Methods of Environmental Social Impact Assessment

Environmental and social impact assessment (ESIA) is an important and often obligatory part of proposing or launching any development project. Delivering a successful ESIA needs not only an understanding of the theory but also a detailed knowledge of the methods for carrying out the processes required. Riki Therivel and Graham Wood bring together the latest advice on best practice from experienced practitioners to ensure an ESIA is carried out effectively and efficiently. This new edition:

- explains how an ESIA works and how it should be carried out
- demonstrates the links between socio-economic, cultural, environmental and ecological systems and assessments
- incorporates the World Bank's IFC performance standards, and best practice examples from developing as well as developed countries
- includes new chapters on emerging ESIA topics such as climate change, ecosystem services, cultural impacts, resource efficiency, land acquisition and involuntary resettlement.

Invaluable to undergraduate and MSc students of ESIA on planning, ecology, geography and environment courses, this internationally oriented fourth edition of *Methods of Environmental and Social Impact Assessment* is also of great use to planners, ESIA practitioners and professionals seeking to update their skills.

Riki Therivel is a partner of Levett-Therivel sustainability consultants, and a visiting professor at the School of the Built Environment, Oxford Brookes University.

Graham Wood is a Reader in Environmental Assessment and Management at the School of the Built Environment, Oxford Brookes University.

The Natural and Built Environment Series

Editor: Professor John Glasson
Oxford Brookes University

Introduction to Rural Planning 2nd edition
Nick Gallent, Iqbal Hamiduddin, Meri Juntti, Sue Kidd and Dave Shaw

Contemporary Issues in Australian Urban and Regional Planning
Julie Brunner, John Glasson

Sustainability Assessment
Alan Bond, Angus Morrison-Saunders and Richard Howitt

Real Estate: Property Markets and Sustainable Behaviour
Peter Dent, Michael Patrick and Xu Ye

Introduction to Environmental Impact Assessment 4th edition
John Glasson, Riki Therivel and Andrew Chadwick

The Environmental Impact Statement after Two Generations
Michael Greenberg

Building Competences for Spatial Planners
Anastassios Perdicoulis

Spatial Planning and Climate Change
Elizabeth Wilson and Jake Piper

Water and the City
Iain White

Transport Policy and Planning in Great Britain
Peter Headicar

Urban Planning and Real Estate Development
John Ratcliffe and Michael Stubbs

Regional Planning
John Glasson and Tim Marshall

Strategic Planning for Regional Development
Harry T. Dimitriou and Robin Thompson

Landscape Planning and Environmental Impact Design
Tom Turner

Public Transport 6th edition
Peter White

Methods of Environmental and Social Impact Assessment 4th edition
Riki Therivel and Graham Wood

Methods of Environmental and Social Impact Assessment

Fourth Edition

Edited by Riki Therivel and Graham Wood

Routledge
Taylor & Francis Group

NEW YORK AND LONDON

Fourth edition published 2018
by Routledge
711 Third Avenue, New York, NY 10017

and by Routledge
2 Park Square, Milton Park, Abingdon, Oxon, OX14 4RN

Routledge is an imprint of the Taylor & Francis Group, an informa business

First edition published by UCL Press 1995
Third edition published by Routledge 2009

Library of Congress Cataloging-in-Publication Data
Names: Therivel, Riki, 1960- editor.
Title: Methods of environmental and social impact assessment / edited by
Riki Therivel and Graham Wood.
Other titles: Methods of environmental impact assessment
Description: 4th edition. | New York : Routledge, 2017. | Series: The
natural and built environment series | Includes bibliographical references.
Identifiers: LCCN 2017010184 | ISBN 978-1-138-64764-0 (hardback) |
ISBN 978-1-138-64767-1 (pbk.)
Subjects: LCSH: Environmental impact analysis--Great Britain. |
Environmental impact analysis--European Union countries.
Classification: LCC TD194.68.G7 M48 2017 | DDC 333.71/4—dc23
LC record available at https://lccn.loc.gov/2017010184

ISBN: 978-1-138-64764-0 (hbk)
ISBN: 978-1-138-64767-1 (pbk)
ISBN: 978-1-315-62693-2 (ebk)

Typeset in Goudy and Akzidenz Grotesk
by FiSH Books Ltd, Enfield

Visit the eResources: www.routledge.com/9781138647671

MIX
Paper from
responsible sources
FSC   FSC® C013985
www.fsc.org

Printed in the United Kingdom
by Henry Ling Limited

This book is dedicated to Peter Morris: friend, colleague and co-editor of the first three editions.

Contents

●●●

Contributors

• •

Polash Banerjee is a research scholar in the Department of Computer Science and Engineering, Sikkim Manipal Institute of Technology, Majitar, Sikkim. His research area is geospatial analysis of the environmental impacts of the transport sector.

Sally-Beth Betts is a fluvial geomorphologist at Credit Valley Conservation, Ontario (Canada).

David S. Brew FGS, CGeol, PhD, BSc is Principal Coastal Geomorphologist at Royal HaskoningDHV.

Martin Broderick is a visiting researcher at Oxford Brookes University and the sustainability adviser on Strategic Projects to Savills UK.

Andrew Brookes is Head of Geomorphology at Jacobs, Reading (UK).

Andrew Chadwick is a former research associate with the School of the Built Environment, Oxford Brookes University.

Amanda Chisholm works at the Scottish Government.

Richard Cottle PhD, BSc works at Ecot Consulting.

Hannah Dalton is a technical associate specialising in air quality at Ramboll Environ.

Bridget Durning is Senior Lecturer in the School of the Built Environment, Oxford Brookes University.

Roy Emberton is Director and member of the Impact Assessment Practice in Ramboll Environ in the UK.

Chris Ferrary is a director with environmental consultants Temple Group Ltd, with 40 years' experience in undertaking EIA.

John Glasson is Professor Emeritus at Oxford Brookes University, and a planning/impact assessment consultant.

Graham Harker leads the air quality discipline at Peter Brett Associates LLP; he is a chartered mechanical engineer and a member of the Institute of Air Quality Management and the Institution of Environmental Sciences.

Kiri Heal is a principal air quality scientist with Peter Brett Associates LLP, and a member of the Institution of Environmental Sciences and of the Institute of Air Quality Management.

Martin J. Hodson is a visiting researcher and former Principal Lecturer in Environmental Biology at Oxford Brookes University.

Júlio de Jesus is a consultant at Júlio de Jesus – Consultores, Lda. (Lisbon, Portugal).

Sian John MSc is Director of Environment, Royal HaskoningDHV.

Katy Kemble is a senior geomorphologist at Jacobs, Manchester (UK).

Rebecca Knight BSc DipLA MA CMLI is a landscape architect, an associate director of LUC and a Chartered Member of the Landscape Institute.

Florence Landsberg is an ecosystem services expert, working for NatureBenefits LLC., Washington DC, USA.

Hugh Masters-Williams is a divisional director at Jacobs UK Ltd, Chartered Engineer, Chartered Environmentalist and an honorary lecturer at Cardiff University.

Garry Middle PhD, is Adjunct Senior Research Fellow at Curtin University, environmental planning consultant, independent member of the Western Australian Planning Commission, and member of the International Association of Impact Assessment.

Phill Minas is Associate Director in the EIA team at Amec Foster Wheeler.

Marla Orenstein is President of Habitat Health Impact Consulting and President-elect of the International Association for Impact Assessment.

Luis E. Sánchez is a full professor at Escola Politécnica, University of São Paulo, Brazil.

Eddie Smyth is Director of Intersocial Consulting, a boutique consulting firm that specialises in resettlement. He is also undertaking a PhD through the University of Groningen.

Chris Stapleton is HS2 Ltd lead on agriculture and soils and visiting lecturer at Oxford Brookes University.

Riki Therivel is a visiting professor at Oxford Brookes University and a partner at Levett-Therivel sustainability consultants.

Jo Treweek is the Director of eCountability Ltd, Chancery Cottage, Kentisbeare, Cullompton, Devon, EX15 2DS, UK.

Frank Vanclay is Professor of Cultural Geography in the Faculty of Spatial Sciences at the University of Groningen. His specific interest is in managing the social issues of large projects.

David P. Walker is Director of Environment at Peter Brett Associates.

Richard J. Wenning is a principal and global leader for ecological services at Ramboll Environ, Portland, Maine, USA.

Elizabeth Wilson is Reader in Environmental Planning at Oxford Brookes University; she is author with Jake Piper of *Spatial Planning and Climate Change* (Taylor and Francis, 2010).

Graham Wood is Reader in Environmental Assessment and Management at Oxford Brookes University's School of the Built Environment.

1 Introduction

Riki Therivel and Graham Wood

1.1 Environmental and Social Impact Assessment (ESIA) and the aims of the book

This book aims to improve the practice of ESIA by providing information about how ESIAs are, and should be, carried out. This introductory chapter (a) summarises the process and current status of ESIA, (b) explains the book's structure, and (c) considers some trends in ESIA methods.

ESIA is an extension of what has traditionally been the process of Environmental Impact Assessment (EIA), which was concisely defined by the United Nations Economic Commission for Europe (UNECE 1991) as 'an assessment of the impacts of a planned activity on the environment'. The definition adopted by the International Association for Impact Assessment (2009) is 'the process of identifying, predicting, evaluating and mitigating the biophysical, social and other relevant effects of proposed development proposals prior to major decisions being taken and commitments made'. The World Bank, a key international driver of both EIA and subsequently ESIA, explains that

> The key process elements of an ESIA generally consist of
> (i) initial screening of the project and scoping of the assessment process;
> (ii) examination of alternatives;
> (iii) stakeholder identification (focusing on those directly affected) and gathering of environmental and social baseline data;
> (iv) impact identification, prediction, and analysis;
> (v) generation of mitigation or management measures and actions;
> (vi) significance of impacts and evaluation of residual impacts; and
> (vii) documentation of the assessment process (i.e., ESIA report).
>
> (IFC 2012)

ESIAs involve assessments of aspects of the environment and society (e.g. landscape, heritage, air, soil, biodiversity, Indigenous people) that are likely to be significantly affected by a proposed project. This book focuses on practical assessment methods and techniques used in the part of the ESIA process concerned with analysing a development's impacts on these *environmental and social components*.

The overall ESIA process is explained and discussed in this book's 'sister volume', *Introduction to Environmental Impact Assessment* (Glasson et al. 2012). This fourth edition of *Methods of Environmental and Social Impact Assessment* is a departure from the first three editions (previously entitled *Methods of Environmental Impact Assessment*) in that it aims to be of wider international relevance, rather than focused on the UK and Europe. It also, for the first time, uses the International Finance Corporation's (2012) performance standards on environmental and social sustainability as its 'backbone'. This has involved major revisions to the content, including the addition of several new chapters. We have used the terms 'ESIA' and 'ESIA report' where possible, but in cases where the literature or practice relates explicitly to EIA, we have kept the terms EIA and EIS *(Environmental Impact Statement)*.

1.2 The ESIA process

1.2.1 Introduction

Figure 1.1 summarises the main ESIA procedures followed in the assessment of any environmental or social component. The figure assumes that the developer has conducted feasibility studies, and that *screening* to determine the need for ESIA (see Glasson et al. 2012) has already been carried out – and these assumptions are made in the chapters. The model illustrates the stepwise nature of ESIA, but also the requirement for continuous reappraisal and adjustment (as indicated by the feedback loops). The model, and the ESIA steps discussed below, form the broad structure for the book's subsequent chapters.

Figure 1.2 shows how ESIA's ability to influence a project is highest in the early stages of project development, decreasing over time. The accuracy of assessment, instead, increases as more is known about the project, but may be very limited early in project development.

1.2.2 Scoping and baseline studies

Scoping is an essential first step in the assessment of a component. The main aims of scoping are:

- to consider and agree at an early stage (when the project type, location and/or design are still relatively amenable to modification): key *receptors*, impacts and project *alternatives*; the scope of the ESIA in time and space; the methodologies to use; and whom to consult;
- to ensure that resources and time are focused on important impacts and receptors;

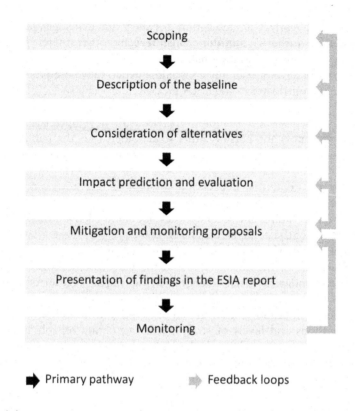

Figure 1.1

Procedures in the assessment of an environmental or social component for an ESIA

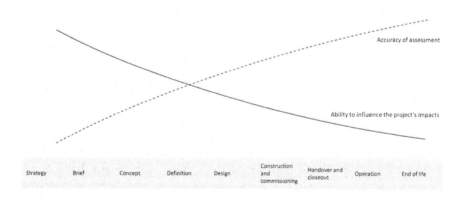

Figure 1.2

Stages of project development and dimensions of ESIA (based on BSI 2016)

- to establish early communication between the developer, consultants, *consenting authority, statutory consultees* and other interest groups who can provide advice and information;
- to warn the developer of any constraints which may pose problems if not discovered until later in the ESIA process.

Scoping is carried out in consultation between the developer, the consenting authority, regulators and possibly other stakeholders (e.g. the public, NGOs). The scoping process should conclude with preliminary agreement on:

- the project's **potential impacts** on component receptors, estimated from the project description (including its size, construction requirements, operational features and secondary developments such as access roads) and the nature of the components and receptors;
- the **impact area/zone and time** within which impacts are likely to occur, estimated from the impact types and the nature of the surrounding area and environmental/social components, e.g. impacts on air or water may occur at considerable distances from the project site;
- the project **alternatives** that should be assessed in the ESIA;
- the **methods and levels of study** needed to obtain reliable information that can be used to evaluate the baseline conditions, make accurate impact predictions, and formulate adequate mitigation measures and monitoring procedures. The selection of methods should involve consideration of:
  - the impact and component receptors on which the studies will focus, and the accuracy and precision required;
  - the most appropriate methods for collecting, analysing and presenting information;
  - the resource requirements and timing considerations, especially for field surveys;
  - constraints such as the time and resources available;
- possible **mitigation measures**;
- the need and potential for **monitoring**.

Depending on the jurisdiction, developers may request a **scoping opinion** from the consenting authority. The findings of the scoping process are typically documented in a **scoping report** that is made available to the developer, consultants, consenting authority and possibly the other stakeholders.

Table 1.1 shows some commonly used aids in ESIA. Two of these, checklists and matrices, are useful scoping tools, particularly for tasks such as identifying key impacts and receptors, and selecting appropriate consultees and interest groups. Lack of detailed information at the scoping stage means that scoping estimates and decisions should be reassessed in the light of baseline information gained as the ESIA progresses.

Baseline studies form the backbone of component assessments. It is only when they provide sound information on the socio-economic or environmen-

Table 1.1 Commonly used ESIA methods

Method	Attributes
Checklists	Useful, especially during scoping, for identifying key impacts and ensuring that they are not overlooked. Can include information such as data requirements, study options, questions to be answered, and statutory thresholds – but are not generally suitable for detailed technical analysis.
Matrices	Can have various uses, e.g. (a) to identify impacts and cause–effect links between impact sources (plotted along one axis) and impacts or receptors (plotted along the other axis); (b) to link features such as magnitude and extent (e.g. localised or extensive, short or long term); and (c) to derive estimated impact significances from receptor sensitivities and impact magnitudes.
Flowcharts and networks	Useful for identifying cause–effect links/pathways: between impact sources; between sources and impacts; and between primary and secondary/tertiary/etc. impacts. However, normally they do not quantify the magnitudes of impacts/effects.
Mathematical/ statistical models	Are based on mathematical or statistical functions, which are applied to calculate deterministic or probabilistic quantitative values from numerical input data. They range from simple formulae to sophisticated models that incorporate many variables. They need adequate/reliable data, can be expensive, and may not be suitable for 'off the peg' use.
Maps	Are often essential. They can indicate features such as impact areas/zones, and locations and extents of receptor sites and/or features within these. Overlay maps can combine and integrate two or three 'layers', e.g. for different impacts and/or environmental components or receptors. Where the maps are part of a Geographical Information System, they can also be used in conjunction with an external tool (such as an expert system or simulation model) as a means of analysing quantitative data and modelling outcomes.

tal systems in the impact area that valid impact predictions can be made, and effective mitigation and monitoring programmes formulated.

The distinction between baseline studies and scoping is not clear cut because (a) consultation should be ongoing, and (b) scoping includes gathering information, much of which is effectively baseline material that can at least form the starting point for more detailed studies. In both stages, it is usually possible to compile some of the required information by means of a desk study.

A thorough search should be made because (a) gathering existing information is generally less expensive and time-consuming than obtaining new data, and (b) it is pointless to undertake new work that merely duplicates information that already exists. However:

- both scoping and baseline studies will usually require brief site visits (e.g. for reconnaissance or to confirm features identified on maps) – perhaps including walkover surveys. Such initial visits are best undertaken by several members of the ESIA and design team, so that relationships between components can be identified;
- in most cases, existing baseline data will be inadequate or out of date, and it will be necessary to obtain new information by some form of field survey.

In order to predict the impacts of a development it is also necessary to consider changes in the baseline conditions that may occur in its absence (a) prior to its initiation, which can be several years after the ESIA report is produced, and (b) during its projected lifetime. These can be assessed in relation to the current baseline conditions and information on past, present and predicted conditions and trends.

The **description and evaluation of baseline conditions** should include:

- a clear presentation of methods and results;
- indications of limitations and uncertainties, e.g. in relation to data accuracy and completeness;
- an assessment of the value of key receptors and their sensitivity to impacts.

1.2.3 Alternatives

Alternatives (or *options*) are increasingly being considered within ESIAs. For instance the 2014 amendments to the European EIA Directive (EU 2014) require:

> A description of the reasonable alternatives (for example in terms of project design, technology, location, size and scale) studied by the developer, which are relevant to the proposed project and its specific characteristics, and an indication of the main reasons for selecting the chosen option, including a comparison of the environmental effects.
>
> (Annex IV.2)

Some of the alternatives listed by the IFC (2012) include alternative project locations, project designs, forms of project operation and access routes; renewable or low carbon energy sources; emission reduction technologies; use of alternative water supplies; alternative options for waste disposal; and water consumption offsets to reduce total demand for water resources. Some of the more technical alternatives can also be considered as mitigation measures once a project's impacts are better understood: this is how alternatives are primarily considered in the chapters of this book. The 'no action' of not proceeding with the project is another alternative, one which is explored in the analysis of the future baseline without the project as part of the baseline description.

The 'hierarchy of alternatives' (adapted from ODPM et al. 2005) can help to structure the consideration of alternatives. The higher up the hierarchy the decisions are made, the more sustainable they tend to be:

Need or demand: Is the project necessary? Can the need or demand be met without the project? (e.g. can the demand for electricity be reduced through improved appliances rather than a new coal-fired power station?)

↓

Mode or process: How should it be done? Are there technologies or methods that can meet the need with less environmental damage than 'obvious' or traditional methods? (e.g. wind farms rather than a new coal-fired power station)

↓

Location: Where should it go? (e.g. more environmentally robust location, minimizing transport requirements)

↓

Timing and detailed implementation: When, in what form and in what sequence, should the project be developed? What details matter, and what requirements should be made about them?

The impacts of the alternatives are typically assessed in parallel with those of the proposed project; compared against the proposed project; and a justification is provided for why the proposed project is being put forward, especially when an alternative is likely to have fewer significant negative social or environmental impacts.

1.2.4 Impact prediction and evaluation

Impact prediction and evaluation is fundamental to ESIA, and the likely impacts of a project should be considered for all 'scoped in' environmental and social components. Most of the relevant baseline information will be drawn from desk study although comparison of field survey findings with previous data can help to elucidate recent trends.

Impact prediction should include assessment of:

- **Direct/primary impacts** – that are a direct result of a development;
- **Indirect/secondary impacts** – that may be 'knock-on' effects of (and in the same location as) direct impacts, but are often produced in other locations and/or as a result of a complex pathway;
- *Cumulative impacts* – that accrue over time and space from a number of developments or actions, and to which a new project may contribute.

An additional possibility is **impact interactions** – between different impacts of a project, or between these and impacts of other projects – that result in one or more additional impacts, e.g. $(A + B) \rightarrow C$. For instance, the interaction of population and air pollution may cause health effects.

All impacts may be: **positive** (beneficial) or **negative** (adverse); **short-**, **medium-**, or **long-term**; **reversible** or **irreversible**; and **permanent** or **temporary**.

Ideally, impact prediction requires:

- a good understanding of the nature of the proposed project, including project design, construction activities and timing, and decommissioning where appropriate;
- adequate information about the relevant receptors, and knowledge of how these may respond to changes/disturbances;
- knowledge of the outcomes of similar projects and ESIAs, including the effectiveness of mitigation measures;
- knowledge of past, existing or approved projects which may cause interactive or cumulative impacts with the project being assessed;
- predictions of the project's impacts on other environmental and social components that may interact with that under study.

Methods of impact prediction vary both between and within ESIA components. The assessment of impact **magnitude** (scale or severity) may be qualitative or quantitative. Qualitative assessments usually employ ratings such as neutral, slight, moderate, large – applied to both negative and positive impacts. They are typically used where quantitative assessments are difficult or impossible, for instance in landscape, archaeological, and ecological assessment. Quantitative assessments involve the measurement or calculation of numerical values, e.g. of the level of a *pollutant* in relation to a statutory threshold value.

The evaluation of **impact significance** brings together 1. an impact's magnitude and 2. the value, sensitivity/fragility and recoverability of the relevant receptor(s). Figure 1.3 shows a typical impact evaluation matrix. Many of the chapters in this book give criteria for determining the sensitivity of receptors and the magnitude of impacts, which can be inserted into a matrix such as that in Figure 1.3. Whereas impact assessment is an objective exercise, impact evaluation is a subjective exercise requiring expert judgement.

Impact prediction and evaluation is often poorly addressed, perhaps because it is the most difficult step in ESIA. Direct impacts are usually relatively easy to identify, but accurate prediction of *indirect, cumulative* and *synergistic impacts* can be much more problematic. Whatever methods are employed, there are bound to be uncertainties (that can sometimes be expressed as ranges) which should be clearly stated in the ESIA report.

Value or sensitivity	Magnitude of impact			
	Imperceptible/ no change	Small	Medium	Large
Very high	Negligible	Moderate/major	Major	Very major
High	Negligible	Moderate	Moderate/major	Major
Medium	Negligible	Minor/moderate	Moderate	Moderate/major
Low	Negligible	Minor	Minor/moderate	Moderate

Figure 1.3

Typical impact evaluation matrix

1.2.5 Mitigation

Mitigation measures aim to avoid, minimise, remedy or compensate (in that sequence) for the predicted adverse impacts of the project. They can include:

- selection of project locations, alignments (of linear projects), size or layout;
- modification of the methods and timing of construction;
- modification of design features, including site boundaries and features, e.g. landscaping;
- use of alternative production techniques and/or technologies to minimise operational impacts (e.g. pollution and waste);
- specific measures, perhaps outside the development site, to minimise particular impacts;
- measures to compensate for losses, e.g. of amenity or habitat features.

Much of the environmental damage caused by developments occurs during the **construction phase**, which may be contracted to a construction company who will not necessarily have participated in the ESIA process, and over whom the developer may have little control. Consequently, there is a need to provide a Construction Environmental Management Plan, ideally as part of an overall project Environmental and Social Management Plan (see Chapter 20). In addition, because project specifications frequently change between publication of the ESIA report and the start or completion of construction (often for unforeseeable reasons), developers sometimes employ site environmental managers to ensure (a) that such modifications take account of environmental and social considerations, and (b) that construction phase mitigation measures are carried out.

Different mitigation measures will be needed in relation to specific impacts on different environmental and social components. The ESIA should propose detailed measures for each impact, indicate how they would actually be put in place, and propose how they might be modified if unforeseen post-project impacts arise. A primary consideration is the likely significance of post-mitigation *residual impacts*, and care is needed to ensure that a mitigation measure does not generate new impacts, perhaps on receptors in other environmental components.

Best practice dictates that the *precautionary principle* should be applied, i.e. that mitigation should be based on the possibility of a significant impact even though there may not be conclusive evidence that it will occur. Similarly, on the basis that preventive action is preferable to remedial measures, and that environmental damage should be rectified at source, the best mitigation measures involve modifications to the project rather than containment or repair at receptor sites, or compensatory measures such as habitat creation – which should normally be considered only as a last resort.

Although the chapters of an ESIA, like the chapters in this book, are presented as separate entities, in practice the individual environmental component assessments should be integrated, and be part of the wider process of project planning. Clearly, an ESIA must involve a **team of experts** on the various components, and in many cases on different aspects of a given component. Close coordination is needed to avoid duplication of effort, while ensuring that important aspects are not omitted. This is particularly important for interrelated components such as soils, geology, air, water and ecology; and historical, landscape and cultural issues. In addition, the ESIA report must be an integrated document in which relationships between components are clearly explained.

It follows that there must be an **ESIA coordinator** who will ensure that (a) cross-component consultation is carried out throughout the ESIA process, and (b) appraisals are conducted to consider aspects such as components' relative importance, the relative significance of different impacts, interactions between impacts, possible conflicts of interest, and distributional effects. For example:

- One sector of the community, or part of the impact area, may be particularly affected by multiple developments, or by the concentration of a project's impacts. For instance, lower socio-economic groups are more likely to suffer from traffic accidents, air pollution and noise (Lucas and Simpson 2000). Identification of the groups/areas most strongly affected can be facilitated by use of geographical information systems (GIS) or simply by a table listing receptors (e.g. particular socio-economic groups, sensitive sites) on one axis, and the main impacts of a project on the other axis. A more equitable distribution of impacts may then be sought, or strongly affected groups may be compensated in some way.
- Mitigation measures proposed for different environmental components should be consistent with those for other components, and should not

themselves cause negative impacts. For instance, earth bunds which reduce visual impacts could have beneficial side effects for noise, but could intrude on archaeological remains.

In addition to mitigation, opportunities for **environmental enhancement** (improvement of current environmental conditions and features) should be sought in ESIA. For instance, projects may seek to reduce flood risk or enhance biodiversity.

1.2.6 Presentation of findings and proposals in the ESIA report

The information presented in the ESIA report must be clear and, at least in the non-technical summary, should be in a form that can be understood by non-experts without compromising its integrity. It should also be transparent, e.g. in relation to limitations and uncertainties. Presentation methods vary between components, but can include the use of maps, graphs/charts, tables and photographs.

The ESIA report must be an integrated document, and this will necessitate assessing one component in relation to others, e.g. to evaluate its relative importance, and ensure that potential conflicts of interest have been addressed.

An *audit trail* can be particularly beneficial if further ESIA analysis is needed because the project changes substantially between the time when it is approved and when it is built. Ideally, final assessment should result in the preparation of a list of proposed planning conditions/obligations and an Environmental and Social Management Plan (ESMP) for the proposed development, to be included in the ESIA report or presented in a separate document.

1.2.7 Monitoring

Monitoring can be defined as the continuous assessment of environmental or social variables via the systematic collection of specific data in space and time. It can be continuous, e.g. using recording instruments, but more commonly involves periodic repeat data collection, usually by the same or similar methods as in baseline surveys. Monitoring in ESIA can include:

- **Baseline monitoring** – which may be carried out over seasons or years to quantify ranges of natural variation and/or directions and rates of change that are relevant to impact prediction and mitigation. This can avoid the frequent criticism that baseline studies are only 'snapshots' in time. However, time constraints in ESIA usually preclude lengthy survey programmes, and assessments of long-term trends normally have to rely on existing data.

- **Compliance monitoring** – which aims to check that specific conditions and standards are met, e.g. in relation to emissions of pollutants.
- **Impact and mitigation monitoring** – which aims to compare predicted and actual (residual) impacts, and hence to determine the effectiveness of mitigation measures.

Unless otherwise specified, "monitoring" in ESIA normally refers to impact and mitigation monitoring, which is also sometimes called auditing. Considerable uncertainty is often associated with impacts and mitigation measures, and it is responsible best practice to undertake monitoring during both the construction and post-development phases of a project. Monitoring is essential to learn from both successes and failures. For example:

- it is the only mechanism for comparing predicted and actual impacts, and hence of checking whether mitigation measures have been put in place, testing their effectiveness, and evaluating the efficiency of the project management programme;
- if mitigation measures are amenable to modification, it should still be possible to reduce residual impacts identified during monitoring (feedback loop in Figure 1.1);
- it can provide information about responses of particular receptors to impacts;
- it is the only means of ESIA report evaluation and of identifying mistakes that may be rectified in future ESIAs. For example, it will provide information that can be used to assess the adequacy of survey and predictive methods, and how they may be improved. Thus, a principal aim of monitoring should be to contribute to a cumulative database that can facilitate the improvement of future ESIAs.

(Morrison-Saunders and Arts 2012)

Three requirements are essential for successful monitoring: (a) good quality baseline data; (b) funding to carry out the monitoring survey work; and (c) sufficient contingency funds to enable modifications to mitigation measures to be made, or faults to be rectified, if necessary.

Monitoring is not strictly part of the ESIA process, is not legally required in many countries, and can be expensive. Consequently, lack of monitoring is a serious deficiency in current ESIA practice, although preparation of an Environmental and Social Management Plan (see Chapter 20) is becoming increasingly widespread.

1.3 The broader context for ESIA

Since the first ESIA system was established in the United States in 1970, ESIA systems have been set up worldwide and have become a powerful environmental and social safeguard in the project planning process. Until relatively recently, the focus of ESIAs was on environmental issues, i.e. EIA. However there has been more recently a strong trend towards considering social as well as environmental issues in a more comprehensive and balanced ESIA. In particular, international lending institutions such as the World Bank, which play a significant role in promoting impact assessment practice in developing countries, have been requiring ESIAs. This is consistent with the World Commission on Environment and Development's principle of *sustainable development*, which was further elucidated at the UN Conference on Environment and Development (UNCED 1992) – the 'Rio Earth Summit'.

A greater recognition of the links between different environmental and social issues is also emerging. This is reflected in the IFC's (2012) environmental and social performance standards (Box 1.1) which consider, for instance, air, water and soil pollution as interlinked phenomena; and consider landscape and biodiversity together.

Box 1.1 IFC performance standards on environmental and social sustainability (IFC 2012)

1   Assessment and management of environmental and social risks and impacts
2   Labor and working conditions
3   Resource efficiency and pollution prevention
4   Community health, safety and security
5   Land acquisition and involuntary resettlement
6   Biodiversity conservation and sustainable management of living natural resources
7   Indigenous peoples
8   Cultural heritage

The concept of *ecosystem services* – the services provided to people by the environment – is a different way of expressing these links and considering them in ESIA. Ecosystem services include supporting services that are necessary for the production of all other ecosystem services (e.g. primary production, nutrient recycling); provisioning services that produce food, water, minerals and other products; regulating services such as carbon sequestration and water purification; and cultural services such as recreational and cultural experiences (Millennium Ecosystem Assessment 2005). Use of the ecosystem services approach within ESIA can lead to very different identified impacts and mitigation measures from those that would be developed through 'traditional' ESIA. This is discussed in Chapter 8.

Projects are not planned, built, operated and decommissioned in isolation, but within regional, national and international processes of change which include other projects, programmes, plans and policies. The aim of assessing cumulative impacts is to take these into account as far as possible in relation to a single development project. However, some projects are so inextricably related to other projects, or their impacts are so clearly linked, that a joint ESIA of these projects should be carried out. For instance if a gas-fired power station requires the construction of a new pipeline and gas reception/processing facility to receive the gas, and new transmission lines to carry the resulting electricity, these projects should be considered together in an ESIA.

Other projects are 'growth-inducing', i.e. necessary precursors to other projects. For instance a new road may, directly or indirectly, trigger the construction of new towns with associated infrastructure requirements; or the infrastructure provided for one project may make a site more attractive, or may present economies of scale, for further development. Although it is probably not feasible to consider induced impacts in detail in an ESIA, the ESIA should at least acknowledge the possibility of these further developments.

The broadening of ESIAs' remit to encompass other projects may allow trade-offs to be made between impacts and between projects. For instance, an environmentally beneficial **shadow project** may be proposed to compensate for the negative impacts of a development project. An example of this is the creation of a new waterfowl feeding ground on coastal grassland as compensation for the loss of tidal mudflat feeding grounds caused by the Cardiff Bay Barrage. However, shadow projects need to be treated with caution. For instance, it can be argued that the provision of a coastal grassland area does not effectively compensate for the loss of tidal mudflats because it is a different habitat supporting different wildlife communities. Compensatory like-for-like creation of valuable habitats is generally much more difficult (see Chapters 6 and 7).

Project ESIAs also need to be set in the context of **strategic environmental assessment** (SEA) of sectoral or spatial policies, plans and programmes. SEAs can, in theory, reduce the time and cost of ESIA, and even eliminate the need for certain types of ESIA, although not much evidence of this currently exists. SEA can also provide background information about the local policy context, baseline environmental conditions, and existing environmental problems in the project area. SEA is required in many countries, including all of the European Member States as a result of the 'SEA Directive' (European Commission (EC) 2001).

ESIA and SEA should be, and are increasingly being, linked to other related techniques. For example:

- Project design is increasingly being influenced by environmental concerns e.g. minimising resource use in building construction and use, and greater application of techniques such as sustainable drainage systems, passive solar heating, photovoltaics and greywater recycling (Chapter 17). *Life cycle analysis* can help to identify the impacts of buildings from the production of the materials used to build them through to their ultimate dismantling and disposal.
- There is increasing use of environmental risk assessment and risk management (Chapter 18). Risk assessment is particularly relevant to the

prediction of impacts from accidents, which are of increasing concern following events such as the Bhopal chemical leak, Fukushima and Chernobyl nuclear events, and the Deepwater Horizon oil platform blowout. Similarly, the **resilience** of projects to likely future change (e.g. climate change) and shocks (e.g. earthquakes or floods), and resilience of communities to the changes brought about by projects is of increasing interest.

- Participatory approaches to ESIA are continually developing. This can include village/community mapping exercises which can help to identify features that are particularly valued by the community; assessment and comparison of the impacts of different project alternatives in community workshops; identification of ecosystem services that are valued by local residents; and ESIA monitoring of impacts by the community.
- In Europe, *appropriate assessment* is required where a project may have a significant 'in combination' impact on the integrity of a Special Protection Area, Special Area of Conservation, or Ramsar site. Appropriate assessment of projects has been carried out for more than 20 years in response to the Habitats Directive (EC 1992), but systematic appropriate assessment of plans has only been carried out more recently.

New tools, techniques and approaches are being developed which complement and support the ESIA process. For example:

- mapping software and GIS now allow much more effective analysis and presentation of information than in the past;
- there is a rapid expansion in the range and availability of information databases, including remote sensed data and other digital data suitable for use in GIS;
- the internet now provides ready access to a wealth of information, including legislation, other publications, databases, software and ESIA examples;
- in ecology and landscape analysis, although legislation and government guidelines still focus on protecting designated areas, there is a shift from 'save the best and leave the rest' to consideration of the 'wider countryside' and characterisation of areas, with the aim of promoting their uniqueness and joint diversity;
- more emphasis is being placed on environmental enhancement, not just mitigation of negative impacts.

Finally, concern about wider distributional impacts – for instance about who wins and loses as a result of development, and whether some countries are 'importing' sustainability at the cost of making environmental conditions in other countries unsustainable – is likely to lead to more evolved forms of public participation and political negotiations, and ultimately to a more equitable approach to development and the environment.

1.4 Book structure

The book broadly moves from environmental (biophysical) to social impacts, with some concluding cross-sectoral chapters. Chapters 2–5 deal with the basic environmental components: water, soil, air and the climate. Chapters 6–8 cover different dimensions of **biodiversity**, flora and fauna: terrestrial ecology, coastal ecology and geomorphology, and ecosystem services. Chapters 9 and 10 deal with, respectively, noise and transport, whilst Chapter 11 considers landscape and visual impacts. Chapters 12–16 cover different human-related impacts: cultural heritage, economic and social impacts, land acquisition and involuntary resettlement, and health. Chapters 17–20 discuss cross-sectoral and integrative approaches: resource efficiency, risk assessment, cumulative impacts and Environmental and Social Management Plans.

Table 1.2 shows how the chapters correspond to the IFC performance standards, and Table 1.3 shows how they correspond to the environmental components listed in European EIA Directive 2014/52/EU. The book includes

Table 1.2 The book's coverage of the topics covered in the IFC (2012) performance standards

IFC performance standard	Chapter number and topic
1 Assessment and management of environmental and social risks and impacts	[most of the chapters] 18. Risks
2 Labor and working conditions	*
3 Resource efficiency and pollution prevention	2. Water 3. Soils, land and geology 4. Air 5. Climate and climate change 10. Transport 17. Resource efficiency
4 Community health, safety and security	9. Noise 16. Health
5 Land acquisition and involuntary resettlement	15. Land acquisition, resettlement and livelihoods
6 Biodiversity conservation and sustainable management of living natural resources	6. Ecology 7. Coastal ecology and geomorphology 8. Ecosystem services
7 Indigenous peoples	*
8 Cultural heritage	11. Landscape and visual 12. Cultural heritage

Note: * Not directly covered, although some chapters indirectly cover this

Table 1.3 The book's coverage of the environmental and social components listed in Annex IV of Directive 2014/52/EU

Environmental or social component	Chapter number and topic
Population, human health	9. Noise 14. Social 15. Land acquisition, resettlement and livelihoods 16. Health
Biodiversity (for example fauna and flora)	6. Ecology 7. Coastal ecology and geomorphology 8. Ecosystem services
Land (for example land take)	3. Soils, land and geology
Soil (for example organic matter, erosion, compaction, sealing)	3. Soils, land and geology 7. Coastal ecology and geomorphology
Water (for example hydromorphological changes, quantity and quality)	2. Water 7. Coastal ecology and geomorphology
Air	4. Air
Climate (for example greenhouse gas emissions, impacts relevant to adaptation)	5. Climate and climate change 2. Water
Material assets	10. Transport 13. Economic 17. Resource efficiency
Cultural heritage, including architectural and archaeological aspects	12. Cultural heritage
Landscape	11. Landscape and visual

some components not specifically listed in the Directive but often discussed in practice, namely noise, transport, geology and geomorphology.

Each chapter discusses the main ESIA steps for the assessment of an environmental or social components outlined at Figure 1.1:

- introduction;
- definitions and concepts;
- legislation and guidance;
- scoping and baseline studies;
- impact prediction and evaluation;
- mitigation and monitoring.

The chapters discuss legislation, guidance and case studies from a range of countries, with an emphasis on current and emerging good practice.

A book of this size cannot cover these subjects in depth, and each chapter aims to provide an accessible overview and entry point to the subject. Terms that may not be familiar to some readers are shown in the text in bold italics and defined in the **glossary**. This reduces repetition in different chapters.

The book does not include chapters on some specialist topics which are found in only a limited number of ESIAs. Groundwater is covered in Box 2.1. Odour is briefly considered in Chapter 4, and vibration is briefly considered in Box 9.2. Impacts on daylight and sunlight are covered in Box 11.4.

Acknowledgements

We are very grateful to the many authors who contributed to this edition and previous editions of this book. In many cases, the authors from the third edition, who are UK experts, teamed up with a colleague from another country to provide an international perspective. Most of the new chapters were written by acknowledged international experts. As such, in each case either an existing chapter was completely rewritten or a new chapter was written. Thank you to everyone involved.

References

BSI (British Standards Institution) 2016. *PAS 2080: 2016 Carbon Management in Infrastructure*. London: BSI.

EC (European Commission) 1992. Council Directive 92/43/EEC on the conservation of natural habitats and of wild fauna and flora. *Official Journal L 206, 22/07/1992 P. 0007 – 0050*. http://eur-lex. europa.eu/LexUriServ/LexUriServ.do?uri=CELEX:31992L0043: EN:HTML.

EC 2001. Directive 2001/42/EC on the assessment of the effects of certain plans and programmes on the environment (The SEA Directive). Brussels: EC. http://eur-lex.europa.eu/LexUriServ/LexUriServ.do?uri=OJ:L:2001:197:0030:0037:EN:PDF.

EU (European Union) 2014. Directive 2104/52/EU on the assessment of the effects of certain public and private projects on the environment. Brussels: EU.

Glasson J, R Therivel and A Chadwick 2012. *Introduction to Environmental Impact Assessment*, 4th Edn. London: Routledge.

IFC (International Financial Corporation, World Bank Group) 2012. Performance standards on environmental and social sustainability. www.ifc.org/wps/wcm/connect/ 115482804a0255db96fbffd1a5d13d27/PS_English_2012_Full-Document.pdf?MOD= AJPERES.

International Association for Impact Assessment 2009. *What is Impact Assessment?* Fargo, ND: IAIA.

Lucas K and R Simpson 2000. Transport and accessibility: the perspectives of disadvantaged communities. Research paper for the Joseph Rowntree Foundation. London: Transport Studies Unit, University of Westminster.

Millennium Ecosystem Assessment 2005. *Ecosystems and Human Well-being*. Washington DC: Island Press.

Morrison-Saunders A and J Arts 2012. *Assessing Impact: Handbook of EIA and SEA Follow-up*. London: Earthscan.

ODPM (Office of the Deputy Prime Minister), Scottish Executive, Welsh Assembly Government and Department of the Environment for Northern Ireland 2005. *A practical guide to the Strategic Environmental Assessment Directive*. London: ODPM.

UNCED 1992. *United Nations Conference on Environment and Development* (UNCED), Rio de Janeiro, 3–14 June 1992. www.un.org/geninfo/bp/enviro.html.

UNECE (United Nations Economic Commission for Europe) 1991. *Policies and Systems of Environmental Impact Assessment*. Geneva: UNECE.

2 Water

Andrew Brookes, Katy Kemble and Sally-Beth Betts

2.1 Introduction

This chapter concerns project impacts on the surface water environment: water quantity, water quality and fluvial geomorphology. Surface water is a rare resource, constituting only approximately 1 per cent of the Earth's free water. Water is an important resource for domestic, industrial, agricultural and recreational purposes. It is also a receptor that can be impacted by developments and activities such as power stations, chemical works, quarries, hydropower facilities, waste-water treatment, groundwater abstraction for water supply, reservoirs, urban land use, inland navigation and intensive agriculture.

Surface water can also cause significant flooding issues, particularly where towns and cities impinge on the natural floodplain of a river system. Developments that have the potential to impact floodplains and, therefore, flood flows and storage, include railways, roads and airports. The number of people living and working in floodplains is expected to increase, with a consequent increase in flood risk, as more floodplains are built on. This is expected to worsen under climate change.

Conversely, developments that lead to low flows (for instance through abstraction) can also affect people and the environment. Reduced flows can impact the availability of water as a resource downstream and also aquatic ecology and sediment transport. This is increasingly being recognised as an issue, and guidelines regarding environmental flows are being established around the world to address it. Changes to the nature of a catchment, a watercourse or the flows that arrive at a watercourse can also lead to channel adjustment in the form of erosion or deposition. This can lead to impacts on rivers themselves and their associated habitats, and put existing infrastructure at risk.

Hydrology is the study of the occurrence, distribution, movement and properties of water and its relationship with the environment. Hydrological systems are highly dynamic and planning any development that may affect them requires an understanding of variations in the storage and flow of water (water quantity) and of the materials it carries (water quality). The study of the behaviour of water in physical systems – or the *hydraulics* – describes how water moves from one point to the next. Fluvial geomorphology concerns processes shaping river channels and the resultant landforms, and is therefore essential to understanding how channels might respond to development activities. Some effects of development on water quantity, quality and fluvial geomorphology can extend well beyond the project's development area, especially in the downstream direction.

Surface water has a pivotal role in environmental systems and the water assessment in an Environmental and Social Impact Analysis (ESIA) is bound to overlap with many other components. For example, water can have high landscape and recreational values, and water and the *sediments* it carries can shape the landscape. Navigable waters have economic importance. River floodplains have long been a focus for settlement, so they often contain important archaeological features. The link between hydrology and ecology is particularly strong in freshwater ecosystems, and increasingly fluvial geomorphology is incorporated into surface water assessments to address this.

Specialists that contribute to a surface water chapter in an ESIA typically include a hydrologist, a geomorphologist and a water quality expert. In addition, a mathematical modeller may be needed to run hydraulic, sediment transport or contaminant transport models.

Many books and guides are concerned with surface water issues. For example, hydrology is covered by Chow (1959), Linsley et al. (1982), Atkinson (1999), Hendriks (2010), Brooks et al. (2012), Bedient et al. (2012) and Ward et al. (2015); water quality aspects in Tchobanoglous and Schroeder (1987), Novotny (2002), Chapra (2008), Viessman and Hammer (2008), Chin (2012), Begum (2015) and Boyd (2015); and fluvial geomorphology in Knighton (1998), Thorne (1998), Kondolf and Piegay (2002), and Sear et al. (2010).

This chapter focuses on processes affecting the land surface and does not cover coastal processes, or impacts on the estuarine or coastal environment that could result from connectivity to a land surface by rivers and streams: these are discussed in Chapter 7. It briefly introduces groundwater at Box 2.1, and discusses it in the text where that topic links with the surface water environment.

2.2 Definitions and concepts

2.2.1 Water quantity

The hydrological cycle describes the movement and storage of water between the earth's liquid and solid components, the atmosphere and the biosphere. This global circulation of water is a closed system with no significant gains or losses. Water is stored in the atmosphere, oceans, rivers, lakes, soils, glaciers and snowfields and as groundwater. Ninety-eight per cent of the world's water is stored in the oceans, with the remaining 2 per cent being freshwater. Eighty-seven per cent of the world's freshwater is stored as ice, 12 per cent is stored in groundwater and only 1 per cent is stored in rivers and lakes.

Water moves between these 'stores' through processes which are either atmospheric or land-based (see Figure 2.1). Precipitation and **evapotranspiration** bring about the interchange of water between the atmospheric and land-phase water. **Runoff** and groundwater flow form the land-based interchanges. Oceans provide most of the evaporated water in the atmosphere. Ninety-one per cent of this water returns to the oceans by way of precipitation, and only the remaining 9 per cent is transported to land where climatological factors include the formation of precipitation. Runoff and groundwater flows correct the resulting imbalance between rates of evaporation and precipitation over the oceans and land. When precipitation exceeds evapotranspiration, there is a water surplus which is discharged as runoff. When precipitation is less than evapotranspiration, there is a water deficit leading to a reduction in stored water and runoff. In the long term, precipitation to evapotranspiration ratios are a function of the local or regional climate.

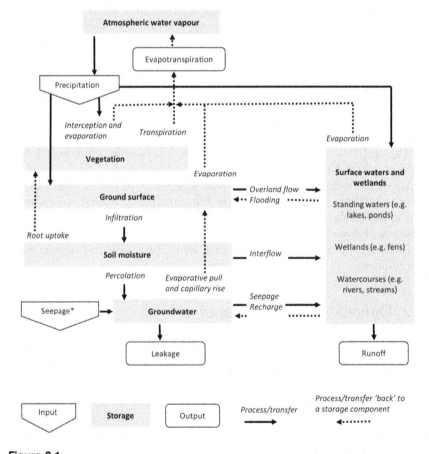

Figure 2.1

Catchment processes and storage components

Catchment approach and the water balance

A water assessment usually adopts a **catchment** (drainage basin) approach. Catchments are separated by a watershed boundary which is the natural division line along the highest points of land. Precipitation falling into the area of a catchment eventually reaches the same river unless it evaporates beforehand or becomes groundwater through infiltration. Catchments can be divided into sub-catchments, also along lines of elevation.

Hydrological catchments vary widely in size and contain different combinations of freshwater bodies (surface water and groundwater). The movement of water through the phases of the hydrologic cycle varies in time and space. This can lead to surplus water and a rise in water levels and potential flooding, or to a deficit of water potentially leading to drought conditions. Figure 2.1 depicts the various storage components and fluxes in a catchment.

For any hydrologic system, a **water balance** (or budget) can be estimated, to predict or account for various flow pathways and storage components in the system. Water budgets are usually calculated for a period of time and are given in depth (m). They can be presented as a volume by multiplying the depth by the surface area of the watershed. A commonly used method to estimate water budgets is by Thornthwaite and Mather (1955):

$$P - R - G - ET = \Delta S$$

where
P = precipitation
R = surface runoff
G = groundwater flow
ET = evapotranspiration
ΔS = change in storage in a specified time period

Table 2.1 summarises these parameters and the process of calculating the water budget where data/information is available.

Water budgets may be repeated for a range of conditions at the development site. For example, for a site with sensitive receptors such as a wetland, it may be desirable to assess 'wet' and 'drought' conditions.

Numerous development activities requiring an ESIA can affect the water budget, including:

- removal of forest cover (e.g. affecting evapotranspiration);
- dam building (affecting flood pulse);
- draining wetlands and marshes (increasing runoff from the land);
- agriculture (increasing runoff);
- water transfers (movement of water between catchments for drinking water or irrigation).

Table 2.1 Components of the hydrological cycle and collection of relevant data for a water budget calculation

Component	Description and types of data/information
Precipitation	Precipitation is the main input to the hydrologic cycle. It can be in the form of rain, snow or hail and is typically a result of atmospheric moisture. Precipitation data for a normal year is used to evaluate long-term effects of projects. Data should ideally be tabulated by month, possibly using an Excel spreadsheet.
Surface runoff	Apart from overland flow, which is normally transitory, surface waters can be divided into standing waters (lakes, reservoirs, ponds, etc.) in which there is little lateral flow, and watercourses (streams, rivers etc.) in which there is appreciable flow. This is not normally an important component in water budget calculations unless, for example, a wetland is dependent on runoff to prevent it from drying out. Surface runoff can be roughly calculated from runoff curves (in the USA, derived from the US Department of Agriculture 1956, 2004). The curves were developed from combinations of hydrologic soil grouping, land use and class of treatment (such as straight ploughing or contouring). Daily runoff values can be collected using a device such as a funnel-shaped runoff collector. For detailed assessment requiring more accurate results, hydrologic models can be used to estimate surface water runoff.
Groundwater flow	See Box 2.1 for general information about groundwater. The distribution of **hydraulic head** through an aquifer determines where groundwater will flow. For unconfined aquifers the hydraulic head is the water table, but for confined aquifers it is not.
Evapotranspiration	**Evapotranspiration** is the combined loss of water vapour from the surface of plants (**transpiration**) and the evaporation of moisture from the soil and other surfaces such as lakes, rivers and oceans. Soil and plant types will determine the amount of potential evapotranspiration. Methods exist for estimating actual evapotranspiration (e.g. Thornthwaite and Mather 1957). Inputs required are precipitation, temperature and latitude.

Infiltration can be considerably altered through land-use change. For instance, converting a forested area into a paved urban development increases the volume and rate of runoff, and reduces recharge to groundwater. Stormwater basins (or stormwater management ponds) can help to mitigate these flashy urban flows, but they can also alter the flow regime and may impact low flows, sediment transport and water temperature. The water budget may also be affected by anthropogenically influenced inflows or outflows, for instance abstraction for industry or farming, or discharge from a waste-water treatment plant.

Box 2.1 Groundwater (based on Kelday et al. 2009)

The subsurface system can be divided into an unsaturated zone that normally has air-filled spaces, and a saturated zone in which all available spaces are filled with groundwater. Soil is an important component of the unsaturated zone, and the properties of a soil, especially its texture and structure, affect its ability to retain water and to allow water to move through it. Within the saturated zone, groundwater is usually held in strata of porous rock called **aquifers**. An unconfined aquifer will have a free water table, and a well sunk into it will fill to the water table level. A confined aquifer lies beneath an impervious confining stratum and fills from one or more unconfined areas: it is artesian (i.e. under pressure). The storage capacity of an aquifer depends largely on its dimensions and porosity. Many strata (e.g. clays and shales) are not usually classed as aquifers because the porous material is thin (< 50 m), and the groundwater supply tends to be small and unreliable during droughts.

Aquifer storage levels (and associated water table levels) normally follow a seasonal cycle. In most Northern countries, storage is depleted during the summer, when output to springs and rivers continues, but input is minimal because there is a meteorological water deficit; and abstraction demands increase. Groundwater recharge occurs mainly during winter, when there is a meteorological water surplus. Consequently, groundwater droughts are mainly caused by a lack of winter rainfall rather than dry summers, and serious droughts occur when (a) a dry summer follows a very dry winter, or (b) winter recharge is below average for several years.

Groundwater and surface water levels are often intimately linked, and groundwater is responsible for river baseflows which continue when there has been little rainfall for some time. In these linked systems, river low flows during droughts can be partly due to over-abstraction of groundwater. Groundwater is also often important in supporting wetland ecosystems which are therefore threatened by groundwater depletion.

Porous rock has a filtering effect as water moves through it; so groundwater is generally much cleaner than surface water, and often requires little or no treatment before use. Chemicals are not completely removed however, and there is increasing concern about groundwater pollution from both urban and rural sources, and from **point source** or **non-point source pollution**. Groundwater pollution can impact surface water supplies, river ecosystems and wetlands. Because groundwater moves very slowly, pollutants take a long time to disperse naturally, and deep groundwater can remain contaminated for centuries. Remedial measures, such as pollutant removal or degradation, are difficult and expensive; so it is particularly important to focus on pollution prevention such as groundwater protection zones.

The two most important aspects of groundwater quantity are storage and flow. For example, if a project is likely to affect soil drainage, it may be important to consider the soil moisture levels, water retention and flow properties of local soils. Soil moisture data may be available from, for instance, the relevant meteorological office, but can otherwise be measured. If the texture of a soil is known, its water retention properties (such as **saturation capacity** and **field capacity**), and its saturated **hydraulic conductivity**, can be estimated using a soil texture triangle calculator (LandIS 2016).

If the project may have a significant impact on groundwater abstraction rates, it will be necessary to consider the local aquifer's storage capacity and storage level patterns. It may be important also to know its specific yield: the volume of water that can be withdrawn under the influence of gravity. Brassington (2006) gives indicative values of specific yield for a range of geological materials.

Methods of monitoring groundwater are described in Brassington (2006) and Nielson (2005). Groundwater hydraulics can be studied using (a) pumping tests in which water is pumped from wells, and groundwater flow rates are calculated from observed recharge rates, and (b) models based on the properties of the aquifers. Groundwater models are discussed in Kresic (2006). These can be complex, but they often incorporate a simple formula known as Darcy's Law:

$$V = K \frac{\Delta H}{L}$$

where: V = velocity, i.e. the distance that water can be expected to flow (m/day)
K = hydraulic conductivity (m/day)
ΔH = the difference in hydraulic head (liquid surface elevation) between two points in the aquifer (m)
L = the distance between the two points (m)
$\Delta H/L$ = the groundwater slope

Typical hydraulic conductivity values are given in Atkinson (1999) and Brassington (2006). The groundwater slope can be determined from aquifer maps or from field measurements of water table levels (as explained below). The simple application of Darcy's Law has limitations, e.g. it assumes aquifer homogeneity (with a single hydraulic conductivity throughout) which is rarely the case.

Groundwater storage levels can be monitored by measuring water level changes in wells. Drilling new wells is expensive, but most areas contain existing monitored and/or unmonitored wells. In wetland sites where the water table is normally near the surface, lengths of plastic waste pipe can be inserted into the ground to act as mini-wells. Water level measurements can be made using continuous recorders, or more simply by weekly or monthly observations using a 'dipper'. This consists of an electric probe attached to a graduated cable, and a visual or audible signal that is activated when the probe contacts water. Because of weather-related fluctuations in water levels, monitoring should be continued for at least a year.

Measurements taken at a network of wells can also provide information on groundwater contours, and hence on likely flow patterns. Recorded water-level depths are subtracted from the relevant ground level altitudes to calculate absolute water table elevations. A water table contour map can then be produced to show the groundwater slope(s) and hence the likely direction(s) of flow. Such information may be useful for assessing the vulnerability of a wetland to impacts such as pollution or water abstraction; and for estimating the site's water budget, and in particular the relative importance of precipitation, surface water recharge and groundwater recharge. However, a site water budget can only be calculated if all but one of the variables in the budget equation (§2.2.1) can be measured or neglected – and requires measurements taken over at least a year.

Typical mitigation measures for groundwater pollution are to site development away from any groundwater protection zones; use buffer zones between the site and

Surface water flow (flooding and low flows)

Apart from overland flow, which is normally temporary, surface waters include streams and rivers (or watercourses) and lakes, ponds and artificial reservoirs (or standing waters). Water can also flow in man-made ditches and canals. Unless standing waters are 'off line' from a watercourse they will have direct inflow and outflow to watercourses. Typically there is also movement between the surface body of water and groundwater; this is referred to as baseflow. The contribution of baseflow to river flow varies significantly with the geology and topography of a catchment, and with season. Standing waters also receive water by direct precipitation and overland flow, and lose appreciable amounts by evaporation (Figure 2.1).

Streamflow in rivers is the gravitational movement of water in a channel. Water velocities and depth can be measured at a channel cross section through time allowing the calculation of a streamflow versus time curve, commonly referred to as the hydrograph. The rate of flow is a function of channel slope, cross-sectional area/shape and hydraulic roughness of the channel boundary conditions. River discharge is related to the effective precipitation over a catchment. Once precipitation occurs, direct runoff begins to increase as a rising limb on the hydrograph, levels off at the peak flow, and eventually runoff from storage adds to the overall response of the catchment. After the event, the hydrograph recedes to a low value that is either baseflow or no flow. A proposed development that includes impervious areas can impact the hydrograph by increasing the rate of the rising limb as well as the peak flow. This can exacerbate flooding and erosion risks. Flooding occurs where watercourses do not have the capacity to convey excess water. As natural channels have limited capacity, the water may rise above the bankfull level and spill out onto an adjacent floodplain. These points are illustrated in Figure 2.2.

Flooding may occur from:

- excessive levels of precipitation over a long period of time (causing saturation of the soil and increased runoff);
- intense precipitation over a relatively short period of time (perhaps a few hours) such that the ground cannot soak up the rainfall (with more water reaching the river than would normally occur);
- snow melting while the soil is still frozen, such that infiltration capacity is reduced.

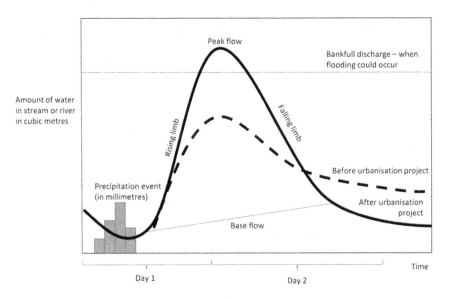

Figure 2.2

Example hydrograph for two hypothetical flood events: before and after an urbanisation project

People have a long history of living on and developing floodplains, from settlements associated with utilising watercourses for transport to large cities, to farming fertile soils, and the construction of roads and railways on the relatively flat floodplain lands. However, these activities affect flow regimes and tend to restrict the capacity of rivers to accommodate large storm events, especially in urban areas where river channels and natural floodplains may be modified, reduced or restricted. This has increased the incidence of serious flooding and has led to further human intervention in the form of flood alleviation/management measures, such as channel modification (e.g. widening and deepening) or embankments (Brookes 1988), in turn further disconnecting the river from its floodplain and altering flow regimes.

In the UK, one in six properties (5.2 million people) are at risk of flooding (Environment Agency 2009). In Italy, 11 per cent of the population (6.7 million people) live in the floodplain and in Hungary it is 18 per cent of the population (1.8 million people, European Environment Agency 2016). In India, on average 4.84 million people are affected by river floods every year (World Resources Institute 2015). Clearly, flood risk and its potential impact on human health and river systems are key reasons for considering stream flow in ESIAs.

Conversely, low flows can be an issue. Climatic factors can cause prolonged dry periods, reducing flows. Low flows can also be caused by over-abstraction either directly from the river itself or from connected groundwater bodies; and from storm water attenuation features such as ponds. This impacts the availability of water as a resource downstream and also impacts aquatic ecology and sediment transport. This is increasingly being recognised as an issue, and guidelines regarding environmental flows are being established around the world including in the USA and Europe (Arthington 2012, European Environment Agency 2010). Environmental flows are the quality, quantity, and timing of water flows required to maintain the components, functions, processes, and resilience of aquatic ecosystems, providing goods and services to people.

Return period or recurrence interval is the most common means of describing the likelihood of a flow or rainfall event. It is a statistical analysis based on measured historic data (gauged flows or weather station data). A return period describes the probability of exceedance in a year, so a two-year return period event has a 50 per cent likelihood of being exceeded in any one year, and a 25-year return period event has a 4 per cent probability of being exceeded in a year. The 100-year event is often used for flood risk mapping. The ESIA process may have to demonstrate that a proposed development will not increase the likelihood of a particular return period event, and therefore will not increase flood risk to properties downstream. Return period flows can also be used to aid in channel design (as a surrogate for bankfull flows) and in low-flow assessments.

2.2.2 Water quality

Surface and groundwater quality is determined by assessing three classes of attributes: chemical, physical and biological. Many countries have set standards for each of these attributes. Aspects of water quality that are usually most relevant in ESIAs are briefly discussed below and at Table 2.2. Further information can be found in the texts listed in §2.1.

Chemical: Water is never pure and (naturally) always contains at least some dissolved chemicals (solutes) originating from the atmosphere, soils or the weathering of bedrock. These include nitrates, phosphates, metals and chlorides. Water chemistry varies widely and depends largely on the geology and climate of the catchment. Freshwater systems have temperature regimes to which the aquatic life is adapted. Temperatures above the normal range can lead to oxygen starvation in freshwater ecological communities because increasing temperature, (a) promotes oxygen consumption by increasing rates of animal and microbial respiration, and (b) reduces the amount of dissolved oxygen held by water.

Physical: Natural waters also vary in the amount of particulate material present. Rivers typically carry large quantities of particulates. Fine suspended particles will not normally settle in a river unless the flow is slow and even in standing waters silt particles may remain in suspension for some time. River bed particulates are called the bedload and can range in size from silts to coarse sands, gravels and boulders. Bedload materials can move, but only when the flow of water exceeds a threshold which is related to both the channel slope and discharge.

Table 2.2 Some key water quality indicators in surface waters

Indicators	Type	Relevance in ESIA
Chlorides	Chemical	Increased chloride concentration in surface water and groundwater is a major concern for the ecological health of sensitive aquatic species and drinking water supplies. The main chloride ion contributor is winter road maintenance (anti-icing and de-icing agents). Other anthropogenic sources include processing of agricultural products, industrial production of chemicals and food, water softening, wastewater treatment plant effluent, septic system effluent, and leachate from landfills. Chlorides do not biodegrade, readily precipitate, volatilise, or bioaccumulate. The persistence of chlorides in the environment does not allow easy or economical treatment of contaminated water.
Metals	Chemical	Manganese (Mn), zinc (Zn) and copper (Cu) are essential to biochemical processes that sustain life. However these metals (and others) are toxic above critical doses. The toxicity and bioavailability of many metals is dependent on the state of oxidation and the form in which they occur.
Nutrients (especially nitrogen and phosphorus)	Chemical	When the surface water environment becomes enriched with nutrients then the process of **eutrophication** may occur, leading ultimately to the development of **algal blooms** which can have a highly detrimental effect on water quality. As the algae grow they use up all available oxygen in a waterbody. They may also shade the water body. Some algae are toxic. Nutrient levels are typically enhanced when fertilisers are used in farming. In many areas, nitrate levels in waterbodies used for drinking water (particularly rivers and aquifers) are now sufficiently high to cause concern.
Dissolved oxygen	Chemical	Dissolved oxygen in water is essential for aquatic life. Levels tend to be highest in flowing rivers because turbulence enhances absorption of oxygen from the atmosphere. Still waters such as ponds and slow-flowing ditches have highly variable oxygen levels. Natural stream purification processes require adequate levels of dissolved oxygen to provide for aerobic life forms. Oxygen depletion can occur through pollution, mainly by organic matter from sources such as sewage, soils, and agricultural or industrial effluents. If dissolved oxygen falls below 5.0 mg/l then aquatic life is put under stress. The lower the concentration then the greater the stress. Large fish kills can arise if levels remain below 1-2 mg/l for a few hours. Reduced oxygen levels can in turn lead to increased levels of potentially harmful chemicals (e.g. ammonia, methane, hydrogen sulphide, and heavy metals) by increasing their production or solubilities.

Table 2.2 continued

Indicators	Type	Relevance in ESIA
Temperature	Chemical	Thermal pollution is the introduction of water that is warmer than the river or stream into which it flows. The most common physical assessment of water quality is temperature measurement. Temperature affects both the chemical and biological characteristics of surface water. It affects the dissolved oxygen content of water, and the photosynthesis of aquatic plants. It also impacts aquatic organisms; some fish species can only survive at temperatures maintained by cold-water baseflow. Sources of thermal pollution include power stations, urban surfaces, disconnection from baseflow sources and removal of the forest canopy which provides shade.
Hydrocarbons	Chemical	Polycyclic aromatic hydrocarbons (PAHs) are found generally as complex mixtures of thousands of compounds which have varying toxicity to aquatic organisms. Toxicity is highest in the aqueous form.
Sediments	Physical	Sediment can be harmful to fish spawning. There is also the potential for sediment to adsorb chemicals. The carbon content of sediments can be used as an indicator of organic matter, including hydrocarbons.
Pathogens	Biological	Microbial pollutants such as bacteria, viruses and protozoa may present considerable risk to human health, aquatic ecosystems and biodiversity. Viral pathogens tend to have a limited host range, so sources are usually limited to waters containing human wastes such as wastewater and agricultural and animal wastes. In many countries pathogens are now regarded as the leading pollutant.

Biological: Biological water characteristics are used to describe the presence of microbiological organisms and waterborne pathogens. Many organisms can lead to illness if consumed by humans and animals. Micro-organisms and waterborne pathogens enter water bodies either naturally or through the release of untreated or partially treated sewage. Pollution sources can be divided into point source and non-point source (diffuse) pollution.

The likely effects of a development or activity on water quality will depend not only on the type, but also on the characteristics and quality of the receiving waters. For example, rivers export most of their pollutants downstream, so the effect at any one point may be transitory, but polluted water and silts may be carried considerable distances before they are sufficiently degraded or diluted to have no effect. Standing waters such as lakes and ponds are sediment sinks and the turnover rate of water is usually slow, so sediment and pollutants tend to accumulate, and impacts intensify with time.

2.2.3 Fluvial geomorphology

Fluvial geomorphology refers to the processes of water and sediment movement in rivers and their floodplains, together with the forms produced by those processes. It can be applied to both natural environments and those altered by humans. Fluvial geomorphology is 'the study of sediment sources, fluxes and storage within river catchments and channels over short, medium and longer timescales and of the resultant channel and floodplain morphology' (Newson and Sear 1993). An understanding of the processes of water and sediment movement in river catchments, across floodplains and within channels, together with the forms produced by those processes, is essential to the assessment and design of sustainable fluvial projects (Maas and Brookes 2010, Sear et al. 2010).

Within Europe, the Water Framework Directive has introduced the much broader term 'hydromorphology' (that encompasses aspects of fluvial geomorphology) with an emphasis on impact assessment and mitigation. This specifically includes:

- Quality and dynamics of water flow
- Connection to groundwater bodies } Process
- River continuity

- River depth and width variation
- Structure and substrate of the river bed } Form
- Structure of the riparian zone

Under natural conditions, processes result in diverse and dynamic habitats, both instream and across riparian and floodplain environments (Anderson et al. 1996, Petts and Amoros 1996). Central to the understanding of fluvial geomorphology is the concept of the geomorphological threshold: 'a threshold of landform stability that is exceeded either by intrinsic change of the landform itself, or by a progressive change of an external variable' (Schumm 1979). Schumm argued that fluvial systems achieve a state of dynamic equilibrium through complex responses. For example, a river that meanders across a floodplain is stable over time within the floodplain (the erosion and deposition processes are in balance), but it is not static.

Table 2.3 summarises the external and internal controls that can affect the geomorphology of a river channel or floodplain. It is also important to distinguish between dynamic equilibrium, which is a fluctuation around an average, and metastable equilibrium, which is where external controls carry a system over a threshold into a new regime. An example of this is where, in periods of lower to moderate flows, lateral migration across a floodplain may be the dominant process, but in large flood events cut-offs occur leading to a change in behaviour. This has important implications when the calculated rate of channel migration to inform a proposed development is based on snapshots of historical information.

Table 2.3 External and internal controls on a river channel or floodplain

External controls	Internal controls
Catchment form	Stream gradient
Drainage network	Mode of adjustment
Flora and fauna	Cross section morphology
Land use	Bed and bank conditions
Modification	Floodplain connectivity
Management practices	River continuity
Geology	Flow regime

Table 2.4 shows a simple geomorphological classification of channels based on cohesiveness. Another classification system (Montgomery and Buffington 1993) is based on a categorization of the landscape and the channel that provides a foundation for interpreting channel morphology and condition, and predicting response to a disturbance. Many other publications list principles for undertaking geomorphological assessments (see Sear et al. 1995, 2010; Maas and Brookes 2010). However, river types vary widely and an impact or mitigation solution applicable in one geography may be totally inappropriate for another.

Geomorphology is a specialist discipline, typically requiring specialist consultants to undertake the ESIA. A fluvial geomorphologist will understand and collect evidence on the controls on relevant water bodies (Table 2.3), drawing on the outputs from other assessments such as hydraulic models and inputs from ecology and geology to supplement or validate their assessment. They will interpret how a proposed development will impact the energy and sediment available to a river system and how it might adjust. They will determine how a development may lead to changes in flow patterns and what the channel response will be. This is useful to determine the impacts of a development on the watercourse as a receiver, but also to assess the risk posed to infrastructure as a result of a channel adjustment. For instance, a geomorphologist can advise on potential erosion rates and direction of channel migration to inform the selection of the location and span (width) of a

Table 2.4 Simple classification of channel types (Knighton 1998)

Primary type	Secondary type
A Cohesive	A1 Bedrock channels
	A2 Silt-clay channels
B Non-cohesive	B1 Sand bed channels
	B2 Gravel bed channels
	B3 Boulder bed channels

proposed bridge. They can help to ensure resilient, self-sustaining systems, functioning aquatic and riparian habitats, and minimal maintenance and management costs.

2.3 Key legislation, guidance and standards

2.3.1 Legislation

Water is an integral component in ESIA. For instance, the European EIA Directive 2014 requires consideration of 'water (for example hydromorphological changes, quantity and quality)'. Water is also a supporting factor to many other ESIA components, for example biodiversity, soil and human health.

Water quality (including groundwater protection) and geomorphological standards are usually set by governments and regulated by government bodies and agencies. Typically in a country, the most stringent standards are set for drinking water (see WHO 2011). Standards usually exist for regulation of waste water from domestic and industrial sources. Restrictions may also be placed on the volume of water abstraction in specific periods of time. This is particularly the case for low-flow periods as well as flood standards. Specific flows may be required in a stream or river to support aquatic fauna and flora and one method for calculating the requisite flows is the Instream Flow Incremental Methodology. This is regarded as one of the best available methods for determining the instream flows that should be protected and preserved to provide adequate fish habitat. It is commonly used in the USA (Bovee 1982). Other types of water-related assessments may also be required for specific consents and licences. This will vary dependent on the country, but in European countries they may include:

- Water Framework Directive (WFD) Compliance Assessment to determine if there is likely to be any deterioration of ecological status as a result of development activities. The WFD (European Commission 2000) makes 'hydromorphic condition' a key building block in the assessment of compliance with the regulations. Hydromorphic condition is typically described as the interrelationship between flow regime and the channel perimeter. Geomorphological baseline information is common to both the ESIA and WFD Compliance Assessment and any associated mitigation. Typically a WFD Compliance Assessment forms a Technical Appendix to a surface water ESIA chapter.
- Flood Risk Assessment, to minimise the likelihood of new projects being subject to flooding and/or increasing flooding elsewhere.

Parallel tracking of these processes with the ESIA may be an efficient and effective way to proceed.

2.3.2 Policies

The World Bank requires ESIA of projects proposed for bank financing through its Operational Policy 4.01 (World Bank 1999) and Environmental Assessment Sourcebook (World Bank 1991). Policy OP4.07 (World Bank 1998) is concerned specifically with water resources management. Bank involvement in water resources management includes support for flood control, potable water, sanitation facilities and water for productive activities in a manner that is economically viable, environmentally sustainable and socially equitable. The Bank assists borrowers in key priority areas, including consideration of cross-sectoral impacts in a regional setting (e.g. a river basin), restoration and preservation of aquatic ecosystems, and avoidance of the water-logging and salinity problems associated with irrigation investments. The World Bank (1998) specifies pollution and abatement measures and emission levels that are normally acceptable to them. The ESIA needs to take account of any variation on pollution prevention and abatement measures that may be requested locally in a specific country.

2.4 Scoping

The scoping stage involves determining the types of activity likely to be involved in the development from construction, through operation to decommissioning (if relevant); the length of each of these stages; and any alternatives considered to date. A clear understanding of how the proposed project will operate relative to the local surface water environment is necessary.

All that is usually required at this stage is an early identification of potential impacts based on professional judgement informed by qualitative sources. A wealth of published information obtained from an internet search of specific projects can be used to scope projects and to identify potential impacts. Specific guidance produced by regulatory bodies can also be found (e.g. Environment Agency 1996). Figure 2.3 gives a generic indication of how water quantity, water quality and fluvial geomorphology parameters can be changed by various development activities. This has been simplified from the original work by Douglas (1983) to show potential increases and decreases (changes) as a result of urbanisation. No attempt is made here to delineate whether these changes are likely to lead to positive or negative effects. Each change is likely to be site specific and assessed locally as required.

Table 2.5 shows an example scoping checklist for dam operations in Pakistan, where many such developments have been taking place (a similar checklist exists for dam construction). Dam development has progressed over the last few decades as there is a greater understanding of the potential impacts, both operational and during construction. Table 2.5 shows that water-related impacts of dam operation are typically quite extensive, ranging from hydrological to water quality and sediment/morphological issues; and covering the reservoir area and kilometres downstream.

		Modification of land cover or landform						Construction												
		Removal of plant cover	Bulldozing of land	Gravel extraction	Clay extraction	Cut and fill	Terracing	Foundation works	Unsealed roads	Paved roads	Kerbing	Guttering	Isolated buildings	Gardens	Mass housing	Car parks	Office buildings	Warehouses and factories	Storage yards	Airports
Water quantity	Precipitation														•	•	•	•	•	•
	Interception	○	○	○	○															
	Surface runoff	●	●	◦	◦	◦	◦		●	●	●	●	•	◦	●	●	●	●	●	●
	Infiltration	○	○	◦	◦	•	•						◦	•	◦	○	◦	◦	◦	○
	Throughflow	•	◦	•	•	•	•	◦					◦	•	◦	○	◦	◦	◦	○
	Water table level	•	◦	•	•	•	•	◦					◦		◦	○	◦	◦	◦	○
	Flood height	■	■	▫	▫	▫	▫	▫	■	■	■	■			■	■	■	■	■	■
	Flood duration	▫	▫	▪	▪	▪	▪		▫	▫	▫	□			□	□	□	□	□	□
	Base flow	▫	▫	▪	▪	▪	▪		▫	▫	▫	□			□	□	□	□	□	□
	Evaporation	•	•	•	•	•	•	•	•	●	●	●	•	◦	●	●	●	●	●	●
	Transpiration	○	○	○	○	○	•		◦	○	○	○	◦	●	○	○	○	○	○	○
Water quality	Sediment concentration	●	●	●	●	•	◦		●	●	•	•		•	•	•	•	•	●	•
	Solute concentration	•	•	•	•	•	•		•	•	•	•	•	•	•	•	•	•	•	•
	Organic concentration	•	•	•	•	•	•		•	•	•	•	•	•	•	•	•	•	•	•
	Trace elements	•	•	•	•	•	•		●	●	●	●	•	•	•	●	•	●	●	●
	Dissolved oxygen	◦	◦	•	•	•	•		◦	◦	◦	◦			◦	◦	◦	○	◦	◦
Fluvial geomorphology	Channel stability	□	□	□	▫	▫	▪		□	▫	▫	▫			▫	▫	▫	▫	▫	▫
	Bank erosion	■	■	■	▪	▪	▫		■	▪	▪									
	Channel extension	●	○	◦	◦				●	•										
	Gully erosion	●	○	◦	◦				●	•										
	Channel aggradation	■	■	■	▪				■	▪	▪	▪			▪	▪	▪	▪	▪	▪
	Silt deposition	●	■	■	■	•	●		■	▪	▪	▪								

		Water supply					Waste disposal					Channel modification						
		Sinking wells	Surface storage	Major impoundments	Allogenic water supply	Municipal wells	Sanitary fill	Septic tanks	Sewer systems	Return of treated wastes	Industrial waste water	Road culverts	Bridges	Channel realignment	Flood protection works	Dumping of debris in channels	Stormwater drainage	Canalisation of rivers
Water quantity	Precipitation																	
	Interception																	
	Surface runoff		○	○													●	
	Infiltration																◦	
	Throughflow	○					•	•										
	Water table level	○	•	●	•	◦	•	●	•									
	Flood height		▫	□										■	■	■	■	■
	Flood duration		▪	▪											■			□
	Base flow		■	■	■	□	▪	▪	■	■	▪				▫		□	▫
	Evaporation		●	●	•													
	Transpiration																	
Water quality	Sediment concentration		▫	□							•					●	•	
	Solute concentration		•	•	•		•	•	•		●					•	•	
	Organic concentration		•	•	•		●	●	●		●					•	•	
	Trace elements		•	•	•						●					•	●	
	Dissolved oxygen		•	•	•		○	○	○	●	•					◦		
Fluvial geomorphology	Channel stability		▫	□								□	□	●		□		●
	Bank erosion		▪	■								■	■	○				○
	Channel extension																	
	Gully erosion																	
	Channel aggradation											■	■	■		■		
	Silt deposition											■	■			■		

Key: type of effect

	increase		decrease	
	major	minor	major	minor
on site	●	•	○	◦
downstream	■	▪	□	▫

Figure 2.3

Potential changes to the surface water environment (without mitigation) caused by activities involved in urbanisation (modified following Douglas, 1983)

Table 2.5 Scoping checklist for potential water-related operational impacts, with particular reference to dam schemes (adapted from Meynell and Zakir 2014)

Issues	Sources of impact	Potential impacts
Surface water hydrology	Storage reservoir	Reduction in peaks in high flows
	Run-of-river power (hydropower on a small scale, usually without large water storage capacity)	Increase in flows in the low-flow season
		Little seasonal change in downstream flows, creating a more uniform flow regime
	Daily peaking operation	Flows downstream will vary over 24 hours altering flow regime
	Base load operation	No daily variations in flow
River hydrology	Flow diversion/ alteration of flows downstream	Dewatering
		Altering flow regime and subsequent impacts on morphological processes and features
		Change in habitats and biota
		Altering the fish species and productivity of the river
		Disturbance to recreational users
		Potential changes to erosion patterns (i.e. exacerbating erosion problems)
Surface water quality	Reservoir – standing water	Breakdown of residual vegetation and organic matter could lead to oxygen depletion (primarily in lower level reservoirs)
		Thermocline changing, bring poor quality anoxic water to the surface
		Algal blooms from nutrient loading
	Downstream watercourse	Altering chemical composition by changing sediment sources and water sources
Erosion and sedimentation	Accumulation of sediment	Shorten reservoir life
		Starve downstream catchment of sediment
		Delta formation raising river bed level causing potential flood risk
		Removal of nutrient input downstream (those carried by the sediment)
		Downstream erosion due to altered sediment regime
	Changes to hydrology	Erosion of reservoir edges from wetting and drying

Table 2.5 continued

Issues	Sources of impact	Potential impacts
Geomorphology	Presence of a dam	Changes in flow regime and downstream morphological processes and features
		Scouring of the river bed and banks
		Channel bed created by clearwater emerging from the dam and eroding away smaller sediment sizes, leaving larger particles (or an armoured layer)
Aquatic ecosystems	Reservoir and dam	Direct fluvial habitat loss
		Removal of migration routes upstream and potentially spawning habitats

An alternative or complementary approach is the source–pathway–receptor model. The source of an effect is a specific development activity. The pathway is the mechanism by which that source can affect a receptor (such as via hydraulic effects). Receptors can include physical, human or fauna/flora generally. A watercourse can then be considered as a pathway and a receptor.

Potential physical effects to the water environment from development include:

- river channel modification (e.g. by dredging) to increase depth to facilitate arterial drainage;
- river channel modification to increase channel capacity for conveyance of flood waters;
- river channel modification to facilitate navigation;
- river channel diversion to facilitate development;
- maintenance dredging;
- elimination of freshwater lakes by development;
- building a new dam (leading to loss of a water body and downstream impacts on water and sediment discharge);
- discontinuance of a dam or a weir with potential to cause erosion and lead to sediments being released downstream.

Potential water quantity effects include:

- replacement of 'natural' vegetated or bare earth surfaces with impermeable paved areas and man-made drainage systems leading to decreased infiltration to groundwater (potentially decreasing aquifer levels), increased flood runoff and potential for downstream flooding and erosion;

- abstraction from rivers which can cause reduced flows, leading to deposition of fine silt and (in some instances) channel narrowing;
- new development which can increase the burden on the water supply (from rivers/aquifers) and increase the volume of waste water (from treatment works).

Potential water quality effects include:

- artificial discharges to the water environment which can lead to changes of water chemistry (and ultimately pollution when the receiving environment can no longer absorb the chemicals). Pollutants from new road surfaces (for example) can include heavy metals, rubber, paint, hydrocarbons and salts;
- artificial cooling water discharges to the environment from a power station leading to thermal changes (affecting ecological and human receptors);
- direct point-source inputs such as from industry, agriculture and wastewater treatment works (e.g. increased nutrients leading to eutrophication and algal blooms);
- diffuse pollution (including fine sediment inputs) from ploughed fields;
- leaching of pollutants from previously developed sites which are proposed for redevelopment;
- detriment to ecological status of a water body by increased pollutants issuing from an urban area (such as oils).

Potential geomorphological effects include:

- increased channel erosion in streams and rivers below newly paved areas;
- effects on sediment conveyance by construction of a new dam (leading to downstream channel adjustment);
- changes to erosion and sedimentation patterns caused by in-channel structures such as flood walls, weirs, outfalls and bridge piers;
- changes to flows below an urban area or dam affecting the morphology of a channel;
- in-channel mining leading to localised erosion upstream and in tributary channels;
- diffuse point sources of fine sediment entering the channel from adjacent farmland, adversely affecting the channel morphology;
- channel straightening leading to slope adjustment and incision;
- changes to the sediment and flow regime caused by river channel restoration.

The scope of works for a modelling study needs to be determined by a competent modeller. The scope needs to be proportionate to the ESIA and to the stage of design. The accuracy of the model needs to be stipulated – for example whether water levels, volumes and flows are accurately represented.

The extent of the study area (upstream and downstream catchment) needs to be defined. The boundary conditions of a model should be defined by a known physical discontinuance such as a weir or known water level. The scoping process can also assist in identifying any legal standards and guidance for the country in question, and to gain an indication of how current conditions are with respect to these standards. If information is not available on standards from that country, it may be necessary to base the ESIA on generic criteria developed from similar or nearby geographic areas.

2.5 Baseline

This stage involves describing existing baseline conditions for water quantity, water quality and fluvial geomorphology for the area potentially impacted by a project. Table 2.6 summarises typical data collection methods for the surface water environment. Different data collection standards apply to different countries, and professional judgement informed by some data/information may be all that is available. Again, the use of expert consultants is recommended.

Collection of field data for hydrological parameters can be difficult, time consuming, and require sampling over extended periods. Resource and time constraints in ESIA often impose severe limitations on the range and depth of field survey work that can be undertaken, and it is important to maximise the use of existing data by means of the desk study.

It is usually obligatory for the relevant environmental protection agency to provide the developer (on request) with any relevant ESIA-related information in their possession. Other useful sources of information include local authorities, angling clubs, local universities, previous ESIAs, studies and scientific papers. Historical information may also be relevant as it may contain information on geology and soils quality.

Where baseline surveys and monitoring is required, these should be carefully designed. Baseline flow and water quality studies may be needed, together with a geomorphological walkover survey to characterise the channel. Limited data are often misleading, and surveys should aim to ensure validity in terms of accuracy of measurements, number of samples, length of sampling period and frequency of sampling.

The baseline includes information on likely future conditions without the development. This may involve looking at empirical information on the known impacts of developments elsewhere (e.g. along the same river system). Only in a relatively few cases is the impact study area confined to the project site and its immediate surroundings. Impacts on the surface water environment are likely to be much more widespread, especially in the downstream direction. This can hinder accurate determination of the impact area, and initial estimates may have to be changed in the light of information obtained during the assessment process.

Table 2.6 Typical sources of information and data for surface water

Discipline	Desk study	Field study
Water quantity	Catchment boundary/area and drainage patterns, and information on geomorphology (especially slopes, geology, soils and land use) determined from base maps, geological and soil survey maps. Use of digital terrain models.	Temporary stream gauging to determine flow regime (for high flows and low flows) is needed but this can be expensive. Stream gauging methods are discussed in most hydrology texts, e.g. Boiten (2000), Gordon et al. (2004) and Herschy (1999).
	Procurement of existing flow records or estimation through use of rainfall-runoff models (e.g. to estimate flood risk).	If an assessment is needed of a length of river, this is normally divided into reaches (sections of fairly uniform morphometry and flow), which are then used as study units (can be obtained from desk study or field survey).
	The limits of a river floodplain may already exist based on topography and hydraulic modelling. This could be the approximate extent of floods with a 1% annual probability of exceedance (1-in-100 year flood) or the highest known level.	Detailed survey data to input to hydraulic models.
Water quality	Use of existing information (if available). Environmental protection agencies in many parts of the world hold data on chemical pollutants and on biological surveys used to assess water chemistry.	Surveys of river chemistry (e.g. nitrates, oxygen, pH etc.) initiated for the purposes of ESIA often have limited value as time constraints usually mean resorting to spot checks rather than a long programme of continuous monitoring. This is often an expensive process. However biological or fisheries surveys are of considerably more value as they give a reliable indication of the long-term health of the water environment. Most groups of freshwater organisms have been used as indicators of given pollutants; but macroinvertebrate families (not species) are by far the most widely used taxa in Europe and North America (for example). Many aquatic vascular plants are sensitive to water and sediment nutrient concentrations, and methods for
	Routine monitoring may be available for human faecal contamination of water. In some countries bathing waters, and surface waters used for extraction of water intended for human consumption, are also monitored for faecal streptococci and Salmonella.	

Table 2.6 continued

Discipline	Desk study	Field study
		assessing eutrophication in rivers using plants exist. Biological monitoring methods are now available in some parts of the world for still waters (lakes, ponds, canals, ditch systems). A disadvantage of these methods is that it is not possible to determine the exact pollutant impacting a system.
Fluvial geomorphology	Channel change over time determined from historic and contemporary maps and aerial photographs (using GIS overlay if possible). Desk study using existing survey information including geologic and topographic maps. Use of topographical (cross-section) information and digital elevation models to calculate stream power.	Fluvial audit or stream reconnaissance survey to collect information on channel size and slope, erosion or deposition, and features such as pools and riffles, and to characterise the channel in terms of existing impacts of modifications, etc. Survey of the channel to obtain accurate slope and bankfull characteristics and to record geomorphological features. Sediment analysis might include bedload and bank samples and/or pebble counts. Increasingly, the use of unmanned aerial vehicle technology to complete aerial surveys obtaining, for instance, elevation data and aerial photograph imagery.

2.6 Impact prediction and evaluation

Table 2.7 summarises potential direct impacts on the water environment arising from different development types, and Table 2.8 shows indirect impacts. These tables should not be used without consideration of local data and information, informing expert judgement. Where possible (or where required by a regulator) predictions should be quantified and tested against any legal standards.

Table 2.7 Impacts from direct manipulation or utilisation of hydrological systems (based on Kelday et al. 2009)

Sources	Potential impacts
River engineering/ manipulation	
Resectioning/channelisation (widening, deepening, realigning/ straightening), e.g. to increase channel capacity for flood defence or drainage, or to facilitate project layout.	Loss of channel and bank habitats. Enhanced erosion and hence silt production (especially during construction, when pollution risks also increase). Increased flood risk and siltation downstream. Lowering of floodplain water table caused by deepening.
Embanking and bank protection (e.g. with concrete) usually for reasons as above.	Floodplain inundation and siltation prevented, with consequent risk of soil drought and loss of wetlands. Drainage from floodplain inhibited (unless sluices installed) with consequent waterlogging. Potential for failure leading to flooding and erosion if flows exceed design criteria. Installation of hard engineering such as concrete blocks and steel sheet piling can lead to erosion downstream as energy is transferred.
Clearing bank vegetation.	Loss of wildlife habitats and visual/amenity value.
Fluvial dredging and deposition of dredgings, e.g. to maintain/ enhance flood capacity or navigation.	Damage to channel morphology, habitats and biota at dredging sites. Increased sediment load and hence turbidity and smothering of downstream benthic and marginal ecosystems.
Diversion, e.g. to increase water supply to receptor area, or as a flood relief channel.	Decreases supply in donor area. Channelisation and evaporative loss from open channels. Risk to habitats in main river corridor.
Culverting.	Direct loss of open channel length, often associated with an increase in velocities and potentially back up of high flow events. Can lead to erosion and flooding.
Development on river floodplains	
Use of floodplain area. Construction of flood defences. Laying of impermeable surfaces.	Increased flood risk upstream and downstream. Reduced groundwater recharge and river baseflows. Loss of ecological, heritage and visual/amenity/ recreational features.
Reservoirs and dams	
General.	Loss of terrestrial habitats/farmland/settlements. Local climate change and rise in water table. Visual impacts of retaining walls. Water-borne pathogens. Earthquake/landslip/failure risks.

Table 2.7 continued

Sources	Potential impacts
On-stream dams: above dam siltation.	Loss of river section; changes in flow regime.
On-stream dams: below dam.	Reduced flows, oxygen levels and floodplain siltation.
On-stream dams: barrier effects.	Migration of fish and invertebrates blocked. Impacts of sediment transport.
Off-line dams (not on a main channel).	Changes in groundwater recharge, levels and flow directions.
Irrigation	Water abstraction (often from rivers). Increased evapotranspiration and local runoff. Risk of waterlogging and salination.
Drainage schemes (possibly involving channelisation)	Increased soil drought risk and oxidation of organic soils. Water table lowered and wetlands lost. Increased flood/erosion risk downstream.
Water abstraction	Water resources depleted. Water table lowered. Risks of river low flows, loss of wetlands, soil droughts and subsidence.
Sewage treatment works	Increases in silts, nutrients (especially if treatment is poor), heavy metals, organics, and pathogens, e.g. faecal coliforms.
River crossings (e.g. bridges, pipelines, cables)	Constriction of flow, scour, localised erosion and requirement for installation of bed and bank protection.

Impact prediction can be qualitative (typically informed professional judgement). It can be assisted by the uses of checklists, matrices, flow charts and networks analyses, but these tools do not assess the nature, magnitude or significance of the impacts. More complex mathematical modelling may be required, for instance to determine the impact of a development's activities on flood risk through hydraulic modelling. Laboratory methods are also used in certain situations to determine changes to water chemistry from activities (e.g. leaching from dredged materials). A simple form of hydraulic analysis – hydraulic flood routing – involves the mass balance equation of unsteady flow. More commonly, hydraulic modelling is carried out. Quantification is recommended wherever possible to avoid subsequent challenge and also to assist in developing appropriate and proportional mitigation.

Numerous models have been developed for simulating, and predicting changes in, hydrological systems. Reviews are provided in many hydrology texts

Table 2.8 Impacts indirectly associated with manipulation or utilisation of hydrological systems (based on Kelday et al. 2009)

Sources	Potential impacts
Roads	Changes in drainage systems, e.g. due to gradient changes, bridges, embankments, channel diversion or resectioning. Drawdown by dewatering when deep cutting. Increased runoff from impermeable surfaces, with risks of flash floods and erosion. Increased sediment loads from vehicles, road wear, and erosion of cuttings and embankments. Pollution of watercourses by organic content of silt, other organics (e.g. oils, bitumen, rubber), de-icing salt (and impurities), metals (mainly vehicle corrosion), plant nutrients and pesticides from verge maintenance, and accidental spillages of toxic materials.
Urban and commercial development	Changes in drainage due to landscaping. Abstraction. Drawdown/changes in groundwater flow, e.g. when dewatering deep foundations. Reduced groundwater recharge, and increased runoff velocities and volumes (with flood and erosion risks from rapid stormflows and flashy hydrographs) due to impermeable surfaces. Pollution of watercourses and groundwaters by a wide range of pollutants which are rapidly transported to receiving waters by increased runoff. Increased sewage treatment (Shaw 1993, Walesh 1989).
Industrial development	As above but with: greater runoff effects (from a higher proportion of hard surfaces); higher pollution levels and a wider variety of pollutants including metals and micro-organics from heavy industry and refineries, pesticides from wood treatment works, and nutrient-rich or organic effluents from breweries, creameries etc. Thermal pollution from power plants.
Mineral extraction	*Operation phase* – Removal/realignment of watercourses. Loss of floodplain storage/flow capacity. Drawdown and reduced local streamflows caused by dewatering for dry extraction, or increased runoff from process wash water or extraction methods involving water use. Increased siltation and chemical pollution downstream, e.g. from spoil heaps/vehicles/machinery/stores. *Restoration/aftercare phase* – see landfill.
Landfill	Increased runoff from raised landforms, especially if clay-capped. Reduced groundwater recharge and river baseflows if clay sealed. Pollution of groundwater and near-surface runoff by **leachates** and by fertilisers and pesticides from restored grassland.
Forestry and deforestation	Reduced evapotranspiration and infiltration after felling, with consequent (a) decreased groundwater recharge, (b) increases in runoff, soil erosion, stream-sediment loads and siltation. Pollution by pesticides, especially herbicides used to prevent re-growth after clear felling.

Table 2.8 continued

Sources	Potential impacts
Intensive agriculture	Increased runoff and erosion from bare soils. Drainage or irrigation impacts. Pollution of surface and groundwaters by fertilisers; pesticides; organics from soil erosion, silage clamps and muck spreading; heavy metals from slurry runoff; pathogens in animal wastes.

and the use of models in ESIA is discussed by Atkinson (1999). Physical models are sometimes used, but most modelling involves the mathematical and statistical analysis of input data. Some calculations can be made using a hand calculator or computer spreadsheet (e.g. Thompson 1998, Wanielista et al. 1997), but more detailed modelling is carried out using software packages, many of which can be run on PCs. Digital data are becoming increasingly available and may facilitate the use of GIS and/or hydrological and hydraulic models. Table 2.9 shows selected government software, much of which is free. Commercial packages are also available. Many of these approaches were developed for specific geographies and may therefore not be more widely applicable.

The use of models has limitations, especially in relation to the time and resource restrictions common in ESIA. Some software is expensive, although much of that available from US agencies, for example, is free. Most models need expert input by a hydrologist/hydraulic engineer and even simple models should be used only under supervision by a competent hydrologist. Models can only be as good as the input data, and inadequate data can be a major source of error. The current capabilities of models are often limited by incomplete understanding of hydrological systems and even complex models 'necessarily neglect some factors and make simplifying assumptions about the influence of others' (Schwab et al. 1993). Finally, all predictions have a degree of uncertainty and should be validated throughout the life of a project.

2.6.1 Water quantity

Typical questions that should be considered in relation to potential surface water quantity impacts of a project are:

- Is a river channel/corridor, standing water or wetland feature likely to be significantly affected by the project through activities such as river works or realignment of a watercourse?
- Is flood risk likely to be increased through project activities constricting a river channel, reducing floodplain storage or increasing surface water runoff?
- Are surface water and/or groundwater levels likely to be decreased (resulting in low flows risk)?

Table 2.9 Examples of hydrological, hydraulic modelling and water quality software provided by government agencies

Type of software	Description	Web address
Hydrologic (watershed model) – USA	**HEC-HMS** (Hydrologic Engineering Center Hydrologic modelling system). Simulates the complete hydrologic processes.	www.hec.usace.army.mil
Hydrologic (flood estimation) – UK	**FEH CD-ROM** Digital descriptors (e.g. boundaries, drainage paths) for catchments ≥ 0.5 km^2; rainfall depth-duration-frequency (DDF) data for catchments and 1 km grid points; a user-interface for selection of catchments/points; facility to compute design rainfalls, or estimate rainfall event rarity, from DDF data.	www.ceh.ac.uk
Hydrologic (flood estimation) – UK	**WINFAP** – given annual maximum flood data for a site, can estimate probable events, e.g. the magnitude of an event in a given return period, or the return period of a flood of given magnitude. **WINFAP-FEH** – Flood frequency analysis methods using FEH CD-ROM.	www.ceh.ac.uk
Hydrology – USA	**WinTR-55** – US Department of Agriculture. Urban hydrology for small watersheds.	www.nrcs.usda.gov/wps/portal/nrcs/detailfull/national/water/?cid=stelprdb1042901
Hydrology – Canada	**WATFLOOD** – used to forecast flood flows.	www.watflood.ca
Hydrology-hydraulic water quality – Canada	**EPA SWMM** – a dynamic hydrology-hydraulic water quality simulation model used for analysis and design of stormwater infrastructure.	www.epa.gov/water-research/storm-water-management-model-swmm
Hydraulic (flood flow analysis) – USA	**HEC-RAS** (Hydrologic Engineering Center River Analysis System). Computer model of the hydraulics of water flow through natural rivers and other channels.	www.hec.usace.army.mil
Hydraulic – France	**TELEMAC-2D** – used to simulate free surface flows in two dimensions of horizontal space.	www.opentelemac.org/index.php/presentation?id=17

Table 2.9 continued

Type of software	Description	Web address
Hydrodynamics, sediment transport, morphology and water quality – Holland	**DELFT 3D** – for use in fluvial, estuarine and coastal environments. Calculates non steady flow and transport phenomena.	http://oss.deltares.nl/web/delft3d
Hydraulics/ Geomorphology – USA	**SIAM** (Sediment Impact Assessment Model) – used to evaluate the impact of a range of catchment erosion control measures on sediment delivery to downstream sensitive sites.	www.hec.usace.army.mil
Geomorphology – USA/UK	**STREAM POWER SCREENING TOOL** Flood Risk Management Research consortium. Simple tool for calculating stream power in any river and related potential for change (e.g. of sediment movement).	https://web.sbe.hw.ac.uk/frmrc/downloads/UR9%20signed%20off.pdf
Water quality – USA	**BASINS** – GIS/model for pollutants from point and nonpoint rural and urban sources. **QUAL2E** – max. daily chemical streamloads in relation to dissolved oxygen.	www.epa.gov
Water quality – USA	**WATERSHEDSS** – online package to assist in formulating mitigation/ management for non-point source pollution. Includes information on pollutants and sources, and a linked GIS/water quality model.	www.epa.gov/exposure-assessment-models/watershedss
Water quality – UK	**HAWRAT** (Highways Agency Water Regulation Assessment Tool) – assesses the risk of pollution to surface waters from highway runoff. Looks at soluble pollutants (e.g. zinc, copper) and sediment-bound pollutants.	http://infrastructure.planningportal.gov.uk/wp-content/ipc/uploads/projects/TR050002/2.%20Post-Submission/Application%20Documents/Environmental%20Statement/5.2%20ES%20Appendix%208.4%20140409_HAWRAT_Outputs-Catchment%20E%20(M1%20Junction).pdf

Table 2.9 continued

Type of software	Description	Web address
Water quality – Australia	**AusRivAS** (Australian River Assessment System) – an adaptation of the River Invertebrate Prediction and Classification Scheme allows rapid sampling to be used for predictive models of macroinvertebrate communities using a reference site database. Comparisons are made between 'predicted' and 'actual' taxonomic compositions to indicate a site's ecological health. Can assess the biological responses to changes in water quality and/or habitat condition in rivers and be integrated with the existing network of physico-chemical water quality monitoring sites.	http://ausrivas.ewater. org.au
Water quality – USA	**NAWQA** (National Water Quality Assessment) – determines status and trends in contaminant, nutrient and sediment loads and concentrations and related trends in ecological conditions. Uses ground monitoring (including water quality monitoring sites and aquatic ecological conditions) and integrates this with modelling and other scientific tools to extend the water quality knowledge to unmonitored sites.	http://water.usgs.gov/ nawqa
Water quality – worldwide	**World Water Quality Assessment** – UNEP assessment that aims to identify current and future problem areas of freshwater quality in surface waters, and to propose and evaluate policy options to address problems. Involves 1. A bottom-up analysis of data sources, and 2. Top-down modelling of water quality on the continental scale.	www.unep.org/dewa/ Portals/67/pdf/ WWQA_24_March_ 2014.pdf

Physical, hydraulic and computer modelling are all used to predict the impacts of river works. Hydraulic modelling is the calculation of water surface elevations and velocities along a watercourse. It can be used to estimate return period flows or continuous flow as well as changes to those flows as a result of a development. It is important to understand from an examination of the catchment the importance of key flow paths and complicated hydraulics. Key flood risk areas need to be identified and mapped, and it may be necessary for a walkover survey to ground truth the results. The likely interactions of fluvial, pluvial, and tidal sources of flooding should also be defined.

All types of surface water models describe the physical system using mathematical equations (typically, partial differential equations). Trying to represent nature by equations introduces errors, since all models simplify the physical processes in the real world. There are one, two and three dimensional models (1D, 2D and 3D) as well as 1D/2D coupled models. 1D models are the simplest and most widely used: flow is calculated in one direction (generally upstream to downstream) and an average velocity in the channel cross section and water surface elevation is derived. HEC-RAS (see Table 2.9) is a relatively simple tool not suited to complex flow situations but it can be used to assess the impacts of bridges and channel works on river flows and downstream flood levels. 2D models can be used for more complicated projects, for instance where secondary flow circulation or overland flows are an important consideration. These models use a square grid digital elevation model to estimate the flow regime in the channel and floodplain. 3D models are more widely used for sediment transport analysis.

Typical steps in developing a hydraulic model (for flood risk purposes) are:

1 Develop the scope of works (based on a catchment understanding).
2 Collect existing data (e.g. flow data, historic flood levels, reports and photographs of historic flood events to calibrate the model).
3 Collect other existing data (e.g. mapping data, aerial photography, digital elevation data; topographical survey of the channel and floodplain; existing models; information on existing levels of protection).
4 Collect new data sets (e.g. updated and detailed topographic survey and watercourse crossing structures).
5 Quality assure and manage the data collected.
6 Build the model (starting with a conceptual model and choice of the level of modelling required).
7 Represent the river in the model (schematisation) – this links in-channel to floodplain flows and storage and considers flood routes.
8 Model boundary conditions (e.g. hydrologic inputs (inflows) to the hydraulic model).
9 Introduce channel sections, hydrology, input structures and out of bank areas of flow to the model.
10 Undertake model simulations.
11 Complete sensitivity analysis.
12 Undertake model calibration.
13 Complete verification and validation of the model.
14 Identify modelling risks and mitigation measures.

2.6.2 Water quality

Techniques and methods for predicting changes in water quality are discussed by for example Kiely (1997) and Singh (1995). They range from using macro-invertebrates and data registers to infer conditions, to specified pass/fail mechanisms (see Table 2.9).

Point-source pollution is relatively straightforward to predict and all point-source pollutants discharged to controlled waters are likely to require a consent licence from the relevant environmental protection agency. In considering the application, the agency will examine potential discharges in relation to the relevant water quality objectives and standards, including those for designated waters. If a proposed development does not require a consent licence, but could still pose a threat (e.g. through accidental spillage), the same criteria can be applied. If adequate data can be obtained, a mathematical model may be employed.

Bespoke methods exist in individual countries. In the UK, the HAWRAT (Highways Agency Water Assessment Risk Assessment Tool) assesses both short-term risks related to the intermittent nature of road runoff and long-term risks of chronic pollution. The assessment is informed by details of the highway catchment draining to an outfall, the flow rate of the receiving watercourse and its physical dimensions, to calculate the available dilution of soluble pollutants and potential dispersion of sediments:

- Step 1 allows an initial check to assess the quality of the direct highway runoff against the toxicity thresholds assuming no in-river dilution and no treatment or attenuation. If Step 1 shows that the toxicity is acceptable, then no further assessment is necessary.
- Step 2 takes account of the diluting capacity of the watercourse which receives the road runoff and the likelihood and extent of sediment deposition. If failures are highlighted by Step 2, Step 3 allows assessment of in-river impacts post-mitigation (for example, treatment, flow attenuation and settlement).

In Canada and the USA, a significant threat to surface water and groundwater quality is chloride contamination from the application of road salts. Road salts, particularly sodium chloride, represents the largest chemical loading to Canadian surface waters. Road salts can impact drinking water supplies as well as aquatic ecosystems. A method for calculating the chloride loading to salt vulnerable areas was developed by Betts et al. (2015). This uses readily available GIS data to identify areas vulnerable to road salts. The methodology quantifies the risk to identified areas to prioritise implementation of best management practices, and can be used to guide development options.

Estimating the amount and effect of non-point-source pollution is generally more difficult. There are relatively few methods and they tend to have limited capability. Commonly used methods include the Unit Load Method, the

Universal Soil Loss Equation, and the Concentration Times Flow Method. Walesh (1989) reviews the applications and drawbacks of these and other methods, some of which are incorporated in computer models. As a precaution against the uncertainty inherent in many of these methods, more than one should be employed wherever possible. An additional problem is that many projects will not in themselves cause significant impacts, but may contribute to the cumulative impacts, e.g. from an existing urban area.

2.6.3 Fluvial geomorphology

Failure to address geomorphological issues prior to undertaking activities or direct interventions in rivers frequently leads to undesirable consequences. A proposed development may introduce an instability to the fluvial system, impacting channel equilibrium. Many river and riparian/floodplain environments have already been modified (Brookes and Shields 1996, Wohl 2011) with detrimental effects. Adjusting a channel flow or sediment load can lead to changes in cross-section width and depth, the slope of the channel bed, roughness (resistance to flow), velocity and sediment transport. In turn, this can cause channel adjustment in the form of erosion or deposition until the system reaches a new dynamic equilibrium. Perhaps one of the most ubiquitous and marked adjustments is to channel slope caused by straightening (Brookes 1988). Straightening a watercourse reduces its length, meaning that more energy is available to do geomorphological work. This can lead to downcutting of the channel bed (incision) as the channel adjusts to its reduced length. Impacts are not always confined to the modified reach but may be transmitted upstream and downstream and along tributaries.

For some channel types, studies and guides are available on the nature of effects on channel processes and form that might originate from a particular activity. However, it is dangerous to follow a 'cook book' approach. A typical geomorphological assessment will begin with a desk study, pulling together and analysing available information such as topographic and geological maps, historic aerial photographs, existing hydrology data and hydraulic models, and any relevant existing reports about the study site to help characterise the river system, for instance aquatic ecology reports (Maas and Brookes 2010, Sear et al. 2010).

A field assessment is almost always required for geomorphological input to ESIAs. Depending on the river system and the proposed development, the walkover survey(s) may be supplemented with more quantitative data collection. This can include a geomorphological survey of the channel to obtain accurate slope and bankfull characteristics, and to record geomorphological features such as riffles, pools, knickpoints and field indicators of channel forming flow. Sediment analysis might include bedload and bank samples which would be analysed in a laboratory to determine grain size distribution, or pebble counts (typically following the Wolman (1954) pebble count method) may be completed.

Once the desk and field data has been collected, analysis of various geomorphological parameters will be completed. Equations such as stream power, bed-shear stress and erosion rates can identify potential channel change. Some examples of types of assessments are given below.

The stream power tool can characterise the potential for change arising from particular development activities. Stream power is the rate at which the energy of flowing water is expended on the bed and banks of a channel: it is the potential for flowing water to perform geomorphic work. Stream power can be calculated relatively cheaply from topographical information for the channel section and long profile. Overlays of maps of different dates are also used to determine if the platform of a channel has changed over time or is likely to do so in the future in response to development activities.

Geodynamics assessment uses desk-based information combined with a walkover survey to characterise a channel in terms of bare and vegetated bars, mature deposits and islands, eroding and stable banks and changes in slope, and to assess likely future change. The volume of sediment sources (natural and artificial) and sediment sinks (areas of deposition) are also sometimes estimated to develop a basic 'sediment budget'. If there are significantly more sinks (e.g. as permanent or semi-permanent channel bars) than sources, this implies that the sediment supplied to the channel is stored rather than being transferred through the system. Sediment deposition could be an indication of an artificially over-wide channel reach, which is attempting to regain a more natural equilibrium through narrowing.

Hydraulic and sediment modelling can be used to enhance geomorphological approaches, although these are often time consuming and costly. Typically, a one-dimensional sediment model is employed. These more sophisticated approaches are often triggered by an initial geomorphological assessment but can be undertaken in parallel. For instance, a geomorphologist might be required to input to the design of a proposed stormwater management pond (or basin) to ensure that the flow out of the pond into a river maintains sediment transport under low flows but does not lead to excessive erosion downstream under higher flows. In this case, they might calculate critical velocity and/or shear stress thresholds for sediment transport based on the sediment analysis. They would then use a continuous flow model to determine an appropriate outflow from the pond that has a minimal impact on sediment transport in the river channel. This is known as erosion threshold analysis.

2.6.4 Evaluation of impact significance

ESIA practitioners are familiar with combining the magnitude of impact with the value of an individual receptor to derive levels of significance (see Figure 1.3). There have been attempts over the years to standardise this approach, including for the water discipline. Typically a matrix is used and can be applied to the development activities 'without' and 'with' mitigation (the latter being an assessment of residual impacts). Understanding regulations and standards for the jurisdiction in which the development lies is an integral part of determining whether impacts are likely to be significant or not (for that locale). Professional judgement is a key component when populating a significance matrix.

The matrix will need to be bespoke according to the overall ESIA and in line with local circumstances. Tables 2.10 and 2.11 show, respectively, examples of criteria for assessing the magnitude of impacts on the water

Table 2.10 Examples of parameters for assessing potential magnitude of impact on the water environment (HA 2008)

Magnitude of impact	Example of typical descriptor
Major (adverse or beneficial)	• Results in loss/gain of attribute and/or quality and integrity of the attribute - increase or reduction in peak flood level (1% annual probability) > 100 mm. • Loss/gain of habitat due to scheme. • Loss/gain or extensive change to a fishery. Loss/gain or extensive change to a designated Nature Conservation Site. • Major permanent or long-term change to groundwater quality or available yield. Existing resource use irreparably impacted upon. Changes to quality or water table level would have an impact upon local ecology.
Moderate (adverse or beneficial)	• Results in an effect upon integrity of attribute, or loss/gain of part of attribute – increase/decrease in peak flood level (1% annual probability) > 50 mm. • Partial loss/gain or damage/improvement to habitat due to modifications. Replacement of the natural bed and/or banks with artificial/natural material. • Partial loss/gain in productivity of a fishery. • Changes to the local groundwater regime predicted to have a slight positive/negative impact on resource use. Minor impacts/improvements on local ecology could result.
Minor (adverse or beneficial)	• Results in some measurable change in attributes quality or vulnerability – increase/decrease in peak flood level (1% annual probability) > 10 mm. • Slight change/deviation from baseline conditions or partial loss/gain or damage to habitat due to modifications. • Changes (adverse or beneficial) to groundwater quality, levels or yields not representing a risk to existing resource use or ecology.
Negligible	• Results in an effect on attribute, but of insufficient magnitude to affect the use or integrity – negligible change in peak flood level (1% annual probability) < +/- 10 mm. • Very slight change from surface water or groundwater baseline conditions, approximating to a 'no change' situation.

Table 2.11 Examples of parameters for assessing sensitivity of the water environment (HA 2008)

Value (sensitivity)	Example of typical descriptor
Very high	• Attribute has a high quality and rarity at a regional or national scale. • Floodplain or defence protecting more than 100 residential properties. • A watercourse that appears to be in complete natural equilibrium and exhibits a natural range of morphological features (such as pools and riffles). There is a diverse range of fluvial processes present, free from any modification or anthropogenic influence. • European Commission (EC) Designated Salmonid/Cyprinid fishery. Site protected/designated under EC or UK habitat legislation (Special Area of Conservation, Special Protection Area, Site of Special Scientific Interest (SSSI), Water Protection Zone, Ramsar site, salmonid water) and or species protected by EC legislation. • Watercourse widely used for recreation, directly related to watercourse quality (e.g. swimming). • Principal aquifer providing a valuable resource because of its high quality and yield, or extensive exploitation for public and/or agricultural and/or industrial supply. Designated sites of nature conservation dependant on groundwater.
High	• Attribute has a high quality and rarity at a local scale. • Floodplain or defence protecting between 11 and 100 residential properties or industrial premises. • A watercourse that appears to be in natural equilibrium and exhibits a natural range of morphological features (such as pools and riffles). There is a diverse range of fluvial processes present, with very limited signs of modification or other anthropogenic influences. • Major cyprinid fishery. Species protected under EC or UK legislation. Watercourse used regionally for recreation. • Aquifer capable of supporting water supplies at a local scale and forming an important source of base flow to significant surface waters. • Local areas of nature conservation known to be sensitive to groundwater impacts.
Medium	• Attribute has a medium quality and rarity at a local scale. • Floodplain or defence protecting 10 or fewer industrial properties from flooding. • A watercourse showing signs of modification and recovering to a natural equilibrium and exhibiting a limited range of morphological features (such as pools and riffles). The watercourse is one with a limited range of fluvial processes and is affected by modification or other anthropogenic influences. • Watercourse not widely used for recreation or limited local use, or recreation use not directly related to watercourse quality.

Table 2.11 continued

Value (sensitivity)	Example of typical descriptor
	• Aquifer with poor groundwater quality and/or low permeability make exploitation of groundwater unlikely. Changes to groundwater not expected to have an impact on local ecology.
Low	• Attribute has a low quality and rarity at a local scale. • Floodplain with limited constraints and a low probability of flooding of residential and industrial properties. • A highly modified watercourse that has been changed by channel modification or other anthropogenic pressures. The watercourse exhibits no morphological diversity and has a uniform channel, showing no evidence of active fluvial processes and not likely to be affected by modification. • Highly likely to be affected by anthropogenic factors. Heavily engineered or artificially modified and could dry up during summer months. Fish sporadically present or restricted; no species of conservation concern. Not used for recreation purposes. • Very poor groundwater quality and/or very low permeability make exploitation of groundwater unfeasible. No known past or existing exploitation of this water body. Changes to groundwater are irrelevant to local ecology.

environment, and the sensitivity of the receiving environment, for road development in the UK. Clearly different criteria will be relevant for different projects, but the broad approach is universally relevant.

2.7 Mitigation

Most developments can have potentially negative impacts on the water environment and mitigation of some type is therefore usually needed. What is needed is likely to vary between project type, geographical location and jurisdiction. Mitigation measures should be an inherent part of the design process but bespoke measures and enhancements may be needed to reduce the magnitude of surface water impacts or provide compensation measures. The following are some examples (see also Environment Agency 2013):

- Incorporate measures to reduce construction impacts (e.g. reduce or prevent pollution downstream). Such measures may be part of a construction environment management plan.
- Create flood storage areas to reduce the rate of runoff from new surfaces, reducing downstream flood risk.

- Use sustainable drainage systems to reduce rate of runoff and to provide water quality treatment.
- Promote water conservation and waste-water treatment and reuse at the site.
- Manage source and non-point-source sediment pollution (e.g. by use of sediment traps).
- Use cooling ponds to mitigate thermal pollution from point sources.
- Use different operational modes for the development (for example within a reservoir allow for enhanced aeration).
- Re-naturalise a previously modified channel, increasing the morphological and ecological value.
- Create traps to collect sediments and liquid or dissolved pollutants that have adhered to the sediment particles themselves.

Table 2.12 outlines other mitigation and enhancement measures commonly adopted in relation to various water-impact issues.

Sustainable drainage systems (SuDS or 'low impact development') is a departure from the traditional approach to draining a site. SuDS aim to mimic natural drainage, to reduce the effect on the quality and quantity of runoff from developments, and provide amenity and biodiversity benefits by storing runoff and releasing it slowly (attenuation); allowing water to soak into the ground (infiltration); slowly transporting water on the surface; filtering out pollutants; and allowing sediments to settle out by controlling the flow of the water. When specifying SuDS, early consideration of their potential multiple benefits and opportunities will help deliver the best results (Lanarc Consultants et al. 2012, Susdrain 2012).

2.8 Monitoring

The sensitivity of the water environment and the inevitable uncertainties associated with impact predictions mean that environmental monitoring is an important consideration. Monitoring may be required for the construction, operation and/or decommissioning phases of a project. In the USA for example, a 'targeted monitoring programme' can be required to cover specific project types, baseline environmental sensitivity and types of impact. Such monitoring can utilise baseline survey methods, and may justify the use in the baseline study of techniques and sampling programmes which may otherwise be excluded by time constraints. Monitoring is often specified as part of a parallel consenting or licensing regime.

Monitoring of the water environment is frequently hindered by the difficulty of isolating the effects of different projects. However, it can set the aspirations for the success of a project, i.e. to prevent, reduce and/or offset adverse effects. Monitoring strategies for the water environment need to be proportional and to take into account the scale and probability of potential

Table 2.12 Typical water-related mitigation and enhancement measures (based on Kelday et al. 2009)

Damage to riparian features, and/or change in channel morphology, caused by river works etc.
Use project management and restoration techniques to minimise and repair damage. Create new features such as pools and riffles. Use dredgings positively, e.g. for landscaping or habitat creation (Brookes 1988).

Increased sediment loads and turbidity caused by river channel works
Select appropriate equipment and timing, e.g. construct new channels in the dry and allow vegetation to establish before water is diverted back in (as above).

Impacts on floodplains
Avoid floodplains, wetlands and high-flood-risk areas. Ensure that new flood defences do not increase flood risk elsewhere. Take compensatory measures, e.g. floodways and flood storage areas/reservoirs to provide flood storage and flow capacity. Allow for failure/overtopping of defences, e.g. by creating flood routes to assist flood water discharge. Promote enhancement, especially where (as in many urban sites) existing conditions are poor, e.g. use river corridor works to restore floodplain (by removing inappropriate existing structures), enhance amenity and wildlife value, and create new floodplain wetlands (Smith and Ward 1998). The deliberate setting back of flood embankments to widen the floodplain is known as managed realignment or retreat (DEFRA/Environment Agency 2002).

Impacts of mineral workings, especially on floodplains
Operational phase – Carefully manage the use and storage of materials/spoil, and runoff from spoil heaps/earthworks. Use siltation lagoons. Route dewatering flow into lagoons, wells or ditches to recharge groundwater, and/or watercourses to augment streamflows. **Restoration phase** – Careful backfill and aftercare management. Enhancement, e.g. of amenity/wildlife value.

Impacts of new roads and bridges, or road improvement schemes
Use: careful routing; designs to minimise impacts on river corridors (not just channels); and measures to control runoff, e.g. routed to detention basins or sewage works, and not into high quality still waters. If construction imposes river re-alignment, create new meandering channel with vegetated banks.

Impacts of dams and reservoirs
Adjust size or location (avoid sensitive areas). Minimise height and slope of embankments, and plant with trees.

Water depletion by abstraction
Promote infiltration and hence groundwater recharge in urban areas (see below). Minimise water use, e.g. metering and the installation of water-efficient equipment/appliances.

Increased runoff from urban and industrial developments
Use sustainable urban drainage schemes with (a) efficient piped drainage and sewer systems and (b) runoff source control measures, i.e. at or near the point of rainfall – to promote infiltration and/or delay runoff before it reaches piped systems or watercourses – e.g. porous surfaces (car parks, pavements etc.), soakaways (gravel trenches, vegetated areas); flow detention measures (grass swales, vegetated channels, stepped spillways, detention/balancing ponds/basins/storm reservoirs, and project

---

Table 2.12 continued

layout/landscaping to increase runoff route) (Ferguson 1998, Schwab et al. 1993, Shaw 1993, Walesh 1989).

Increased runoff and pollution (including sediments) from construction sites
Minimise soil compaction and erosion. Ensure careful storage and use of chemicals, fuel etc. Install adequate sanitation. Guard against accidental spillage, vandalism and unauthorised use.

Chemical pollution from built environments, e.g. roads, urban/industrial areas
Control runoff (as above). Use: oil traps; siltation traps/ponds/lagoons; vegetated *buffer zones* and wetlands, e.g. constructed reed beds.

Increased sewage and/or sewage-pollutant content
Increase capacity and/or *sewage treatment level*, e.g. from primary to secondary, or secondary to tertiary.

Chemical pollution from an accidental spillage
Effective contingency plans. Use booms and dispersants.

---

effects. Methods range from photography (simple) to full resurveys of the topography of a channel and floodplain, hydraulic analysis, and ecological, water quality and sediment sampling.

World Bank (1999) Operational Policies require that an environmental management plan (EMP) is produced consisting of a set of mitigation and monitoring measures for construction and operation, to eliminate adverse impacts or to reduce them to acceptable levels. EMPs are considered essential by the World Bank for projects that have a significant impact on the environment. Chapter 20 discusses EMPs in more depth.

2.9 Conclusions

The surface water environment is one of the most complex in terms of legislation, regulations and procedures for individual countries. The methods presented in this chapter can be used to inform other licences and consents related to the water environment that may be required by regulatory bodies. The steps also have applicability to more strategic forms of assessments for plans and policies.

Many of the adverse impacts covered by this chapter could be avoided or reduced by avoiding development in the floodplain wherever possible. However, where development is essential then SuDS can be incorporated into the design at an early stage of project planning. As many developments occur in already urbanised areas, there are opportunities for restoration of previously degraded river channels using the tools developed for urban river restoration (see Brookes and Shields 1996).

References and further reading

Anderson MG, DE Walling and PD Bates (eds.) 1996. *Floodplain Processes*. Chichester: John Wiley & Sons.

Arthington A 2012. *Environmental Flows – Saving Rivers in the Third Millennium*. Jackson, TN: University of California Press. www.ucpress.edu/ebook.php?isbn= 9780520953451.

Atkinson S 1999. Water impact assessment. In *Handbook of Environmental Impact Assessment*, Vol. 1, J Petts (ed.), 273–300. Oxford: Blackwell Science.

Bedient PB, WC Huber and BE Vieux 2012. *Hydrology and Floodplain Analysis*, 5th Edn. Oxford: Pearson.

Begum L 2015. *Water Pollution: Causes, Treatments and Solutions*. North Charleston, South Carolina: Create Space Independent Publishing Platform.

Betts AR, B Gharabaghi and EA McBean 2015. Salt vulnerability assessment methodology for urban streams. *Journal of Hydrology* 517, 877–888.

Boiten W 2000. *Hydrometry*. Rotterdam: Balkema.

Bovee, KD 1982. *A Guide to Stream Habitat Analysis Using the Instream Flow Incremental Methodology*. Instream Flow Information Paper 12, U.S. Fish and Wildlife Service. FWS/OBS-82-26.

Boyd CE 2015. *Water Quality: An Introduction*. New York: Springer.

Brassington R 2006. *Field Hydrogeology*, 3rd Edn. Chichester: John Wiley & Sons.

Brooks KN, PF Ffolliott and JA Manger 2012. *Hydrology and the Management of Watersheds*. Chichester: John Wiley & Sons.

Brookes A 1988. *Channelized Rivers: Perspectives for Environmental Management*. Chichester: John Wiley & Sons.

Brookes A and FD Shields 1996. *River Channel Restoration – Guiding Principles for Sustainable Management*. Chichester: John Wiley & Sons

Chapra SC 2008. *Surface Water Quality Modelling*. Long Grove, IL: Waveland Press Inc.

Chin DA 2012. *Water Quality Engineering in Natural Systems: Fate and Transport Processes in the Water Environment*. Chichester: John Wiley & Sons.

Chow VT 1959. *Open-channel Hydraulics*. New York: McGraw-Hill.

DEFRA and Environment Agency 2002. *Managed Realignment Review*, Project Report (Policy Research Project FD 2008), Flood and Coastal Defence R&D Programme. London: DEFRA/EA.

Douglas I 1983. *The Urban Environment*. London: Edward Arnold.

Environment Agency 1996. *Environmental Assessment, Scoping Handbook for Projects*. London: Her Majesty's Stationery Office.

Environment Agency 2009. *Flooding in England: A National Assessment of Flood Risk*. Bristol: Environment Agency.

Environment Agency 2013. *Water Framework Directive Mitigation Measures Manual*. Bristol: Environment Agency. http://evidence.environment-agency.gov.uk/FCERM/en/SC060065.aspx.

European Commission 2000. *The Water Framework Directive – Integrated River Basin Management for Europe*. http://ec.europa.eu/environment/water/water-framework/index_en.html.

European Environment Agency 2010. *The European Environment – State and Outlook Water Resources: Quantity and Flows – SOER 2010 Thematic Assessment*. Copenhagen, Denmark.

European Environment Agency 2016. *Flood Risks and Environmental Vulnerability: Exploring the Synergies Between Floodplain Restoration, Water Policies and Thematic Policies*. EEA Report No.1/2016. Copenhagen, Denmark.

Ferguson B 1998. *Stormwater: Concept, Purpose, Design*. New York: John Wiley & Sons.

Gordon ND, TA McMahon, BL Finlayson, CJ Gippel and RJ Nathan 2004. *Stream Hydrology: An Introduction for Ecologists*, 2nd Edn. Chichester: John Wiley & Sons.

HA (Highways Agency) 2008. *Design Manual for Roads and Bridges*. London: Highways Agency. www.standardsforhighways.co.uk/ha/standards/dmrb/vol11/index.htm.

Hendriks M 2010. *Introduction to Physical Hydrology*. Oxford: Oxford University Press.

Herschy RW 1999. *Hydrometry Principles and Practices*, 2nd Edn. Chichester: John Wiley & Sons.

Kelday S-B, A Brookes and P Morris 2009. Water, Ch. 10 in Morris P and R Therivel (eds). *Methods of Environmental Impact Assessment*, 3rd Edn. London: Routledge.

Kiely G 1997. *Environmental Engineering*. London: McGraw-Hill.

Knighton D 1998. *Fluvial Forms and Processes*. London: Edward Arnold.

Kondolf M and H Piegay 2002. *Tools in Fluvial Geomorphology*. Chichester: John Wiley & Sons.

Kresic N 2006. *Hydrogeology and Groundwater Modeling*, 2nd Edn. Boca Raton, FL: CRC Press.

Lanarc Consultants Ltd., Kerr Wood Leidal Associates Ltd. and Goya Ngan 2012. *Stormwater Source Control Design Guidelines 2012*. Vancouver: Metro Vancouver. www.metrovancouver.org/services/liquid-waste/LiquidWastePublications/Stormwater SourceControlDesignGuidelines2012.pdf.

LandIS (Land Information System) 2016. *Soil Texture Triangle*. Cranfield: Cranfield University. www.landis.org.uk/services/tools.cfm.

Linsley RH, JL Paulhus and MA Kohler 1982. *Hydrology for Engineers*. London: McGraw-Hill.

Maas S and A Brookes 2010. *Fluvial Design Guide*, Chapter 3 Background. Environment Agency. http://evidence.environment-agency.gov.uk/FCERM/en/FluvialDesignGuide/Chapter_3_Background.aspx.

Meynell P-J and N Zakir 2014. *ESIA Guidelines for Large-scale Hydropower in Pakistan*. Islamabad: IUCN Pakistan. http://cmsdata.iucn.org/downloads/niap_large_scale_hydropower.pdf.

Montgomery DR and JM Buffington 1993. *Channel Classification, Prediction of Channel Response, and Assessment of Channel Condition*. Washington State Timber/Fish/Wildlife Agreement.

Newson MD and DA Sear 1993. River conservation, river dynamics, river maintenance: contradictions? in White S, J Green and MG Macklin (eds). *Conserving our Landscape*. London: Joint Nature Conservancy.

Nielson DM (ed.) 2005. *Practical Handbook of Environmental Site Characterization and Groundwater Monitoring*, 2nd Edn. Boca Raton, FL: CRC Press.

Novotny V 2002. *Water Quality: Diffuse Pollution and Watershed Management*, 2nd Edn. Chichester: Wiley.

Petts GE and C Amoros 1996. *Fluvial Hydrosystems*. London: Chapman and Hall.

Schumm SA 1979. Geomorphic thresholds: the concept and its applications. *Transactions of the Institute of British Geographers* 4, 485–515.

Schwab GO, DD Fangmeier, WJ Elliott and RK Frevert 1993. *Soil and Water Conservation Engineering*, 4th Edn. New York: John Wiley & Sons.

Sear DA, MD Newson and A Brookes 1995. Sediment-related river maintenance: the role of fluvial geomorphology. *Earth Surface Processes and Landforms* 20(7), 629–647.

Sear DA, MD Newson and CR Thorne 2010. *Guidebook of Applied Fluvial Geomorphology*. London: Thomas Telford Press.

Shaw E 1993. *Hydrology in Practice*, 3rd Edn. London: Chapman and Hall.

Singh VP 1995. *Mathematical Modelling of Flow in Watersheds and Rivers*. Colorado: Water Resources Publications.

Smith K and R Ward 1998. *Floods: Physical Processes and Human Impacts*. Chichester: John Wiley & Sons.

Susdrain 2012. www.susdrain.org.

Tchobanoglous G and ED Schroeder 1987. *Water Quality: Characteristics, Modelling, Modification*. Oxford: Addison-Wesley Publishing Inc.

Thompson SA 1998. *Hydrology for Water Management*. Rotterdam: Balkema.

Thorne CR 1998. *Stream Reconnaissance Handbook*. Chichester: John Wiley & Sons.

Thornthwaite CW and JR Mather 1955. The water balance. *Publications in Climatology* 8(1), 1–104.

Thornthwaite CW and JR Mather 1957. Instructions and tables for computing potential evapotranspiration and the water balance. *Publications in Climatology* 10(3), 185–311.

US Department of Agriculture Soil Conservation Service 1956. *National Engineering Handbook (NEH-4)* The Soil Conservation Service Curve Number (SCS-CN) method (developed in 1954).

US Department of Agriculture Natural Resources Conservation Service 2004. *National Engineering Handbook*, Estimation of Direct Runoff from Storm Rainfall, Chapter 10.

Viessman W and MJ Hammer 2008. *Water Supply and Pollution Control*. New York: Harper and Row.

Walesh SG 1989. *Urban Surface Water Management*. Chichester: John Wiley & Sons.

Wanielista MP, R Kersten and R Eaglin 1997. *Hydrology: Water Quantity and Quality Control*, 2nd Edn. Chichester: John Wiley & Sons.

Ward AD, SW Trimble, SR Burckhard and JG Lyon 2015. *Environmental Hydrology*, 3rd Edn. Boca Raton, FL: CRC Press.

WHO (World Health Organization) 2011. *Guidelines for Drinking-water Quality*, 4th Edn. Geneva: WHO.

Wohl E 2011. *A World of Rivers: Environmental Change on Ten of the World's Great Rivers*. Chicago: University of Chicago Press.

Wolman MGW 1954. A method of sampling coarse river bed material. *EOS, Transactions American Geophysical Union* 35(6), 951–956.

World Bank 1991. *Environmental Assessment Sourcebook*. Washington DC: World Bank. http://elibrary.worldbank.org/doi/abs/10.1596/0-8213-1843-8#.

World Bank 1998. *Pollution Prevention and Abatement Handbook 1998: toward cleaner production*. Washington DC: World Bank. http://documents.worldbank.org/curated/en/1999/04/442160/pollution-prevention-abatement-handbook-1998-toward-cleaner-production.

World Bank 1999. *Environmental Assessment. Operational Policies 4.01*. Washington DC: World Bank. http://siteresources.worldbank.org/INTFORESTS/Resources/OP401.pdf.

World Resources Institute 2015. *Aqueduct Global Flood Analyser*. http://floods.wri.org.

3 Soils, land and geology

Chris Stapleton, Hugh Masters-Williams and
Martin J. Hodson

3.1 Introduction

Soils, land and geology are important components of Environmental and
Social Impact Assessment (ESIA) that can also play a vital role within other
impact chapters in an ESIA Report (e.g. water, ecology, ecosystem services,
livelihoods, health, and resource efficiency). While the International Finance
Corporation (IFC) Performance Standards mention soils and land issues
(primarily within Performance Standard 3 Resource Efficiency and Pollution
Prevention; see IFC 2012), their conservation and management is given a
more prominent status within the United Nations Sustainable Development
Goals (SDGs), which entered into force in January 2016 (UN 2015). Soils and
land appear in several of the SDGs, with Goals 2, 12 and 15 being particularly
relevant here (see Box 3.1). Importantly, the intention is that these goals
should apply to all countries, not just developing countries.

With the exception of mining projects, less emphasis is generally given to
geology within ESIA, since relatively few types of development have
significant impacts on geology. This chapter therefore concentrates on the
assessment of land and soil impacts (including contaminated land), although
some important geological and geomorphological aspects are described briefly.

3.2 Definitions and concepts

3.2.1 Geology and geomorphology

Geology is a vast and complex subject, and only a few aspects of key relevance
to ESIA will be considered here. Keller (2010) provides a good introduction to
environmental geology covering the topics of interest in this context. Surface
geology concerns superficial deposits (e.g. drift, glacial deposits, river gravel)
while solid geology only concerns pre-superficial formations. The three main
groups of rock are igneous, sedimentary and metamorphic. Many different
igneous rocks have formed as a result of volcanic activity. They are characteristically hard and crystalline, and have crystallised from magma, a silicate melt.
Sedimentary rocks are formed from pre-existing rocks by processes of erosion

Box 3.1 Sustainable Development Goals and targets linked to soils and land, to be achieved by 2030 (UN 2015)

SDG2 on sustainable agriculture has two targets:

- Double agricultural productivity.
- Ensure sustainable food production systems and implement resilient agricultural practices that increase productivity and help maintain ecosystems.

SDG12 on sustainable consumption and production patterns has the following target:

- Sustainable management and efficient use of natural resources.

SDG15 on sustainable use of terrestrial ecosystem has four targets:

- Ensure the conservation, restoration and sustainable use of terrestrial ecosystems.
- Promote the implementation of sustainable management of forests and halt deforestation.
- Combat desertification, restore degraded land and soil.
- Integrate ecosystem and biodiversity values into national/local planning and development processes.

and sedimentation. They are relatively soft and easily eroded and include limestones, coal, *evaporites* and sedimentary iron ores. Sedimentary rock strata are often important as *aquifers*, and many are rich in fossils. Metamorphic rocks are formed as the result of heat, pressure and chemical activity on pre-existing solid rock, and examples include quartzite, marble and slate.

A number of aspects of geology are of direct importance in ESIA, including the conservation, protection and management of fossils, stratigraphy, minerals, or other geological features of interest. The underlying geology also has engineering and construction implications, and affects both geochemistry and geophysics (Keller 2010).

Some geological aspects are of indirect importance in ESIA. For example, both the storage and movement of ground and surface waters, and water geochemistry will be affected by the hard geology of an area (see Chapter 2). The geology and hydrogeology of a site influences the potential for on-site and off-site *pollution* as a result of development, and pathways for any pollution that may have occurred in the past. Finally, competition between mineral extraction and other land uses may be critical in some circumstances. For example, Cuba et al. (2014) investigated overlapping claims between mineral extraction projects and river basins, agricultural land use and protected areas in Ghana and Peru.

Geomorphology includes the study of topography (the terrain), and the factors that have moulded the land to the present form. This includes the nature of the rock and soils in relation to the erosion and deposition caused by glaciers, rivers and wind. Human impacts can include landscape/visual aspects (Chapter 11), but also consequences such as erosion, slope failure, subsidence, and *sedimentation* in aquatic systems. Some aspects of geomorphology, such as soil erosion, overlap with soil studies.

3.2.2 Land and soils

Land and soil are often considered to be the same thing, but there are important differences between these terms. The quality of land for agriculture and forestry is determined by the combined physical properties of climate, topography and soil. The value placed on land also has social and economic dimensions, which are influenced by the potential uses of land and also its location in relation to settlements.

Land is a more complex concept than soil, and because these terms are practically indivisible it is more helpful to consider the relationship between land and soil than to seek separate definitions. Land is the terrestrial part of the earth's surface, and people think of land in terms of what they can do with it. Economists consider land to be one of the basic factors of production, along with labour and capital. Land is an area or location where we can live and carry out activities for subsistence or wealth-creating employment, leisure activities and other lifestyle choices. People need land for economic activity and to produce things, whether it is within an office or factory, or within parcels of agricultural land.

Where soil occurs, it is the top layer of the terrestrial part of the earth's surface (i.e. land). Soil contributes towards terrestrial ecosystem services that people rely upon when they use land for food, shelter and biomass production (Chapter 8) and these are central to social, economic and environmental sustainability. Soil is a component/subsystem of terrestrial ecosystems, providing a growing medium for flora and a habitat for fauna. From the human perspective, soil provides raw materials (e.g. gravel) and is also the basis of agricultural and forestry production for food, wood and textiles. It provides a platform for human activities, including foundations for construction. Soil is one of the main environmental *receptors* of development impacts. Avoiding significant impacts upon soil ultimately protects the whole of an ecosystem from degradation.

The productive value of soils is determined by a number of important physical and chemical properties. An appreciation of a development's impacts on soils requires an understanding of basic soil properties which are summarised in Table 3.1. The coverage of soil science here is necessarily brief, and the reader is referred to Ashman and Puri (2002), and Brady and Weil (2007), for further information.

Table 3.1 Summary of key soil properties (adapted from ASRIS 2011)

Soil property	Significance
Texture	Affects most chemical and physical properties. Indicates some processes of soil formation.
Clay content	As for texture.
Coarse fragments	Affects water storage and nutrient supply.
Bulk density	Suitability for root growth. Guide to permeability.
pH	Controls nutrient availability and many chemical reactions. Indicates the degree of weathering.
Depths to A1, B2, impeding layers, thickness of solum and regosols	Used to calculate volumes of water and nutrients (e.g. plant available water capacity, storage capacity for nutrients and contaminants).
Volumetric water content	Used to calculate water availability to plants and water movement.
Plant available water capacity	Primary control on biological productivity and soil hydrology.
Saturated hydraulic conductivity	Indicates likelihood of surface runoff and erosion. Indicator of the potential for waterlogging. Measure of drainage.
Electrical conductivity	Presence of potentially harmful salt. Indicates the degree of leaching.
Aggregate stability	Guide to soil physical fertility. Potential for clay dispersal and adverse impacts on water quality.
Sum of exchangeable bases	Guide to nutrient levels. Indicates the degree of weathering.
Cation exchange capacity	Guide to nutrient levels. Indicates the degree of weathering. Guide to clay mineralogy (when used with clay content).
Exchangeable sodium percentage	Indicator of dispersive clays and poor soil physical properties.
Substrate type	Control on soil formation, landscape hydrology, groundwater movement, nutrients and solutes.
Substrate permeability	Affects landscape hydrology and groundwater movement.

3.2.3 Soil composition and texture

There are two major types of soil: organic and mineral. Organic soils include peat and are classified as histosols (see §3.2.4). They contain >20–30 per cent organic matter, depending on the clay content, although that figure can be much higher (Brady and Weil 2007). Typically mineral soils have four major components: mineral particles, usually derived from **weathering** of parent rock (about 45 per cent of the volume); organic matter (about 5 per cent); water (about 25 per cent); and air (about 25 per cent). Organic matter is an important component of the soil that is derived mainly from decomposing vegetation. It combines with inorganic particles and cements like iron oxides and calcium carbonate to create stable structural aggregates. The nature of the organic matter (mainly concentrated in topsoils) varies according to the vegetation cover and environmental conditions. In warmer areas, the organic matter decomposes more completely to form stable complex compounds that are collectively known as humus. Most arable agricultural topsoils contain 2–6 per cent organic matter, and structural stability is impaired at lower levels.

The inorganic component of soils consists of particles that are classified into standard size ranges (gravel, clay, silt and sand). There are a number of national classifications of these particles, but the United States Department of Agriculture's (USDA) (1993a) classification now seems to have gained wide acceptance, and has been adopted by the Food and Agriculture Organization of the United Nations (FAO):

Gravel – particle size over 2.0 mm
Sand – between 0.05 and 2.0 mm
Silt – between 0.002 and 0.05 mm
Clay – less than 0.002 mm

These categories are known as separates, and their proportions in a soil define its texture. Sandy soils contain at least 70 per cent sand, and less than 15 per cent clay; clays usually have no less than 40 per cent clay; and loams have more equal proportions of clay, silt and sand. The texture of a soil is of great practical importance. Together with the humus content, texture influences the development of **soil structure**, which is the degree of aggregation of the separates, the size and shape of aggregates/structures and both the range and total volume of pore spaces. Soil structure has a major influence on:

- the soil's aeration properties;
- the capacity of the soil to retain moisture, and its *hydraulic conductivity* (and hence drainage properties);
- the soil biota and plant root growth.

Texture also affects the behaviour of the soil at different moisture contents (its consistency). Thus clay soils tend to have less well-developed structures and are less well-drained than sandy and loamy soils. They may be waterlogged in winter, show poor infiltration, and have a plastic consistency for much of the year. They are described as 'heavy' as they are difficult to cultivate. Sandy soils are described as 'light'. They are very friable ('crumbly') and easy to work, but prone to drought. Loams are generally thought to have the most favourable textures and structures for agriculture. Soil textures often vary with depth, as a result of the mixing and redistribution of parent materials, e.g. during fluvial and glacial erosion and deposition, and subsequent soil-forming processes.

3.2.4 The soil profile and soil classification

Clearly, it is important to know what type of soil is present in a study area. In most places a topsoil lies above a subsoil (containing a lower organic matter content) and these layers overlie weathering rock, although in some places topsoils lie directly over hard rocks. A pit dug into an undisturbed soil will reveal the topsoil and subsoil layers. Such a vertical section is called a soil profile, and each individual layer is called a horizon. Two different soil profiles are shown in Figures 3.1 and 3.2. Not all of the subsoil horizons are always present, and the horizons are frequently subdivided. Pedological classifications of soils are concerned with natural horizons that are created as a result of soil-forming processes. Most natural soils have an organic-rich topsoil which contains humus. A and E horizons are eluvial upper horizons in which the inorganic particles have become depleted of nutrients and other minerals as a result of the **leaching** effect of precipitation as it percolates through the profile to groundwater and watercourses. In contrast, illuvial B horizons are often

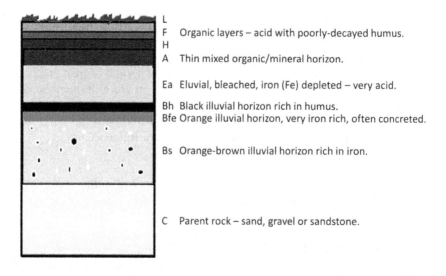

Figure 3.1

Profile of a typical spodosol. These temperate zone soils and their aquic (waterlogged) variants occur extensively over relatively cold and wet higher ground on wide and relatively flat interfluves and in low-lying receiving sites. They can occur in some freely drained sandy parent materials low in nutrients in lowland areas. Their typical vegetation cover is coniferous forest or heath (figure redrawn from Bridges 1978).

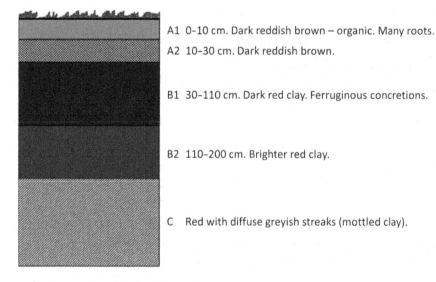

A1 0–10 cm. Dark reddish brown – organic. Many roots.

A2 10–30 cm. Dark reddish brown.

B1 30–110 cm. Dark red clay. Ferruginous concretions.

B2 110–200 cm. Brighter red clay.

C   Red with diffuse greyish streaks (mottled clay).

Figure 3.2

Profile of a typical oxisol. These soils are highly weathered, forming in tropical zones with hot and wet/moist climates. They are typical soils of tropical rainforests. Oxisols have low-nutrient status, but are well-drained and suitable for agricultural production where fertiliser inputs are available and can be managed in a way that maintains ecosystem services (figure redrawn from Bridges 1978).

enriched with nutrients, iron, clays or organic matter which have been leached from above and deposited in the lower subsoils. The C horizon is the weathering parent material or rock.

The differentiation of horizons within the soil profile is the main criterion used in soil classifications. There are many such classifications, and we will concentrate on two that are widely used in an international context, namely: the US *Soil Taxonomy* (USDA 1999); and the *World Reference Base for Soil Resources* (WRB 2014), produced by the United Nations Food and Agriculture Organisation (FAO). Both the USDA and WRB classifications are used within data sources that are highlighted in §3.4.2. Table 3.2 compares the two classifications, but note that they do not precisely match. The 12 USDA orders are compared with the 32 Reference Soil Groups of the WRB. Many countries have their own distinctive soil classification systems (e.g. the British system described by Avery 1990), which may contain elements of the USDA and WRB classifications.

We will now briefly consider each of the 12 USDA orders, starting with the least weathered and naturally developed soils, and working towards those that are highly weathered and more fully developed.

Table 3.2 A comparison between the World Reference Base (WRB) for Soil Resources, and the USDA (1999) soil taxonomies

USDA classification	WRB classification	Notes
Entisols	Regosols, Leptosols	Thin recent soils with little profile development.
Psamments	Arenosols	Sandy soils.
Fluvents	Fluvisols	Soils on recent alluvial deposits.
Inceptisols	Cambisols, Umbrisols	A wide range of soils showing the beginning of B horizon development.
Andisols	Andosols	Soils usually formed on volcanic ash.
Gelisols	Cryosols	Soils affected by permafrost and frost churning.
Histosols	Histosols	Organic soils with no permafrost.
Aridisols		Arid zone soils.
Durids	Durisols	Arid zone soils with a cemented secondary silica layer forming a hardpan within 100 cm of the soil surface.
Salids	Solonchaks	Saline soils of arid and semi-arid zones and coastal areas.
Gypsids	Gypsisols	Desert soils with a substantial secondary accumulation of gypsum.
Calcids	Calcisols	Arid and semi-arid zone soils with secondary accumulation of calcium carbonate.
Natric soils	Solonetz	Develop on flat lands with a hot dry summer. Soils contain free sodium carbonate and have high pH.
Vertisols	Vertisols	Heavy clay soils mostly in warm regions.
Mollisols	Chernozems	Dark soils of grasslands.
Ustolls and Xerolls	Kastanozems	Dry grassland soils.
Udolls and Albolls	Phaeozems	Wet grassland and forest soils of continental climates.
Alfisols	Lixisols, Luvisols, Retisols	Temperate and boreal forest soils. Characterised by an argillic clay horizon and medium to high bases.
Ultisols	Acrisols, Alisols	Warm to tropical climates. Characterised by an argillic clay horizon and low bases.
Spodosols	Podzols	Acid soils of temperate zones often associated with coniferous forest and heathland.
Oxisols	Ferralsols, Nitisols	Highly weathered soils mostly in the tropics.

Table 3.2 continued

USDA classification	WRB classification	Notes
Plinthaquox	Plinthosols	Plinthite soils, iron rich (laterite), most common in the tropics. Subject to repeated wetting and drying.
Soils influenced by standing water		
Aquic suborders	Gleysols, Stagnosols, Planosols	Water lies in the profile for at least part of the year causing reducing conditions.
Human-influenced soils		
Anthropic suborders	Technosols	Soils dominated by human-made material.
	Anthrosols	Soils affected by human activity.

The 12 USDA-NRCS orders are in bold with some suborders in italics. The table shows the relationships between the USDA-NRCS orders/suborders and the 32 Reference Soil Groups (RSGs) of the WRB.

Entisols are thin soil types where the parent rock is the dominant feature, and represent an early stage in soil development processes. One of the best-known entisols is the calcareous rendzina (a leptosol in the WRB classification), which develops over chalk or limestone. In a typical rendzina, the A horizon, which is generally fairly thin, rests directly on the parent C horizon. The soil is very dark brown or black in colour and is alkaline (pH 7.5–8.4). In contrast, rankers (also leptosols) are young, acidic soils that develop over non-calcareous rocks such as sandstones. After the entisols, the next most developed soils are the inceptisols, where the B horizon is just beginning to form. Andisols are also recent soils that form on volcanic ash and cinders. The gelisols are another young soil order with little profile development because of extremely cold temperatures for much of the year.

Histosols, or peat soils, are an important soil type in some parts of the world, but cover a relatively minor fraction of the land surface globally (about 1 per cent). Pure peat is partly decayed organic (mainly plant) material that accumulates in areas associated with high water tables and waterlogging, where a lack of oxygen inhibits the activity of microbial decomposer organisms. Mires (peatland ecosystems) occur where there is near-permanent waterlogging and consequent peat accumulation. They provide valuable wildlife habitats, and are also important 'carbon sinks'. Lowering of water tables (e.g. by agricultural drainage schemes and/or water abstraction) can seriously damage peatland ecosystems and lead to soil loss by oxidation and erosion.

As the name suggests, the aridisols are soils of hot arid zones. They are quite variable, and hence five of the WRB Reference Soil Groups are contained within this order: durisols, solonchaks, gypsisols, calcisols and solonetz. In the USDA classification these are denoted as suborders. Vertisols are heavy clay soils that mostly occur in humid and semi-arid warm regions. They are usually grassland soils. The mollisols are more widely distributed grassland soils, and are typical of the Great Plains in the United States and the steppes of Russia, where chernozems are very common. Mollisols are very productive soils often associated with high crop yields.

Alfisols are generally found in areas originally covered by deciduous forest, although some are under grasslands and mixed forest (e.g. the retisols within the WRB classification have most similarities with alfisols). They are generally located in temperate climates, and are fairly fertile, with pH 4.5–6.5. Most of the original forest that grew on these soils has been cleared for agriculture. They have topsoils that extend to relatively uniform depths over subsoils, largely as a result of agriculture, with a gradual transition into weathering parent material. Ultisols are typical of warm or tropical climates, usually under forests. Like the alfisols they are characterised by an argillic clay horizon, but they are generally low in nutrients.

Spodosols (Figure 3.1) are typical of northern areas of Europe and North America where they are associated with the boreal coniferous forest and heaths, and the climate is characteristically cold and wet. These soils are highly leached and acidic (pH often 3.0–4.5). They are little used for agriculture, but are very important for forestry and **heathland** habitats. Spodosols develop best on permeable sands and gravels. Oxisols are highly weathered soils mostly found in the tropics (Figure 3.2). They are often iron rich (lateritic), and are the major soils of tropical rainforests. They have low natural fertility, and nutrients are either rapidly cycled through rainforest systems, or lost from the soil and hence they are not ideal for agriculture.

The USDA classification does not include an order specifically for those hydromorphic soils that are waterlogged for at least part of the year. Rather, these are assigned to the aquic suborders. For example, an alfisol affected by standing water for part of the year is an aqualf, and an oxisol thus affected is an aquox. In contrast, the WRB assigns such soils to three Reference Soil Groups: gleysols, stagnosols, and planosols. With waterlogging, water saturates the soil and rises to the ground surface, filling most of the pore spaces and driving out air. Any remaining oxygen is soon used up by micro-organisms, causing the development of anaerobic conditions and gleying.

Many soils have been influenced by human activity to some extent. The USDA classification assigns such soils to anthropic suborders, while WRB gives two Reference Soil Groups: technosols, which are soils dominated by human-made materials; and anthrosols, soils affected by human agricultural activity.

3.2.5 Soil structure

In most soils, the soil particles or separates are organised into aggregates. Soil structures called peds vary in size and shape, and standard types are described in Brady and Weil (2007). Each soil horizon in a soil type usually contains a type of texture and one shape and size of structure, but structure frequently varies with depth. For example, angular and mainly subangular blocky structures in loams become coarser (larger) with depth. In clays, there is frequently a transition from coarse angular and subangular blocky to prismatic structures with increasing depth. Sandy soils may have weakly developed angular and subangular structures in the upper subsoils, but sand particles lack cohesion, and such soils are usually devoid of structures (i.e. they are apedal) in the lower subsoil. In addition to drainage channels, soil structure provides air spaces or pores within the aggregates or peds. These provide the space for plant roots, and the air and water necessary to sustain plants.

3.2.6 Soil colour

Field observations of colour can be a clue to soil composition, parent material and soil drainage status. Soil charts (Munsell Color Co 2000) provide standard examples of the normal range of soil colours. A black or grey-brown subsoil is likely to have high humus content. Predominant yellow or red-brown subsoil colours are due to the presence of iron oxides. A white soil may contain abundant silica, aluminium hydroxide, gypsum or calcium carbonate. Well-drained soils tend to have uniform brown, yellow-brown or red-brown soil colours. Colour is often inherited from the parent material. In poorly drained soils, the drainage channels and pore spaces are saturated and air is largely absent. Under these anaerobic conditions, iron compounds are reduced from the ferric (Fe^{3+}) to the ferrous (Fe^{2+}) state. The ferric compounds are characterised by ochreous colours and the ferrous compounds are characterised by blue-grey colours. Occasional waterlogging gives soils a mottled ochreous and grey appearance, while more permanent waterlogging at greater depths leads to predominantly grey soil colours. These colours are known as gley morphology, and are indicative of impeded soil drainage. In general, the greater the depth at which gleying occurs, the better the drainage status and quality of the soil.

3.2.7 Soil fertility

This is a vast topic and the reader is referred to Brady and Weil (2007), and Troeh and Thompson (2005) for more details. Two major soil chemistry issues that are of importance in an ESIA are low soil fertility and toxicity, both of which will lead to poor plant growth. Low soil fertility is due either to low levels of nutrients (e.g. nitrogen, phosphorus, potassium and magnesium) in the soil, or their being made unavailable for plant uptake. Soil toxicity is caused by high levels of toxic elements or compounds being present in the soil, usually as a result of human activity such as the spraying of pesticides, deposition of industrial waste, fuel spillage and the spreading of farm manure, slurries and sewage sludge. The source of toxic materials may not be on the affected land, and 'pollutant linkages' within a source–pathway–receptor model should be investigated.

Atmospheric deposition and movement in solution in groundwater may also be significant. Some elements (e.g. copper and zinc) which are essential **micronutrients** for plant growth can be toxic at high concentrations (Hodson 2012). Soil pH per se rarely affects plant growth, but it strongly influences the availability of plant nutrients. Aluminium and nearly all of the **heavy metals** are much more available for plant uptake and entry to food chains in acid soils than in neutral or alkaline soils.

High levels of plant **macronutrients**, especially nitrogen and phosphorus, stimulate plant growth. However, the plant communities of semi-natural habitats, such as heathlands and 'unimproved' grasslands, are adapted to low-nutrient levels – and their value for biodiversity can be degraded by soil **eutrophication** that increases nutrient levels and favours species such as vigorous grasses at the expense of ericoids and forbs.

In developed countries, it is assumed that nutrient deficiencies can be remedied by fertiliser applications under normal standards of land management. In developing countries, land evaluation methodologies have to consider whether this soil limitation can be effectively addressed. For example, in addressing nutrient deficiencies in Tigray, northern Ethiopia, Kraaijvanger and Veldkamp (2015) found that trial plots at experimental stations do not adequately reflect field conditions. Experiments should therefore be conducted on farms and in fields selected by farmers with local knowledge. There is no simple relationship between nutrient applications and yield responses because there are other potential non-soil constraints on crop yields, most notably altitude, annual rainfall, the types of crops grown and local management practices. At the farm level these other non-soil factors may have more influence on levels of soil nutrients and crop yields, and definitive advice on good farming practice cannot necessarily be extended beyond a study area.

3.2.8 Land evaluation

'Land quality' has two meanings in current ESIA practice. Firstly, it relates to the quality of undisturbed land and natural soils, their value for agriculture and forestry, and the provision of terrestrial ecosystem services. Secondly, it is concerned with the degree to which soils have been degraded and polluted by disturbance and contamination arising from human activity, and the necessary remedial measures.

Land evaluation methodologies for the assessment of natural land quality concentrate on physical properties that cannot be changed by land management. For land-use planning purposes the focus has, until recently, been on determining the relative productive value of different areas of land for agriculture and forestry. The concept of sustainable development has introduced the need to protect the other functions of soils within the hydrological and carbon cycles, supporting habitats and biodiversity, and maintaining terrestrial ecosystem services (see Chapter 8).

A physical land evaluation methodology

The Land Capability Classification (LCC) of the United States Department of Agriculture (USDA 1961) is a well-established physical land evaluation methodology which classifies land quality according to long-term climatic, topographic and soil limitations to agriculture. This is in contrast to the more recently developed multidisciplinary approach of the FAO evaluation methodology (described below), which also considers social and economic factors, and the overarching concept of sustainability.

Combined physical and social/economic land evaluation

The FAO has published generic guidelines (Verheye et al. 2008) which can be adapted for the evaluation of land in the context of major settlement or development projects, national land-use planning and local land-management plans. In this approach, the analyst works back from the intended social, economic and environmental objectives of development projects to address environmental challenges, protect terrestrial ecosystem services and support traditional patterns of sustainable land management. This generally involves the specification of suitable land-management regimes, and the selection of crops with due consideration of social and economic factors. Land evaluation then seeks to identify locations where land has the suitable physical attributes for this objective.

The FAO approach starts with the identification of specific crops and their requirements. These are then matched against the physical attributes of land within the study area. At the end of the evaluation procedure the land is given a preliminary suitability class for the land utilisation type under consideration. The exercise can be repeated in an iterative process for a number of different land uses before a final choice is made. Land-suitability classes are finally determined after field validation of the productive potential of the land. Thus the FAO approach recommends land uses, crops and management that are considered to be sustainable within a study area and socio-economic context, given the physical attributes of land and soil. The evaluation might lead to a choice between agriculture, forestry, ranching or other uses.

The FAO approach is intended to be universally applicable and scale-independent, but the level of input detail determines the scale of maps produced, and the accuracy and reliability of the interpretation of the results. The FAO has identified the need for more quantified data in land evaluation, but the integration of physical and social/economic criteria remains challenging. For instance, while the physical attributes of land remain relatively fixed over time, social and economic factors are much more changeable in respect of availability of labour, land tenure systems, market fluctuations, technological innovations, economic developments and political decisions.

3.3 Key policy and legislation

3.3.1 Geology

Policy and legislation to protect sites of geological significance (i.e. sites important for their fossils, minerals or other geological/geomorphological interest), so-called 'geodiversity', has lagged behind that intended to protect biodiversity. The first country to introduce legislation in this area was the United States in 1906. Other countries followed, although it was not until 1949 that the United Kingdom had such provision under the National Parks and Access to the Countryside Act (Matthews 2014). At an international level, geological heritage was formally recognised by the International Union for Conservation of Nature (IUCN) at the 5th World Conservation Congress in 2012 (IUCN 2012). Moreover, support was given for collaboration with UNESCO and the International Union of Geological Sciences (IUGS) to extend the inventories for the Global Geosites Programme and other regional and international sites of geological interest. On 17 November 2015 the 195 Member States of UNESCO ratified the creation of a new designation, the UNESCO Global Geoparks (UNESCO 2015). These are sites of international geological significance, and in 2015 there were 120 parks in 33 countries. They are important for the conservation, protection and management of their fossils, stratigraphy, minerals or other geological interest. Geoparks have scientific and amenity value, and include exposures of value to wildlife (e.g. rocky shores, shingle structures, cliffs, screes and limestone pavements).

3.3.2 Soil protection and conservation

Most countries have laws and regulatory procedures for the protection of land, particularly if a development displaces soil and converts land from agriculture or forestry to urban use or infrastructure. Here we briefly consider the wider international agenda set by the United Nations, before outlining examples of policy and legislation in Europe and the UK.

United Nations

Agenda 21 arose from the 1992 Earth Summit (UN Conference on Environment and Development) in Rio de Janeiro, Brazil. It is a UN action plan for voluntary implementation by multilateral organisations and individual governments, and it can be delivered at local, national and global levels. Agenda 21 is concerned with programmes for the integration of environment and development to achieve sustainable development. For land and soil,

Agenda 21 has a number of objectives of relevance including the management of land resources and fragile ecosystems; combating deforestation and desertification; and promoting sustainable agriculture. Subsequently, in September 2000 the UN Millennium Declaration was adopted (UN 2000), incorporating time-bound targets – the Millennium Development Goals (MDGs) – that included 'environmental sustainability'. Building on the MDGs and to tackle 'unfinished business', the 2016 Sustainable Development Goals include several goals that refer more specifically to agriculture, land and soils (see Box 3.1).

European Union and the United Kingdom

In 2006, the European Commission adopted a Soil Thematic Strategy (EC 2006) to protect soils across the European Union. A proposal for a Soil Framework Directive for the sustainable management of soils was subsequently withdrawn in 2014, but the Commission remains committed to the protection of soil. Few European member states have specific legislation on soil protection. For example in England, indirect protection is provided, mainly through guidance documents. Defra (2009a) set out the government's wider approach to the sustainable use and protection of soil. This introduced the concept of protecting soil functions (see §3.2.2) which are valuable in respect of a wider range of environmental objectives beyond only the protection of soils on high quality agricultural land. The National Planning Policy Framework (DCLG 2012) refers to the protection of good quality agricultural land within the land-use planning system.

Annex IIA of the most recent European EIA Directive (EC 2014) states that any development requiring EIA must describe the significant effects associated with 'the use of natural resources in particular soil, land, water and biodiversity'. The UK has comprehensive guidance on soil handling for land restoration, applicable within cool temperate climate zones (DoE 1996a; MAFF 2000; Defra 2009b,c).

3.3.3 Contaminated land

Within the context of contaminated land in ESIA, development that introduces new receptors (such as occupants of proposed buildings), or which creates or blocks pathways (such as when soils are removed or replaced), must be assessed to determine whether the changes introduced by the development will be significant. Development has the potential to convert contaminated land to uncontaminated land and vice versa without actually changing the chemistry of the ground at depth. There can also be an overlap between the remediation of contaminated land and waste management, particularly in relation to the management of excess excavated material, including soil. An ESIA for a road or rail tunnel, for example, would consider the risk of excavated material being contaminated, and how the material should be reused or disposed of appropriately.

The IFC Performance Standards address contaminated land issues under the heading of 'Pollution Prevention' within PS3:

> The client will avoid the release of pollutants or, when avoidance is not feasible, minimize and/or control the intensity and mass flow of their release. This applies to the release of pollutants to air, water, and land due to routine, non-routine, and accidental circumstances with the potential for local, regional, and transboundary impacts. Where historical pollution such as land or ground water contamination exists, the client will seek to determine whether it is responsible for mitigation measures. If it is determined that the client is legally responsible, then these liabilities will be resolved in accordance with national law, or where this is silent, with GIIP [good international industry practice].
>
> (IFC 2012)

In the United States, legislation relating to land contamination was introduced through the Comprehensive Environmental Response, Compensation, and Liability Act (Environmental Protection Agency 1980). Specific legislation on contaminated land in other countries has followed, and Land Quality Management Ltd (2013) provides details for 14 countries. Within the IFC Performance Standards, GIIP stands for 'Good International Industry Practice' and it is in this context that examples from Australia and the UK are briefly outlined below.

In Australia, regulatory authorities in each state have slightly different approaches and the level of assessment is highly dependent upon the nature of the proposed development. However the identification and management of contaminated land always involves a risk-based approach. In Queensland a Contaminated Land Act was introduced in 1991. Guidance from the Queensland Government (2014) on the typical content of an Environmental Impact Statement (EIS) includes a section on contaminated land. Similarly, following the introduction of the Contaminated Land Management Act in 1998 in New South Wales, it is very common to see dedicated EIS sections for contaminated land. The need for remediation and perhaps some remedial goals may be identified in the EIS, but detailed design and implementation is not normally undertaken until after the development application is approved. This means that the risk assessment undertaken can range from a simple generic screening approach through to a detailed and site-specific quantitative risk assessment.

In the UK, the regulation of land contamination is now devolved to the governments in England, Scotland, Wales and Northern Ireland. Relevant legislation and guidance on the assessment of contaminated sites is shown in Box 3.2. UK legislation on contaminated land applies the principle of a 'suitable for use' approach, whereby remedial action is only required if there are unacceptable risks to health, property or the environment, taking into account the use of the land and its environmental setting. Local authorities are

Box 3.2 Guidance and planning policy documents in the UK

England	Environmental Protection Act 1990: Part 2A Contaminated Land Statutory Guidance 2012
	National Planning Policy Framework 2012
	Guidance – Environmental Impact Assessment, ID: 4, Department for Communities and Local Government, 2015
	Guidance – Land affected by contamination, ID: 33, Department for Communities and Local Government, 2014
	Water Act 2003
Scotland	Planning Advice Note PAN 33, Development of Contaminated Land, October 2000
Wales	Contaminated Land Statutory Guidance for Wales, WG19243, Welsh Government, 2012
	Planning Policy Wales, Edition 7, July 2014, WG26105, Welsh Government, 2014
Northern Ireland	Part 3 of the Waste and Contaminated Land (Northern Ireland) Order 1997 (not in force)
	Development Control Advice Note 10 (Revised 2012) Environmental Impact Assessment, Department of the Environment Northern Ireland, September 2012

responsible for determining whether land meets that definition. All of the related statutory guidance documents enshrine a risk assessment methodology based on the concept of 'pollutant linkages' within a source–pathway–receptor model of the application site. For land to be defined as 'contaminated', there must be a source of contamination, a receptor which can be affected by the contaminant, and a pathway which may connect the two. If a potential pollutant linkage is established, then the risk assessment considers whether significant harm is, or could be, caused to any relevant receptor.

In the UK, planning and pollution control systems are separate, but complementary. Both are designed to protect human health, property and the environment from potential harm caused by development and operations. Historic land contamination is a material planning consideration that must be taken into account in the planning process, irrespective of the need for an ESIA, when considering proposals for land-use changes. A local authority may require remediation to be undertaken as part of the redevelopment of a site to make it suitable for the intended use. These works usually encompass a desk study, site investigation, risk assessment and a plan for any remediation works that may be required as a result of the risk assessment. A verification report confirming that the remediation works have been completed is also required.

3.4 Scoping and baseline studies

3.4.1 Introduction

During scoping it is necessary to decide whether a desk study, a reconnaissance field survey, or possibly a detailed field survey and laboratory analysis of soils is required. Scoping site visits will normally be brief but may involve walkover surveys, which ideally should be undertaken with other members of the ESIA team, to identify interactions between subject areas. For example, an understanding of geology, geomorphology, land use and soils may be relevant to the assessment of landscape/visual impacts, water and ecology. Coordination at an early stage is important to achieve an integrated approach, avoid duplication of effort and ensure that key aspects are not omitted.

The most important scoping considerations are whether the geological or soil resources within a project's impact area are likely to be significantly affected, and if there are any practical measures which can be undertaken to mitigate the anticipated impacts. Where a significant impact on land or soils is anticipated, a Land Capability Classification or preferably a land evaluation should be performed, ideally using the FAO approach (see §3.2.8).

Where soils need to be conserved for land restoration (i.e. at mineral extraction sites), or where the developer wishes to make beneficial use of this resource on the development itself (e.g. for landscaping), the field survey should also include an assessment of the volumes of topsoil and subsoil available. Where contamination is anticipated an appropriate site investigation will be required (see §3.4.2 and §3.4.3).

3.4.2 Desk study

The desk study should interpret existing information on geology, geomorphology, soils, land quality and associated aspects such as site history and local climate. It should also consider the potential for contaminated land on or adjacent to the proposed site.

Geology and geomorphology

Geological maps range from those depicting whole continents (at perhaps 1:20,000,000 scale), through to regional maps (between around 1:1,000,000 and 1:250,000 scale), down to more detailed (1:10,000 scale) maps for areas of special interest to geologists: see Table 3.3. When used in conjunction with geological and soil maps, topographical data and digital terrain models (DTMs) can serve to give a general idea of geomorphology.

'Solid' maps show only pre-Quaternary rocks, and 'drift' maps also show superficial Quaternary deposits that have been laid down more recently after being moved by wind, water or ice. Lithology (the general physical characteristics of rocks) has a big influence on soil types through the mineralogical

Table 3.3 Examples of geological maps and Digital Terrain Model (DTM) data sources

Coverage	Name	Web link
International	One Geology Portal	portal.onegeology.org
	US Geological Survey – Mineral Resources and Geology	http://mrdata.usgs.gov/general/global.html
	ASTER Global Digital Elevation Map (DTM)	https://asterweb.jpl.nasa.gov/gdem.asp
	Shuttle Radar Topography Mission (SRTM GL1) (DTM)	https://catalog.data.gov/dataset/shuttle-radar-topography-mission-srtm-gl1-global-30m
Australia	Australian Geoscience Information Network	www.geoscience.gov.au
Canada	Geological Survey of Canada	www.nrcan.gc.ca/earth-sciences/science/geology/gsc/17100
India	Geological Survey of India	www.portal.gsi.gov.in
South Africa	Council for Geoscience (CGS)	www.geoscience.org.za/index.php
United Kingdom	British Geological Survey	www.bgs.ac.uk/data/maps/home.html
	Ordnance Survey OS Terrain 50 (DTM)	www.ordnancesurvey.co.uk/business-and-government/products/terrain-50.html
United States	US Geological Survey	www.usgs.gov

composition and texture of the weathered rock. However, because of the erosion, mixing and redistribution of surface rocks and weathered materials, drift maps tend to give the most informative indication of the soil parent materials in a survey area.

Land and soils

The starting point for the evaluation of land and soils is to determine the usefulness of the information available. The scope and quality of existing data at the national and local scale can vary considerably. Published sources are not always directly applicable to new projects, but can indicate the land and soil likely to be found in a study area, and assist in planning field surveys (see §3.4.3). Generalised, medium to small-scale mapping (i.e. 1:50,000 to 1:100,000) may be appropriate for use in studies involving large areas (e.g. in pipeline or electricity transmission line routing), but would be less useful where a very detailed and precise picture is required for a small, confined area.

Table 3.4 lists a range of land cover and soil map data sources. For example, the ISRIC World Soil Information shows the distribution of individual soil types using either the USDA or World Reference Base classification, plus other soil characteristics including pH, bulk density, organic carbon, and sand/silt/clay content. However, the value of published sources of information for a specific ESIA is determined by the area covered, and the precision with which the survey results can be interpreted.

It should be recognised that land and soil evaluations carried out at national and regional levels rely on soil association maps and broadly defined regional climatic information. The results of such studies are shown on small-scale maps which cannot be interpreted in detail, even if GIS systems allow the user to 'zoom in' to a larger scale. Remote sensing models set up relationships between defined categories of land cover and land use, and relate these to image colours, tones and textures, for the assessment being undertaken. This bespoke approach means that the results of different studies may not be directly comparable, but they will contribute towards a knowledge base within a given climate/topographic zone. This contrasts with local studies at a large scale, which can delineate soil types in homogeneous map units within micro-climatic zones, and which can be interpreted more precisely.

Remote sensing supported by focused field survey is considered to be a cost-effective approach to the evaluation of land and soils in large, relatively inaccessible areas within developing countries. However, the accuracy of predictions about land cover, land-use categories and associated patterns of soil

Table 3.4 Examples of land cover and soil map data sources

Coverage	Name	Web link
International, national, regional	Global Land Cover by National Mapping Organizations (GLCNMO)	www.iscgm.org/gm/glcnmo.html
	FAO Soils Portal	www.fao.org/soils-portal/en
	ISRIC SoilGrids1km (web-GIS)	https://soilgrids.org
Africa	ISRIC AfSoilGrids250m (web-GIS)	www.isric.org/data/AfSoilGrids250m
Australia	Australian Soil Resource Information System (ASRIS)	www.asris.csiro.au
Europe	European Soil Data Centre (ESDAC)	esdac.jrc.ec.europa.eu
United States	Soil Survey Geographic (SSURGO)	www.nrcs.usda.gov/wps/portal/nrcs/detail/soils/survey/geo/?cid=nrcs142p2_053627

types based on remotely sensed imagery depends on field survey at representative sample locations selected to provide the necessary ground truth. This can be determined by stratified sampling of ground-truth observations, the use of detailed aerial photographs and topographic maps, together with statistical analysis to evaluate the findings. The figure of 85 per cent accuracy was initially set by the US Geological Department (Anderson et al. 1976), and contemporary studies have found this is a suitable benchmark (Kurt 2013).

Contaminated land

The possibility that a site may be contaminated is usually first assessed on the basis of its land-use history, which may be derived from historic maps (where available). In some countries this material is commercially available alongside other relevant environmental data from regulatory bodies, including geological and hydrogeological maps. This information is important in developing an initial conceptual site model of the site. The New South Wales Government (2011) in Australia and Defra and Environment Agency (2004) in the UK have produced guidance on carrying out a preliminary survey of a potentially contaminated site.

The type and extent of any contamination on a site will depend upon the previous uses of the site, the operational processes undertaken and the effectiveness of environmental protection measures. In addition, pollution of the subsurface on adjacent sites can migrate into the study area. It is important, therefore, to ascertain as far as is possible the activities and management practices that occurred on-site, and on adjacent sites, and the type of chemicals used. However, uncertainty will always remain and this should be recognised and recorded in the desk study report (e.g. gaps in the map record of land use).

Information on the type of materials and chemicals that were used in a wide range of commercial activities can be gathered from published data (e.g. GOV.UK 2015a), and professional bodies such as the Society of Chemical Industry. The Construction Industry Research and Information Association (CIRIA 1995–1998) has published 12 volumes covering all aspects of remedial treatment for contaminated land. Although dated, the first of these is a useful guide and introduction to all aspects of this topic. In the United States, procedures for gathering information concerning previous land uses and potential contamination are outlined by the Environmental Protection Agency (2012).

3.4.3 Field survey

Baseline information on land and soils should be acquired from field surveys of landforms, soil types and land quality, ground cover/crops, land ownership and tenure. Ideally this information should be compiled into a GIS suitable for spatial and statistical analysis. In addition, the influence of land management for farming and forestry should be determined by engaging with people who use land in the study area.

The USDA (1993a) *Soil Survey Manual* is being updated to include changes in methodologies, terminology and technologies used in carrying out soil surveys under the National Cooperative Soil Survey Program. It refers to the soil survey methods, standards and procedures of the National Soil Survey Handbook (USDA 1993b). *The Soil Survey of England and Wales Soil Survey Handbook* (Hodgson 1997) also sets out a methodology, terminology, definitions and criteria for the description of soils in the field.

As soils are observed only where samples are taken, the sample network and density have to be designed to be representative of the variation in soil types within the survey area. Soil survey methods are discussed in Tan (2005), and a detailed statistical account is provided in Webster and Oliver (1990). Where geology and geomorphology are complex, particularly where soil parent materials have been redistributed (e.g. during Ice Ages), soils can vary considerably over short distances, and this can be a major challenge for soil surveyors. Field observations are made by using a soil auger to take samples from successive horizons within a soil profile.

Generalised soil surveys of large areas are carried out by a physiographic (or free survey) technique to ensure that samples are representative of the geomorphology within the survey area. The results are shown on maps at medium to small scales ranging from 1:50,000 to 1:100,000 (see §3.4.2). For detailed surveys of development sites on undisturbed and uncontaminated agricultural land in a cool temperate climate zone where glaciation has occurred, most practitioners favour a grid sampling pattern and a minimum density of one sample per hectare. Supplementary samples are taken as necessary to accurately delineate soil boundaries. Soil pits are dug to observe the soil structures and the extent of crop rooting in each of the main soil types. For surface mines, topsoil and subsoil resource maps are derived from the information collected during soil survey. These maps indicate the areas, thicknesses and volumes of the topsoils and subsoils that are available for conservation and land restoration. Detailed soil maps are usually shown at a scale of 1:10,000, and they can be interpreted with a reasonable degree of precision. It is important to note, however, that soil survey is a field technique and not an exact science.

During the field survey, some soil properties like soil depth are easily measured, and other properties are either estimated by eye, or assessed using a quantified standard technique, depending on the degree of precision required. For example, stone content can be estimated or measured using a sieve and weighing scales, and applying the known density of the rock. Determining soil texture by hand requires experience, occasional calibration with standard samples, and laboratory analysis may also be necessary.

Surveys should identify soil resources that should be conserved and retained for land restoration. Such surveys provide information on the depths, volumes and physical characteristics of available topsoil, subsoil and soil-forming materials, together with a description of the original landform and drainage.

If the desk-study findings indicate the possibility of contaminated land, then a walkover survey will help define the extent of any intrusive ground investigation required. Such investigations will be essential for scoping studies, but may not be necessary at the initial project planning stage, provided the desk study and walkover reports are sufficient for a qualitative assessment of any risks. Where potential for contamination has been identified, investigation targeted at confirming the presence and extent of any potential sources will be necessary and this will form part of the baseline description of the site. International ESIA best practice requires that land, soil and water are considered as natural resources and this inherently recognises the inseparable nature of the ground and the groundwater that exists within it. The investigation should recognise that contamination exists in three phases – solid (soil), liquid (ground and surface waters) and vapour (ground gases). For the UK, guidance on the scope and extent of any investigation can be found in BS10175 (BSI 2011). Further guidance on the appropriate layout of exploratory holes, locations and depths of sampling and analysis are discussed by DoE (1994) and Defra/EA (2004). International guidance on soil quality and sampling can be found in the ISO 18400 series of documents (BSI 2017).

All land contamination studies, investigations and risk assessments should be undertaken by suitably qualified and experienced practitioners who possess the knowledge required to address the health, safety and environmental risks (Chapter 18) of potentially contaminated sites.

3.4.4 Laboratory work

Some physical land evaluation methodologies set quantified thresholds between the grades and classes of land for selected climatic, topographic and soil variables. In cases where the field observations indicate a marginal classification, samples should be analysed in the laboratory for greater precision and the definitive grading of land quality. For example, this can apply to the analysis of soil texture and pH.

Laboratory analysis can be expensive, and it is usually undertaken on samples of soils for specific verification purposes only. Soil samples may be analysed at a number of stages in the ESIA process: during baseline studies for land evaluation, and to inform mitigation measures, including the treatment of contamination. Soils may also be analysed during project construction and operation for monitoring and management purposes. While a wide range of analyses are undertaken, only the more common are described below.

During baseline studies and the evaluation of undisturbed agricultural land, topsoil texture is often analysed in the laboratory for a definitive grading. Analytical methods differentiate between the mineral fractions of soils on the basis of particle size. This involves sedimentation of mineral particles in a water column. This takes several days, and at current (2016) prices each determination will cost about £35 in the UK, although prices in the United

States seem to be lower at around \$12. The British Standards Institute (1994) specification for topsoil refers to a number of qualifying threshold levels in soil texture and other variables. Soil chemistry variables include pH, organic matter content, electrical conductivity, available phosphorus, potassium, magnesium and total nitrogen. Rowell (1994), Tan (2005) and FAO (2016) provide detailed methods of soil analysis.

Surface mine excavations may be preceded by soil stripping and storage, and followed by the reinstatement of the soils. The moisture content at which soils may be moved is an important variable. Restricting soil handling to moisture levels below the *plastic limit* of soils reduces damage to soil structures. It can be determined in the field by hand, but the moisture content may have to be determined with more precision in the laboratory. Most analysts use gravimetric analysis, which involves taking a sample of soil from the field and weighing it before and after heating in an oven. For soil surveys on well-defined sites at a given time, the gravimetric method yields good results, giving information identifying where soils are too wet to be moved. For larger study areas where monitoring is required over a longer period of time more sophisticated machinery can be used (e.g. neutron probes or time domain reflectometry) (Brady and Weil 2007).

When mineral sites have been restored, a period of aftercare is instituted to recreate favourable soil conditions for a range of beneficial uses, including agriculture, forestry and wildlife and amenity planting. As a part of this rehabilitation process, soil samples may be taken to determine bulk density and plant nutrient status. Bulk density is the mass of dry soil per unit of bulk volume (g/cm^3), including the air space. Before and after comparisons can be used to assess damage to soil structure and reduced porosity caused during soil handling. It is usually measured directly with the use of a volumetric corer. Essentially, a pipe is pushed into the ground to extract a core of soil on which measurements can be made. In ESIA, bulk density is a very useful measure if soil compaction is likely to occur. The results of these tests can be used to guide subsequent remedial cultivations (like subsoiling) and fertiliser applications.

Measurement of contaminants

For a meaningful assessment it is essential that the soils are analysed for all of the contaminants potentially in the ground within the site. Investigations of contaminated sites are often hindered by an incomplete understanding of the polluting activities that have taken place, and it would be excessively expensive to test for every possible contaminant. Where there are no clearly 'anticipated' contaminants on a site, it is common to analyse soils for pH, arsenic, cadmium, chromium, lead, mercury, nickel and selenium, as well as cyanide, the 16 most common polycyclic aromatic hydrocarbons (PAH) and total petroleum hydrocarbons (TPH). Often the concentration of phenol in the ground is assessed to determine whether this could permeate plastic water supply pipes. Other specific analyses will depend upon the history of the site, and may include contaminants such as pesticides, PCBs, chloride, mineral oils, elemental sulphur, organic acids, and the components of landfill gas.

In the UK, methods of testing and assessment for contaminants is standardised, with agreed generic guideline maximum mean levels of contaminants ('soil guideline values') for particular site uses (Defra/EA 2004). Certain suites of contaminants associated with the main industrial processes and mining operations are provided by GOV.UK (2015a, b). The US Environmental Protection Agency (2012) covers similar material.

3.5 Impact prediction

3.5.1 Geological and geomorphological impacts

Potential impacts on UNESCO Global Geoparks and other sites of conservation interest are likely to be direct and hence relatively easy to predict. Surface mining often has significant impacts because it removes the geological resource and may also affect the local hydrogeology. Apart from possible benefits of new rock exposures (e.g. revealing fossil beds) the impacts are negative, particularly in view of the finite nature of mineral resources, and the consequential loss of agriculture and wildlife habitats. However, an ESIA takes place within a statutory context, and governments may consider mineral extraction to be a valid component of sustainable development (e.g. to provide a localised supply of aggregates to the construction industry). Worked-out mineral sites can be restored to a beneficial use, and may be used for waste disposal. If not carefully managed, this landfill phase can result in groundwater pollution by **leachates**.

Seismic risk is a significant problem in some parts of the world, and may require consideration. For example, hydraulic fracturing ('fracking') can potentially cause significant geological problems that an ESIA for a fracking operation would need to assess. Fracking involves pumping liquid under pressure into rock formations to force shale gas out. The main geological risks are that the expelled gas might contaminate underground aquifers, and the possibility of earthquakes. Earthquakes caused by fracking are usually small, but associated waste-water disposal by injection into deep wells can induce larger earthquakes (Ellsworth 2013). For example, a fracking-induced magnitude 5.7 earthquake in central Oklahoma in November 2011 destroyed 14 homes and injured two people.

In some parts of the world, a section of the ESIA should be devoted to volcanic risk. Subsidence and slope stability are also factors that should be considered. Subsidence is caused by underground mining and is usually associated with traditional coalfield areas, where the subsidence extends for considerable distances around collieries (Schrefler and Delage 2010). It can also occur as a result of the underground extraction of salt and chalk, and in limestone and chalk areas where natural chemical dissolution has occurred. Natural slope stability is a more widespread problem, and the objective of ESIA should be to avoid the construction of new developments in unstable areas, particularly when the development might make the area even less stable by concentrating runoff. Information concerning planning issues involving subsidence and slope stability is available from DCLG (2014) in the UK, and much of the material presented is applicable in other countries.

Road developments can have a direct impact on geology and geomorphology by displacing rocks and changing landforms. Indirect effects may be felt through alterations to hydrogeology (e.g. diverting streams or affecting the recharge of aquifers). The major impacts of such developments are, however, likely to be in respect of damaging or introducing geological exposures, fossil beds, stratigraphy and geomorphological systems (PAA 1994). Many sites of geological interest have been created as a result of human activity, and road developments can create new exposures that may be of scientific interest. It is more difficult to preserve geomorphological features, and the best of these (e.g. stream systems, glacial forms) should be avoided by a proposed development.

3.5.2 Impacts on land and soils

Human impacts on land and soil occur due to activities such as deforestation, unsustainable intensive farming, and poor planning and management of urban and industrial development. These activities lead to problems including soil erosion and degradation due to the loss of organic matter, compaction and destruction of soil structures, and salinisation. Poor environmental management of polluting materials causes land degradation by contamination. Other significant human impacts occur where inappropriate development causes the avoidable loss of land and soils, particularly where they are permanently displaced.

ESIA methodology

The ESIA of development projects should consider impacts on soil functions and terrestrial ecosystem services (Chapter 8). Assessing the impacts of a development project should begin with the identification of a study area within which significant impacts are likely to occur, and then determine the appropriate temporal scope (i.e. the duration of the construction, operation and restoration phases). ESIA then has to predict the magnitude of the potential permanent and temporary impacts of land-take and the displacement of soils. A further consideration is the sensitivity of land and soil as environmental receptors. This is determined by the quality of the land or soil, the value placed on its terrestrial ecosystem services and its resilience to soil handling.

The assessment is based on selected physical criteria (climatic, topographic and soil), together with social and economic criteria. The evaluation methodology should set out a range of quantified values as class limits for magnitude and sensitivity categories to determine the overall significance of change. Class limits can be adjusted to match the scale of the development project.

There is no single recognised evaluation methodology for predicting and assessing the impacts of development projects on land and soils. However, the FAO guidelines for land evaluation provide a set of principles and a methodological framework for assessing impacts in compliance with the UN programmes for sustainable development. USDA (2013a) also outlines how to carry out an ESIA, and rate the residual (post-mitigation) significance of anticipated impacts for a range of environmental receptors, including land and soil.

Temporary land-take and soil displacement

Land is subject to development pressure from urban and industrial expansion, resource extraction and infrastructure. Development potentially takes land out of agricultural production, removes habitat and displaces soils – effects that can be temporary or permanent. Insufficient consideration is given to the conservation of soils displaced by development. The temporary displacement of soils should be followed by their reinstatement and land restoration.

Permanent land-take and soil displacement

Permanently displaced soils have to be reinstated in another location identified and secured within the development proposals. The challenge is to find suitable land (e.g. where soils are degraded or absent) in a nearby location (to reduce transport costs and impacts) where the displaced soils can be reinstated. The loss of permanently displaced soils is an unsustainable practice that has yet to be effectively addressed, even in developed countries (EEA 2006).

Development often changes land use from agriculture and forestry to a more urban or industrial character. In many cases indirect changes also occur (e.g. tree planting on agricultural land for landscaping and for the creation of wildlife habitats) to meet the wider environmental management objectives of development projects. These changes are another form of land-take from agriculture where soils are not permanently displaced. Land-take is superimposed on patterns of land ownership or tenure. Security of tenure and access to land resources and ecosystem services may also be critical in more remote regions, e.g. where subsistence pastoralism occurs and rangeland resources are managed as collective common property. The important parameters are: patterns of land resource usage; land holding boundaries; the proportion of a holding affected by land-take; the degree to which land management within the holding is affected by the loss of land; severance (particularly by linear infrastructure development); fragmentation; and the loss of farm buildings and infrastructure.

Soil erosion

Soil erosion by wind and water is a serious threat to the soil resource worldwide (Montanarella et al. 2016). Most soil erosion occurs as a result of agricultural land-management practices which are not subject to development planning controls, and which are normally beyond the scope of ESIA. However, almost all ESIA developments are also likely to lead to some soil erosion unless suitable mitigation procedures are adopted, including:

- construction phase impacts, for example vegetation removal and the creation of unstable slopes;
- increased runoff from impermeable surfaces associated with urbanisation;
- creation of unstable and unvegetated surfaces due to poor quality land restoration following construction and mineral extraction.

There are two major types of erosion: by water and by wind (Lal 2001, Morgan 2005, Wang 2015). Where soils are not protected from the force of raindrops by crops or other vegetation, during the rainy season their structures are broken down into particles, which are moved downslope by surface runoff. This occurs when rainfall exceeds the infiltration capacity of the soil. In periods of heavy rainfall or melting snow this can lead to sheet erosion over areas, and rill erosion which can concentrate runoff into linear channels to form gullies. Displaced soil particles form suspended sediment which, together with stones, can have a further erosive effect. The factors that most influence erosion by water are: mean annual rainfall; storm frequency and intensity; slope; soil infiltration capacity; and vegetation cover. Rain and overland flow cause some natural erosion in most environments, but this is insignificant compared with accelerated erosion resulting from human activities such as the disturbance or removal of vegetation (e.g. for agriculture, mineral extraction or development). When water erosion occurs, damage will often not be restricted to the terrestrial environment because the removed soil often causes increases in *turbidity*, siltation and soil nutrient levels in nearby watercourses. It is not uncommon for the levels of certain nutrients (particularly nitrate) to exceed legal limits in streams and rivers as a result (see Chapter 2).

Only dry soil is subject to wind erosion and so rainfall must be relatively low for it to occur (< 250–300 mm). Steady prevailing winds are generally found on large, fairly level landmasses, which can be susceptible to wind erosion. For example, the Loess Plateau in China is particularly susceptible to wind erosion, which occurs when the force of air movement at the ground surface picks up and carries away loose particles of soil. This removes the most fertile part of the soil, namely the clays and organic matter. Wind erosion is, therefore, a major source of land degradation in arid and semi-arid regions where it is often initiated or accelerated by deforestation and inappropriate agricultural practices.

Soil scientists have developed a number of predictive equations that can be used to predict soil loss due to erosion (Lal 2001). The standard model for most water erosion assessments and conservation planning is the empirically based Universal Soil Loss Equation (Wischmeier and Smith, 1960, 1978). This predicts the long-term average annual soil loss from the erosive nature of rainfall, soil erodibility, topography (steepness and length of slope) and crop/vegetation ground cover. More recently the USDA (2013b) has developed a Wind Erosion Prediction System for predicting the effects of land-management practices and crop rotations on wind erosion for an individual field. Several models also predict soil loss at a regional scale (de Vente et al. 2013).

Damage to soil structure

Soil stripping, storage and subsequent reinstatement operations at land-restoration sites can damage soil structure. This is due to the use of inappropriate methods and machinery and moving soil when it is too wet and plastic. Vehicles driving over soil will compact it, destroying soil structure and increasing bulk density. Topsoils also tend to become mixed with the less favourable subsoils when they should be stripped, and stored separately to facilitate the restoration of a natural soil profile upon reinstatement. Soils which have been damaged in this way lack the natural drainage channels and porosity which normally absorb precipitation, and transfer it to groundwater reserves. Compaction by machines reduces infiltration, increases runoff and erosion risk, and inhibits root penetration. Damaged soils also have a reduced capacity to retain moisture and to make it available to plant roots, with the result that plants are prone to severe drought. In response to this threat, the UK Government has published three guides to good practice (DoE 1996b, c; MAFF 2000). The United States Government Publishing Office (2010) has also published detailed legislation and guidance concerning surface coal mining and reclamation.

Soil pollution

Two types of situation exist where soil pollution is an important factor in an ESIA: (i) where the site is already contaminated and a clean-up operation is required prior to development, and (ii) where pollution may be caused by the project itself.

If the baseline survey shows that the ground below the site contains contaminants, a risk assessment based on the pollutant linkage (source–pathway–receptor) model should be carried out to determine whether, in the context of the project, the contamination is significant (see §3.3.3). Significance may be assessed in terms of risks to receptors (e.g. human health, surface and groundwaters, property and ecosystems). The baseline survey should identify whether the contaminated soils are causing any adverse impacts to identified receptors before the commencement of the project, since the proposed work may have beneficial impacts by breaking the pollutant linkage.

In simple situations, a qualitative risk assessment can be undertaken to establish whether a pollutant linkage exists, although some form of numerical modelling will be necessary to quantify the potential adverse impact upon an identified receptor. The basis for modelling contaminant transportation to groundwater is considered by Adriano et al. (1994). Modelling techniques include hydrogeological-flow models such as AQUA3D (SSG 2008), MODFLOW (USGS 2007), ConSim (EA 2003), the *Remedial targets worksheet* (EA 2006); and air dispersion models such as ADMS (CERC 2016). Risks to human health can be assessed using a range of modelling tools such as the contaminated land exposure assessment tool (GOV.UK 2015c), and in the USA, Risk-Based Corrective Action (GSI Environmental 2014). Risk assessments must also consider gaseous contaminants including: vapours from hydrocarbon spillages; subsurface methane generation resulting from the biological degradation of organic materials in the soils and from coal workings; and from naturally occurring radon emanating from geological sources.

During the operational phase, airborne emissions (Chapter 4) may also begin to impact on the local soils. **Acid deposition** arises mainly from sulphur dioxide emitted from power stations, and nitrogen oxides from vehicles. Emissions have had major effects on soil pH in some locations; for example, Hallbäcken and Tamm (1986) found that soil pH in southwest Sweden dropped by 0.3–0.5 units between 1927 and 1984, although more recently there have been signs of recovery (Löfgren et al. 2011). Lowered soil pH increases available soil aluminium and heavy metal levels, and causes increased leaching of soil nutrients. There are often many sources of acid rain and these create a cumulative and dispersed impact that is felt some distance from the sources.

Policy to reduce emissions of key pollutants across Europe and North America has led to some improvements. For example, sulphur emissions have fallen by 95 per cent and nitrogen oxide emissions by 69 per cent between 1970 and 2014 in the UK (Defra 2015). Nonetheless, nitrogen oxides remain a significant problem, particularly in causing **eutrophication** of terrestrial systems, which can lead to changes in vegetation and biodiversity (Emmett 2007). Moreover, sulphur and nitrogen oxide emissions in China and India have increased significantly since 2000, as they industrialised. China subsequently cut its sulphur emissions by 50 per cent between 2012 and 2014, but emissions from India are continuing to increase (Krotkov et al. 2016).

Critical loads provide a means to estimate the vulnerability of land to atmospheric pollution (especially acid, nitrogen and sulphur deposition) in relation to receptor soils, geology, freshwaters and vegetation. The United Nations Economic Commission for Europe has specified critical loads for deposition onto sensitive ecosystems (APIS 2016), and a critical loads database for the United States is now available (Blett et al. 2014).

All of the above impacts can have serious effects on soils, but the soil types outlined in §3.2.4 will be affected to different extents by each. Spodosols are already acidic and have a low buffering capacity (the greater the buffering capacity, the more acid rain will be needed to change the pH of a soil). Consequently, these soils are vulnerable to acid precipitation. Spodosols also suffer disruption by disturbance and soil mixing because they have distinctive layers. Aquic soils are vulnerable to changes in soil hydrology; as are histosols which are extremely sensitive soils, especially to erosion and compaction. Susceptibility to erosion is also a feature of many sandy soils. The tropical rainforest soils, oxisols and ultisols, often have low-nutrient status and are easily eroded when their tree cover is removed.

3.6 Mitigation

It is not possible to mitigate the permanent loss of land arising from a development project. Therefore it is necessary to ensure that the smallest area of land is lost, consistent with the sustainable operation of the proposed development, and to avoid more highly valued land and soils where practicable. Site boundaries and the corridors of land affected by linear developments can be adjusted during the project design phase to achieve this. Development layouts can be configured to locate hard development (e.g. structures and impermeable surfaces) on less valued land, and to maintain the viability of adjacent agricultural land.

Where land degradation occurs as a result of unsustainable farming practices on marginal land, land restoration involves introducing a management regime to increase levels of organic matter and nutrients in the topsoil, alleviate soil compaction and develop subsoil structures to increase moisture retention. Further measures (e.g. the introduction of vegetation cover and windbreaks) may also be required to control erosion. Land restoration also refers to the conservation and reinstatement of soils temporarily and permanently displaced by development (also known as 'soil salvation'). The identification, conservation and sustainable reuse of temporarily displaced soil resources (i.e. the reinstatement of soil profiles) has the potential to effectively mitigate the temporary land-take impacts of development projects. Finding a suitable location and beneficial reuse for permanently displaced soils and retaining their natural functions is a greater challenge.

Soil conservation depends on effective project design and well-managed construction operations that conform to the good practice guidelines for soil handling referred to in §3.3.2 (DoE 1996a; MAFF 2000; Defra 2009b, c). Pre-development surveys (§3.4.3) will have identified soil resources that should be conserved and retained for land restoration. Mitigation stage plans should show soil stripping, storage and reinstatement areas, and haulage routes. The stripping, storage and reinstatement of topsoils, and all significantly different subsoil horizons present within the development site, should be done using specialist machinery under appropriate weather and soil-moisture conditions. If the soils are to be stored for any length of time, they may need to be grassed over to prevent erosion and colonisation by weeds. Land restoration should be followed by a period (approximately five years) when the land is managed to rehabilitate the soil and return it to its original state, rather than to maximise agricultural production.

The guidance on soil handling for land restoration within a cool temperate climate zone (DoE 1996a; MAFF 2000; Defra 2009b,c) emphasises the need to handle soils when they are relatively dry and friable (i.e. within plastic limits largely determined by soil texture and moisture content) to avoid compaction and damage to soil structures. Some of this guidance has to be adjusted for conditions in drier climate zones where, for example, there is more concern over avoiding soil erosion during soil handling. However, many of the other basic principles set out in the cool temperate guidance are universally applicable. In the USA, guidance on land restoration in wetter and drier conditions is provided by individual states. For example, Wright (2002) and the Utah Department of Natural Resources: Oil, Gas and Mining (2008) set out restoration considerations more relevant to conditions in a drier climate zone, including soil salinity, electrical conductivity, organic matter content and available water capacity. Further information on legislation and guidance relating to the reclamation of surface and underground coal-mining sites in a range of countries is available from Sloss (2013).

Contaminated land

Where pre-existing contamination sources are affected by the proposed development, mitigation approaches may have the beneficial effect of clearing up pre-existing pollution or they may leave the source in place and isolated, and ensure that no new pathways linking the source to receptors are introduced by the development. In simple terms, the mitigation or remediation should break the pollutant linkage and/or prevent any new linkages being formed. This can involve source removal, pathway removal, or in some cases receptor removal. For example, it may be possible (and cost effective) to relocate a small population of a protected species to a suitable location if it is potentially impacted by the movement of contaminated soils.

Techniques used to clean up contaminated land include:

- **Removal of the contamination for off-site disposal**. Historically this was the most commonly used technique, but is not sustainable, and usually requires pre-treatment of the waste. Landfill taxes have also rendered other approaches more cost effective. The transportation of contaminated material along public highways also increases adverse noise and air quality impacts.
- **Excavation and on-site reuse/disposal**. This approach avoids off-site transport impacts, but regulatory approval is likely to be needed. Modelling and risk assessment are required to demonstrate that the material will have no adverse impacts in the new location. Verification of the planned work and post-construction monitoring are likely to be

essential. Disposal on-site (as a waste) is likely to require a permit or licence for a fully engineered containment facility. This will require long-term monitoring and maintenance, which introduces financial commitments.

- **Containment of the contamination to break the pathway.** This approach includes the provision of a cap of suitable thickness which can comprise combinations of different materials (concrete, paving, aggregates, membranes and soils) over the contaminated materials. The cap breaks the pathway to the surface. It may have to be of low permeability, be designed to prevent the osmotic rise of soluble contaminants to the surface, and in some cases it may have to be capable of collecting and dissipating gas. Where groundwater is the pathway or vector of the contaminant movement (rather than the receptor) the cap will prevent the infiltration of precipitation. A vertical barrier may also be required. This can be either a physical enclosing wall constructed in the ground, or a hydraulic barrier created by manipulating groundwater flow by pumping so there is no outward flow. *In situ* physical walls can be constructed from a range of materials including concrete, sheet piles, and bentonite (with or without a high-density polyethylene membrane within it).
- **Thermal treatments.** These methods can be used to remove and, in some cases, destroy the contaminants. Often, the soil will need to be excavated prior to treatment, and although some *in situ* treatment methods exist they may not be viable for treating large quantities of soil. The optimum method will depend on the nature of the contaminants present (Stegmann et al. 2013). The temperature at which the treatment is carried out will govern the chemical and physical properties of the treated material. The products of these processes are inevitably sterile and unsuitable for use as a growth medium.
- **On-site stabilisation by additives such as cement or bentonite.** These techniques immobilise pollutants.
- **In situ and ex situ bio-remediation.** These methods encourage natural micro-organisms to break down organic pollutants in soil and groundwater, generally by adding oxygen and nutrients to optimise their activity.
- **Soil washing.** Acid or solvent washing of soils is commonly undertaken, and is very effective at removing contaminants and minimising material for disposal. However, its use is dependent on the soil being mostly granular in nature.
- **Air sparging, vacuum extraction and pump and treat methods.** These are effective at removing a range of contaminants from groundwater. Sparging (stripping of contaminants using air) and vacuum extraction are *in situ* approaches which are most effective where volatile and semi-volatile contaminants are present. The action of drawing air through the soil has the added benefit of increasing micro-organism activity and increasing the breakdown of organic material. Pump and treat methods can recover the contaminants which are dissolved in the groundwater and any free-phase

product which may be present. The treatment occurs at the surface and can include a range of methods appropriate to the contaminants present.

Further guidance on how to select the best practicable environmental option is given in CIRIA (1995–1998). Extensive information on remediation techniques and methods is also available from the US Environmental Protection Agency (2012), and BSI (2017) provides some useful case studies. The choice of technique will depend upon the type of pollutant(s) present, the geology and hydrogeology, the development type and the sensitivity of the surrounding environment. Even where contamination is found, it does not automatically mean that some form of remediation is required.

3.7 Monitoring

Measures undertaken to mitigate the impacts of development should be monitored to determine their effectiveness. Such monitoring is generally a project requirement for programme funding and can be linked to Environmental and Social Management Plans (Chapter 20).

In respect of land and soil, the main aims of monitoring are to:

- determine the permanent loss of land in hectares, and ensure that a development project does not take more land than was proposed;
- ensure that residual land adjacent to the development continues to provide ecosystem services;
- ensure that temporarily displaced soils are conserved and reinstated to their original quality and the proposed end-uses within the site;
- ensure that permanently displaced soils are conserved and reinstated for beneficial uses on suitable land in a convenient location nearby.

Monitoring can take the form of on-site supervision of soil handling during the construction phase, maintaining a soil audit based on the volumes of soils identified on soil resource maps before disturbance, and a soil survey of restored land to determine whether all soil resources have been conserved. For sites where contamination has been contained in some way, monitoring will be required to ensure that leakage is not occurring.

3.8 Conclusions

Soils, land and geology are important components of ESIA and have recently been given a more prominent status within the Sustainable Development Goals (SDGs). This chapter begins by defining the subject area and describing some of the key features of soils, land and geology. We then outline the policies and legislation that have been developed to protect geological features and soils, and the legislative framework for dealing with contaminated soils. The details of such legislation vary from country to country, with some being more advanced than others. For an ESIA, a decision has to be made on whether a desk study or a field survey and laboratory analysis of soils is needed. Consideration is then given to the likely impacts of developments on geology, geomorphology, soils and land. Finally, mitigation of those impacts and monitoring of a site after development is briefly covered. The geology, soils and land aspects of an ESIA have markedly changed during the time we have been writing the four editions of this book, and they will certainly continue to do so.

Acknowledgement

The authors are grateful to Roy Emberton and Kevin Hawkins for their work on previous editions of this chapter.

References

Adriano DC, AK Iksandar and IP Murarka (eds) 1994. *Contamination of groundwaters*. Boca Raton, FL: CRC Press.

Anderson J, E Hardy, and JT Roach 1976. *A land use and land cover classification system for use with remote sensor data*. US Geological Survey Professional Paper 964.

APIS (Air Pollution Information System) 2016. Site-relevant critical loads and source attribution. www.apis.ac.uk/srcl.

Ashman MR and G Puri 2002. *Essential soil science*. Oxford: Blackwell.

ASRIS (Australian Soil Resource Information System) 2011. www.asris.csiro.au.

Avery BW 1990. *Soils of the British Isles*. Wallingford, Oxon: C.A.B International.

Blett, TF, JA Lynch, LH Pardo, C Huber, R Haeuber and R Pouyat 2014. FOCUS: A pilot study for national-scale critical loads development in the United States. *Environmental Science & Policy* **38**, 225–236.

Brady NC and RR Weil 2007. *The nature and properties of soils*, 14th Edn. Upper Saddle River, NJ: Prentice Hall.

Bridges EM 1978. *World soils*, 2nd Edn. Cambridge: Cambridge University Press.

BSI (British Standards Institute) 1994. BS 3882. *Specification for topsoil*. London: BSI.

BSI 2011. BS 10175. *Investigation of potentially contaminated sites – Code of practice*. London: BSI.

BSI 2017. ISO 18400 Series. http://shop.bsigroup.com/SearchResults/?q=18400.

CERC (Cambridge Environmental Research Consultants) 2016. www.cerc.co.uk/environmental-software.html.

CIRIA (Construction Industry Research and Information Association) 1995–1998. *Remedial treatment for contaminated land, Volumes I–XII* (SP164). www.ciria.org/CMDownload.aspx?ContentKey=c6c21523-a8d3-475e-a747-9465b6ca86dd&ContentItemKey=ca0f121d-a701-4dd7-bf13-dae4e95b8bbc.

Cuba N, A Bebbington, J Rogan and M Millones 2014. Extractive industries, livelihoods and natural resource competition: Mapping overlapping claims in Peru and Ghana. *Applied Geography* **54**, 250–261.

DCLG (Department of Communities and Local Government) 2012. *National planning policy framework*. http://planningguidance.communities.gov.uk/blog/policy.

DCLG 2014. *Land stability*. Planning practice guidance. http://planningguidance. communities.gov.uk/blog/guidance/land-stability/land-stability-guidance/

de Vente J, J Poesen, G Verstraeten, G Govers, M Vanmaercke, A Van Rompaey, M Arabkhedri and C Boix-Fayos 2013. Predicting soil erosion and sediment yield at regional scales: where do we stand? *Earth-Science Reviews* **127**, 16–29.

Defra (Department for Environment, Food and Rural Affairs) 2009a. *Safeguarding our soils: A strategy for England*. London: Defra. www.gov.uk/government/uploads/system/ uploads/attachment_data/file/69261/pb13297-soil-strategy-090910.pdf

Defra 2009b. *Construction code of practice for the sustainable use of soils on construction sites*. London: Defra.

Defra 2009c. *Code of good agricultural practice: protecting our water, soil and air*. London: Defra.

Defra 2015. *National statistics release: emissions of air pollutants in the UK, 1970 to 2014*. www.gov.uk/government/uploads/system/uploads/attachment_data/file/486085/Emissi ons_of_air_pollutants_statistical_release_2015_-_Final__2_.pdf.

Defra and EA (Environment Agency) 2004. *Model procedures for the management of contaminated land*. Report CLR11. Bristol: EA. www.gov.uk/government/publica- tions/managing-land-contamination

DoE (Department of the Environment) 1994. *Sampling strategies for contaminated land*. CLR4. London: HMSO.

DoE 1996a. *Mineral planning policy guidance note 1 (MPG 1): general considerations and the development plan system*. London: HMSO.

DoE 1996b. *The reclamation of mineral workings to agriculture*. London: HMSO.

DoE 1996c. *Guidance on good practice for the reclamation of mineral workings to agriculture*. London: HMSO.

EA (Environment Agency) 2003. *Contamination impacts on groundwater: simulation by Monte Carlo method*, ConSim release 2, Environment Agency R&D Publication 132. Nottingham: Golder Associates (UK) Ltd. www.consim.co.uk.

EA 2006. *Remedial targets methodology: Hydrological risk assessment for land contamination*. Bristol: EA

EC (European Commission) 2006. *Soil thematic strategy*, COM-2006, 231.

EC 2014. Directive 2014/52/EU of the European Parliament and of the Council of 16 April 2014 amending Directive 2011/92/EU on the assessment of the effects of certain public and private projects on the environment Text with EEA relevance. http://eur- lex.europa.eu/legal-content/EN/TXT/?uri=celex%3A32014L0052.

EEA (European Environment Agency) 2006. *Urban sprawl in Europe. The ignored challenge*. EEA Report No 10/2006.

Ellsworth WL 2013. Injection-induced earthquakes. *Science* **341**, 1225942.

Emmett BA 2007. Nitrogen saturation of terrestrial ecosystems: some recent findings and their implications for our conceptual framework. *Water, Air, and Soil Pollution: Focus* **7**, 99–109.

EPA (Environmental Protection Agency) 1980. *Summary of the Comprehensive Environmental Response, Compensation, and Liability Act (Superfund)* 42 U.S.C. §9601 *et seq.* (1980). www.epa.gov/laws-regulations/summary-comprehensive-environmental- response-compensation-and-liability-act.

EPA 2012. *Brownfields road map to understanding options for site investigation and cleanup*, 5th Edn. Washington DC: EPA.

FAO (Food and Agriculture Organization of the United Nations) 2016. *Sampling and laboratory techniques*. FAO soils portal. www.fao.org/soils-portal/soil-survey/sampling-and-laboratory-techniques/en.

GOV.UK 2015a. *Contaminated land*. www.gov.uk/contaminated-land

GOV.UK 2015b. *Land contamination: Department of Environment (DOE) industry profiles*. First published 1 January 1995; last updated 27 May 2014. www.gov.uk/government/publications/department-of-environment-industry-profiles.

GOV.UK 2015c. *Contaminated land exposure assessment (CLEA) tool*. Version 1.071. www.gov.uk/government/publications/contaminated-land-exposure-assessment-clea-tool.

GSI Environmental 2014. RBCA tool kit for chemical releases. www.gsi-net.com/en/software/rbca-software-tool-kit-for-chemical-releases-version-2-6.html.

Hallbäcken L and CO Tamm 1986. Changes in soil acidity from 1927 to 1982–1984 in a forest area in south-west Sweden. *Scandinavian Journal of Forest Research* **1**, 219–232.

Hodgson JM (ed.) 1997. *Soil survey field handbook*, 3rd Edn. Cranfield, Beds: Cranfield University.

Hodson, MJ 2012. Metal toxicity and tolerance in plants. *The Biochemist* **34**(5), 28–32.

IFC (International Finance Corporation) 2012. Performance standard 3, Resource efficiency and pollution prevention. Washington DC: World Bank Group. www.ifc.org/wps/wcm/connect/25356f8049a78eeeb804faa8c6a8312a/PS3_English_2012.pdf?MOD=AJPERES.

IUCN (International Union for Conservation of Nature) 2012. *Resolution WCC-2012-Res-048-EN: valuing and conserving geoheritage within the IUCN programme 2013–2016*. https://portals.iucn.org/library/sites/library/files/resrecfiles/WCC_2012_RES_48_EN.pdf.

Keller EA 2010. *Environmental geology*, 9th Edn. New Jersey: Prentice Hall.

Kraaijvanger R and A Veldkamp 2015. The importance of local factors and management in determining wheat yield variability in on-farm experimentation in Tigray, Northern Ethiopia. *Agriculture, Ecosystems and Environment* **214**, 1–9.

Krotkov NA, CA McLinden, C Li, LN Lamsal, EA Celarier, SV Marchenko, WH Swartz, EJ Bucsela, J Joiner et al. 2016. Aura OMI observations of regional SO_2 and NO_2 pollution changes from 2005 to 2014, *Atmospheric Chemistry and Physics* **16**, 4605–4629, www.atmos-chem-phys.net/16/4605/2016/.

Kurt S 2013. Land use changes in Istanbul's Black Sea coastal regions between 1987 and 2007. *Journal of Geographical Sciences* **23**(2), 271–279.

Lal R 2001. Soil degradation by erosion. *Land Degradation & Development* **12**, 519–539.

Land Quality Management Ltd 2013. *SP1004 International processes for identification and remediation of contaminated land*. Defra Report No.: 1023-0.

Löfgren S, M Aastrup, L Bringmark, H Hultberg, L Lewin-Pihlblad, L Lundin, GP Karlsson and B Thunholm 2011. Recovery of soil water, groundwater, and streamwater from acidification at the Swedish integrated monitoring catchments. *Ambio* **40**(8): 836–56.

MAFF (Ministry of Agriculture, Fisheries and Food) 2000. *Good practice guide for handling soils*. London: MAFF. http://webarchive.nationalarchives.gov.uk/20090306103114/http://www.defra.gov.uk/farm/environment/land-use/soilguid/index.htm.

Matthews TJ 2014. Integrating geoconservation and biodiversity conservation: theoretical foundations and conservation recommendations in a European Union context. *Geoheritage* **6**, 57–70.

Montanarella L, DJ Pennock, N McKenzie, M Badraoui, V Chude, I Baptista, T Mamo, M Yemefack, MS Aulakh et al. 2016. World's soils are under threat. *Soil* **2**, 79–82.

Morgan RPC 2005. *Soil erosion and conservation*, 3rd Edn. Oxford: Blackwell.

Munsell Color Co. 2000. *Soil color charts*. Newburgh, NY: Macbeth Division, Kollmorgen Instruments Group.

New South Wales Government 2011. *Guidelines for consultants reporting on contaminated sites*. State of NSW and Office of Environment and Heritage.

PAA (Penny Anderson Associates) 1994. *Roads and nature conservation: guidance on impacts, mitigation and enhancement*. Peterborough: English Nature.

Queensland Government 2014. *Information guideline for an environmental impact statement*. Department of Environment and Heritage Protection. www.ehp.qld.gov.au/management/impact-assessment/eis-processes/eis-tor-support-guidelines.html#eis_specific_content.

Rowell DL 1994. *Soils science: methods and applications*. London: Addison Wesley Longman.

Schrefler B and P Delage (eds) 2010. *Environmental geomechanics*. Chichester: Wiley-ISTE.

Sloss L 2013. *Coal mine site reclamation*. IEA Clean Coal Centre. www.usea.org/sites/default/files/022013_Coal%20mine%20site%20reclamation_ccc216.pdf.

SSG (Science Software Group) 2008. *AQUA3D: 3-dimensional finite-element ground-water flow and transport model*. Sandy, UT: SSG. www.scisoftware.com/environmental_software/product_info.php?products_id=31.

Stegmann R, G Brunner, W Calmano and G Matz G (eds) 2013. *Treatment of contaminated soil: fundamentals, analysis, applications*. Berlin: Springer.

Tan KH 2005. *Soil sampling, preparation and analysis*, 2nd Edn. London: Taylor & Francis.

Troeh FR and LM Thompson 2005. *Soils and soil fertility*, 6th Edn. Oxford: Blackwell Publishing.

UN 2000. *United Nations Millennium Declaration 55/2*. www.un.org/millennium/declaration/ares552e.htm.

UN 2015. *2030 Agenda for sustainable development*. Sustainable Development Goals (SDGs). UN Department of Economic and Social Affairs. https://sustainabledevelopment.un.org.

UNESCO 2015. *UNESCO global geoparks*. www.unesco.org/new/en/natural-sciences/environment/earth-sciences/global-geoparks.

USDA (US Department of Agriculture) 1961. *The land capability classification (LCC)*. Washington DC: USDA.

USDA 1993a. *Soil survey manual (Handbook 18)*. Washington DC: Soil Conservation Service.

USDA 1993b. *430–VI National soil survey handbook*. Washington DC: USDA-Natural Resources Conservation Service.

USDA 1999. *Soil taxonomy: a basic system of soil classification for making and interpreting soil surveys*, 2nd Edn. Washington DC: USDA – Natural Resources Conservation Service.

USDA 2013a. *Environmental assessment worksheet* (Revised). Washington DC: USDA, Rural Development.

USDA 2013b. *NRCS wind erosion prediction system (WEPS) information*. www.nrcs.usda.gov/wps/portal/nrcs/detail/national/technical/tools/weps/software/?cid=nrcs144p2_080196.

US Government Publishing Office 2010. *Title 30. Mineral lands and mining. Chapter 25. Surface mining control and reclamation*. www.gpo.gov/fdsys/pkg/USCODE-2010-title30/pdf/USCODE-2010-title30-chap25.pdf.

USGS (US Geological Survey) 2007. *MODFLOW and related programs*. Denver, CO: USGS http://water.usgs.gov/nrp/gwsoftware/modflow.html.

Utah Department of Natural Resources: Oil, Gas and Mining 2008. *Guidance for the management of topsoil and overburden.* https://fs.ogm.utah.gov/pub/MINES/ Coal_Related/soilsguide.pdf.

Verheye W, P Koohafkan and F Nachtergaele 2008. *The FAO guidelines for land evaluation. Land use, land cover and soil sciences.* Vol. II: UNESCO Encyclopaedia of Life Support Systems.

Wang H 2015. *Soil erosion.* New Delhi: ML Books International.

Webster R and MA Oliver 1990. *Statistical methods in soil and land resource survey (Spatial information systems).* Oxford: Oxford University Press.

Wischmeier WH and DD Smith 1960. A universal soil-loss equation to guide conservation farm planning. *Transactions of the 7th International Congress of Soil Science,* 418–425.

Wischmeier WH and DD Smith 1978. *Predicting rainfall erosion losses: A guide to conservation planning.* Agriculture Handbook No. 537. Washington DC: USDA/Science and Education Administration.

WRB (World Reference Base for Soil Resources) 2014. *International soil classification system for naming soils and creating legends for soil maps. Update 2015.* Rome: FAO.

Wright MA (ed) 2002. *The practical guide to reclamation in Utah.* Utah: Office of Surface Mining Reclamation and Enforcement, Division of Oil, Gas, and Mining. https://fs.ogm.utah.gov/pub/MINES/Coal_Related/RecMan/Reclamation_Manual.pdf.

4 Air

● ●

David P. Walker, Hannah Dalton, Graham Harker
and Kiri Heal

4.1 Introduction

A proposed development that will change the concentration of pollutants in the atmosphere from the baseline situation may result in effects on people, plants, animals, materials and buildings (Harrop 2002). These effects can occur at the local, regional or even global scale. Major developments, such as power stations, oil refineries, waste incinerators, chemical processing plants and roads, pose obvious potential pollution problems. Other developments (e.g. edge-of-town shopping or leisure complexes), once complete, may give rise to additional vehicle emissions as people travel to use them.

Many developments also have the potential to create a local dust nuisance due to earth-moving and materials-handling operations during the construction stage, especially during periods of dry weather (although with appropriate mitigation these effects can be managed successfully). For developments such as opencast mines and quarries, the potential exists for dust generation during the entire life cycle of the development, and with it the potential for contaminants (such as heavy metals, silicates and other irritants) to be deposited in the surrounding area.

Developments involving combustion may give rise to both routine and non-routine pollutant emissions. For example, they may normally use one type of fuel (e.g. natural gas), but may occasionally need to switch to an alternative (e.g. fuel oil) due to a lack of availability of the primary fuel source. The alternative fuel may lead to different pollutant emissions and therefore the short-term environmental effects of these need to be considered. For example, natural gas is essentially sulphur-free whereas fuel oil can contain varying amounts of sulphur (depending on the specification). Non-routine emissions could also include accidental releases at a proposed development giving rise to the risk of the release of hazardous substances, for example through the rupturing of a tank, or an explosive failure of part of a process or plant due to excess pressure.

As well as pollutants and dust, certain types of development might release odour to the atmosphere causing a response among the local community. The nature of the response is likely to vary considerably depending upon the type of odour released, e.g. a bakery is less likely to attract adverse comment than a development which releases strong odours from waste-processing activities. Any odour can be offensive if it is continuous and intrusive and therefore the potential effects of releases should be assessed carefully.

Finally, developments may cause a *perception* of emitting pollutants to the air when in actual fact they are not causing any significant releases at all. A typical example might be the release of steam from a vent stack which might cause concern to onlookers as well as resulting in a visual effect (Chapter 11). In some instances, this visual impact can be a material consideration – for example, emissions of water vapour from stacks in the vicinity of airports may have the potential, under certain meteorological conditions, to cause a plume of mist which disrupts important sightlines.

4.2 Definitions and concepts

4.2.1 Air pollutants and human health

Air *pollutants* can affect the health of a person during inhalation and exhalation as the pollutants inflame, sensitise and even scar the airways and lungs in extreme conditions or following chronic exposure. On reaching deep inside the lungs, they may enter the bloodstream, thus affecting organs other than the lungs, and they can take up permanent residence in the body. In addition, some pollutants affect health through contact with the skin and through ingestion of contaminated foods and drinks. Pollutants affect health in varying degrees of severity, ranging from minor irritation through to serious illness to premature death in extreme cases. They may produce immediate (acute) symptoms as well as longer term (chronic) effects. Health effects depend upon the type and amount of pollutants present, the duration of exposure, and the state of health, age and level of activity of the person exposed (Harrop 2002; see also Chapter 16).

Epidemiological studies of community groups and laboratory-based toxicological experiments using human volunteers provide assessments of the health effects of pollutants. Based on these findings, various national and international organisations have identified levels of air pollution concentrations (air quality standards) which should not be exceeded in order to protect human health (see §4.3.1).

4.2.2 Air pollutants and the natural and built environment

Pollution damage to plants and animals is caused by a combination of physical and chemical stresses that affect the receptor's physiology. Pollutants can affect crops by causing leaf discoloration, reducing plant growth and yields, or by contaminating a crop, so making it unsafe to eat. Effects on terrestrial and aquatic ecosystems can occur locally or even regionally in the case of pollutants that contribute to *acid deposition* ('acid rain'), especially where soils and lakes lack substances to neutralise or buffer the acidic inputs (see Chapters 2 and 3). Deposition of nitrogen-based compounds (such as ammonia) on sensitive ecosystems (e.g. upland grasslands, sensitive watercourses and lakes, or ancient woodlands) can also disrupt the nutrient balance of these habitats and compromise their long-term integrity.

The effects of air pollution on vegetation and ecosystems can be related to a pollutant concentration (termed *critical level*), or the deposition of a pollutant (termed *critical load*). Critical levels are defined as 'concentrations of pollutants in the atmosphere above which direct adverse effects on receptors, such as human beings, plants, ecosystems or materials, may occur according to present knowledge' (UNECE 1996). Critical loads are defined as a 'quantitative estimate of exposure to one or more pollutants below which significant harmful effects on specified sensitive elements of the environment do not occur according to present knowledge' (UNECE 1996). The critical load relates to the quantity of pollutant deposited from air to the ground, unlike the critical level, which is the gaseous concentration of a pollutant in the air.

Pollution problems for buildings can be short-term and reversible such as soiling by smoke (which can be removed by cleaning), whereas the effects of acid deposition can be cumulative and irreversible by causing erosion and crumbling of the stone.

4.2.3 Odours

Odours tend to be associated with mixtures of chemicals that interact to produce what is detected as a smell. Odour is an olfactory response to a chemical or chemicals and this is the key response of concern within an Environmental and Social Impact Assessment (ESIA), not the potential health effects of the chemicals themselves. Occasionally odour can be associated with a single substance with a particularly strong odour potential (e.g. hydrogen sulphide), but more commonly odour is associated with a mixture of substances. Odours of concern are those that are objectionable or offensive and these can lead to dis-amenity, annoyance, nuisance or complaints. Odour is subjective to the individual and therefore a response can vary significantly from person to person, which makes assessing the significance of the effect particularly challenging.

Before an adverse effect can exist, individuals must experience exposure to odour. For odour exposure to occur there must be an odour source and a pathway, and the scale of the exposure can be determined by reference to parameters known as FIDOL (frequency, intensity, duration, offensiveness and location). The first four of these factors relate to the magnitude or scale of the exposure, and the latter factor to the sensitivity of a particular receptor population) (Bull et al. 2014).

4.3 Key international guidelines and standards

4.3.1 Ambient air quality guidelines and standards for human health

Levels of pollution set within air quality standards are sometimes advisory, such as the World Health Organisation guideline values, while others are mandatory and backed by legislation, such as the US standards and the EU limit values. Concentrations are expressed either as mass of the substance per unit volume of air (e.g. micrograms per cubic metre, abbreviated to $\mu g/m^3$) or as volume of the substance to the volume of air (e.g. parts per million or parts per billion, abbreviated to ppm and ppb respectively). These units can be converted from one to another using published conversion factors, which can vary slightly due to standardisation at different atmospheric pressure and temperature conditions (Defra 2014).

World Health Organisation (WHO)

The WHO guideline values (see Table 4.1; WHO 2000, 2005) are based on the lowest concentration a pollutant has been shown to produce adverse health effects or the level at which no observed health effect has been demonstrated, plus a margin of protection to safeguard sensitive groups within the population. Sensitive groups include people with asthma, those with pre-existing heart and lung diseases, the elderly, infants, and pregnant women and their unborn babies. Some pollutants, notably carcinogenic pollutants (e.g. arsenic, benzene, chromium, polycyclic aromatic hydrocarbons (PAHs) and vinyl chloride) have not been given a guideline value. Instead, exposure-effect information is provided, giving guidance to risk managers about the major health impact for short- and long-term exposure to various levels of this pollutant.

The WHO guideline values are based on health considerations alone, and do not consider the technical feasibility or the economic, political and social dimensions of attainment (European Parliament 2014). This explains why air quality standards vary nationally around the world and why they are often not as strict as the WHO guidelines.

International Finance Corporation (IFC)

The IFC Performance Standards provide an international benchmark for environmental and social risk management. The Environmental, Health and Safety (EHS) Guidelines (IFC 2016) provide technical advice with general and industry-specific examples of 'good international industry practice' (GIIP) to meet specified Performance Standards. Under Infrastructure Planning Commission (IPC) Performance Standard 3 (Resource Efficiency and Pollution Prevention), commercial clients/investees are required to integrate

Table 4.1 Comparison of air quality assessment levels

Pollutant	Averaging time	World Health Organisation	US National Ambient Air Quality Standards	EU air quality limit values for the protection of human health
Nitrogen dioxide	1 h	200 µg/m³	100 ppb[a] (188 µg/m³)	200 µg/m³[b]
	Annual	40 µg/m³	53 ppb (100 µg/m³)	40 µg/m³
Sulphur dioxide	10 min	500 µg/m³	–	–
	1 h	–	75 ppb[c] (196 µg/m³)	350 µg/m³[d]
	3 h	–	0.5 ppm (1,310 µg/m³)	–
	24 h	20 µg/m³	–	125 µg/m³[e]
PM$_{2.5}$	24 h	25 µg/m³[f]	35 µg/m³[g]	–
	Annual	10 µg/m³	12 µg/m³[h] / 15 µg/m³[h]	25 µg/m³
PM$_{10}$	24 h	50 µg/m³[f]	150 µg/m³[i]	50 µg/m³[j]
	Annual	20 µg/m³	–	40 µg/m³
Carbon monoxide	15 min	100 mg/m³	–	–
	30 min	60 mg/m³	–	–
	1 h	30 mg/m³	35 ppm (40 mg/m³)	–
	8 h	10 mg/m³	9 ppm (10 mg/m³)	10 mg/m³[k]
Ozone	8 h	100 µg/m³	0.07 ppm (137 µg/m³)[l]	120 µg/m³[k m]
Benzene	Annual	–	–	5 µg/m³
	UR/ lifetime[n]	6x10^{-6} (µg/m³)$^{-1}$	–	–
Dichloromethane	24 h	3 mg/m³	–	–
Formaldehyde	30 min	0.1 mg/m³	–	–
PAHs°	UR/ lifetime[n]	8.7x10^{-5} (ng/m³)$^{-1}$	–	–
	Annual	–	–	1 ng/m³
Styrene	1 week	0.26 mg/m³	–	–
Tetrachloroethylene	Annual	0.25 mg/m³	–	–
Toluene	1 week	0.26 mg/m³	–	–

Table 4.1 continued

Pollutant	Averaging time	World Health Organisation	US National Ambient Air Quality Standards	EU air quality limit values for the protection of human health
Trichloethylene	UR/ lifetime[n]	4.3×10^{-7} $(\mu g/m^3)^{-1}$	–	–
Arsenic	UR/ lifetime[n]	1.5×10^{-3} $(\mu g/m^3)^{-1}$	–	–
	Annual	–	–	6 ng/m³
Cadmium	Annual	5 ng/m³	–	5 ng/m³
Chromium	UR/ lifetime[n]	0.04 $(\mu g/m^3)^{-1}$	–	–
Lead	Rolling 3 months	–	0.15 µg/m³	–
	Annual	0.5 µg/m³	–	0.5 µg/m³
Manganese	Annual	0.15 µg/m³	–	–
Mercury	Annual	1.0 µg/m³	–	–
Nickel	UR/ lifetime[n]	3.8×10^{-4} $(\mu g/m^3)^{-1}$	–	–
	Annual	–	–	20 ng/m³

Notes: The values in parentheses have been converted from their original metric to simplify comparison with the WHO and EU guidelines.

[a] 98th percentile of 1-hour daily maximum concentrations, averaged over 3 years.
[b] No more than 18 exceedances per year.
[c] 99th percentile of 1-hour daily maximum concentrations, averaged over 3 years.
[d] No more than 24 exceedances per year.
[e] No more than 3 exceedances per year.
[f] 99th percentile, averaged over 3 years (3 days/year).
[g] 98th percentile, averaged over 3 years.
[h] Primary and Secondary objectives. Annual mean, averaged over 3 years.
[i] Not to be exceeded more than once per year on average over 3 years.
[j] No more than 35 exceedances per year.
[k] Maximum daily 8-hour mean.
[l] Annual fourth-highest daily maximum 8-hour concentration, averaged over 3 years.
[m] No more than 25 exceedances per year averaged over 3 years.
[n] UR = excess risk of dying from cancer following lifetime exposure. Thus for benzene, 6 people in a population of 1 million will die as a result of a lifetime exposure of 1 µg/m³; for PAHs 87 people in a population will die from cancer following lifetime exposure to 1 ng/m³.
[o] Specifically benzo[a]pyrene.

pollution prevention and control technologies and practices (as technically and financially feasible as well as cost-effective) into their business activities. The guidelines state that projects with significant air emissions, and potential for significant impacts to ambient air quality, should prevent or minimise impacts by ensuring that:

- Emissions do not result in pollutant concentrations that reach or exceed relevant ambient quality guidelines and standards by applying national legislated standards, or in their absence, the current WHO Air Quality Guidelines or other internationally recognised sources (e.g. the United States National Ambient Air Quality Standards and EU air quality limit values).
- Emissions do not contribute a significant portion to the attainment of relevant ambient air quality guidelines or standards. As a general rule, the Guidelines suggest less than 25 percent of the applicable air quality standards to allow additional, future sustainable development in the same air-shed.

United States National Ambient Air Quality Standards

In the United States, the Clean Air Act, last amended in 1990, requires the Environmental Protection Agency (EPA) to set National Ambient Air Quality Standards for pollutants considered harmful to public health and the environment. The Clean Air Act identifies two types of National Ambient Air Quality Standards. Primary standards provide public health protection, including protecting the health of 'sensitive' populations such as asthmatics, children and the elderly. Secondary standards provide public welfare protection, including protection against reduced visibility and damage to animals, crops, vegetation and buildings. The EPA has set standards for six principal pollutants, called 'criteria' pollutants (Table 4.1; EPA 2016a).

European Union air quality limit values

The EU began setting air quality standards in the form of mandatory health-based limit values and more stringent non-mandatory guide values to protect the environment from 1980 onwards. Guide values are intended to be long-term objectives which, when met, will protect vegetation as well as aesthetic aspects of the environment such as long-range visibility and soiling of buildings.

As part of the European Community's Framework Directive on Ambient Air Quality Assessment and Management (96/62/EC), agreed in September 1996, the EU has set limit values for a number of pollutants. The values are specified in a series of 'Daughter Directives', with the first one being agreed in 1999 and covering SO_2, particulate matter (PM_{10} and $PM_{2.5}$), NO_2 and lead (Pb). Subsequent Daughter Directives refer to ozone (O_3), benzene and carbon monoxide (CO), polycyclic aromatic hydrocarbons (PAHs), cadmium, arsenic, nickel and mercury. The most recent Directive 2008/50/EC combines most of the existing legislation into a single document (EC 2008). This introduced new air quality objectives for $PM_{2.5}$ and permits certain time extensions for the member states to comply with the limit values, in addition to a two-year period to transpose the new legislation (Table 4.1; EC 2015).

4.3.2 Air quality guidelines and standards for vegetation and ecosystems

The effect of air pollution on vegetation and ecosystems is a relatively complex and evolving area that requires cooperation between air quality scientists and ecologists to properly assess the significance of the effects within the context of ESIA.

Critical loads for deposition onto sensitive ecosystems have been specified by the United Nations Economic Commission for Europe (UNECE). Exceedance of a critical load is used as an indication of the potential for harmful effects to occur. The UK Air Pollution Information System (APIS 2016) provides a useful guide to critical loads and levels. Examples of critical levels and the source of the limit value are shown in Table 4.2.

Substances released to air do not need to be assessed for deposition to ground unless they contribute to acidification and eutrophication or they are highly toxic, bio-accumulative or persistent. Critical loads for nutrient nitrogen are set under the Convention on Long-Range Transboundary Air Pollution. They are based on empirical evidence, mainly observations from experiments and gradient studies. In Europe, critical loads are assigned to habitat classes of the European Nature Information System (EEA 2016) to enable consistency of habitat terminology. The values for critical loads are given as ranges (e.g. 10–20 kgN/ha/yr) to reflect variation in ecosystem response.

Deposition of sulphur, as sulphate (SO_4^{2-}), and nitrogen, as nitrate (NO_3^-), ammonium (NH_4^+) and nitric acid (HNO_3^-), can cause acidification and both sulphur and nitrogen compounds must be taken into account when assessing acidification of soils. For the purposes of determining links between critical loads and atmospheric emissions of sulphur and nitrogen, critical loads are further derived to produce a maximum critical load for sulphur (CLmax), a minimum critical load for nitrogen (CLminN) and a maximum critical load for nitrogen (CLmaxN). These components define the 'critical load function' and when compared with deposition data for sulphur and nitrogen, they can be used to assess critical load exceedances.

There are no well-defined critical loads for heavy metals, but researchers have started to develop effects-based critical levels and critical loads for some heavy metals in soils and freshwaters (APIS 2016).

Table 4.2 Critical levels for key pollutants in the UK (APIS 2016)

Pollutant	Receptor	Time period	Critical level
NO$_x$	All	Annual mean	30 µg/m³
NO$_x$	All	24-hour mean	75 µg/m³
SO$_2$	Crops	Annual mean	30 µg/m³
SO$_2$	Forests and natural vegetation	Winter mean (1 Oct to 31 Mar)	20 µg/m³
SO$_2$	Forests and natural vegetation	Annual mean	20 µg/m³
SO$_2$	Sensitive lichens	Annual mean	10 µg/m³
O$_3$ (Ozone)	All	AOT40, calculated from 1h values May–July. Mean of 5 years	18,000 µg/m³.hr (9,000 ppb hours)
O$_3$	Crops	AOT40, May to July	3,000 ppb hours
O$_3$	Forests	AOT40, April to September	10,000 ppb hours
O$_3$	Forests, semi-natural vegetation dominated by perennials	AOT40, April to September (semi-natural) growing season (trees)	5,000 ppb hours
O$_3$	Semi-natural vegetation dominated by annuals	AOT40, May to July	3,000 ppb hours
Ammonia	Lichens and bryophytes (where they form a key part of the ecosystem integrity)	Annual mean	1 µg/m³
Ammonia	Other vegetation	Annual mean	3 µg/m³ (with an uncertainty range of 2–4 µg/m³)

4.3.3 Odour guidelines and standards

There are no internationally recognised standards for the assessment of odour, and while common concepts can be identified, the circumstances in individual countries will differ. The precise definition of odour and how it is assessed in concentration terms varies from country to country.

One approach adopted in the UK involves measurements of odour which are based on the European Odour Unit (OU$_E$). The definition of 1 OU$_E$ represents the threshold whereby 50 per cent of the members of an odour panel can detect an odour in laboratory conditions. The odour unit can be related to a mass unit of pollution such that a concentration level can be set for the acceptability of a particular odour, depending on the particular land use.

When considered as a concentration, the 98th percentile of the hourly average is normally taken to be the assessment level. This allows for the particular concentration threshold to be exceeded for up to 2 per cent of the time (i.e. for up to 175 hours per year), which suggests that tolerance to a certain level of odour in the environment is to be expected or is acceptable. While some research has attempted to relate a specific odour concentration threshold to a level of complaints or annoyance (e.g. Miedema and Ham 1988, Miedema et al. 2000), the results of the assessments are not definitive. This may partly be due to the fact that the 98th percentile concentration is based on a steady state odour whereas emissions and actual concentrations may fluctuate widely.

In terms of the FIDOL parameters, an odour concentration as a 98th percentile takes account of the frequency, intensity and duration elements of the equation. For the locational dimension, different concentration thresholds may be applied to different land uses, i.e. a higher level of odour may be tolerable for industrial or commercial locations than for residential locations. It is then necessary to decide whether the concentration threshold should vary according to the offensiveness of the particular substance. In the UK this is the approach adopted within odour management regulatory guidance (Environment Agency 2011), where odour is classified into three offensiveness ranges: 'most offensive', 'moderately offensive' and 'less offensive', with different concentration criteria suggested for each range (see Table 4.3).

Table 4.3 Odour concentration and offensiveness (Environment Agency 2011)

Offensiveness	Odour concentration C_{98}, 1 hour (OU_E/m^3)	Examples
Most offensive	1.5	Processes involving decaying animal or fish remains Processes involving septic effluent or sludge Biological landfill odours
Moderately offensive	3.0	Intensive livestock rearing Fat frying (food processing) Sugar beet processing Well aerated green waste composting
Less offensive	5.0	Brewery Confectionery Coffee roasting Bakery

The Institute of Air Quality Management (IAQM) guidance (Bull et al. 2014) provides suggested odour-effect descriptors based on receptor sensitivity for ranges of predicted odour concentrations. However, the practitioner is required to use professional judgement in the choice of criteria and must be able to justify their choice. Overall, the acceptable odour concentration is anticipated to lie somewhere in the range 1–10 OU_E/m^3 as a 98th percentile of hourly averages.

4.3.4 Emission limits and standards

Air quality standards refer to the levels of air pollution to which people and ecosystems are exposed. Another type of legislated standard is the emission standard, which specifies the maximum amount or concentration of a pollutant that can be emitted from a given source. Emission standards are usually derived from consideration of the cost and effectiveness of the available control technology.

In 1990, the US Congress passed the Pollution Prevention Act and in 1996 the EU Directive on Integrated Pollution Prevention and Control (IPPC) (96/61/EC) was adopted. IPPC has been superseded by the Industrial Emissions Directive (IED) (2010/75/EU) which builds upon seven previous directives covering industrial emissions from a wide range of industries. The IED aims to achieve a high level of protection of human health and the environment taken as a whole by reducing harmful industrial emissions across the EU, in particular through better application of Best Available Techniques (BAT). Currently around 50,000 industrial installations are required to operate in accordance with a permit granted by the authorities within the EU Member States.

IED permits must take into account the whole environmental performance of the plant, covering e.g. emissions to air, water and land, generation of waste, use of raw materials, energy efficiency, noise, prevention of accidents and restoration of the site upon closure. The permit conditions, including emission limit values, must be based on the use of BAT. In order to define BAT and the BAT-associated environmental performance at an EU level, the European Commission organises an exchange of information with experts from the member states, industry and environmental organisations. This process results in the production of BAT Reference Documents (BREFs), which are available for a wide range of industrial sectors from the Joint Research Centre of the European Commission (JRC 2016). The BREFs contain BAT conclusions which are adopted by the Commission as 'Implementing Decisions'. The IED requires that these BAT conclusions are used as the reference for setting permit conditions. For certain activities (i.e. large combustion plants, waste incineration and co-incineration plants, solvent using activities and titanium dioxide production) the IED also sets EU-wide emission limit values for selected pollutants.

The IFC Environmental, Health and Safety Guidelines (IFC 2016) state that emissions from point sources should be avoided and controlled according to good international industry practice (GIIP). It emphasises the combined application of process modifications and emissions controls, examples of which are provided along with additional recommendations regarding stack height and emissions from small combustion facilities.

Emission standards for vehicles are covered by Federal or State Standards in the US, or at a European level. Standards vary according to the age and type of vehicle, and in Europe, by the fuel (diesel or petrol). As such, the emissions performance of the vehicle fleet will vary over time, which needs to be taken into account in assessing the impacts and effects of road traffic emissions that may be relevant to ESIA.

Emission standards for odour are rarely applied due to the difficulty in measuring odour emissions at source, unless the odour is related to a specific chemical pollutant that can be readily measured. Instead, it is more normal to control odour by reference to the impact (see §4.3.3).

Emission limits for pollutants can also apply in terms of total, national emissions. For example, in Europe, the UNECE *Protocol to Abate Acidification, Eutrophication and Ground-level Ozone* (UNECE 1999) sets national ceilings for four acidifying, eutrophying and ozone-forming air pollutants: SO_2, NO_x, VOCs and NH_3. Stricter national ceilings for these four pollutants are set out in the EU Directive on *National Emission Ceilings (NEC) for Certain Atmospheric Pollutants* (2001/81/EC) which came into force in 2001. Parallel to the development of the EU NEC Directive, the EU Member States together with Central and Eastern European countries, the United States and Canada negotiated the 'multi-pollutant' protocol under the Convention on Long-Range Transboundary Air Pollution (the so-called Gothenburg Protocol, agreed in November 1999). In 2012 the parties to the Gothenburg Protocol agreed new emission reduction commitments to be achieved by 2020. The restrictions imposed by national emissions ceilings are only likely to impact on the very largest industrial developments that fall under the remit of ESIA.

4.3.5 Regulations for hazardous chemicals

In the case of a proposed development that involves materials that could be harmful to people in the event of an accident, the ESIA should include an indication of the preventative measures to be adopted, so that such an occurrence is not likely to have a significant effect. Recent amendments to the European EIA Directive 2014/52/EU include, among other things, reference to the effects of major accidents or disasters on the environment when undertaking EIAs (EC 2014). In the case of major accidents, the standards or guidelines against which the pollutant concentrations are assessed are often different to those used for routine releases as they relate to acute, short-term exposure. Appropriate reference will therefore need to be made to the toxicology of the particular chemicals that could be released in the event of an accident or disaster.

A new directive on the control of major accident hazards involving dangerous substances known as Seveso III was published on 24 July 2012 by the European Commission (EC 2012). It amended and subsequently repealed earlier versions and came into force on 1 June 2015. The directive lays down rules for the prevention of major accidents which involve dangerous substances in order to limit the consequences of a major accident.

4.4 Scoping and baseline studies

4.4.1 Introduction

During scoping, the consideration of air quality impacts in an ESIA should be confirmed through discussions with the regulatory authorities, including which pollutants to consider and the need to address odour impacts.

Depending upon the development project, a wide range of atmospheric pollutants may potentially be relevant to ESIA (Table 4.4). For example, the existence of international objectives and limit values clearly indicate the need to consider NO_2, SO_2, fine particulates, CO, Pb, benzene, 1,3 butadiene, O_3, polycyclic aromatic hydrocarbons (PAH), cadmium, arsenic, nickel and mercury. In addition, many other health-threatening pollutants, some of which have been given WHO guideline values and others which have not (simply because of insufficient evidence to be able to define an appropriate safe level), should be considered. These latter pollutants include polychlorinated biphenyls (PCBs), dioxins (PCDDs), furans (PCDFs), toxic chemicals (e.g. ammonia, fluoride, chlorine) and toxic metals (e.g. chromium, manganese, platinum). Ionising radiation (radionuclides) released from certain medical facilities and nuclear power plants should also be considered for these developments.

In scoping the ESIA, the need to consider a particular atmospheric pollutant will depend on the nature and scale of the development, the sensitivity of the receiving environment, the baseline conditions and an initial view of the potential changes that may occur in relation to this. It is also important to establish whether the baseline conditions are likely to change in the future, irrespective of the planned development. This is essential because even if a development is likely to add only small amounts of pollution to the area, it could lead to air quality standards being exceeded if the baseline air quality is already poor, or may become poor in the future. In cases where the pollutant is scoped out of detailed consideration through experience of the likely level of the effect, an assessment of non-routine or emergency releases may still be necessary if the potential exists for large scale releases of hazardous substances. As well as differing sensitivities among buildings (for example hospitals, houses, industrial premises etc.), other considerations may also influence scoping, for example the proximity of the development to ecologically sensitive sites.

Developments such as sewage treatment works, chemical plants, paint works, food processing factories, brick works and commercial kitchens, could

Table 4.4 Key air pollutants and their anthropogenic sources

Pollutant	Anthropogenic sources
Nitrogen oxides (NO_x: NO, NO_2)	Coal-, oil- and gas-fired power stations, industrial boilers, waste incinerators, motor vehicles
Sulphur dioxide (SO_2)	Coal- and oil-fired power stations and industrial boilers, waste incinerators, diesel vehicles, metal smelters, paper manufacturing
Particulates (dust, smoke, PM_{10}, $PM_{2.5}$)	Coal- and oil-fired power stations and industrial boilers, waste incinerators, domestic heating, many industrial plants, diesel vehicles, construction, mining, quarrying, cement manufacturing
Carbon monoxide (CO)	Fuel combustion
Volatile organic compounds (VOCs) e.g. benzene	Petrol engine vehicle exhausts, leakage at petrol stations, paint manufacturing
Toxic organic micropollutants (TOMPS) e.g. PAHs, PCBs, dioxins	Waste incinerators, coke production, coal combustion
Toxic metals e.g. lead, cadmium	Metal processing, waste incinerators, oil and coal combustion, battery manufacturing, cement and fertiliser production
Toxic chemicals e.g. chlorine, ammonia, fluoride	Chemical plants, metal processing, fertiliser manufacturing
Greenhouse gases e.g. carbon dioxide (CO_2), methane (CH_4)	CO_2: fuel combustion, especially power stations; CH_4: coal mining, gas leakage, landfill sites
Ozone (O_3)	Secondary pollutant formed from VOCs and nitrogen oxides
Ionising radiation (radionuclides)	Nuclear reactors and waste storage, some medical facilities
Odours	Sewage treatment works, landfill sites, chemical plants, oil refineries, food processing, paintworks, brickworks, plastics manufacturing, commercial kitchens

give rise to offensive odours. Odours often generate great annoyance when residents are subjected to them in their gardens and homes. Development which places an additional load on treatment facilities (for example new residential development generating more waste water requiring treatment by a local sewage works) might also result in releases of odour from the sewage works if it is not of sufficient capacity to handle the additional loading.

4.4.2 Pollution data availability

Before the impact of a proposed development can be predicted, information on the ambient levels of the pollutants of concern at one or more locations in the study area is needed to assess the amount of pollution present. Using information from existing pollution monitors is the simplest and least expensive approach to obtaining baseline data. In many countries, national or local monitoring networks exist, and monitoring may also be undertaken by universities and other organisations. Pollution data from national networks are often freely available online, for example the Automatic Urban and Rural Network in the UK (Defra 2015), the US Air Quality System (EPA 2016b), or the World Air Quality Index (World Air Quality 2016), or they can be requested from the relevant authorities.

Pollution monitoring sites are often classified by type of location and therefore, in the absence of a monitoring site in the vicinity of the proposed development, the data may be considered as indicative of what may be experienced at similar sites elsewhere (Defra 2009). Expert opinion obtained from environmental consultancies and universities can advise on the validity of using pollution data in this way. Alternatively, the data can be modified to reflect the location of interest by using established empirical relationships, for example the relationship between annual mean and one-hour objectives for NO_2 and the annual mean/24-hour mean for PM_{10} (Defra 2016a). In some cases empirical relationships enable the levels of one pollutant to indicate the likely levels of another pollutant.

When the impact of an existing odour source on a proposed development needs to be considered (e.g. placing residential areas close to a sewage treatment works) there may be data on the complaints history of the site (from the regulatory authority) or specific modelling information from the operator of the installation. Complaints data may be analysed to indicate whether an existing odour source is likely to be an issue for a new development by consideration of the differences in wind speed, direction and the nature of the pathway between the odour source and the receptor.

4.4.3 On-site pollution monitoring

If pollution data are not available or are insufficient, on-site baseline monitoring will be required. This should be planned and initiated during the scoping phase of an ESIA and needs to consider: (a) what pollutants to monitor; (b) what type of monitoring equipment to employ; (c) the number and location of sampling sites; (d) the duration of the survey; and (e) the time-resolution of sampling.

Selecting the equipment to measure air pollution concentrations depends upon (a) the intended use of the data; (b) the budget allocated to purchase or hire the equipment; and (c) the expertise of personnel available to set up and maintain the equipment and, in some cases, to undertake laboratory analyses of collected samples. Setting up an automatic pollutant analyser can be costly, so hiring the equipment may be more appropriate. It is important that the equipment selected for monitoring is fit for purpose so that the data collected can be compared with relevant air quality standards.

A number of air quality monitoring options exist, ranging from relatively simple and inexpensive equipment such as passive diffusion tubes to the more expensive automatic analysers, which continuously measure pollutant concentrations. Passive diffusion tubes absorb the pollutant on to a metal gauze placed at the bottom of a short cylinder open at the other end to the atmosphere. After exposure, the tubes are sent for laboratory analysis. They can provide useful information for a range of pollutants including ammonia, benzene, CO, hydrogen sulphide, NO_2, O_3 and SO_2. In areas of high pollution concentrations they can produce results for daily or even three-hourly exposures, although in areas with low concentrations they are usually exposed for at least two weeks at a time. Monthly exposure readings from these tubes can provide estimates of the annual mean concentrations.

When using diffusion tubes, it is always preferable to correct the readings obtained from a specific batch against more sophisticated equipment and it is good practice to 'co-locate' some diffusion tubes from the same batch as those exposed in the study area at an automatic pollution analyser. Provided the automatic analyser is operating correctly and has been calibrated, any difference between the two data sets is likely to be due to a bias in the measurement of the diffusion tubes and a 'bias correction factor' can be calculated from this differential which is then applied to all results obtained from diffusion tubes exposed over the same period from the same batch. Alternatively, there may be databases of national factors for the same tube manufacturer/preparation method that could be used to obtain a bias adjustment factor, e.g. the UK's national bias adjustment factor database (Defra 2016b).

The duration of baseline monitoring for an ESIA will depend upon the pollutant to be tested and the standard which is to be assessed. Pollution concentrations vary from hour to hour, day to day, and month to month and are influenced by a range of external variables such as wind speed and direction, sunlight, temperature, precipitation, humidity, etc. As a result, monitoring for a short period of time is unlikely to provide a satisfactory indication of baseline conditions. A minimum monitoring period of three months should be used, and six months is preferable. For this reason, early commencement of monitoring in the ESIA process is essential if the results obtained are to be useful in informing the assessment process before it is complete. Where monitoring is undertaken for periods of less than one year and an estimate of the annual average concentration is required, then it will be necessary to annualise the data by reference to historic pollution measurements over the same time periods in the same region of the assessment.

Pollution bio-indicators, types of plant that are sensitive to pollution levels (e.g. lichen for SO_2, tobacco plants for O_3) may provide supplementary information on pollution levels (Mulgrew and Williams 2000). Soil and vegetation analyses can also provide long-term levels of pollutants such as metals.

When siting monitoring equipment, it is necessary to consider: (a) the need to protect against vandalism; (b) access to the site; (c) the avoidance of pollution from indoor and localised sources which may make the data unrepresentative of the wider area; and (d) the availability of a power supply (if needed). In circumstances where monitoring is undertaken in order to verify the predictions made by modelling it is necessary to place the monitoring where it will pick up the influence of the source being modelled. For example, when verifying road traffic modelling the monitoring should be placed in close proximity to the road network near to the proposed development site.

If data is not available on a particular odour source it may be necessary to undertake sampling and chemical analysis/olfactometry in order to determine an appropriate odour-source term. For industrial installations this requires the cooperation of the operator and may not always be possible. Generic odour emission factors exist for certain types of industrial process such as sewage treatment works (e.g. EPA 1995). Such generic emission factors must be treated with caution when being used to model specific scenarios, and appropriate explanatory text must be included with any assessment based upon them to make this clear to the reader. Nonetheless, such data does have its uses in providing an opportunity to complete sensitivity tests or to provide an initial indication as to whether or not a proposed installation or industrial process is likely to result in the need for mitigation or emission control in order to prevent adverse effects.

4.4.4 Projecting the baseline forwards

Having established current baseline pollution levels, it is then necessary to consider how these levels are expected to change in the future, irrespective of the possible effects of the proposed development. If emission sources and strengths are not expected to change, then current pollution levels may be considered to be representative of pollution levels in the near future. However, changes in population and activity patterns, new industrial developments or closures, changes in technology, the type of fuel used, or stricter emission standards can change the future baseline against which the assessment is made. Weather conditions that favour a build-up of pollutants (e.g. periods of calm or light winds, higher temperatures promoting increased evaporative emissions) may alter too but, in practice, these are not usually considered.

The implications of significant changes to emission rates and patterns for future pollution concentrations need to be assessed. Local, district and county authorities can usually supply information on new developments under construction as well as details of likely population and land-use changes. A judgement will then have to be made as to how these and other changes (e.g. relevant legislation) will alter emissions in the area and consequently alter baseline pollution levels.

If there is insufficient pollution data available in the study area it may be necessary to compile an **emissions inventory**. Taking into account the factors that may affect emissions in future years may enable emission sources and rates to be approximated for future years. These emission data can then become the input into a suitable numerical dispersion model (see §4.5.2) in order to predict future pollution concentrations in the area. Emission inventories are sometimes compiled on a national scale and include reports of existing emissions from qualifying industries and sources and estimates of future year emissions and future year background concentrations at a particular geographic scale. Examples include the National Pollutant Inventory in Australia (Department of Environment and Energy 2016), AP-42 in the US (EPA 1995) and the UK National Atmospheric Emissions Inventory (UK NAEI 2016).

4.5 Impact prediction and evaluation

4.5.1 Physical models and expert opinion

Physical (scale) models using wind tunnels or computer graphics are employed occasionally in situations involving complex hilly terrain or where numerical models suggest uncertainty concerning the possible effects of nearby buildings on the dispersion of pollution emissions.

Predictive methods also include the use of expert opinion, providing it is backed up with reasoned justification which support that opinion, such as comparison with similar existing developments or planned projects for which prediction has already been undertaken. Expert opinion is valuable throughout the ESIA process, but is particularly beneficial at the early stages of project design, when design concepts and options are being considered. Experienced expert opinion is therefore of value for any type of project with a potential impact on air quality, whether relating to industrial process, mineral extraction or transport related schemes, and should be considered before the detail of dispersion modelling, stack height analysis, fuel type, operational cycles, etc. are considered. The use of expert opinion can be justified readily in terms of cost when a number of similar projects are proposed in different locations.

4.5.2 Numerical dispersion models: overview

The type of model used most frequently in predicting air pollution is the numerical dispersion model. It calculates how specified emission rates are transformed by the atmospheric processes of dilution and dispersion (and sometimes chemical and photochemical processes) into pollution concentrations at various distances from the source(s). Models are available for predicting pollution concentrations for emissions from a single point source (e.g. industrial stack or vent) as well as for emissions from a large number of point sources simultaneously. Dispersion models have also been developed for line source emissions (e.g. road transport networks) and those with an area source component (e.g. lagoons, settling tanks, open cast mines or quarries).

The basic model can be improved in accuracy by taking into account complications appropriate to the specific location under study, such as type of terrain (e.g. flat or hilly), surface roughness (e.g. urban or rural conditions), coastal influences (e.g. effects of a sea breeze) and the presence of nearby buildings which may cause building wake effects. Models are also available for area sources (e.g. construction sites, car parks, motorway service stations, industrial processes with numerous vents, urban areas, county regions, storage lagoons), line sources (e.g. open roads, street canyons, railways) and volume sources (area sources with a vertical depth, e.g. leaking gases from a group of industrial processes, take-off and landing activities at an airport).

The effects of odour sources can also be predicted by point-source dispersion models, either based on the emissions of a pure substance (e.g. hydrogen sulphide) or in terms of emissions of odour units. However, when modelling of odour has been undertaken as the predictive element of the assessment and the odour source is already in existence, then observational data should be used to support the conclusions of the modelling.

Both simple and complex (advanced) versions of numerical dispersion models exist. Simple models (known as 'screening' models) are designed for use as an inexpensive tool to identify whether or not air quality warrants further investigation. Screening models employ grossly simplified assumptions about the behaviour of pollutants in the atmosphere and can be used with limited meteorological input data (e.g. minimum and maximum ambient air temperatures and minimum wind speed) to calculate the worst-case pollution concentrations. If a screening model predicts that emissions from a proposed development will produce air pollution concentrations far below an air quality standard this would indicate that it may not be necessary to obtain a more accurate estimate of the predicted concentrations using a complex model. However, if the screening model predicts that pollution concentrations are likely to approach or exceed air quality standards then a more rigorous investigation using a complex model is needed. For major developments, regardless of how small an increase in pollution levels are caused by their emissions, the use of a complex model may be appropriate for an ESIA.

Examples of screening models include the US EPA's SCREEN3 and AERSCREEN. Examples of complex models include the EPA's CALPUFF and AERMOD dispersion models and the UK's ADMS dispersion modelling suite of programmes (CERC 2016). The US EPA models in particular have been widely used internationally (e.g. in Australia, New Zealand, Canada and South Africa), facilitated by open access to the source code via the Support Center for Regulatory Atmospheric Modeling (SCRAM 2016). Commercial versions of these models with user-friendly input and output routines can be purchased from specialist software companies, typically requiring an annual licence fee, e.g. software such as CALPUFF View (Lakes Environmental 2016).

The choice of dispersion model also depends on the specific requirements of the assessment. In many cases where industrial emissions are being modelled there may be a requirement to apply for a permit to operate. The regulatory authorities may therefore have specific requirements regarding the type of modelling software to use, the input data and modelling procedures. In general, the larger and more complex the installation, the more likely it is that a complex dispersion model will be required.

4.5.3 Principles of air pollution dispersion modelling

For many years numerical prediction models have been developed based upon Gaussian assumptions. For a point source of pollution (e.g. a chimney stack) the basic **Gaussian model** assumes that the pollutant emissions spread outwards from a source in an expanding plume aligned to the wind direction, in such a way that the distribution of pollution concentration decreases away from the plume axis in horizontal and vertical planes, according to a specific Gaussian mathematical equation, a symmetric bell-shaped distribution. Although a plume may appear irregular at any one moment, its natural tendency to meander results in a smooth 'cone-shaped' Gaussian distribution after ten minutes of averaging time. The horizontal axis of the plume does not normally coincide with the height of the stack or point of emission, as the density and momentum of the emissions quickly carries the plume to a higher elevation, known as the 'effective release height' (sometimes many times higher than the stack or point of emission).

Basic Gaussian models assume the rate of dispersion of the plume is a function of atmospheric turbulence which is based around simple meteorological parameters involving solar radiation, cloud cover and mean wind speed, and expressed in the form of six or seven Pasquill stability categories. Stability categories range from class A (very unstable) occurring during hot, sunny conditions with light winds through category 4 (neutral) to class F or G (both very stable) occurring during cold, still nights with clear skies. The new generation of Gaussian plume air dispersion models (e.g. AERMOD or ADMS-5) incorporate improved understanding of turbulent flows in the atmospheric boundary layer (particularly the Monin-Obukhov length) in place of Pasquill stability classes. The maximum ground-level concentration experienced from a pollution plume is where the plume touches the ground. Numerous situations can give rise to high pollutant concentrations, such as meteorological conditions, the existence of nearby buildings and the local topography.

The highest ground-level concentrations from an elevated source tend to occur close to the source during light winds when the atmosphere is very unstable with substantial vertical mixing, such as happens on hot summer days. It can also be seen that during light winds the peak concentration is found further from the source during conditions of increasing atmospheric stability. It is therefore important to take into account local meteorological conditions at the site.

Where tall buildings lie adjacent to a tall stack, strong winds can on occasions give rise to high ground-level concentrations. This happens because buildings cause eddies to form that make the plume touch the ground much closer than would otherwise be expected. It is generally considered that building downwash problems may occur if the stack height is less than 2.5 times the height of the building above which it protrudes. Similarly problems may occur if adjacent buildings are within about five stack heights of the release point. Where buildings could have an effect on stack emissions these should therefore be included in the modelling.

Other situations giving rise to high pollution concentrations may be when plumes impact directly on hillsides under certain meteorological conditions, or when valleys trap emissions during low-level inversions (Harrop 2002).

4.5.4 Road traffic air pollution modelling

Line source models are used to predict the contribution from the source, or sources, at receptors in the surrounding locality. Several models have been developed specifically to predict pollution concentrations arising from emissions from road vehicles, e.g. the UK's ADMS-Roads model (CERC 2016) and the US EPA's CAL3QHC and CALINE 4 models (both of which are available as open source code). Other proprietary models exist, and some traffic-flow modelling software provides the option to predict total vehicle emissions (of NO_x, PM_{10}, etc.) across the network being modelled to inform mass emission calculations from traffic, e.g. EnViVer (PTV Group 2016).

The difference in the predicted concentrations for each scenario being modelled (such as a 'do nothing' and 'with development' scenario) allows the contribution of the development-generated traffic to total pollution levels, at a given location, to be predicted, and the impact of that development-generated traffic to be understood.

As with point-source models, appropriate meteorological data needs to be used in a road traffic model. However, a more significant influence on the results of the modelling will be how the model is set up in terms of the influence of queueing traffic, the treatment of street canyons and the choice of vehicle emissions. In contrast to point-source models, it is normally necessary to specifically validate the results of a road traffic model by comparing the results of the baseline model against local monitoring data.

4.5.5 Modelling inputs: data requirements

Emissions data

All numerical dispersion models require emissions data either in the form of a specified emission rate for the source (e.g. the amount of pollutant released per unit of time) or a measure of the level of activity of the source (e.g. amount of fuel consumed) together with the corresponding emission factor (e.g. the quantity of pollutant emitted per unit of activity). Emission rates need not necessarily be exact, as the likely impact of a planned development could be assessed by using the highest likely emissions, such as the maximum emission limits defined for the operational permit. If the emission rate for a proposed development is not already specified in the plant design then an estimate may be based upon expected fuel consumption and characteristics of the fuel, or by obtaining 'surrogate' information from another similar plant or process elsewhere. Information on emissions and emission factors are available from the UK National Atmospheric Emissions Inventory (UK NAEI 2016), the European Monitoring Evaluation Programme/European Environment Agency Air Pollutant Emission Inventory Guidebook (EMEP-EEA 2013) and AP-42 in the US (EPA 1995). Emission factors (F) are described in terms of, for example, grammes of NO_x per kilowatt fired for boilers, and grams of NO_x per tonne of nitric acid product for a nitric acid works. Emissions would then be calculated as $M \times F$ where M is a measure of the level of activity.

Typical emission rates can be used when calculating long-term pollution concentrations but for short-term models a number of worst-case scenarios may be needed (e.g. periods of intensive activity, during start-up, and the operation of emergency release vents). Complex models applied to a point source will require input information about the release conditions of the emissions. This will include the stack height and internal exit diameter as well as the flue-gas exit temperature and exit velocity (or volumetric flow rate).

In the case of road traffic models, vehicle emission rates for a specific section of road are calculated by the model itself from input data such as vehicle flow (e.g. vehicles per day, peak hourly value), average vehicle speed, vehicle mix (e.g. fraction of heavy goods vehicles, fraction of petrol/diesel engine cars) and vehicle emission factors (Defra 2016c). If the model is being used to predict pollution concentrations for a future year, then input forecast data not only for future traffic flow, speed and mix are needed but also the likely change in emission factors. Emission factors for future years, which take into account the expected effects of phasing in of cleaner technologies and fuels, are embedded in some models, although care needs to be taken to ensure that the latest version of the emissions database is being used.

Meteorological data

All dispersion models utilise meteorological data; some use simple statistical data sets (e.g. Pasquill stability classes), whereas others use more complicated data sets such as hourly sequential data. The data used should be appropriate to the location being modelled and the type of assessment being undertaken. As weather patterns vary from year to year, when modelling large-scale industrial installations it is common practice to use three or five years' worth of meteorological data and compare the results. Other factors to take into account in the choice of meteorological data are the relative height of the meteorological station and assessment site, and the potential for coastal effects.

Topography influences turbulence at, and close to, ground level, which in turn will influence the dispersion of pollutants between source and receptor. The more sophisticated dispersion models therefore consider the terrain and topography, typically utilising imported data from proprietary sources (such as the Ordnance Survey in the UK), to create a three-dimensional map of the modelling area. This allows for landscape features (e.g. hills, mountains, valleys) and even man-made structures such as motorway embankments, to be considered by the modelling algorithms. Thus, the potential for updrafts and downdrafts, microclimate effects and other impacts (such as the potential for a plume emitted from a stack in the bottom of a valley to ground at some point on the side of the valley) can be evaluated by the model.

While these features undoubtedly have the potential to add to the model's performance, they can significantly increase run times on all but the fastest computers, and an understanding of the manner in which the models handle this data is essential to ensure that the results are correctly interpreted, and the model is not simply treated as a 'black box'.

4.5.6 Model outputs

The most appropriate model outputs that should be incorporated in an ESIA are predictions of short-term pollution impact (e.g. highest or 'worst-case' hourly mean concentration) and long-term impact (e.g. annual mean concentration). Outputs need to be compared with the appropriate air quality standards, objectives and guideline values and any locations which approach or exceed these concentrations must be identified. In some cases a model may not calculate pollution concentration over the averaging period used to define an air quality standard. In this situation it is necessary to use empirical relationships to decide whether the air quality standard is exceeded or not. Complex air dispersion models use meteorological data that is based upon one-hour averages – so any predictions presented by the model for periods of less than one hour are mathematical calculations performed within the model and will be based on empirical relationships.

One of the key benefits of using more complex models is that they allow the user to generate graphical outputs such as *isoline plots* which can greatly assist the reader of the ESIA in understanding the results being presented (Figure 4.1). Many of the latest models will also interface with Geographical Information Systems, allowing population counts to be overlain with model pollution isolines, etc. so that the change in exposure to pollution from a given source (or a new source) can be presented.

4.5.7 Model limitations

Within the ESIA there should be specific acknowledgement of the limitations of the assessment methodology and, where possible and helpful, information

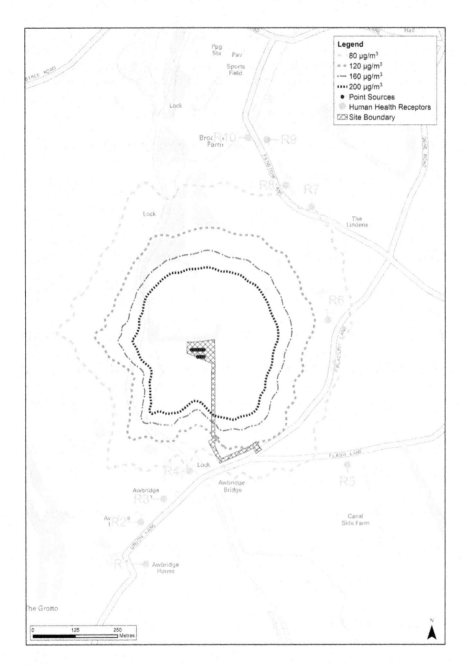

Figure 4.1

Predicted one-hour 99.79th percentile nitrogen dioxide concentrations, modelled using ADMS5. The development is an emergency standby electricity generation facility in the West Midlands, UK.

should be provided on the accuracy of the predictions being made. All predictions have an element of uncertainty and it is important to acknowledge this and not treat the model as a 'black box' by concentrating only on the results produced. Models are simplifications of reality and their limitations, accuracy and confidence levels should be recognised. The quality of the input data will clearly affect the accuracy of a model. Even if accurate input data were available, the algorithms employed in the model to represent the behaviour of pollutants released into the atmosphere contain many uncertainties. Confidence in the accuracy of a model is gained by assessing its ability to predict the current baseline conditions in the study area since the results can be verified using monitored pollution data. For example, current UK guidance on modelling local air quality sets out methodologies that can be applied to verify road traffic air pollution modelling results against monitoring data (Defra 2016d). The predictions made by modelling software for point sources are normally validated during the development of the model, and it is rare to attempt to validate the results of a point-source modelling study by monitoring.

4.5.8 Odour effects

In assessing odour effects, both predictive and observational methods can be used. Predictive methods can involve: (a) a risk-based assessment of the source, pathway and receptor to come to a conclusion regarding the likely significance of an odour source; (b) semi-quantitative assessments using screening models or look-up tables to predict odour concentrations; or (c) dispersion modelling whereby odour concentrations are predicted and compared to benchmark values. Observational methods can involve: (a) monitoring of odour in the environment, i.e. undertaking 'sniff tests'; or (b) using the local community to provide evidence, i.e. by the use of odour diaries or community surveys or by analysing historic complaint records.

For new sources of odour in an existing environment it is likely that a predictive method will need to be used. Where the effect of existing odour sources is being evaluated, combining both predictive and observational methods is likely to provide the best means of assessment.

4.5.9 Assessing significance

The level of significance of the likely pollution impacts of a proposed development is assessed by comparing the predicted changes in the area to air quality standards, objectives or guideline values, and determining whether these are likely to be exceeded at any locations, after taking into account the existing and predicted baseline pollution levels. If the planned development is predicted to increase pollution levels in excess or close to the air quality standard, then mitigation measures may be required. If the predicted concentrations are well below the standard or limit value, it is useful to express the increase in ground-level pollution concentrations in a meaningful way. For example, an ESIA may conclude that a proposed development is expected to increase the annual average NO_2 concentration at the location worst affected (5 km downwind) by only 3 per cent and that this increase is well within the year-to-year variability of annual average concentration produced by meteorological fluctuations. Even when a development is likely to add only small amounts of pollution to the area, it is important that an ESIA makes specific assessment of what effect (perhaps negligible) this will have on any nearby sensitive receptors such as residential areas, schools, nature reserves or ecological sites of importance and historical buildings. A number of organisations have published guidance to assist in the assignation of significance to the results of air dispersion modelling. These include the Environment Agency guidance for point sources (Environment Agency and Defra 2016) and Environmental Protection UK and IAQM (Moorcroft et al. 2015).

Human health and odour receptors for ESIA purposes are chosen on the basis that people will be present for the averaging period that the standard or limit value applies over. There are no grades of sensitivity for human health and odour receptors; they are all regarded as being equally sensitive to the pollution being assessed (this is because the standards and guidelines are determined for vulnerable individuals in the population as a whole, which are deemed to apply to everyone).

The impacts at individual receptor locations can be judged in terms of the change in pollutant concentration at that location and whether a standard or limit value is approached or breached. Greater weighting in terms of significance is generally given to the situation where a standard or limit value is either just met or breached, rather than when there is considerable headroom. Given that pollution disperses rapidly from the source, the magnitude of change in concentration often varies markedly between receptor locations such that the significance of the effect also varies at each receptor. Within the ESIA it is necessary to come to a conclusion about whether the development overall has a significant effect, and this requires professional judgement to balance the effects at each individual receptor location.

For ecological receptors it is possible to distinguish the sensitivity and importance of the receptor based on the level of designation (i.e. an internationally or nationally designated site would be a more significant receptor than a locally designated site). The judgement regarding the significance of an effect on an ecological receptor is further complicated as the assessment criteria (the critical levels and loads) are only indicative of the chance that an adverse air quality effect may occur. The results of the air quality assessment in terms of the critical level or critical load must also be considered in the context of other factors that could affect the health of the ecological receptor and therefore in these circumstances the judgement of significance passes from the air quality professional to the ecologist.

Assessing the significance of odour effects requires each element of the assessment to be given appropriate weight, i.e. for an existing odour source, where there is observational information this should be given significant weight and the results of the predictive assessment should be considered in light of the observational results and not used to try to disprove that an existing adverse odour effect exists. Ultimately it will be necessary to judge whether a significant effect on amenity exists, which may also need to take into account regulatory and statutory considerations.

4.6 Mitigation and enhancement

4.6.1 The need for mitigation measures

Mitigation measures should aim to avoid, reduce or remedy any significant adverse effects that a proposed development is predicted to produce. At one extreme, the prediction and evaluation of likely impacts may indicate such extreme adverse effects that abandonment or complete redesign of the proposed development is the only effective mitigating measure. More likely, modifications to the development can be suggested in order to avoid or reduce potential impacts.

Some mitigation measures may be required by law for new – though not for existing – developments (e.g. fitting of specific types of pollution-control devices), but the use of others depends upon the significance of the predicted impacts. For regulated industrial processes, mitigation through design is always promoted ahead of the application of 'end of pipe' processing to 'clean up' emissions. This is an example of embedded mitigation, which is in place before the impacts and effects are determined within the ESIA. Other examples would include the incorporation of travel plans to reduce the road traffic generation from a scheme. As mitigation feeds back into design, measures proposed to minimise adverse impacts can be incorporated as alternatives in the project description. Subsequent development proposals can also make use of the information contained in a previous ESIA in order to incorporate appropriate mitigation measures at the outset, rather than wait for its own ESIA to identify potential problems.

Where several alternative mitigation measures exist to solve a potential problem it is important to assess the likely effectiveness of each measure, as well as the costs of implementation. Whatever mitigation measures are proposed, it is important to ensure that they do not create problems of their own. For example, increasing the height of a chimney stack may help to increase the dispersion of emissions, but would also increase the potential visual impacts.

General mitigation measures such as the provision of electric vehicle charging points are sometimes required as a matter of policy by the regulatory authorities. In this case, evaluation of the effects of the mitigation on pollutant concentrations may not be possible, and, perhaps, all that could be said is that 'emissions have been reduced below levels that would otherwise have occurred'. In other words, the beneficial impact is achieved not in terms of an absolute, or quantifiable impact improvement (or 'reduced worsening') of air quality associated with a particular scheme undergoing an ESIA, but in terms of compliance with a local or regional planning policy aimed at improving air quality over a wider area.

4.6.2 Mitigating adverse pollution impacts

If a planned development is likely to exceed, say, maximum hourly pollution standards only during periods of poor atmospheric dispersion, then one possible mitigation measure would be to keep a cleaner standby fuel for use during those periods, or to reduce emissions by reducing production output in the case of an industrial process. In many cases the type and amount of pollutants emitted are a function of the fuel burned, so alternatives such as fuel oil with a very low sulphur content or natural gas may be appropriate substitutes.

Traffic-generated pollutants decrease rapidly away from roads, and therefore the primary mitigation measure for road traffic impacts is to increase the distance between the road and the receptor. An alternative would be to install mechanical ventilation within affected buildings, with the air intakes high up or further from the road to provide the building with 'cleaner' air, or to use filtration to reduce the pollutant concentration entering the building (i.e. NO_x or PM_{10} filtration).

Mitigation against odour impacts is also likely to be provided by means of separation distance. Increased vegetative screening between the source and the receptor may promote dispersion and thereby reduce odour concentrations, although the effectiveness of this is difficult to ascertain and any trees would need to be evergreen to remain effective throughout the year. There is also some evidence that the perception of odour reduces if the source of the odour cannot be seen (Tyndall 2008). It may be possible for the internal areas of residential and commercial buildings to use mechanical ventilation with filtration on the air intake to reduce the odour concentration entering the building. However, this leaves the outside areas of the properties vulnerable to adverse odour effects and this is not normally acceptable for residential properties unless there are fundamental advantages to the development which outweigh the effects.

The construction stage of most projects has the potential to cause localised windblown dust problems, either when excavation is taking place or when materials are being transported and stored in stockpiles. Careful design of construction operations including the selection of haulage routes into the site and the location of stock piles can help to minimise dust problems in nearby sensitive areas. Mitigation measures can include: (a) frequent spraying of stockpiles and haulage roads with water; (b) regular sweeping of access roads; (c) covering of lorries carrying materials; or (d) enclosing the site with hoardings. A standard suite of mitigation measures can be applied to mitigate construction dust effects with the resultant effect generally not being significant.

Commonly, the management of construction-related effects is dealt with in fairly generic terms in the ESIA Report, because the details of construction methods, timings, phasing, etc. might not always be sufficiently developed or known at this stage in the development application process. One solution in these circumstances is to commit to providing further detail in a Construction Environmental Management Plan (CEMP), which will be prepared and submitted to the appropriate regulatory authority at an appropriate time, when the project programme has advanced further, but before work starts (see also Chapter 20). A CEMP allows more detailed management of air quality effects to be presented, but also crucially allows the opportunity of the interaction of potential effects with other technical disciplines to be considered (for example dust deposition from construction activities into sensitive watercourses, or on the leaves of sensitive woodlands or grasslands in proximity to the site).

4.7 Monitoring

Numerical prediction models contain uncertainties, so monitoring may need to be continued after completion of the development to compare predictions with those that actually occur. However, unless the effect is due to a dominant source, i.e. a large stack emission or a nearby road, it may not be possible to measure or attribute the effect of a particular development on the environment and therefore identify the need for additional mitigation. This is particularly true where cumulative impacts from many smaller developments occur and road traffic emissions are the principal sources of pollution.

When modelling is completed to provide information in support of a regulatory permit application (for an industrial process), the permit may stipulate the need for ongoing monitoring at specific locations (for example at the site perimeter, or a nearby sensitive location, such as a designated ecologically sensitive site). When dispersion modelling has been carried out using emission factors set out in the relevant process guidance notes (in the absence of operational information to work from), monitoring requirement may focus on stack emissions, to ensure that the emissions do not exceed the limits set out in the permit, hence resulting in potentially different dispersion impacts from those modelled to inform the permit.

4.8 Conclusions

Assessment of the effects of a proposed development project on local air quality is important from the perspective of regulatory compliance, to ensure that relevant standards and objectives for human health or the protection of ecosystems are not compromised. Additionally, air quality is of increasing concern to the general public, and the popular press often carries articles about the impact of air pollution on public health. This increased awareness means that development proposals supported by an air quality assessment are coming under intense scrutiny from a wide range of stakeholders.

The assessment of potential impacts can be approached in a variety of ways, from basic spreadsheet calculations to the use of complex computer models that take account of variables including meteorology, atmospheric chemistry, terrain, adjacent buildings, traffic speed/volume, and different operating conditions for industrial processes. Such models have the ability to provide detailed predictions of the potential impacts, with the added advantage that these can be presented in an easy to understand graphic form, showing concentrations of pollution overlaid onto maps and plans. This makes the results of some complex calculations and assessment processes accessible to a wider, less technical audience, which is fundamental to the ESIA process.

However, it is important that the practitioner undertaking the assessment has a thorough understanding of the entire process, including the scenario(s) being modelled (whether based around traffic flows in future years or elements of an industrial process), the variables being used to inform the assessment, the standards against which the assessment is being completed and the way any model used for the assessment works. Providing this is the case, the air quality assessment process is a robust and reliable way to ensure that any potential impacts are clearly presented and understood by readers of the ESIA Report.

Acknowledgements

The authors are grateful to Professor Derek M. Elsom for his work on previous editions of this chapter.

References

APIS (UK Air Pollution Information System) 2016. Critical Loads and Critical Levels – a guide to the data provided in APIS. www.apis.ac.uk/overview/issues/overview_Cloadslevels.htm.

Bull M, A McIntyre, D Hall, G Allison, J Redmore, J Pullen, L Caird, M Stoaling and R Fain 2014. IAQM Guidance on the assessment of odour for planning. London: Institute of Air Quality Management. www.iaqm.co.uk/text/guidance/odour-guidance-2014.pdf.

CERC (Cambridge Environmental Research Consultants) 2016. www.cerc.co.uk/environmental-software.html.

Defra (Department for Environment, Food and Rural Affairs) 2009. Local air quality management technical guidance LAQM.TG(09). London: Defra

Defra 2014. Conversion Factors between ppb and g/m³ and ppm and mg/m³. https://uk-air.defra.gov.uk/assets/documents/reports/cat06/0502160851_Conversion_Factors_Between_ppb_and.pdf.

Defra 2015. Automatic Urban and Rural Network (AURN). https://uk-air.defra.gov.uk/networks/network-info?view=aurn.

Defra 2016a. Local air quality management: technical guidance (TG16). London: Defra.

Defra 2016b. National Bias Adjustment Factors. http://laqm.defra.gov.uk/bias-adjustment-factors/national-bias.html.

Defra 2016c. Emissions Factors Toolkit (EFT). http://laqm.defra.gov.uk/review-and-assessment/tools/emissions-factors-toolkit.html.

Defra 2016d. Local Air Quality Management (LAQM) Support. http://laqm.defra.gov.uk.

Department of Environment and Energy 2016. National Pollutant Inventory (NPI). Canberra: Australian Government. www.npi.gov.au/npi-data/latest-data.

EEA (European Environment Agency) 2016. European Nature Information System (EUNIS). http://eunis.eea.europa.eu.

EC (European Commission) 1996. Council Directive 96/61/EC of 24 September 1996 concerning integrated pollution prevention and control. http://eur-lex.europa.eu/legal-content/GA/TXT/?uri=CELEX:31996L0061.

EC 2001. Directive 2001/81/EC of the European Parliament and of the Council of 23 October 2001 on national emission ceilings for certain atmospheric pollutants. http://eur-lex.europa.eu/legal-content/EN/TXT/?uri=CELEX%3A32001L0081.

EC 2008. Directive 2008/50/EC of the European Parliament and of the Council of 21 May 2008 on ambient air quality and cleaner air for Europe. http://eur-lex.europa.eu/legal-content/EN/TXT/?uri=CELEX:32008L0050.

EC 2010. Directive 2010/75/EU of the European Parliament and of the Council of 24 November 2010 on industrial emissions (integrated pollution prevention and control) Text with EEA relevance. http://eur-lex.europa.eu/legal-content/EN/TXT/?uri=CELEX:32010L0075.

EC 2012. Directive 2012/18/EU of the European Parliament and of the Council of 4 July 2012 on the control of major-accident hazards involving dangerous substances, amending and subsequently repealing Council Directive 96/82/EC. Text with EEA relevance. http://eur-lex.europa.eu/legal-content/EN/TXT/?uri=celex:32012L0018.

EC 2014. Directive 2014/52/EU of the European Parliament and of the Council of 16 April 2014 amending Directive 2011/92/EU on the assessment of the effects of certain public and private projects on the environment. Text with EEA relevance. http://eur-lex.europa.eu/legal-content/EN/TXT/?uri=celex%3A32014L0052.

EC 2015. Air Quality Standards. http://ec.europa.eu/environment/air/quality/standards.htm.

EMEP-EEA (European Monitoring Evaluation Programme and European Environment Agency) 2013. *Air Pollutant Emission Inventory Guidebook.* Luxembourg: Publications Office of the European Union. www.eea.europa.eu//publications/emep-eea-guidebook-2013.

Environment Agency 2011. *H4 odour management: How to comply with your environmental permit.* Bristol: Environment Agency.

Environment Agency and Defra 2016. Air Emissions Risk Assessment for your Environmental Permit. www.gov.uk/guidance/air-emissions-risk-assessment-for-your-environmental-permit.

EPA (Environmental Protection Agency) 1995. AP-42: Compilation of Air Pollutant Emission Factors, 5th Edn. www3.epa.gov/ttnchie1/ap42.

EPA 2016a. National Ambient Air Quality Standards (NAAQS). www.epa.gov/criteria-air-pollutants/naaqs-table.

EPA 2016b. AirData – Access to monitored air quality data from the EPA's Air Quality System Data Mart. www3.epa.gov/airdata.

European Parliament 2014. EU Air Quality Policy and WHO Guideline Values for Health. www.europarl.europa.eu/RegData/etudes/STUD/2014/536285/IPOL_STU(2014)536285_EN.pdf.

Harrop O 2002. *Air quality assessment and management: A practical guide*. London: Spon Press.

IFC 2016. *Environmental, health and safety guidelines*. Washington DC: IFC. www.ifc.org/wps/wcm/connect/topics_ext_content/ifc_external_corporate_site/ifc+sustainability/our+approach/risk+management/ehsguidelines.

JRC (Joint Research Centre of the European Commission) 2016. Reference Documents under the IPPC Directive and the IED. http://eippcb.jrc.ec.europa.eu/reference.

Lakes Environmental 2016. CALPUFF View™ Graphical Interface for the US EPA Approved Long Range Transport Model. www.weblakes.com/products/calpuff/resources/lakes_calpuff_view_release_notes.pdf.

Miedema HME and JM Ham 1988. Odour Annoyance in Residential Areas. *Atmospheric Environment* **22,** 2501–2507.

Miedema HME, JI Walpot, H Vos and CF Steunenberg 2000. Exposure-annoyance relationships for odour from industrial sources. *Atmospheric Environment* **34,** 2927–2936.

Moorcroft S and R Barrowcliffe et al. 2015. *Land-use planning and development control: Planning for air quality*. London: IAQM.

Mulgrew A and P Williams 2000. *Biomonitoring of air quality using plants*. Air Hygiene Report 10. Berlin: WHO Collaborating Centre for Air Quality Management and Air Pollution Control.

PTV Group 2016. Emissions Modelling with EnViVer. http://vision-traffic.ptvgroup.com/en-uk/products/ptv-vissim/use-cases/emissions-modelling.

SCRAM (Support Center for Regulatory Atmospheric Modeling) 2016. www3.epa.gov/ttn/ scram.

Tyndall J 2008. The use of vegetative environmental buffers for livestock and poultry odour management. *Natural Resource Ecology and Management Conference*. Iowa State University.

UK NAEI (National Atmospheric Emissions Inventory) 2016. UK Emissions Data. http://naei.defra.gov.uk/data.

UNECE (United Nations Economic Commission for Europe) 1996. *Manual on methodologies and criteria for mapping critical levels/loads and geographical areas where they are exceeded*. Texte 71/96, Berlin: Umweltbundesamt.

UNECE 1999. Protocol to Abate Acidification, Eutrophication and Ground-level Ozone. www.unece.org/fileadmin/DAM/env/lrtap/full%20text/1999%20Multi.E.Amended.2005.pdf.

WHO (World Health Organisation) 2000. *Air quality guidelines for Europe*, 2nd Edn. Copenhagen: WHO Regional Office for Europe. www.euro.who.int/__data/assets/pdf_file/0005/74732/E71922.pdf.

WHO 2005. *Air quality guidelines global update 2005: Particulate matter, ozone, nitrogen dioxide and sulfur dioxide*. Copenhagen: WHO Regional Office for Europe. www.euro.who.int/__data/assets/pdf_file/0005/78638/E90038.pdf.

World Air Quality 2016. World Air Quality Index. https://aqicn.org/map/world/

5 Climate and climate change

•••

Elizabeth Wilson and Phill Minas

5.1 Introduction

Climate change is often referred to as the most serious environmental challenge facing humanity. Indeed, as pointed out by Stern (2015), it is also a political, social and economic challenge which requires action at global, national, local and individual scales. The United Nations Intergovernmental Panel on Climate Change (IPCC) Fifth Assessment Synthesis Report 2014 (AR5) states that "Human influence on the climate system is clear, and recent anthropogenic emissions of greenhouse gases are the highest in history". AR5 says "In recent decades, changes in climate have caused impacts on natural and human systems across all continents and across the oceans", with changing precipitation, snow and ice altering hydrological systems and affecting water resources; terrestrial, freshwater and marine species shifting their ranges, seasonal activities, migrations and interactions; and negative impacts on crop yields. "Continued emission of greenhouse gases will cause further warming ... increasing the likelihood of severe, pervasive and irreversible impacts for people and ecosystems". AR5 concludes that "Limiting climate change would require substantial and sustained reductions in greenhouse gas emissions which, together with adaptation, can limit climate change risks" (IPCC 2014).

The first legally binding agreement to take urgent global action to address climate change was reached at the UN Climate Conference in Paris in December 2015, where 195 countries committed to:

> holding the increase in global temperature to well below 2°C above pre-industrial levels and to pursue efforts to limit the increase to 1.5°C above pre-industrial levels, recognizing that this would significantly reduce the risks and impacts of climate change, and increasing the ability to adapt to the adverse impacts of climate change and foster climate resilience and low greenhouse gas emissions development, in a way that does not limit food production.
>
> (UN 2015 Art 2, 1a and 1b)

The Paris Agreement also recognised that more needed to be done to limit global warming to below 2°C by the end of this century, and so countries agreed to meet every five years to set more ambitious targets and report on their progress in implementation.

Environmental and Social Impact Assessment (ESIA) can play an important role in helping to meet the commitments made in Paris. It can help to ensure that new projects do not exacerbate climate change and that they adapt to be more resilient to the consequences. During the last decade, many regulatory and funding agencies such as the World Bank, the Organisation for Economic Co-operation and Development, and the European Union have issued regulations or guidelines requiring or promoting the consideration of climate change in ESIA procedures, and professional bodies such as the International Association for Impact Assessment have published best-practice principles. An earlier review of Environmental Impact Assessment (EIA) application and effectiveness by the Commission of the European Communities had noted that

> Any review of the impacts of climate change is often limited to CO_2 and other greenhouse gas emissions from industry and from increases in transport as part of air quality studies or as indirect impacts. The EIA assessment will often not go beyond evaluating existing emissions and ensuring that ambient air quality standards are met. In addition, the effects on global climate, the cumulative effects of an additional project and adaptation to climate change are not sufficiently considered within the EIA.
>
> (CEC 2009)

One of the reasons for this lack of systematic consideration was that climate change is a complex, multiscale and multifaceted issue: for instance, it is hard to attribute any particular risk of exacerbating climate change through a causal pathway to any specific project (Ohsawa and Duinker 2014, CEQ 2014). Nevertheless, the climate change policy community (such as the IPCC 2014) argues that climate change needs addressing simultaneously on two fronts:

1 *Climate change mitigation* – that is the reduction of the causes of climate change (e.g. lowering carbon emissions);
2 *Climate change adaptation* – that is ensuring that developments will be resilient to unavoidable climate change.

Moreover, in order to promote positive synergies between mitigation and adaptation, and to avoid negative synergies and maladaptation, it is important also to assess the interactions between mitigation and adaptation measures (IPCC 2014, Berry et al. 2015).

We can therefore see that addressing climate change through ESIA raises difficulties in four major assessment areas which have already proven problematic for practitioners (Glasson et al. 2012):

1   Cumulative impacts
2   Risk and uncertainty
3   Time-horizons and lifetimes
4   Interactions.

The first two topics are largely addressed in Chapters 18 and 19, so this chapter focuses on the latter two.

5.2 Definitions and concepts

It will be clear from §5.1 that the climate change policy community refers to some terms somewhat differently from the impact assessment community: mitigation, for instance, is not the same as for other chapters in this book. For clarity and consistency, this chapter adopts the definitions employed by IPCC AR5 (IPCC 2014), although these definitions differ in some respects from earlier IPCC reports.

Adaptation: The process of adjustment to actual or expected climate and its effects. In human systems, adaptation seeks to moderate or avoid harm or exploit beneficial opportunities. In some natural systems, human intervention may facilitate adjustment to expected climate and its effects.

Adaptive capacity: The ability of systems, institutions, humans and other organisms to adjust to potential damage, to take advantage of opportunities, or to respond to consequences.

Climate: Climate in a narrow sense is usually defined as the average weather or, more rigorously, as the statistical description in terms of the mean and variability of relevant quantities over a period of time ranging from months to thousands or millions of years. The classical period for averaging these variables is 30 years, as defined by the World Meteorological Organisation. The relevant quantities are most often surface variables such as temperature, precipitation and wind. Climate in a wider sense is the state, including a statistical description, of the climate system.

Climate change: A change in the state of the climate that can be identified (e.g. by using statistical tests) by changes in the mean and/or the variability of its properties, and that persists for an extended period, typically decades or longer. Climate change may be due to natural internal processes or external forcings such as modulations of the solar cycles, volcanic eruptions, and persistent anthropogenic changes in the composition of the atmosphere, or in land use.

Climate projection: The simulated response of the climate system to a scenario of future emission or concentration of greenhouse gases and aerosols, generally derived using climate models. Climate projections are distinguished from climate predictions by their dependence on the emission/concentration/radiative forcing scenario used, which is in turn based on assumptions concerning, for example, future socio-economic and technological developments that may or may not be realised.

Greenhouse gases (GHG): Those gaseous constituents of the atmosphere, both natural and anthropogenic, that absorb and emit radiation at specific wavelengths within the spectrum of infrared radiation emitted by the earth's surface, atmosphere and clouds. This property causes the greenhouse effect. Water vapour (H_2O), carbon dioxide (CO_2), nitrous oxide (N_2O), methane (CH_4) and ozone (O_3) are the primary greenhouse gases in the earth's atmosphere. Fluorinated gases were also covered under the Kyoto Protocol.

Impacts: The IPCC (2014) uses the term 'impacts' primarily to refer to the effects on natural and human systems of extreme weather and climate events, and of climate change. Impacts generally refer to effects on lives, livelihoods, health, ecosystems, economies, societies, cultures, services and infrastructure due to the interaction of: 1. climate changes or hazardous climate events occurring within a specific time period and 2. the vulnerability of an exposed society or system. The impacts of climate change on geophysical systems, including floods, droughts and sea-level rise, are a subset of impacts called physical impacts.

Mitigation (of climate change): A human intervention to reduce the sources or enhance the sinks of greenhouse gases.

Resilience: The capacity of social, economic and environmental systems to cope with a hazardous event or trend or disturbance, responding or reorganising in ways that maintain their essential function, identity and structure, while also maintaining the capacity for adaptation, learning and transformation.

Sink (carbon): Any process, activity or mechanism that removes a GHG, an aerosol, or a precursor of a GHG or aerosol from the atmosphere.

Vulnerability: The propensity or predisposition to be adversely affected. Vulnerability encompasses a variety of concepts and elements including sensitivity or susceptibility to harm and lack of capacity to cope and adapt.

5.3 Key legislation, guidance and standards

A number of agencies and regulatory authorities have recently set requirements for climate change to be considered in ESIAs/EIAs. A survey of international EIA practitioners in 2010 found that most considered that climate change would be better handled in EIAs if there were specific regulations requiring it (Sok et al. 2011). However, only three countries – Australia, Canada and the Netherlands – were implementing such regulations at the time (Agrawala et al. 2011), while many others (such as Caribbean countries) were moving towards adopting regulations. Regulatory frameworks are still changing and guidance is continuing to emerge on how to integrate climate change considerations into EIA.

5.3.1 Legislation and regulations

This section provides an overview of some of the legislation that has emerged over the last few years that require ESIAs/EIAs to consider aspects of climate change.

USA: In the USA, the Council on Environmental Quality have confirmed that climate change falls within the scope of the environmental issues that should be addressed under the National Environmental Policy Act (NEPA) (CEQ 2014). However, the quality of the assessments varies substantially: for example,

> some environmental statements contained a detailed inventory of GHG emissions, some provided an aggregate estimate of total emissions, and others merely noted that GHG emissions may occur as a result of the project (without quantifying these emissions or identifying specific sources).
>
> (Wentz et al. 2016)

In the last few years a number of individual states, such as the State of Massachusetts, have amended their EIA laws to take account of climate change (Wentz 2015).

Canada: The Canadian Environmental Assessment Act does not explicitly mention climate change, although it does require that the environmental assessment take into account "any change to the designated project that may be caused by the environment" (CEAA 2012, Factor 19, 1h).

Australia: In Australia, there is currently no specific Commonwealth requirement. However, the Australian Capital Territory requires all Environmental Impact Statements (EISs) to consider climate change (both mitigation and adaptation), and its *Climate Change Vulnerability Assessment Framework* supplements its Risk Management Standards for the planning, development, renewal, maintenance and management of public infrastructure (ACT 2012). Western Australia requires risk-based decision-making in the coastal zone to include an allowance for sea-level rise.

European Union: The original EU Directive on EIA required the consideration of 'climatic factors' but this was rarely interpreted to include climate change and subsequent amendments did not specifically refer to 'climate change'. However, following research on the application and effectiveness of the EIA Directive in 2009 (CEC 2009), proposals for a new EIA Directive were developed and subsequently adopted. One of the most significant changes to the new EIA Directive (CEC 2014) is the increased prominence of climate change. The new Directive establishes that climate change will continue to cause damage to the environment and that as part of the EIA process it is appropriate to assess the impact of projects on climate (for example, GHG emissions) and their vulnerability to climate change. The new Directive therefore places a stronger emphasis on climate change mitigation, adaptation and resilience through the screening, scoping and assessment process.

As the Paris Agreement takes effect, and as EU Member States begin to transpose the new EIA Directive into national regulations, it is likely that other jurisdictions at federal or state level will require their EIA processes to include climate change.

5.3.2 Guidance

Guidance is available from a range of organisations including development co-operation funding agencies, professional impact assessment groups and particular political and national jurisdictions.

International Finance Corporation (IFC) Performance Standards on Environmental and Social Sustainability: Performance Standard 1 (Assessment and Management of Environmental and Social Risks and Impacts) states that "a number of cross-cutting issues such as climate change, gender, human rights and water are addressed across multiple Performance Standards" (IFC 2012). Performance Standard 3 (Resource Efficiency and Pollution Prevention) covers greenhouse gases. Performance Standard 4 (Community Health, Safety and Security) points out that "communities that are already subjected to impacts from climate change may also experience an acceleration and/or intensification of impacts due to project activities", and draws attention to the need to "identify those risks and potential impacts on priority ecosystem services that may be exacerbated by climate change". In other words, the IFC guidance reminds practitioners that climate change is likely to affect most adversely communities who are already vulnerable, and it is therefore important to assess the interaction of physical and socio-economic impacts.

International Association for Impact Assessment (IAIA): The IAIA Best Practice Guidelines on handling climate change in impact assessment (Byer et al. 2012) provide a very useful summary of the key principles, and of how they might be operationalised (such as by commitments by government to prior assessment of climate change before decision-making, by further education of practitioners, and by employment of local and indigenous knowledge).

Institute of Environmental Management & Assessment (IEMA): IEMA has prepared detailed practice guidance on integrating climate change adaptation into EIA (IEMA 2015), and is currently (2016) preparing similar guidance for integrating climate change mitigation into EIA.

European Union (EU): The EU's *Guidance on Integrating Climate Change and Biodiversity into EIA* (CEC 2013) aims to help member states improve the way in which climate change and biodiversity are integrated in EIAs carried out across the EU. The guidance also provides links to useful sources of further information and tools.

Canadian Environmental Assessment Agency (CEAA): The CEAA published guidance for practitioners in 2003 (modified in 2012) on how to incorporate climate change considerations in environmental assessment. The guidance includes issues of particular relevance to projects in northern latitudes (for example, projects that are located within Arctic regions, which may be particularly sensitive to climate change).

Scottish Natural Heritage (SNH): In Scotland, the guidance on EIA (SNH 2013) follows the National Planning Framework's recognition of the value of Scotland's carbon-rich soil as an asset to be protected, and so emphasises the intrinsic value of Scotland's soils and their functional role in the context of climate change (for instance, as a carbon repository or a source of GHG release through damage or poor soil management). This emphasis has been particularly important in the assessment of the impacts of the large number of wind farms constructed in Scotland in fulfilment of its renewable energy targets, and an example of its use in practice is given in §5.5.1. Additional guidance is available on calculating potential carbon savings and losses arising from wind farm developments on peat lands (Scottish Government 2011), indicating key issues useful in establishing the environmental baseline for such projects. It also provides a Carbon Calculator Tool in the form of an Excel spreadsheet (Scottish Government 2014): applications for the development of windfarms (of 50MW or greater) on peatland sites in Scotland are expected to use this carbon calculator as part of the EIA.

5.4 Scoping and baseline studies

Most of the available legislation and guidance requires climate change to be considered early in the EIA process. Indeed, during the scoping stage it will be important to identify the key climate change issues by understanding the policy framework for reductions in carbon at national and global levels (see Box 5.1 for examples) and consulting with key stakeholders to take a view on the current baseline conditions and likely future changes.

Box 5.1 Examples of climate change policy frameworks

* Intended Nationally Determined Contributions (as developed under the Paris Agreement of 2015, see §5.4.2)
* Commitments made in Europe to reducing carbon emissions by 80 per cent by 2050
* Legally binding provisions in the UK (under the Climate Change Act 2008) which set five-yearly carbon budgets and hence interim targets to reduce emissions

Since it is the accumulation of GHGs which is contributing to climate change, and since all major infrastructure projects are likely to contribute to this, arguably all climate change considerations should be scoped into the assessment. However, for minor infrastructure projects where there are no significant climate change concerns, it may be appropriate to scope-out climate change from further, more detailed assessment. This recognises the need to be proportionate and pragmatic in the approach to ESIA, but legislation and guidance on this will vary.

The IAIA lists a set of four key questions to ask of any project – whether and to what extent:

- The proposal will, directly or indirectly, increase or decrease GHG emissions;
- The proposal may be beneficially or adversely affected by, and vulnerable to, climate change either directly or indirectly;
- Climate change may affect other aspects of the environment that are potentially affected by the impacts of the proposal;
- An objective of the proposal is to use impact assessment to identify alternatives and measures to mitigate and/or to adapt to climate change.

(Byer et al. 2012)

5.4.1 Scoping for climate change mitigation

Most of the available guidance (including the Canadian guidance for practitioners and IEMA's principles) emphasises the importance of considering greenhouse gas emissions during the scoping stage. The Canadian guidance suggests that during scoping the focus should be on general considerations, rather than detailed quantitative analysis, and reminds practitioners that these considerations include not just greenhouse gas emissions, but also the impacts on carbon sinks. Table 5.1 provides a checklist of potential questions that could assist practitioners during scoping.

It is worth noting that numerous terms are used within ESIAs to describe GHG emissions, for example: carbon; C; carbon dioxide; CO_2; CO_2e equivalent/eq; carbon dioxide-equivalent; and greenhouse gases. A study by Watkins and Durning (2012) demonstrated that these terms were used with considerable variability, sometimes within the same ESIA report, as well as between ESIA reports produced within the same country. Furthermore, the

Table 5.1 Preliminary scoping of GHG emissions (adapted from CEAA 2003)

Checklist question	Illustrative examples of projects that *may* prompt each question
1 **GHG emissions** – Is the project likely to generate significant volumes of GHG emissions during any phase of the project (including exploration, construction, operation or decommissioning)? This may include relatively short periods of intense emissions, as well as indirect emissions due to increased demand for energy.	• Coal-fired generating plant • Hydro-carbon production • Some large-scale industrial manufacturing (e.g. petroleum refining, cement, pulp and paper, iron and steel, chemical production).
2 **Carbon sinks** – Is the project's construction or lifetime operation likely to adversely affect areas that may serve as carbon sinks for GHG emissions (e.g. forests, crops, peatlands and wetlands)?	• Large-scale deforestation • Large-scale flooding of land • Large-scale changes in land use

study noted that using these terms without explanation or inconsistently can create confusion, lack of transparency, difficulties in comparison and occasionally the provision of inaccurate estimates of emissions. The study suggested that using a single term (or single term and abbreviation) would avoid the confusion seen in many of the ESIA reports and improve the ability of non-specialists to understand assessments.

5.4.2 Baseline studies for climate change mitigation

In some countries with a well-developed and long-standing science base, such as the UK, obtaining information on GHG emissions may be relatively straightforward. For instance, the UK has a Greenhouse Gas Inventory that contains all of the UK's official reported greenhouse gas emission estimates by source and by Local Authority (Ricardo-AEA 2015).

The extent and depth of the baseline information presented within any ESIA report will vary, and may include historical, current and likely future conditions/trends. An understanding of how the baseline environment will change is important, particularly for projects that have very long lifespans. However, it may not always be necessary to undertake detailed quantification of GHG emissions and this will need to be judged on a case-by-case basis.

Under the terms of the Paris Agreement, signatory countries have committed through pledges, or Nationally Determined Contributions, to reduce their emissions by the end of the century. However, there is still a "significant gap between the aggregate effect of Parties' mitigation pledges (in terms of global annual emissions of greenhouse gases by 2020) and aggregate emission pathways consistent with holding the increase in the global average temperature to well below 2°C above pre-industrial levels" (UN 2015). Therefore, further negotiations and emission reduction efforts will be necessary (see Figure 5.1). Baseline studies should therefore refer to the latest version of the Intended Nationally Determined Contributions, but should also "take account of the likely future baseline situation" (IEMA 2010) as these contributions are periodically reviewed.

5.4.3 Scoping for climate change adaptation

Compared with mitigation, adaptation has often been neglected in ESIA. Reasons given are: that it has not been the subject of international commitments to the same extent as mitigation; the impacts of climate change on developments are experienced locally and therefore subject to great variation, and are sometimes seen as the responsibility of local rather than national agencies; the timescales into the future (decades or even centuries) are difficult to handle; or because of emphasis on the uncertainty of climate change projections into the future, depending as they do partly on the levels of mitigation adopted and successfully implemented (Wilson and Piper 2010, Larsen 2014, Wentz 2015).

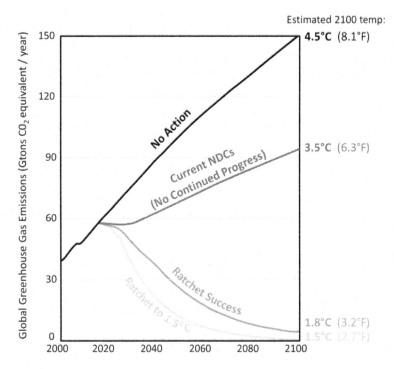

Figure 5.1

UN Climate Pledge Analysis (Climate Interactive 2016)

Project lifetimes and climate change

However, climate change is already impacting human and natural systems, and the climate will continue to change throughout the twenty-first century because of the carbon already in the atmosphere (IPCC 2014). Most major infrastructure projects are likely to have a lifetime of around 60–100 years (CIRIA 2009). However, for some major infrastructure projects this could be significantly longer, e.g. the operational and decommissioning lifetime of a nuclear power station could be 160+ years. It is therefore important that ESIA considers the full lifetime of the plan or project, including the design, construction, operation and decommissioning or abandonment phase. This lifetime may well be longer than the design life originally envisaged by the development's engineers, as many developments remain in place long after the original development has fulfilled its objectives (IEMA 2015). Figure 5.2 shows typical lifetimes of infrastructure alongside climate change timescales.

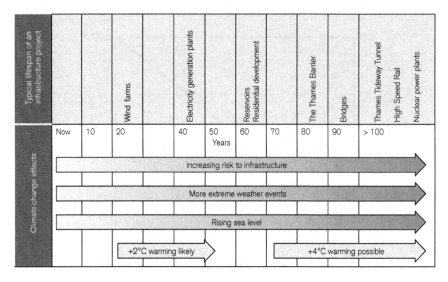

Figure 5.2

Typical lifetimes of infrastructure projects in the UK and climate change effects (based on HMG 2011)

Impacts on project and/or receiving environment

Adaptation should be an essential consideration in all stages of a project's ESIA, and rationales should be provided for the choices made in the assessment. IEMA (2015) suggests that scoping for adaptation should involve:

- agreement with key stakeholders on the most appropriate climate change projection to adopt for the assessment, and any necessary methodological considerations;
- identification of the scale and scope of the project's initial design and its potential impact on the receiving environment, taking into account how this will be affected by a changing climate; and
- engagement with key stakeholders to identify the policy and regulatory regime regarding climate change in the project area.

Issues to consider in scoping for adaptation include:

- applicable regulatory and legislative requirements;
- the nature of the project, its location and resilience to climate change;
- the duration of the project;
- the climate-related parameters likely to influence the project;
- anticipated changes to those climatic parameters over the life of the project;

- how sensitive the environment potentially affected by the project is to those climate parameters (IEMA 2015).

Examples of these climate-related parameters and their consequences are given in §5.4.4.

It will be necessary to consider whether the likely effects of the project on the receiving environment or potentially vulnerable receptors will change over these timescales because of the changing climate (essentially because the baseline environment is changing). This will involve identifying whether the likely impacts of the project will get better or worse, say 50 or 100 years into the future. For example, a climate change scenario might indicate that in 50 years the habitat surrounding the proposed project may be occupied by a protected species that has moved to new "climate-space" and away from less suitable areas. This will be challenging for practitioners and regulators who will have to grapple with high levels of uncertainty and consider how impacts predicted to occur in 50 years can be mitigated and made binding as part of any permission or consent (see §5.6).

Use of scenarios

As part of IPCC's studies, and under the provisions of the United Nations Framework Convention on Climate Change (UNFCCC), the scientific community has undertaken much work to downscale climate change models to regional and national areas, generating a range of climate change scenarios for key climate parameters for time periods to the 2080s, including for developing and least developed countries (see §5.5.2). As indicated, this scoping stage should include agreement with key stakeholders on the appropriate climate change scenario to employ: in some cases, such as for Nationally Significant Infrastructure Projects in England and Wales, national policy advice specifies this. For instance, for major UK energy projects, applicants are expected to use, as a minimum, the emission scenario which the Committee on Climate Change regards as most current, and, for critical parts of energy infrastructure, the high emissions (high impact, low likelihood) scenario; and they are expected to base any adaptation measure on the Climate Change Risk Assessment (CCC 2015) and the latest UK climate change projections (DECC 2011).

5.4.4 Baseline studies for climate change adaptation

Identification of the baseline environment involves taking into account:

a) historic/past climate conditions;
b) current conditions (at the time of the assessment);
c) future climate conditions in the medium (e.g. 15–30 years) and longer term (for instance, 50 years, and at least the lifetime of the project).

Future climate conditions can be identified through available climate change projections. However, it is also important to identify the possibility of extreme weather events (such as extreme rainfall or prolonged periods of drought). As such, the analysis of climatic factors that affect adaptation may include:

- extremes in short-term weather events such as: heat-waves, extreme flooding and freezing conditions, gales and hurricane-force windstorms/ typhoons; storm surges along coastlines;
- extremes in longer-term climatic variability, including precipitation over one or more seasons; variations in average temperature; potential changes in prevailing wind directions;
- changes in average climate norms, resulting in sea-level rise; increases in freezing and thawing; average ambient temperatures; changes in seasonal rainfall patterns (IEMA 2015).

For longer-lasting projects like roads and pipelines that may be in place for more than 100 years "it is probably more useful to define several future baseline environments (the current baseline, then in 30, 50, 70 years' time, and more than 100 years' time)" (IEMA 2015).

The vulnerability and sensitivity of receptors to the existing/prevailing climate should then be determined, employing a classification of:

- high vulnerability: receptor directly dependent on prevailing climatic factors, and reliant on their continuation, or limited variability;
- moderate vulnerability: receptor dependent on some climatic factors, but able to tolerate a range of conditions;
- low vulnerability: climatic factors have little influence on receptors.

A matrix can then be developed which represents the sensitive receptors and the physical location of the project (dimensions such as topography, hydrology, soil conditions, habitats and communities), and considers how the baseline for these dimensions is likely to alter in response to climate change (IEMA 2015). Examples of sensitive receptors might include nomadic populations, marginal farmlands, and coastal habitats and communities.

The ESIA should also consider whether the evolving baseline trends are likely to reach a tipping point (a critical turning point or bottom line) (CEC 2013). This will be an issue where expert judgement such as from key stakeholders will be essential. More details and an example of employing climate change scenarios are given in §5.5.2.

5.5 Impact prediction and evaluation

5.5.1 Impact of the project for climate change mitigation

A project's impacts on climate will depend on the degree to which it:

- generates energy by using fossil fuel;
- consumes energy (most projects do);
- generates energy by using renewables;
- is designed to remove carbon, such as Carbon Capture and Storage projects or protects carbon sinks.

The former two will lead to GHG emissions; renewable energy projects should be GHG-neutral, and the last category will remove GHGs (in practice neutralising some of the effects of fossil-fuel energy generation).

In the context of international and national commitments to lower-carbon energy generation, a large number of projects have as their primary purpose the generation of low-carbon and/or renewable energy, and some of the better efforts at estimating the climate change mitigation aspects of developments under ESIA are for such projects (Watkins and Durning 2012). Box 5.2 shows an example of how carbon losses and savings from a windfarm have been calculated.

However, although carbon calculators such as the Scottish Government's (2014) provide a useful indication of potential carbon losses and savings, they may not always consider the whole life of the development and may be selective with the inclusion/exclusion of certain parameters. Wentz et al. (2016) report that a number of EISs in the US have conducted a lifecycle assessment of GHG emissions from some fossil-fuel related projects, for their extraction, transportation and consumption.

Evaluation

Climate change is a cumulative effect. No one project will 'cause climate change' but jointly almost all projects do. Because of this, it is difficult to evaluate a single project's significance for climate change: even if the project's contribution is deemed not to be significant on a global scale, it may well be significant in relation to locally or nationally determined GHG reduction targets (CEC 2013). The urgent global commitment to reducing these causes implies that at least every major infrastructure project (for example, those projects described in Annex 1 of the EU EIA Directive) should be regarded as having potentially significant effects on climate change and so require mitigation.

Box 5.2 GHG losses and savings from Deuchries wind farm, Scotland

Scotland is committed to a target for renewable energy sources of 100 per cent of annual electricity demand by 2020; but it also has abundant organic soils, containing almost 50 per cent of the UK's stocks of soil carbon. On-shore wind energy makes a large contribution to this target, but as the best location for such installations is often on sites valued for their organic soils, it is important to make a systematic assessment of the project's net carbon contribution (potential savings and losses). One such development which employed this methodology was that for the Deuchries Wind Farm in Aberdeenshire (RSK & Force 9 Energy 2012):

Source of GHG emissions/savings	GHG emissions (tCO$_2$e)
Construction	
Turbine manufacture, construction and decommissioning	5,044
Restoration of borrow pits	44
Back-up	4,586
Reduced carbon fixing potential	534
Loss of soil organic matter	10,120
Leaching of dissolved oxygen content and portable oxygen content	17
Forestry felling	0
Decommissioning	
Improvement of degraded bogs	0
Improvement of felled forestry	0
Removal of drainage from foundations and hardstanding	−135
Subtotal (emissions)	**20,211**
Operation: Annual GHG savings	
Development operation (per year)	−11,007
Sub-total (CO$_2$ savings, per year)	**−11,007**

The EIS stated that the carbon payback period for the project would be 1.8 years (20,211 tCO$_2$e divided by annual savings of −11,007 tCO$_2$e) against a fossil fuel grid mix of electricity. As the development had an expected operational lifetime of 25 years, the net positive GHG effect was approximately 255,000 tCO$_2$e (RSK & Force 9 Energy 2012). One of the benefits of this method is that it helps to value carbon stocks such as organic soils.

5.5.2 Climate change impacts on the project and on receptors

Climate change will exacerbate existing risks for natural and human systems, and cause new risks for the proposed development. The risks are likely to be greater for disadvantaged people and communities (IPCC 2014). In this phase of the ESIA, it is also helpful to consider the sensitivity of elements of the project to climate change parameters, and to include the different stages and supply chains of any project. It is important then in the ESIA to allow for the dynamic nature of the relationship of the project with changing climate and other socio-economic changes over time.

It is not possible here to cover all such risks and vulnerabilities, but many countries have undertaken their own sectoral assessments. For instance, the UK Climate Change Risk Assessment (HMG 2012) assessed the themes of Agriculture and Forestry; Business; Health and Well-being; Buildings and Infrastructure; and the Natural Environment. Table 5.2 summarises the assessment for Buildings and Infrastructure and Health.

Table 5.2 Risks and opportunities of climate change for built environment, and health and well-being (HMG 2012)

Risks	Opportunities
Built environment	
Energy	
• Energy infrastructure at significant risk of flooding • Higher energy demand for cooling • Heat damage/disruption to energy infrastructure such as buckling • Increased water demand for energy generation	• Reduced energy demand for heating
Transport	
• Road, railways and tunnels at significant risk of flooding • Scouring of road and rail bridges (which is caused by the removal of sediment from around bridge abutments or piers)	• Shorter shipping routes and reduced transportation costs due to less Arctic ice
Water	
• Supply-demand deficits	
Buildings	
• Damage to property due to flooding and coastal erosion • Overheating in buildings including homes, schools and hospitals • Increasing impact from the Urban Heat Island effect • Buildings affected by subsidence	

Note: Flooding is identified as the most significant risk, both currently and in the short term across the UK. Water availability and overheating of buildings are assessed to be increasingly significant by the middle of the century, particularly in England. Key adaptation options include green infrastructure – the living network of green spaces, water and other environmental features in both urban and rural areas – which can help to reduce extremes of temperature and manage water flows.

Table 5.2 continued

Risks	Opportunities
Health and well-being	

- Increased summer temperatures may lead to increased risk of mortality and morbidity [illness] due to heat
- Increased flooding would increase the risks of deaths, injuries and people suffering from mental health effects as a result of the impacts of flooding
- Increased ozone levels by the end of the century could lead to an increased risk of mortality and respiratory hospital admissions
- Increased summer temperatures combined with increased periods of time spent outdoors may lead to an increased risk of skin cancer cases and death
- Increased temperatures and changed rainfall patterns may lead to an increased health risk from water, vector and food borne diseases
- Increased sea temperatures may lead to increased marine pathogens and harmful algal blooms

- Increased winter temperatures may lead to decreased levels of mortality and morbidity due to cold
- Increased summer temperatures combined with increased periods of time spent outdoors could increase Vitamin D levels and help to improve physical and mental health of people

Vulnerability assessment at this stage of the ESIA should also include critical interdependencies, as they can lead to cascade-failure, where the failure of one aspect, such as flood defence, can lead to others (e.g. flooded power stations leading to power cuts which in turn affect communication networks (CEC 2013)).

Once possible impacts are identified, IEMA (2015) then recommend a fourfold approach:

- Assess the magnitude of the impacts of the project on baseline conditions under current conditions, and the significance of effects (i.e. conduct the EIA as normal without climate change).
- Identify the effect of climate change on receptors without the project.
- Assess whether the impact of the project will be worse or improved on the future baseline.
- Define if these changes affect the significance of effects identified for the project without climate change.

An example of a risk assessment of climate hazards for the elements of a linear project – high speed rail line, HS2 – is given in Table 5.3.

Sources of information (as mentioned in §5.4.3) on regional climate change scenarios are now available for developed and developing countries: for

Table 5.3 Risks from a number of climate hazards across a range of infrastructure and assets to be addressed through future design – example of high speed rail from London to the north of England (HS2 2013)

Climate hazard	Infrastructure and assets associated with the proposed scheme	Potential climate change impact	Unmitigated risk	Proposed resilience measure
Heat	Human factors	It may be too hot to work or travel	Unmitigated, this is likely to occur with a high consequence of impact	Practical management of heat risk for employees will take place during operation and maintenance. The passenger experience will be addressed during the future design of trains and stations.
Ice and snow/cold	Overhead line (OLE) equipment including overhead lines (OHL)	OLE may fail due to snow overloading	Unmitigated, this is likely to occur with a high consequence of impact	OLE and OHL will be designed to take into account a range of minimum and maximum temperatures and loads under current and future climate conditions.
Heat	Earthworks and landscaping	Increased shrinkage of soil due to decrease in groundwater levels is possible. Planting failures may occur due to drought, and grassland fires are also possible	Unmitigated, this is likely to occur with a medium consequence of impact	Geotechnical issues will be covered in the future design and track-side planting has already been considered in preliminary interim design. Further resilience measures will be considered as necessary within operation and maintenance procedures.

Table 5.3 continued

Climate hazard	Infrastructure and assets associated with the proposed scheme	Potential climate change impact	Unmitigated risk	Proposed resilience measure
Heat	Tunnels	Higher temperatures may require increased climate control within trains.	Unmitigated, this is possible and could have a medium consequence of impact	Sensitivity analyses will be undertaken for a range of possible ambient temperatures during future design to inform the specification of rolling stock
Wind/ flooding	Abstraction, drainage and flood conveyance systems	Windborne debris may cause blockage of railway drainage systems	Unmitigated, this is likely to occur and would have a high consequence of impact	All abstraction, drainage and flood conveyance systems have been specified to allow for climate change and to reduce blockage

instance, Least Developed Countries have employed downscaled climate change models to generate scenarios in their National Adaptation Programmes of Action submitted to UNFCCC 2006–2013, and currently work is being undertaken on their National Adaptation Plans under the Cancun Agreement (UNFCCC 2015). For instance, the National Adaptation Programme of Action for Yemen described a range of scenarios to 2050 for three climate variables (temperature, cloud cover and precipitation) and their impacts on key sectors of water, agriculture and coastal zone/fisheries (Republic of Yemen Environmental Protection Agency 2009). This was followed up by a World Bank Climate Risk and Adaptation Country Profile for Yemen (World Bank 2011), which examined the impact of climate change scenarios for the 2030s, 2050s and 2080s on water resources, agriculture and urban areas.

As part of many countries' climate change risk assessment, information is available for climate change impacts on key sectors. In the UK, for instance, the Climate Change Act 2008 requires infrastructure sectors such as water resources and energy networks to report on risk preparedness. The UK Adaptation Sub-committee, in preparation for the UK's Second Climate Change Risk Assessment, due in 2017, has commissioned studies on projections of flood risk, water availability and the natural environment under medium and high scenarios, and a study on high impact, low likelihood scenarios representing plausible extreme changes (CCC 2015).

Information is also available for assessing the impact of climate change on vulnerable human populations: for instance, the Climate Just mapping tool (www.climatejust.org.uk) displays the geography of England's vulnerability to climate change at a neighbourhood scale, providing data and maps on exposure to heat, on river and surface water flooding, and drivers of social vulnerability such as personal factors (e.g. age and health), exposure (e.g. green space cover), and adaptive capacity (people's ability to prepare/respond and recover, e.g. income).

Box 5.3 shows an example where national climate change scenarios were employed with professional, sector-based knowledge to assess the resilience of elements of an energy-distribution project in Wales.

The susceptibility or resilience of the receptor to climate change must be considered, as well as the value of the receptor. Therefore, a high-value receptor that has very little resilience to changes in climatic conditions should be considered more likely to be significantly affected than a high-value receptor that is very resilient to changes in climatic conditions. The uncertainty of the combined effects needs to be taken into account. If uncertainty about how a receptor will adapt to a changing climate is high, then it is recommended that a conservative threshold of significance is adopted within the evaluation (IEMA 2015). Box 5.4 shows how in-combination impacts upon, and the resilience of, a high-value receptor were assessed.

5.5.3 Interactions between climate change mitigation and adaptation

In adopting climate change mitigation and adaptation measures, in considering alternatives to those measures, and in implementing further actions to reduce any residual negative impacts on those measures, it is important to consider possible trade-offs and synergies between them, especially in the field of spatial planning (Wilson and Piper 2010). The CEC guidance suggests some useful principles for selecting the most appropriate approaches:

- Adopting no-regret or low-regret options that yield benefits under different scenarios;
- Adopting win-win-win options, such as might benefit climate change, biodiversity and ecosystem services, or climate change and poverty reduction;
- Favouring reversible and flexible options that can be modified if significant impacts arise;
- Adding 'safety margins' where possible, without wasteful over-specification, to ensure responses are resilient to a range of future climate impacts;
- Promoting soft adaptation strategies, such as building adaptive capacity to ensure a project is better able to cope with a range of possible impacts (such as through more effective forward planning);
- Delaying projects that are risky or likely to cause significant effects.

(CEC 2013)

Box 5.3 Stages in climate change assessment of a high voltage electricity connection between a proposed wind farm in South Wales and the main electricity network (WPD 2015)

Stage 1: Baseline description (likely future without the project) employing UK Climate Projections (UKCP09) (2009) This used the medium emissions assessment, for Wales to 2050s, with central, low-probability and high-probability estimates, for winter mean temperature, summer mean temperature and summer mean daily maximum.

Stage 2: Assessment of climate risks The energy transmission sector risk assessment for climate change in 2011 to 2080s examined the relationship between current weather conditions and power distribution incidents to understand vulnerability; it then identified three priority risks (extreme events): wind and gales, lightning, and snow and blizzard, as well as temperature increases and vegetation growth. A Further Adaptation Report in 2015 highlighted an increased risk of inter-dependencies, such as the effect of high winds after periods of prolonged rainfall.

Climate change risk: extreme events	2020 Min, mean and max scenario	2050 Min, mean and max scenario	2080 Min, mean and max scenario	Risks
Wind and gales	−11% to +11%	−11% to no change	−11% to no change	Fall of trees on overhead lines: little change
Lightning	No change	No change to +16%	No change to +16%	Increased risk of lightning strike post 2050s to short overhead lines and transformers
Snow, sleet blizzard	−33% to no change	−66% to −33%	−66% to −33%	Decrease in number of days when snow will fall, but considerable uncertainty
	2010–2039 summer	**2040–2069 summer**	**2040–2069 winter**	**Risks**
Temperature increases	high: 2.2–2.6 °C	4.3–5.2 °C	3.3–3.5 °C	Less current (limited by thermal expansion; can cause overhead lines to sag, giving less ground clearance)
	med: 2.3–2.8 °C	3.9–4.7 °C	3.0–3.2 °C	
	low: 2.3–2.7 °C	3.5–4.2 °C	2.7–2.9 °C	
Earthing of overhead lines	Higher temperatures may reduce the amount of moisture in the ground, which can affect the ability to earth an overhead line.			
Flooding	Fluvial, pluvial and tidal flood risk to infrastructure including substations.			
Vegetation growth	It is important that trees and other vegetation near overhead lines are managed to prevent them touching or falling on the line as this			

will lead to faults. Problems occur when there are very strong winds, combined with prolonged rainfall which can make trees more prone to uprooting.

Stage 3: Mitigation measures Alignment and pole positioning especially through forests; management of future vegetation growth; twin poles on exposed land; size of conductor; undergrounding.

Box 5.4 Assessment of resilience of Thames Tideway Tunnel under future climate and population changes (TWUL 2013)

When London's sewerage system was built more than 100 years ago, it served some three million people and was designed for a population of four million. It now serves some six million. The system relies on an integrated network of combined sewer overflows (CSOs). A CSO is an overflow pipe which releases flow through discharge points along the River Thames, and is legally allowed to operate during wet weather. When the sewerage system was built, CSO overflows would happen once or twice a year; they now happen on a weekly basis and no longer require a storm to trigger overflows.

A 25 km waste water storage and transfer tunnel under the Thames (known as the Thames Tideway Tunnel) was proposed to deal with this issue. By intercepting the sewage before it enters the river, the proposed Thames Tideway Tunnel will help prevent the tidal River Thames from being polluted with untreated sewage which can stay in the river for up to three months before the ebb and flow of the tide finally takes it out to sea. The tunnel has a design life of 100 years.

The Resilience Report (which did not form part of the EIS) for the project acknowledged that both the climate and population will change significantly over that timescale, which will influence the project's performance and that of the whole system. Accordingly, the report set out to gather information about these changes and to assess possible impacts on the system. It acknowledged that, given the scale of the uncertainties, it would be overly expensive and inappropriate to design the main tunnel to be able to cope with all possible future conditions. Instead, it considered options for adapting the system to future changes and risks in a proportionate way. National guidance for Climate Resilient Infrastructure requires "climate change impacts in design, build and operation of new infrastructure to be considered" (HMG 2011) and guidance for wastewater infrastructure specified use of the latest UK Climate Projections and the most likely current emissions scenario (Defra 2012).

Current conditions:

- *Frequency and volume of CSO discharges and water quality (as measured by dissolved oxygen):* Assessment of current conditions, and assessment, using hydraulic and water quality modelling including discharge events and volume, of 2020 scenario with planned improvements.

Future changes:

- *Population*: The number and quality of CSO discharges are predominantly affected by rainfall, but also by the volume of foul water flow. Population forecasts for the catchment up to 2030 (by which time the project should be in commission) are available, showing London's population growing faster than the UK; but the EIS scaled national projections to estimate population to 2080s. Based on the best available climate and emission projections for the UK the 10, 50 and 90 percentiles were used to explore the implications of these uncertainties for the 2050s (2040 to 2069) and 2080s (2070 to 2099) time horizons. Mapping showed that the areas likely to experience the greatest population increase lie closest to the River Thames, and are therefore most likely to contribute to CSO discharges.
- *Other change factors*: Sensitivity modelling was used to show that reduced household water consumption (and hence wastewater rates) and estimates of baseflow infiltration (taking account of possible changes to leakage rates or greater provision of *sustainable drainage systems*), were of low significance for CSO discharge volume. These assumptions are somewhat conservative and so tend towards higher potential impacts from future climate and population change, but indicate the scope for impact mitigation through measures to reduce water consumption, repair sewers and reduce run-off.
- *Climate change*: Modified rainfall event data sets were generated, using the UKCP09 Weather Generator for the 2050s and 2080s, and sensitivity analysis was used. Analysis showed that change decelerated between the 2050s and 2080s, and there was little difference between the emission scenarios, so the study focused on the medium emissions scenario for 10th, 50th and 90th percentile for the 2080s, giving a range of CSO spill volume and volume treated. CSO spill volume refers to the amount of overflow sewage that is not able to be treated before discharge.
- *Climatic variables affecting water quality*: Factors such as temperature and river flows were assessed, and a 20 per cent decrease in river flows by 2050s allowed for (as a conservative estimate).
- *Water quality modelling under future climate scenarios*: This showed that there is a possibility of water quality thresholds to be exceeded by the 2080s, but that the exceedances would be far higher without the project.
- *Adapting to changing frequency of CSO spills*: The EIS explored options for adapting the system in the longer term to reduce the risk of exceedances: these included measures for per capita consumption, uptake of sustainable drainage systems, and baseflow conditions. It concluded that these measures offered scope for implementing further adaptation, but that projected climate change, especially temperatures, could cause further deterioration in the tidal Thames.

More positive synergies are typically found in the water and biodiversity sectors, such as through ecosystem-based adaptation or mitigation, or blue/green infrastructure (Berry et al. 2015, IPCC 2014). Many of the methods shown in other chapters of this book will be useful for assessing the possible interactions, but use of a simple mileage chart (Figure 5.3) can ensure that this aspect of addressing climate change is explicitly covered. Examples of the interactions of mitigation and adaptation measures for biodiversity are given in Figure 5.4.

	Mitigation action 1	Mitigation action 2	Adaptation action 1	Adaptation action 2
Mitigation action 1				
Mitigation action 2	✓			
Adaptation action 1	?	✓		
Adaptation action 2	✓	x	x	

Legend: ✓ = consistent   x = inconsistent   ? = uncertain

Figure 5.3

Mitigation and adaptation consistency assessment (Wilson and Piper 2010)

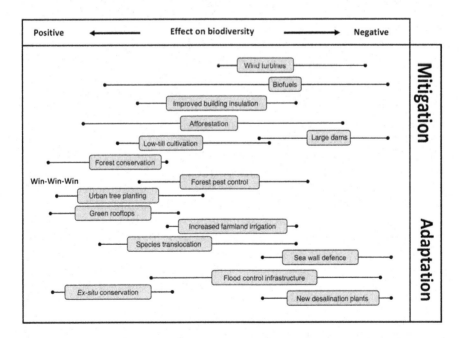

Figure 5.4

Known and potential relationships between mitigation and adaptation measures and their impacts on biodiversity (Paterson et al. 2008)

5.6 Mitigation and monitoring

5.6.1 Mitigation and monitoring for climate change mitigation

Climate change mitigation involves employing the resource hierarchy for energy at all stages of the project (design, construction, operation, and decommissioning), and through its supply chains:

- Adopt alternatives to avoid GHG emissions where possible.
- Minimise energy use though energy-efficient design and energy conservation.
- Employ renewables (non-fossil-fuel sources) and low-carbon materials.
- Employ sources with co-benefits such as combined heat and power stations.

This should minimise project impacts, which will significantly adversely affect climate change. It is important that all projects, whether intended to be energy-generating, energy-using or specifically designed to remove carbon, assess their energy patterns and the possibilities of reducing energy demand (see Chapter 17). Alternatives should be considered at every stage, including 'do nothing'.

However, some residual adverse consequences of the project's direct GHG emissions (through its technologies, materials, supply modes etc.), and its indirect emissions through its energy consumption and transport modes, may require further mitigation. This might include offsetting measures, such as enhancement of natural carbon sinks, and support for carbon management programmes or clean technology, such as renewable energy schemes: either through financial contributions or directly incorporated into the project (CEC 2013).

For many of these measures, provisions should be made and secured in the Environmental and Social Management Plan or similar document (see Chapter 20). Monitoring will require reliable and consistent reporting and evaluation of the project's carbon balance throughout the project's lifetime and decommissioning. It will be useful to tie this in with data required for other accreditations, such as BREEAM, BREEAM Infrastructure, LEED US and Canada etc., and for other corporate reporting such as the Carbon Disclosure Project (2015). Moreover, the effectiveness and efficacy of any offsetting scheme will need secure monitoring with provision for remedy if necessary. The project proponents will also need to monitor the changing global and national commitments to climate change reduction targets under the evolving provisions of the Paris Agreement 2015.

5.6.2 Mitigation and monitoring for climate change adaptation

Adaptation to climate change can include:

- building adaptive capacity and resilience;
- provision and enhancement of green and blue infrastructure;
- water-efficient design and processes;
- flood or drought resistant design and materials;
- fire-resistance or fire-adapted space;
- design of buildings, plant and processes to reduce the need for cooling;
- taking opportunities such as new crops and outdoor activities;
- connectivity for biodiversity (such as corridors, stepping stones and mosaics).

Adaptation should not be a one-off action but a continuous process to ensure the resilience of the system being impacted. Adaptive management involves handling uncertainty, and guidance is available in the European Climate Adaptation Platform CLIMATE-ADAPT: indeed, the CEC guidance argues that EIA may facilitate adaptive management by clearly acknowledging assumptions and uncertainty (CEC 2013).

Adaptive management is exemplified in the much-quoted SEA for the Thames Estuary 2100 flood-management plan, in which a decision-path approach was taken for a replacement for the Thames Barrier, through building-in robustness to climate change uncertainties. Four possible packages of flood-management options – improving the existing defences; tidal flood storage; a new barrier; and a barrier with locks – were tested (in addition to a do-nothing option), and decision points were identified. It was decided to undertake small-scale actions (such as raising some flood defences) at present, and the need for a new barrier will be regularly reconsidered. Monitoring of key indicators (such as water levels, or frequency of barrier closures) allows decision points to be brought forward if necessary (Reeder and Ranger 2011).

Any measures taken to mitigate residual negative impacts of the development should maximise co-benefits and minimise risks of maladaptation or negative synergies (such as increased use of fossil fuels for the cooling of buildings). Possible remedies for risks for water between adaptation and mitigation measures are illustrated in Table 5.4.

5.7 Conclusions

ESIA can help to integrate climate change considerations into new projects. There is evidence that this is increasingly happening, driven by new legislation (such as the new EU EIA Directive which came into force in May 2017), emerging guidelines (such as in the US), and best practice. Various tools for incorporating climate change into ESIA have been developed and are likely to evolve for use within specific sectors (such as renewable energy, nuclear power and water management). This chapter has identified some of the legislative drivers, sources of guidance and methods currently used in ESIA, but the reader is reminded that best practice continues to evolve in this field.

Table 5.4 Risks for water between climate change adaptation and mitigation measures (UNECE 2009)

Examples of climate change mitigation measures and possible impacts on water resources

Mitigation measure	Possible risks for water resources	Possible positive effects	Possible remedies and comments
Large-scale biofuel production	Increased water demand, enhanced leaching of pesticides and nutrients leading to contamination of water, biodiversity impacts, conflicts with food production etc.	Reduced nutrient leaching, soil erosion, runoff and downstream siltation	Appropriate location, design and management are essential
Land management for soil carbon conservation	Enhanced contamination of groundwater with nutrients or pesticides via leaching under reduced tillage	Erosion control, improved water and air quality, increase of food production, reduction of siltation of reservoirs and waterway	Depends on regions and conditions, use of innovative methods of precise agriculture

Examples of climate change adaptation measures in the water sector which can have negative impacts on climate change mitigation

Adaptation measure in water management	Possible negative impacts for GHG emissions	Possible remedies and comments
Desalinisation of saline water for water supply	High energy needs	Mitigation of impact depends on energy source, therefore only use desalination if no other choice, and use renewable energy
Reservoirs/ hydropower plants	GHG emissions as water conveys carbon in the natural carbon cycle; and due to rotting vegetation	Depends on many factors, including depth of reservoir. Multipurpose dams and appropriate location and management
Irrigation	High water and energy needs	Use efficient irrigation techniques, drought-resistant crop varieties

References

ACT (Australian Capital Territory Government) 2012. *Climate Change Vulnerability Assessment Framework for Infrastructure*, Canberra: ACT Policy and Cabinet Division, www.cmd.act.gov.au/__data/assets/pdf_file/0007/346273/Climate_Change_ Vulnerability_Assessment_Framework_for_Infrastructure_August_2012.pdf.

Agrawala S, A Matus Kramer, G Prudent-Richard, and M Sainsbury 2011. *Incorporating Climate Change Impacts and Adaptation in Environmental Impact Assessments: Opportunities and Challenges, OECD Environmental Working Paper No. 24*, Paris: OECD Publishing.

Berry, P, S Brown, M Chen, A Kontogianni, O Rowlands, G Simpson and M Skourtos 2015. Cross-sectoral interactions of adaptation and mitigation measures, *Climatic Change* 128, 381–393.

Byer P, R Cestti, P Croal, W Fisher, S Hazell, A Kolhoff and L Kørnøv 2012. *Climate Change in Impact Assessment: International Best Practice Principles*, Special Publication Series No.8, Fargo, USA: International Association for Impact Assessment.

Carbon Disclosure Project 2015. *Carbon Disclosure Project: The facts*, www.cdp.net/ Documents/CDP-the-facts.pdf.

CCC (Committee on Climate Change) 2015. *Climate Change Risk Assessment 2017.* www.theccc.org.uk/tackling-climate-change/preparing-for-climate-change/climate-change-risk-assessment-2017.

CEAA (Canadian Environmental Assessment Agency) 2003 (modified 2012). *Incorporating Climate Change Considerations in Environmental Assessment: General Guidance for Practitioners*, Federal-Provincial-Territorial Committee on Climate Change and Environmental Assessment. www.ceaa-acee.gc.ca/default.asp?lang= En&n=A41F45C5-1&offset=1&toc=show.

CEAA 2012. *Canadian Environmental Assessment Act 2012* (S.C. 2012, c.19, s. 52). www.ceaa-acee.gc.ca/default.asp?lang=En&n=9EC7CAD2-1.

CEC 2009. *Report from the Commission in the Application and Effectiveness of the EIA Directive (85/337/EEC as Amended by Directives 97/11/EC and 2003/35.EC)*, COM (2009) 378 final.

CEC 2013. *Guidance on Integrating Climate Change and Biodiversity into EIA*, Brussels: CEC.

CEC 2014. *Directive 2014/52/EU of the European Parliament and of the Council Amending Directive 2011/92/EU on the Assessment of the Effects of Certain Public and Private Projects on the Environment*, http://eur-lex.europa.eu/legal-content/EN/TXT/PDF/?uri= CELEX:32014L0052&from=EN.

CEQ (Council on Environmental Quality) 2014. *Revised Draft Guidance for Federal Departments and Agencies on Consideration of Greenhouse Gas Emissions and the Effects of Climate Change in Nepa Reviews*, 79 Fed. Reg. 77802 http://web.law.columbia.edu/ sites/default/files/microsites/climate-change/ceq_revised_draft_guidance.pdf.

CIRIA 2009. Infrastructure longevity, *Evolution*, January, 6–7.

Climate Interactive 2016. Climate Scoreboard (UN Climate Pledge Analysis). www.climateinteractive.org/programs/scoreboard/

DECC (Department for Energy and Climate Change) 2011. *Over-arching National Policy Statement for Energy EN-1*, London: Stationery Office.

Defra (Department for Environment, Food and Rural Affairs) 2012. *National Policy Statement for Waste Water: a framework document for planning decisions on nationally significant waste water infrastructure*, London: Stationery Office.

Glasson, J, R Therivel and A Chadwick 2012. *Introduction to Environmental Impact Assessment*, Abingdon: Routledge.

HMG 2011. *Climate Resilient Infrastructure: Preparing for a Changing Climate*, Cm 8065, London: Stationery Office.

HMG 2012. *UK Climate Change Risk Assessment: Government Report*, London: Stationery Office.

HS2 (High Speed Two) 2013. *London-West Midlands Environmental Statement – Technical Appendices*, Resilience to impacts from climatic conditions, London: HS2.

IEMA (Institute of Environmental Management and Assessment) 2010. *Climate Change Mitigation and EIA*, IEMA Principles Series, Lincoln: IEMA.

IEMA 2015. *IEMA Environmental Impact Assessment Guide to Climate Change Resilience and Adaptation*, Lincoln: IEMA.

IFC 2012. *Performance Standards on Environmental and Social Sustainability*. www.ifc.org/wps/wcm/connect/115482804a0255db96fbffd1a5d13d27/PS_English_201 2_Full-Document.pdf?MOD=AJPERES.

IPCC (Intergovernmental Panel on Climate Change) 2014. *Climate Change 2014: Synthesis Report. Contribution of Working Groups I, II and III to the Fifth Assessment Report of the Intergovernmental Panel on Climate Change* [Core Writing team RK Pachauri and LA Meyer (eds.)], Geneva, Switzerland: IPCC.

Larsen SV 2014. Is environmental impact assessment fulfilling its potential? The case of climate change in renewable energy projects, *Impact Assessment and Project Appraisal* 32(3), 234–240.

Ohsawa T and P Duinker 2014. Climate change mitigation in Canadian environmental impact assessments, *Impact Assessment and Project Appraisal* 32(3), 222–233.

Paterson JS, MB Araújo, PM Berry, JM Piper and MDAR Rounsevell 2008. Mitigation, adaptation and the threat to biodiversity, *Conservation Biology* 22(5), 1352–5.

Reeder T and N Ranger 2011. *How Do You Adapt in an Uncertain World? Lessons from the Thames Estuary 2100 Project*, World Resources Report. www.wri.org/sites/default/files/ uploads/wrr_reeder_and_ranger_uncertainty.pdf.

Republic of Yemen Environmental Protection Authority 2009. *National Adaptation Programme of Action*. http://unfccc.int/resource/docs/napa/yem01.pdf.

Ricardo-AEA 2015. *Local and Regional Carbon Dioxide Emissions Estimates for 2005–2013 for the UK: Report for Department of Energy and Climate Change*, Harwell: AEA. www.gov.uk/government/uploads/system/uploads/attachment_data/file/437426/2005_ to_2013_UK_local_and_regional_CO2_emissions_technical_report.pdf.

RSK & Force 9 Energy 2012. *Deuchries Wind Farm Environmental Statement Volume 1: Environmental Statement Chapter 12 Climate Change*. www.force9energy.com/assets/ uploads/project/deuchries_ES.pdf.

Scottish Government 2011. *Calculating Potential Carbon Losses and Savings from Wind Farms on Scottish Peatlands*. Technical Note version 2.0.1. www.gov.scot/Resource/ Doc/917/0121469.pdf.

Scottish Government 2014. *Full Carbon Calculator for Windfarms on Peatlands Version 2.9.0*. www.gov.scot/Topics/Business-Industry/Energy/Energy-sources/19185/17852-1/CSavings/CC2-9-0.

SNH (Scottish Natural Heritage) 2013. *A Handbook on Environmental Impact Assessment: Guidance for competent authorities, consultees and others involved in the EIA process in Scotland*, Edinburgh: SNH.

Sok V, BJ Boruff, and A Morrison-Saunders 2011. Addressing climate change through environmental impact assessment: international perspectives from a survey of IAIA members, *Impact Assessment and Project Appraisal* 29(4), 317–326.

Stern NH 2015. *Why are we waiting? The logic, urgency and promise of tacking climate change*, Cambridge, MA: MIT Press.

TWUL (Thames Water Utilities Limited) 2013. *Thames Tideway Tunnel. Application for*

Development Consent. Resilience to Change. London: TWUL.

UK Climate Projections 2009. *UKCP09,* http://ukclimateprojections.metoffice.gov.uk/23977.

UN (United Nations) 2015. *Adoption of the Paris Agreement.* http://unfccc.int/resource/docs/2015/cop21/eng/l09r01.pdf.

UNECE (United Nations Economic Commission for Europe) Convention on the Protection and Use of Transboundary Watercourses and International Lakes 2009. *Guidance on Water and Adaptation to Climate Change,* UN: New York and Geneva

UNFCCC (United Nations Framework Convention on Climate Change) 2015. *Best Practices and Lessons Learnt in Addressing Adaptation in Least Developed Countries,* Least Developed Countries Expert Group. http://unfccc.int/files/adaptation/application/pdf/leg_bpll_volume3.pdf.

Watkins J and B Durning 2012. Carbon definitions and typologies in environmental impact assessment: greenhouse gas confusion? *Impact Assessment and Project Appraisal* 30(4), 296–301.

Wentz J 2015. *Assessing the Impacts of Climate Change on the Built Environment under NEPA and State EIA Laws: a Survey of Current Practices and Recommendations for Model Protocols.* http://web.law.columbia.edu/sites/default/files/microsites/climate-change/assessing_the_impacts_of_climate_change_on_the_built_environment_-_final.pdf.

Wentz J, G Glovin and A Ang 2016. *Survey of Climate Change Considerations in Federal Environmental Impact Statements 2012–2014,* New York: Sabin Centre for Climate Change Law. https://web.law.columbia.edu/sites/default/files/microsites/climate-change/survey_of_climate_change_considerations_in_federal_environmental_impact_statements_2012-2014.pdf.

Wilson E and J Piper 2010. *Spatial Planning and Climate Change.* Abingdon: Routledge.

World Bank Group 2011. *Climate Risk and Adaptation Country Profile: Yemen: Vulnerability, Risk Reduction and Adaptation to Climate Change,* Global Facility for Disaster Reduction and Recovery. http://sdwebx.worldbank.org/climateportalb/doc/GFDRRCountryProfiles/wb_gfdrr_climate_change_country_profile_for_YEM.pdf.

WPD (Western Power Distribution) 2015. *Brechfa Forest Connection: Environmental Statement Appendix 7.3; Climate Change Resilience Report,* WPD. http://infrastructure.planninginspectorate.gov.uk/wp-content/ipc/uploads/projects/EN020016/2.%20Post-Submission/Application%20Documents/Environmental%20Statement/BFC_Vol_06.4_ES%20Appendix%207.3%20Climate%20Change%20Resilience%20Report.pdf.

6 Ecology

•••

Roy Emberton, Richard J. Wenning and Jo Treweek

6.1 Introduction

Most countries with requirements and procedures for performing an ESIA require the work to include an assessment of the potential for ecological impacts; countries such as Australia (Elliott and Thomas 2009), Canada (Beanlands and Duinker 1983), New Zealand (Dixon 1993), the Netherlands (Heuvelhof and Nauta 1997), the United States (Eccleston and Doub 2012) and the Member States of the European Union, including the United Kingdom (Arts et al. 2012). Most policies and regulations requiring Strategic Environmental Assessment (SEA) also include a requirement to address the implications for biodiversity and ecosystem services. This stems, in part, from growing recognition that human activities influence ecosystem services, which in turn affects environmental health, ecological processes and functions, biodiversity, and the intrinsic cultural values associated with wildlife. Consequently, ecological assessments are becoming more sophisticated, reflecting the desire to demonstrate that proposed development will not irreparably harm the environment or its capacity to provide ecosystem services for current or future generations (Honrado et al. 2013). The right of current and future generations to enjoy a healthy environment and access ecosystem services is implicit in these requirements.

Relative to other forms of capital, the assets embodied in ecosystems are poorly understood, scarcely monitored, typically undervalued and more often than not recognised only upon their loss (MA 2005a, 2005c). Impact assessment alone cannot resolve the global challenges of biodiversity loss and deterioration of ecosystem services, but it can at least make important trade-offs explicit (CEQ 1993, Cowley and Vivian 2007). Specific aspects of ecosystem services are addressed in Chapter 8, but such reviews are typically underpinned by ecological information and understanding gained through ecological assessments such as those described in this chapter, as well as results of social assessments.

Assessing ecological impacts is challenging because of the complex interactions that take place within and between ecosystems and the fact that their spatial and temporal scales are often quite different from those of planned development (EEA 2011). A further challenge is the fact that ecological baselines are dynamic in space and time, making it difficult to accurately distinguish between changes that would occur in the absence of proposed development and those induced by it. This makes it difficult to obtain baseline information sufficient to derive accurate predictions or provide the technical foundation needed to determine how impacts attributable to planned development might be avoided or moderated so that biodiversity, ecosystem services and human well-being are protected and preserved.

Global environmental change introduces further challenges. Climate change, in particular, is altering baseline scenarios for many ecosystems and increasing uncertainty about future trends. It is becoming increasingly difficult to compare and contrast baseline conditions with post-development conditions because the planet's rapidly changing hydrological environment and the increasing frequency of extreme climate events is altering the natural cycle of change in ecosystems (Mawdlsey et al. 2009). The most important causes for the loss of biodiversity and ecosystem degradation worldwide – habitat degradation, climate change, desertification, and displacement by non-native (i.e. invasive) species – can be attributed to human-accelerated natural changes (Juniper 2007, MA 2005b). For example, habitat loss attributed to climate change is an increasingly serious threat to wildlife in the UK, and is believed to be a primary factor for declines in several species and the extinction of approximately 170 species of plants and animals in the UK during the past 100 years (EN 2004a, ZSL 2007, PI 2000). In developing countries, pressure from human population growth and industrialisation has led to similar large declines in wildlife and habitat. For example, Wikramanayake et al. (2002), Hogan (2012) and others report that losses of tropical forest ecosystems range between 5–29 per cent across the Indo-Pacific Region.

Such threats to the resilience of ecosystems and biodiversity, together with accelerating natural resource use, makes it very important in ESIA to consider *cumulative impacts* on ecosystems (EN 2004a, LUC 2005), taking into account baseline trends. However, it is not always easy to secure an appropriate scope in practice, due to resource constraints and time pressure on the decision-making process.

This chapter provides an overview of Ecologically Focused Impact Assessment (hereafter referred to as "EcIA"), and focuses on terrestrial and freshwater ecosystems. Because marine ecosystems differ significantly from those on land, and because there is a close relationship between coastal ecology and geomorphology, these two components are considered together in Chapter 7. Similarly, the major environmental components of terrestrial and freshwater ecosystems – water, soil and air – are largely dealt with in Chapters 2, 3 and 4.

6.2 Definitions and concepts

Ecology is the study of (in decreasing order of size) ecosystems, habitats, communities and species; and it is important to understand what these are and how they are inter-related.

- An ecosystem is a self-sustaining, functional system consisting of physical, chemical and biological attributes. Ecosystems need to be clearly defined in terms of scale. Planet Earth, for example, can be represented as a single global ecosystem, and yet closer inspection reveals innumerable unique smaller marine and terrestrial ecosystems that each have their own set of attributes.
- A habitat refers to the conditions needed to support a species or community of species in a particular location, but the term is also used more globally, to describe the ecological conditions generally needed to support individuals or populations of a particular species wherever they occur. Habitat types are designated in habitat classifications that are widely used in conservation, environmental legislation and EcIA.
- A community of species in a habitat refers to assemblages of different species that have evolved in ways that facilitate their coexistence and regularly and recognisably occur together. The environment typically supporting such a specific, recognisable community is strictly defined as a "biotope". In marine contexts, the term "biotope" is much more prevalent, with a number of "biotope classifications" in use. However, the term "ecosystem" is also used in this context, so in practice all three terms are used interchangeably. When conducting EcIA it is important to define clearly which usage of the term is intended, to avoid confusion.

ESIA and SEA draw on ecological surveys and assessments to identify, assess and evaluate impacts of proposed development projects on biodiversity, including ecosystems and ecosystem services. Treweek (1999) defined ecological impact assessment as the "process of identifying, quantifying and evaluating the potential impacts of defined actions on ecosystems or their components", which, if properly implemented, "provides a scientifically defensible approach to ecosystem management".

This section provides a brief explanation of definitions and concepts that are particularly relevant to EcIA. Further information can be found in a wide range of ecology books such as Townsend et al. (2007), Begon et al. (2005), and Krebs (2001).

EcIA involves identifying ecological "receptors" or features, generally in the scoping phase of an impact assessment, that are considered likely to be sensitive to project-related changes in the environment. Ecological receptors may be ecosystems, species populations or individual species, and ecological features may include unique landscapes or areas of habitat considered essential to the structure and function of the ecosystem associated with the proposed development. Some EcIA guidance and standards also refer to the identification of biodiversity features or values and the ecological processes that are needed to sustain them. In this context, EcIA should establish whether construction activities, mitigation and the completed development itself are compatible with maintaining biodiversity. In general, it is important to take an "ecosystem approach", as well as considering impacts on individual receptors.

6.2.1 Factors underpinning ecosystems

Ecosystems have numerous interacting components and processes, often involving delicate balances and relationships. Their components are generally influenced by, and in turn affect, other components within the same system (e.g. the effects of vegetation on soil development), which can be influenced by soil geological conditions and the activities of soil animals, bacteria and fungi.

Typical interactions within a terrestrial ecosystem are illustrated in Figure 6.1. A change in even a single component such as a species population or an environmental variable, can cause effects that are difficult to predict. An understanding of food chains and food webs is important because the alteration of habitat conditions or an essential environmental attribute can have unintended and dire consequences to one more species in the ecosystem.

Many types of interactions also occur between ecosystems that are capable of influencing ecological processes such as the flow of nutrients or biomass through food chains and the effects of solar radiation on climate conditions. Ecosystems are sustained by fluxes of energy and materials, ultimately driven by an energy source (most often solar radiation). There is strong evidence that photosynthesis is responsible for the generation and maintenance of atmospheric oxygen, and also compensates for the gradual increase in solar energy emission over geological time, by absorbing CO_2 and hence reducing the greenhouse effect.

Energy also drives important processes such as evapotranspiration (whereby plants emit moisture to the atmosphere as a result of transpiration and evaporation). This has a cooling effect and influences the formation of protective cloud screens, especially over tropical rainforests. Large-scale removal of plants to facilitate a proposed development may alter these processes, resulting in higher ambient temperatures and potentially affecting the hydrological cycle of the local landscape.

Climate can also be affected by geomorphology, hydrology and soil conditions at either a global or ecosystem scale. The exchange of volatile elements (in elemental form or in compounds) between an ecosystem and the atmosphere can influence climate stability and change. The main inputs are carbon (from CO_2) assimilated by photosynthesis and absorbed in precipitation as carbonic acid; and nitrogen (from N_2) "fixed" by lightning and nitrogen-fixing bacteria. The main outputs are CO_2 from respiration; and O_2 from photosynthesis (which normally exceeds CO_2 output from the majority of terrestrial and aquatic ecosystems). In addition, N_2 (mainly from nitrates) may also be released by denitrification in waterlogged soils. Volatile elements can also enter or leave ecosystems as solutes in surface and groundwater, but the atmospheric inputs normally prevent depletion of these elements over time.

Other important material and energy flows in an ecosystem are those related to non-volatile elements.

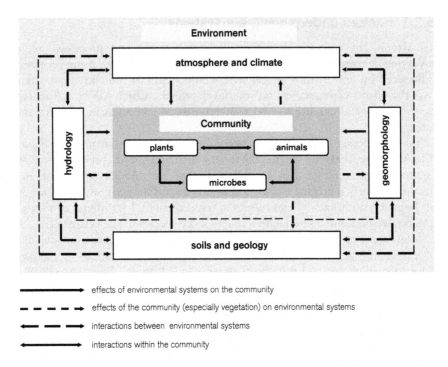

effects of environmental systems on the community

effects of the community (especially vegetation) on environmental systems

interactions between environmental systems

interactions within the community

In any given location, the climate affects the geomorphology, hydrology, soil and all the species of the community. Conversely: **microclimates** and local climates are affected by the other sub-systems; and vegetation and major geomorphological features such as mountain ranges, can affect macroclimates (regional and even global).

The community is also strongly influenced by, and in turn affects, the other sub-systems – particularly the soil, which is an ecosystem in its own right, with a community consisting of plant roots, soil animals, bacteria and fungi.

In addition, there are innumerable interactions within the subsystems, e.g. between abiotic environmental variables and between species.

Figure 6.1

Simple model of interactions between subsystems of a terrestrial ecosystem (Morris and Emberton 2009)

These normally only enter local ecosystems by weathering of bedrock, as airborne particulate matter, or in solution or suspension in water. Water is usually the most important input-output medium and an ecosystem's nutrient budget (Figure 6.2) is strongly influenced by its water budget. Thus, a local ecosystem may receive nutrients in drainage water from higher in its catchment, but may lose nutrients by leaching and erosion. Consequently, if an area has a climatic water surplus, then outputs of non-volatile nutrients are likely to exceed inputs, resulting in gradual nutrient depletion of an ecosystem. Ecosystem nutrient regimes can be markedly affected by human activities,

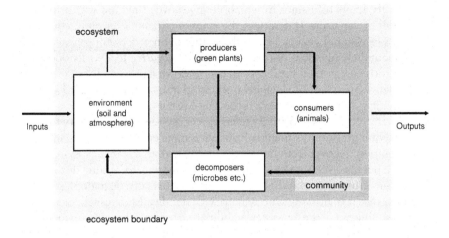

Figure 6.2

Nutrient flows within a terrestrial ecosystem and across its boundaries (Morris and Emberton 2009)

which can result in excessive inputs such as in the case of acid deposition and eutrophication; or nutrient depletion due to enhanced leaching or soil erosion. In addition, many toxic pollutants can enter, and circulate within, ecosystems in the same ways as nutrients.

Several aspects of an ecosystem – energy, biomass and productivity – are often overlooked in EcIA. Traditional approaches have tended to focus more on changes to structural and compositional aspects, neglecting the influence of proposed developments on biodiversity and the future viability of an ecosystem or the availability of natural resources and supplies of important ecosystem services such as climate regulation. Ecosystem structure and function must be considered in EcIA, in addition to the many interactions among species that inhabit the ecosystem, all of which should be considered when ecological impacts are possible or anticipated. Current knowledge of the dynamic nature of ecosystems is limited. Therefore, the potential effects of development activity on ecosystems is often difficult to predict, and then only with high degrees of uncertainty. This likely explains why large non-linear and unexpected environmental changes have been observed as a consequence of undertaking some proposed developments.

6.2.2 Ecosystem resilience

Stability, fragility/sensitivity, resilience/recoverability and recreatability are important attributes of ecosystems from an EcIA perspective because they describe an ecosystem's ability to withstand or recover from impacts.

- Stability refers to a state in which changes over time are within normal bounds of variation. It describes the state of climax vegetation, compared with successional vegetation, or the ability of an ecosystem to resist environmental change. It is normally assisted by **negative feedbacks** operating within the ecosystem: feedbacks that slow down a process. Inertia refers to the delay (time lag) or slowness in the response of a ecosystem to a driver of change, and in its recovery if the cause is removed.
- Conversely, fragility/sensitivity refers to a ecosystem's susceptibility to environmental pressures. A fragile ecosystem is more easily altered and may lose its ability to recover from impacts more easily than a stable ecosystem.
- Resilience can refer to the ability of a ecosystem to absorb disturbance without crossing a threshold to a different (usually degraded) state, in which case it is virtually synonymous with stability. More usually it refers to recoverability, i.e. the ecosystem's ability to return to a pre-disturbance state (and the speed at which it can do so) if the disturbance is removed. In the context of EcIA, this can be particularly important in relation to short-term impacts such as those associated with the construction phase of a project.
- Recreatability refers to the potential for re-establishing an ecosystem of richness and complexity similar to the ecosystem that has been destroyed.

These properties have a bearing on the suitability of different kinds of mitigation. If ecosystems are resilient and likely to recover following perturbations associated with proposed development, no mitigation may be needed. On the other hand, ecosystems that are fragile and lack resilience may have very low capacity to recover, requiring deliberate interventions to restore them to a pre-impact condition. In some cases this may not be technically achievable and irreversible loss is likely inevitable. Distinguishing between these scenarios is an important part of EcIA and requires a good understanding of ecosystem processes.

Similarly, time lags for recovery vary in relation to different drivers. For instance, the impact of overharvesting may be quickly checked or reduced for some species (provided that a threshold has not been exceeded), but much longer time lags apply to recovery from the impacts of drivers such as habitat destruction, nutrient loading, pollution by persistent toxins, severe soil erosion or climate change.

In general, natural and semi-natural ecological systems are more sensitive, less resilient, and less recreatable than highly modified ones. This is important in EcIA because the success of community/habitat restoration as a mitigation method is highly dependent on the complexity of the ecological system prior to development.

6.2.3 Habitats

Several aspects of habitats have a bearing on the ability of the environment to "absorb" impacts associated with a proposed development.

Species composition refers to the numbers and types of species in the landscape. However, meaningful studies require quantitative data on *species abundance*, as well as information describing the dominant species, **keystone** species, and indicator species. Such studies are important in detailed surveys.

Vegetation structure (physiognomy) refers to the physical structure and features of vegetation that influence the behaviour of wildlife populations such as life-form composition (i.e. the types and proportions of plant life that form the physical matrix of vegetation types) and vertical structure (i.e. the stratification of vegetation types). For example, broadleaved woodlands have up to four layers (canopy, under-story/shrub, field, and ground), while heathlands and grasslands rarely have more than two layers of vegetation. Structure is important because it affects the availability of habitat or living space for wildlife populations.

Species richness and species diversity are measures of a community's biodiversity, and are often used in community or site evaluation.

Trophic structure refers to the flows of energy and nutrients through communities and ecosystems (Figure 6.3). A food chain is simply a general route of energy and nutrients and a community's trophic structure consists of a food web, i.e. a network of feeding relationships between species. Some knowledge of food chains and food webs is important for predicting impacts and developing mitigation strategies. This work, however, can be challenging; for example, it took 25 years to fully understand the food web in a small estuary supporting >90 species and nearly 5,500 feeding links (Gorman and Raffaelli 1993).

Community productivity (rate of production) varies widely, largely in relation to environmental temperature, water, and nutrient regimes. For example, tropical rain forests, swamps, estuaries, and beds of marine algae normally have high productivities, while deserts, bogs, and open oceans have low productivities. Highly productive communities have a large **biomass**. Some also have high *species diversities*, although low-productivity ecosystems can be more biologically diverse than many with higher productivities.

Spatial pattern refers to the spatial configuration of communities. While a managed landscape is generally characterised by sharp boundaries, these are mostly man-made; and the spatial pattern of natural communities tends to consist of community gradients rather than discrete entities, with attributes such as species composition adjusting progressively along environmental gradients. Where the environmental gradients are steep, there may be obvious transition zones (ecotones) between adjacent communities, and these are often species-rich because they contain species of adjacent communities. Mosaics of communities may be readily apparent, but less discernible gradients are common, and **semi-natural** vegetation is rarely homogeneous, even within small areas.

Temporal pattern refers to community changes through time. Short-term changes include seasonal variations, intrinsic vegetation cycles (e.g. associated with forest canopy gaps), and environmental perturbations (e.g. fire, storm,

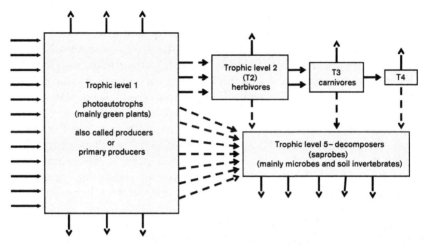

Input of light energy to trophic level 1 (photoautotrophs) by photosynthesis

Transfer of energy (in organic compounds) to higher trophic levels – T2 (herbivores), T3 (carnivores) and T4 (top carnivores) – along the **consumer food chain**

Transfer of energy to trophic level 5 (decomposers) in the form of dead plant and animal remains and animal excretory products. This route is often called the **decomposer food chain**

Loss, from all trophic levels, of energy (mainly heat) generated by respiration.

The sizes of boxes and numbers of arrows indicate the relative amounts of energy entering, leaving and within the various trophic levels – but are not strictly proportional.

Communities need sustained flows of energy and nutrients, and rely on **autotrophs** which synthesise organic compounds using inorganic nutrients and external sources of energy. These are nearly always **photoautotrophs** in which the primary process is photosynthesis of glucose from CO_2 and H_2O using light energy absorbed by chlorophyll. All **heterotrophs** obtain their energy and nutrients from the organic compounds synthesised by autotrophs.

Energy assimilation by photosynthesis is called **primary production** (PP), and the total amount is **gross primary production** (GPP). Plants use c.55% of this; so c.45% (**net primary production** (NPP)) is available for heterotrophs, whose utilization of energy is called **secondary production**. In terrestrial communities, only a small proportion of NPP passes along the consumer food chain – the bulk goes directly to the decomposers as dead plant remains. The decomposers also receive energy in the form of animal remains and excretory products.

All organisms carry out **respiration**, by which organic compounds are broken down to release usable energy (and CO). Much of this energy is lost to the environment as heat; so (a) energy flow through the community must be sustained by PP, and (b) less energy is available to higher trophic levels – which is why there is a **pyramid of decreasing biomass** from trophic levels 1 to 4, and why top carnivore populations are generally small.

Figure 6.3

Simple model of energy and nutrient flow through a terrestrial ecosystem (Morris and Emberton 2009)

flood, drought, and cold). Ecological succession is a progressive process that culminates in the development of a climatic climax community (*biome*) (Figure 6.4).

Precise prediction of succession is difficult because biomes are broad generalisations, within which there is wide variation in relation to local conditions,

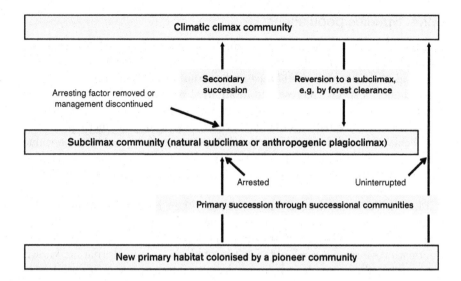

A **primary succession** (or **prisere**) starts from a near-sterile **primary habitat**, e.g. rock (exposed by volcanic activity, glacial retreat, mineral extraction etc.) or new water body (lake, reservoir etc.) which is colonised by a **pioneer community**. This is followed by a series of **successional (seral*) communities** (each replacing the previous one), and ultimately by a **climatic climax community (biome)**. For example, a lake in lowland Britain is likely to be gradually infilled and undergo the following succession:

open water community → swamp → fen/marsh → carr → broadleaved woodland.

Succession can stop at a persistent **subclimax** stage. The arresting factors can be natural, but most 'subclimaxes' (including UK heathlands and grasslands) are semi-natural communities maintained by human activity (including management such as grazing); and because these **anthropogenic climaxes** differ from natural subclimaxes, they are often called **plagioclimaxes**. However, they are much more natural than communities such as 'improved' grasslands. Removal of an arresting factor results in a **secondary succession** (or sub-sere) which can be rapid because features such as soil already exist.

* The terms sere and seral are often used as a synonyms of succession and successional respectively, but a sere is strictly a particular type or example of primary succession. Recognised types include the lithosere (from rock) and the hydrosere (from open water), both of which may eventually culminate in the same climatic climax.

Figure 6.4

Simple model of ecological succession (Morris and Emberton 2009)

and secondary successions are influenced by the "stock" of potential colonisers living in the area and in the soil *seed bank*. In many areas, abandoned land and unmanaged plagioclimax communities will revert (often quite rapidly) to some form of woodland. Under an unchanging climate, climatic climax communities are relatively stable; but they still undergo (usually slow) long-term changes in response to factors such as in-migration, out-migration, evolution and soil development.

6.2.4 Species populations

Species composition describes the numbers and types of species that are represented in a particular area; however, meaningful studies also require information on *species abundance* and measuring not just *species richness* (the number of species represented in an area) but also *species diversity* (the number of species represented and their relative abundance). In addition to overall richness/diversity, individual species may occur that merit particular consideration to understand the conditions required to maintain a viable population. This might be needed for species of conservation concern, or species that are used to indicate environmental change (e.g. a pollution *indicator species*).

Most EcIAs focus primarily on species and their habitat, even though this species-by-species approach can mean inadequate emphasis on the ecological conditions needed to support viable populations. In some EcIAs, the information on species is restricted to lists of plants or animals that are known (perhaps based on a single sighting) or suspected to be present in the vicinity of the development project. This is widely viewed as inadequate because individual sightings provide little insight on the status of the affected population, its *abundance* (which may range from a few individuals to a number of interacting groups), or its distribution in the landscape. "Snapshot" observations also provide little opportunity to address the functionality of the affected ecosystem or identify how it is likely to respond to any change. Meaningful predictions about the future status of a species or population require an understanding of factors that have influenced the population either in the past or in its current condition. This implies the need to define a "dynamic baseline" describing population trends and seasonal or periodic variability over time. Without this, it is not possible to predict likely responses to direct and indirect pressures imposed by an altered landscape or environmental conditions resulting from a project. The viability of a species depends on the presence of a suitable environment with adequate resources, and this involves a set of biotic and abiotic environmental factors that affect population dynamics.

Many species can tolerate short-term environmental variations; and while size, sex, and age-class profiles in a population may fluctuate temporarily, a species often returns to prior stable condition or settles into a new equilibrium supported by the altered environment (Morrison et al. 2012). Species are also capable of responding to slow progressive environmental changes by evolving or changing their geographical ranges; but they may be unable to adjust quickly enough to rapid environmental changes such as those resulting from rapid urbanisation or climate change, especially if their dispersal is inhibited by factors such as habitat fragmentation. Wildlife characterised as specialists (which are adapted to a narrow range of environmental conditions or food sources) are more vulnerable to environmental changes than generalists (which have less specific foraging requirements).

6.3 Key legislation, guidance and standards

Legislation covering EcIA and the protection of biodiversity can be divided into four components:

1   International and multinational agreements and conventions;
2   National and regional legislation;
3   Industry codes and best practice;
4   Special interest groups.

6.3.1 International and multilateral agreements and conventions

Several international conventions and multilateral agreements and conventions affect the assessment and protection of biodiversity and ecosystems, and establish global policy and targets to reduce biodiversity loss. Some, such as the global biodiversity-related conventions, include explicit requirements to consider biodiversity in the planning of development or to include it in ESIA. Others set a broader policy context for reducing rates of biodiversity loss or for conserving it through various protection mechanisms. Several international goals and principles for sustainable development were launched at the Rio Earth Summit in 2002 and have since been refined and advanced at UN World Summits focused on developing international agreements supporting sustainable development and biodiversity. Notable agreements include:

- Convention on Biological Diversity (CBD) 1992 – adopted at the Rio Earth Summit to conserve biodiversity.
- CBD 2010 Biodiversity Target set in 2002 – aimed to achieve significant reductions in the current rate of biodiversity loss by 2010 using ESIA and SEA tools to ensure that biodiversity and its sustainable use are considered in the planning of development.
- CBD Tenth Conference of the Parties (COP 10) Decision x/2 Strategic Plan for Biodiversity 2011–2020 – aimed to achieve several strategic goals to resolve the underlying causes of biodiversity loss; reduce the direct pressures on biodiversity and promote sustainable use; improve the status of biodiversity by safeguarding ecosystems, species and genetic diversity; enhance the benefits to societies from biodiversity and ecosystem services; and enhance implementation through participatory planning, knowledge management and capacity building.
- Bonn Convention on the Conservation of Migratory Species of Wild Animals 1979 – aimed to protect threatened animals that migrate across seas and/or national boundaries using ESIA as a tool for recognising areas needed to maintain viable migration routes.
- Ramsar Convention on Wetlands of International Importance 1971 –

aimed to conserve wetlands of international importance such as at Ramsar sites by using ESIA and SEA tools to consider biodiversity in development planning and ensure wetlands are safeguarded.

- UNESCO Man and the Biosphere Programme 1970 – established Biosphere Reserves to innovate approaches to conservation and sustainable development. These are generally recognised as globally important areas that should be avoided when development is planned.
- UNESCO World Heritage Convention 1972 – aimed to protect natural and cultural areas of outstanding value as World Heritage Sites, including some sites identified because of their biodiversity importance.
- Bern Convention on the Conservation of European Wildlife and Natural Habitats 1979 – aimed to protect endangered species and their habitats, and amended in 1989 and 1996 to include the EMERALD network of Areas of Special Conservation Interest supporting the European Habitats Directive.
- The Pan-European Biological and Landscape Diversity Strategy, developed by the Council of Europe, UNEP and European Centre for Nature Conservation – aims to link European and national protected areas and ecological networks in order to ensure the conservation of Europe's key species, habitats, and ecosystems.

At a European level, several policies and Directives require the establishment of areas to protect biodiversity and planning of development to ensure that threatened species can be maintained at a favourable conservation status. The principal European policies include:

- The EU Biodiversity Strategy (COM(98)42) – which requires the European Commission to produce biodiversity action plans.
- Wild Birds Directive 1979/409/EEC – which aims to protect wild bird species and their habitats, with particular protection of rare species (in Annex I) in Special Protection Areas (SPAs).
- Habitats Directive 1992/43/EEC (and amendments) – which aims to protect habitats and species using measures to maintain or restore their "favourable conservation status", principally by Special Areas of Conservation (SACs) but also through land-use and development policies and landscape management outside SACs; and to safeguard species needing strict protection.
- Water Framework Directive 2000/60/EC – which requires management of aquatic, wetland and terrestrial ecosystems with intent to achieve "good status" by 2015 or, at the latest, by 2027.

6.3.2 National and regional legislation

Most countries have passed national-level ESIA legislation requiring the consideration of ecology and biodiversity in ESIA. In addition, the protection, conservation and management of wildlife within each country is enacted under national legislation, and this will affect the scope of an EcIA and the assessment of effect-significance. National legislation either enacts multilateral agreements, or provides dedicated legislation on biodiversity. An example of the former is the UK Conservation of Habitats and Species Regulations 2010, which follows from the EC Directive on Conservation of Natural Habitats and of Wild Fauna and Flora (EC 92/43/EEC).

National legislation on biodiversity and ecology is either comprehensive or specific. Comprehensive legislation contains measures to protect habitats, species and areas under a single overarching Act. Examples of this include the Wildlife Conservation and Management Act (No:47 of 2013) in Kenya, the Indian Wildlife Protection Act of 1972 (amended in 2006) and the UK Wildlife and Countryside Act 985 (as amended). Specific legislation relates to protection of specific areas, habitats or species such as:

- *Areas*. The most widely used example of this is National Park legislation. Examples of this include the 1980 New Zealand National Parks Act and the 2003 South Africa Protected Areas Act. Areas of international conservation concern such as the Galapagos also have national area protection.
- *Species*. These regulations usually supplement overarching national legislation by adding protection to individual species. An example of this is the 1992 UK Protection of Badgers Act.
- *Habitats/land use*. These regulations relate to specific habitats or land uses but have implications on ecological conservation. Examples include the forest laws of Brazil and India.

Under the terms of the 1993 Convention on Biological Diversity and COP10 of 2010, signatory countries to the agreements must have a national strategic plan for biodiversity for the period 2011–2020 that will achieve the Aichi targets, which are important to consider in the context of EcIA and the assessment of the potential for a development to affect biodiversity.

In countries that adopt a national and regional (state) system, wildlife legislation can occur both under national and regional legislation. An example of this is in Australia, where states such as New South Wales have enacted regional National Park legislation. However, national legislation affecting ecology can be complex as well as comprehensive. A good example of this is Uganda. Whilst primary legislation on ecological protection is under the National Environmental Policy (1994) and the National Wildlife Act (2000), protection of ecology is also affected by: the 1997 Local Government Act, the 2003 National Forestry and Tree Planting Act, the 1995 Wetland Act, the 2008 Oil and Gas Policy for Uganda, the 2000 Fish Act Cap 197, the Animal (Prevention of Cruelty) Act Cap 220, the 2000 Cattle Grazing Act Cap 227, the Plant Protection Act, the 2000 Prohibition of Burning of Grass Act Cap 33, and the 2000 Animal Diseases Act Cap 218 amongst others (Ministry of Tourism, Wildlife and Antiquities 2014). It is important that experts involved in a scoping EcIA have a full understanding of the legislation affecting the protection of wildlife in the subject country.

6.3.3 Codes of practice and best practice

There are numerous ecological codes of practice and best-practice guidance such as the EcIA guidance of the UK Institute of Institute of Ecology and Environmental Management (IEEM 2006). These define the state of best practice for each nation, and where available should be used as the basis for EcIAs undertaken in that nation. Where necessary, these can be supplemented with other, updated, methodologies and specific methods for individual species and habitats.

Internationally, the key Codes of Practice are the International Finance Corporation (part of the World Bank) Environmental and Sustainability Performance Standards (IFC 2012a), and the methods issued by the international development banks such as the European Bank for Reconstruction and Development (EBRD), African Development Bank (AfDB), Asian Development Bank (ADB) and Inter-American Development Bank (IADB). Bilateral funding agencies such as the Danish International Development Agency (Danadia), United States Agency for International Development (USAID) and Japan International Cooperation Agency (JICA) also have requirements related to projects financed under bilateral loans, and these will contain requirements for ecological and environmental studies associated with the loan process.

The International Finance Corporation (IFC) Performance Standards (2012a,b) are used extensively in assessing the potential impacts of projects promoted under international financing agreements, and are increasingly regarded as a benchmark for good practice. Of the Performance Standards, PS1 (Assessment and Management of Environmental and Social Risks and Impacts) and PS6 (Biodiversity and Sustainable Management of Living Natural Resources) are most related to EcIA. PS1 establishes the need to identify and assess potential environmental and social impacts that may occur, to adopt a hierarchy of mitigation to minimise impacts, promote improved environmental and social performance, and undertake adequate engagement with affected peoples. In Paragraph 7, PS1 identifies the need to "consider relevant environmental and social risks" and in Paragraph 8 observes that impacts "will be identified in the context of the projects area of influence", and notes that the assessment should include "(iii) Indirect project impacts on biodiversity or on ecosystem services upon which affected communities depend".

PS6 aims to protect biodiversity, maintain the benefits from ecosystem services and "promote the sustainable management of living natural resources". Where applicable, PS6 requires the client (i.e. the borrower) to adopt and maintain an Environmental and Social Management System to meet the objective mentioned above. PS6 introduces the mitigation hierarchy, and introduces the classification of habitats under the titles "modified land", natural habitat", "critical habitat" and "legally protected and internationally recognised areas". It also contains guidance on invasive alien species.

In most countries, professional institutes and societies, special interest groups and academic organisations provide a range of best-practice approaches to undertaking surveys and assessments of species and habitats, and for their classification. Where possible, best-practice approaches should be adopted in EcIA as they are often accepted by the consenting bodies and those statutory organisations advising them, and often have been tested in public. So long as the approach is scientifically robust and defendable, the use of best-practice approaches is of great advantage in EcIA as it reduces the chances of objection from the consenting organisation and other stakeholders.

6.3.4 Special interest groups

It is good practice to invite review and obtain agreement on EcIA methodologies with the consenting authority (usually a planning authority, ministry of state or delegated authority) and the relevant statutory consultee (statutory nature conservation organisation) who will review the work once complete. The statutory consultee has several important roles and requirements, which typically include:

- It must be notified by the consenting authority about a development application and will assist in the screening and scoping procedures.
- It may hold and provide non-confidential information on local ecology (and sometimes copies of previous ESIA reports undertaken in the local area). This may be in the form of a biological or environmental records office.
- It will employ (or have contacts with) experienced local ecologists, who may give advice on all aspects of the EcIA.
- It must be supplied by the consenting authority with a copy of the ESIA report for comment. In turn, it may provide (a) an appraisal of the ESIA report in terms of its scope, technical competence, validity, and proposed mitigation measures; and (b) an indication of whether it would support or oppose planning consent.

Other relevant government organisations could include the relevant environmental protection agency, especially when there are concerns relating to pollution, contaminated land, freshwater ecosystems, or coastal ecosystems. In most countries, non-governmental organisations often hold significant, relevant, and recent data on the ecology of sites and their surroundings. These can include:

- Local wildlife trusts/groups such as the Naturalist Trusts in the UK, Kenya Wildlife Trust, Wildlife Trust for India, South African Wildlife Trust,

Australian Wildlife Conservancy, Japan Wildlife Conservation Society, and the US Wildlife Habitat Council.
- Special interest and conservation groups for taxa such as bats, butterflies, birds, reptiles, and amphibians. This includes groups such as the Royal Society for the Protection of Birds in the UK and the Audobon Society in the US;
- Local museums and data centres;
- Non-professional special interest groups such as local natural history or wildlife groups or cultural groups.

6.4 Defining the baseline – scoping

The importance of defining an accurate, robust, and comprehensive ecological baseline cannot be overstated. Accurate identification and assessment of impacts can only be undertaken with good knowledge of the ecology of a development site and its interaction with the surrounding area. In screening, the presence of designated ecological sites and particular habitats is used to define the sensitivity of the development site, and contributes to the decision on whether an ESIA is required. For development projects subject to Equator Principles and related International Finance Corporation or regional development bank requirements, an initial project scoping phase must consider the presence of sensitive and highly valued wildlife and natural environments (Tucker et al. 2012, Adeyemi 2014).

Scoping the ESIA (and hence the EcIA) is a critical element in the ESIA process. If undertaken incorrectly, it is expensive and can cause significant delay to projects to rectify, especially where ecological systems are involved due to their seasonal nature. Scoping is more than a simple exercise to define what should be included in the ESIA: it is the opportunity for all relevant parties to agree the study area to be used, survey methodologies, assessment methodologies (especially in the definition of levels of significance) and the key ecological issues to address, as well as those issues that can be put aside and not addressed further. As such, scoping should be undertaken as early in the project as possible; it should be undertaken by experienced, professional ecologists; and include full consultation with both the consenting organisation and statutory nature bodies. In addition, international best practice on scoping recommends the involvement of non-statutory nature organisations, not only for data collection but also to identify aspirations for the development of ecological services in the area of the development site.

The study area (or "impact area" or "project-affected area") is the area over which discernible effects, or impacts (both beneficial and adverse) can be identified on local ecological populations and habitats, and the functioning interfaces between them. It is very rare for this to be confined to the project site itself, and therefore it is important to agree with the consenting authority the study area to be used for the assessment of terrestrial ecology. In the US, the spatial extent or the zone of influence of the potential impact must be specified and addressed in the ESIA (Nash 2014). The work must address impacts that may be site-specific or limited to the project area; locally occurring within the locality of the proposed project; regional and extending beyond the local area; and/or national, and affecting resources on a wide scale. If the spatial extent is believed to have transboundary influences, then the ESIA is also expected to include international considerations.

A new approach in the US for transportation projects that merges impact area and mitigation considerations at the outset is gaining acceptance as an approach useful to focus limited assessment resources (Thorne et al. 2014). Similarly, IEEM (2006) recommend that, rather than stipulating any specific radius from the project site, an assessment should be made of all potential receptors within an estimated zone of influence (or impact area). Using this approach, the impact area could be large, especially where the impact pathway is extensive such as those transmitted in air or water, or where migratory species are involved. IEEM (2006) acknowledges that estimating the spatial extent of the study area may involve professional judgements, which "should be continually reviewed and, if appropriate, amended as the scheme evolves".

Some international practices are more prescriptive, requiring assessment within a fixed width corridor or buffer. For instance, in the UK a minimum 2 km radius from the site boundary is normally considered for non-linear projects; and a corridor at least 1 km wide should be examined along the proposed route of linear transport projects such as roads. Regardless of how the scale of the impact area is determined, either based on prescriptive regulations or site-specific professional judgement, the spatial extent of the impact area will affect the results and accuracy of ESIA studies. Therefore, it is important to define in explicit terms the scale used in ESIA work (João 2002).

The amount of ecological information that could be collected is potentially enormous. Morris and Emberton (2009) advise that it is essential to be proportionate, and to focus resources on the important components of terrestrial ecosystems during scoping and detailed EcIA work. This can be accomplished by identifying **valued ecosystem components** (VECs). These can be species, habitats, or sites that qualify in terms of their ecological/conservation value, or other attributes such as socio-economic value, for instance where ecosystem services are a consideration (Treweek 1999, Tucker 2005, Crossman et al. 2013). It is also important to focus on VECs that are receptors, since there is no point in using resources to study species or habitats that will not be impacted (Wathern 1999).

The most obvious VECs are protected species, habitats, and country-specific priority species/habitats. However, the scoping inventory should include all receptors that may warrant further investigation. Even small areas of habitat and non-optimal habitats can be important to the wider wildlife network. Of particular note are **linear habitats** which, while valuable in their own right, may also act as refuges, stepping stones, **wildlife corridors** or **buffer zones** within an urban or intensively cultivated agricultural landscape. By providing refuges for wildlife, allowing small populations to remain viable and facilitating migration in an otherwise low-diversity area, even small sites with modified communities that are individually of low conservation value, can have a significant role in a "green network" that permeates the area. Ultimately, the baseline survey (and evaluation of baseline conditions) should include all ecological receptors that qualify as VECs.

The scoping stage should also achieve agreements with the consenting authority on the scope of the work that ecologists will undertake in the EcIA, including:

- The methods to be used for surveys for each habitat and species under investigation;
- The number, timing, and location of each survey to be undertaken during the EcIA;
- The methods to be used to assess ecological importance, impact magnitude and impact significance. This may differ for each species and habitat, and also where species are permanently resident or migratory;
- Reporting requirements.

6.5 Defining the baseline – desk studies and surveys

6.5.1 Methods and levels of study

In selecting study methods and levels there is a strong case for adopting a phased strategy for baseline surveys. Preliminary surveys or reviews might identify the main ecosystems, habitats, and species populations that may be affected, as well as providing information on the site context and its connectivity to the wider environment. Subsequent, more detailed surveys may then be carried out, based on the initial findings.

Resource and time constraints often impose severe limitations on the range and depth of field survey work that can be conducted, and it is important to make maximum use of existing information by means of the initial desk study. The desk study should collect as much existing information as possible from third parties. Table 6.1 lists the types of organisation from which data can be sought. Some non-government organisations may be particularly useful sources of local information, but their limited resources may restrict their ability to respond to enquiries within tight timeframes.

While desk study is essential, much of the existing information may be sketchy, out of date, or inaccurate. A detailed review of the robustness of available data is necessary before it can be used in the EcIA (sometimes referred to as a gap analysis) and this should always be undertaken by an experienced ecologist.

Table 6.1 Types of organisation holding suitable ecological data

Ecological and natural history data centres, local history centres
Internet datasets, such as the Global Biodiversity Information Facility, Integrated Biodiversity Assessment Tool, IUCN Red List of Threatened Species, national red lists etc.
National natural history, nature conservation, ecology, or environmental authorities
Local museums and libraries
Academic institutes and universities
Species or habitat professional societies and institutes
National park authorities
Regional administrations and local authorities
Local leaders, land owners, customary land users, and village elders
Public interest groups and societies, natural history charities
NGOs: international, national, and local

6.5.2 Resource requirements and timing

The resources needed for EcIA vary in relation to factors such as the availability of existing information and the need for detailed surveys. Some mapping will be essential, and consideration should be given to the use of GIS (geographic information system), which will facilitate integration of ecological information with other environmental and social considerations in the ESIA.

Ecological expertise is the "resource" of prime importance. It is essential to employ an appropriate number of competent ecologists because scoping and surveys typically require professionals with appropriate licences and certifications to ensure accurate species identification and the application of suitable sampling and data analysis methods. Ecologists involved in work required by financial institutions such as the World Bank, regional development banks, and IFC must also demonstrate an ability to meet certain obligations such as advising the project proponent and other stakeholders on the terms of reference; the key issues and methods for preparing the EcIA; recommendations and findings; implementation of recommendations; and environmental monitoring and management capacity during project implementation (World Bank 2013).

Under a seasonal climate, or where dealing with migratory species, the timing of fieldwork can be a critical factor because it is difficult or impossible to sample many species or communities during much of the year. Similarly, detailed surveys for most species can only obtain appropriate and reliable data during short sampling seasons. Visible constituents of vegetation may vary seasonally, with some plants being inconspicuous or absent for parts of the year – so failure to carry out surveys in all seasons can often lead to error.

Provision of sufficient surveying time and resources to accommodate these seasonality constraints requires careful planning. Detailed field surveys can be started as soon as the need becomes apparent during scoping or the initial survey; however, decisions on their nature and extent are best made after evaluation of the initial survey findings. Ideally, as a rule of thumb, EcIA should start at least one year before the submission date of the ESIA report. Where developers' timescales are inconsistent with this requirement, the developers and their agents must be made aware of the potential consequences. These include conducting surveys "out of season" or during inappropriate weather. If initial surveys are not started until late in the survey season, then surveys for *notable* species may not be possible that year, even if appropriate habitats have been identified.

Such failings are often accompanied by the promise to conduct adequate surveys after development consent or planning permission is granted. This should be avoided due to the possibility that significant risks may not be recognised in time for appropriate mitigation action to be taken. Such an approach also means that the relevant decision-making authorities may lack the information needed to make an informed decision. In some countries, there may be specific requirements and procedures to follow for protected or threatened species, and these can cause significant project delays. In the UK, for example, Circular 06/05 (ODPM 2005) states that all surveys need to be complete and actions for conservation or mitigation in place before planning permission can be granted. Similar requirements are found in several international jurisdictions addressing marine resource development (Warner 2013a, 2014), polar regions (Warner 2013b), and infrastructure projects. As a general rule, EcIA efficiency is improved by "front-loading" of baseline survey or review so that all potential risks are recognised early on.

6.5.3 Initial baseline surveys

The species and habitats present in each region of the world differ, and as a result, national requirements for ecological surveying have developed which are often specific to a country or region. Therefore, the use of local ecological specialists who have experience in the habitats and species potentially present and of the relevant legislation and survey requirements for the subject country is essential. Consistent with IEEM (2007), the aims of the initial survey should be to identify:

- The main ecosystems, habitats and species populations that may be affected and make an initial assessment of their inter-dependencies;
- The potential presence of rare, notable or protected species (flora and fauna);
- Alongside the information collected in the desk study, the potential role of affected features and processes in a wider (landscape) context;
- Important ecosystem services and the different attributes (e.g. biological, chemical and physical materials and energy flows) that support ecosystems;
- Gaps in knowledge and an identification of the need for further work.

The initial survey procedure usually involves ecologists field-walking and undertaking a visual assessment of the development site, and recording and mapping areas occupied by the habitat or vegetation types using a relevant classification system, e.g. the UK Joint Nature Conservation Committee (JNCC) habitat classification, CFG survey monitoring protocols in the state of California (CFG 2009), or the vegetation classification system in the US (Jennings et al. 2009).

Initial spatial mapping of vegetation or habitat types might be done using several different methods (IEEM 2007). The development of hand-held electronic devices and GPS, and the availability of satellite imagery and aerial photography is greatly improving the precision of mapping habitats and other features on-site. The results can then be uploaded into digital form (for instance on GIS) for analysis back in the office. Digital site mapping is particularly useful for large projects (especially linear infrastructure projects such as pipelines, power lines, roads and rail).

There are occasions, however, when more traditional methods might be needed such as the use of the line-intercept method, whereby parallel lines drawn on the maps or aerial photographs are used by ecologists in the field to estimate the spatial distance occupied by each habitat type along a survey transect. This approach provides a measure of the proportion (percentage) of the site covered by each habitat type.

Although initial surveys are likely to focus on habitats, it is also desirable to record other observations such as signs of species-presence that are noticed. These "opportunistic" or "incidental" observations can assist in the identification of more detailed surveys that may be required. The presence of tracks, signs of feeding (bones/carcasses), resting or occupation areas (nests, holes, setts, holts etc.), runnels/tunnels, hairs and feathers, scratches/marks on vegetation and in the soil, and scat/dung/faeces can be important to verifying the presence of wildlife. In addition, naturalist field guides are available in many countries to assist in the survey process (e.g. Richter and Freegard 2009, Liebenberg 1990). Where possible, information on relative abundance should be obtained, not just on species-presence/absence. For example, relative dominance of plants within an area might be described using DAFOR ratings (see Table 6.2) or percentage-cover of the different species within a vegetation type might be recorded.

Where access allows, initial field surveys should cover as wide an area as possible around a proposed development site. The ecologist needs to be able to verify data collected from aerial and satellite photography and third parties, and also to place site-specific information in a wider ecological context. Where the initial survey is used to verify satellite data, this is termed "ground truthing" and for very large projects (e.g. linear projects extending over many miles) can be used to target further initial and detailed surveys.

Where biodiversity on the current site is very poor and the potential for the presence of notable species is low, the initial survey may be sufficient to provide all of the detail necessary for the application for planning/permitting purposes. However, in most cases, additional, more detailed surveys will be required. In

Table 6.2 Species abundance measures

Measure	Description
Semi-quantitative abundance ratings	These are visually estimated using systems such as DAFOR in which: D = dominant; A = abundant; F = frequent; O = occasional; R = rare; with the prefix I (locally) added to any category if required. They are quick to record, but are subjective, approximate, and have reduced potential for analysis and presentation. Consequently, they are generally more suited to initial rather than detailed studies.
Number of individuals	This is a suitable measure for species which have readily discernible individuals that can be counted. It is not usually applicable in community studies because it has little meaning when comparing species of widely differing size. When measured in defined areas, numbers can be expressed as density (number per unit area) and/or as population size (in the study area). There are two counting measures: • Direct counting is only generally valid for plants, near-sedentary animals or small populations of animals within defined areas. Occasionally, whole populations (e.g. of trees or nesting birds) can be counted in small areas. More usually, population estimates are derived from samples, e.g. in quadrats or by plotless sampling. • Indirect counting methods can provide estimates of fairly small populations. Capture-mark-recapture methods (see Krebs 1998, Hill et al. 2005, Sutherland 2006) involve capturing and marking a number of individuals, releasing them, and re-sampling after a suitable time interval. Formulae are used to derive the population estimate from the proportion of marked individuals in the recapture sample. This concept is regularly used for migratory species as a time-based comparative measure.
Cover (usually as a %)	The percentage of ground occupied by the over ground parts of a species. It is usually measured in quadrats by visual estimation along a vertical projection below (and if necessary above) the observer; but there are alternative methods such as the line intercept method of "point quadrating" (see Kent and Coker 1992, Hill et al. 2005, Sutherland 2006). It is suitable for studies of communities which include species of differing size. Visual estimates are prone to observer error (accuracy greater than the nearest 5% is not feasible) and species present as small scattered individuals tend to be under-estimated. The use of cover-abundance scales such as Domin and Braun-Blanquet aim to minimise these errors by grouping % cover values in designated bands and assessing abundance (in the strict sense of numbers) for cover values of less than 4%.

Table 6.2 continued

Measure	Description
Frequency (usually as a %)	The percentage of observations in a sample that contain the species, derived from presence/absence observations, e.g. in quadrats. Limitations are: (a) it is strictly a measure of distribution rather than abundance and does not discriminate between high density and density that is just sufficient for a species to be present in a large proportion of quadrants; (b) it tends to over-represent small species; (c) it increases in value with increasing quadrat size, so results using different-sized quadrats are not strictly comparable, and it is best obtained from a large number of observations using small quadrats. However, frequency can be a cost-effective method for obtaining large representative samples of communities because it is relatively rapid and free from observer errors.

these cases, the initial survey, when allied to the data collected in the desk study, will allow the need and scope of such surveys to be defined, and agreed with the relevant consenting organisation/authorities.

6.6 Detailed surveys and evaluation of baseline conditions

6.6.1 Introduction and sampling options

The purpose of detailed ecological surveys is to supplement the findings of any preliminary surveys and desk studies, which often lack quantitative data. Detailed surveys focus on collecting sufficient quantitative information for the area of influence of a proposed project to generate a robust baseline. These studies need to cover a sufficient spatial scope to understand direct and indirect impacts, and may need to cover more than one survey season. Such studies can be expensive and time-consuming. Therefore, it is important to focus on carefully selected priority objectives, ensure that work on different aspects is coordinated, and that the findings are integrated and clearly presented.

The guidance note accompanying the IFC Performance Standard 6 (IFC 2012b) discusses the need for and scope of baseline surveys:

> For sites with potentially significant impacts on natural and critical habitats and ecosystem services, the baseline should include in-field surveys over multiple seasons and be conducted by competent professionals and external experts, as necessary. In-field surveys and assessments should be recent and data should be acquired for the actual site of the

project's facilities, including related and associated facilities, and the project's area of influence.

It also notes that "in-field" (i.e. detailed) surveys may be required outside the site boundary. These are important general principles for designing appropriate baseline surveys and assessments. Another important consideration is the framework for evaluating impact significance. Baseline assessments are more useful if they can underpin post-ESIA monitoring and provide information on biodiversity receptors that may be used as indicators in the longer term.

Importantly, the IFC guidance (IFC 2012b) recognises the vast range of habitats and species that may be present, along with numerous environmental factors that may affect them and, therefore, provides general advice on the approach to baseline surveys rather than specific advice on methodologies. Recommendations on survey and assessment of biodiversity exist in most nations, prepared by government departments and agencies, professional institutes or academia.

Survey methods for coastal and marine systems are discussed in Chapter 7. However, the interface between terrestrial, wetland, freshwater, coastal and marine systems is dynamic, and so EcIA must effectively cover all ecological habitats in an integrated fashion. Whether surveying individual species or whole habitats, it is important to remember the following:

- Species-conservation value should be checked in a local, national and international sense, especially where the species in question is migratory, or under significant pressure in other parts of its biogeographic range. Many species (and some habitats) are legally protected under national or international law and this needs to be identified as it may impose restrictions on surveying, e.g. any activity likely to involve handling or disturbance may require a licence from the relevant authority.
- Species distributions and habitat requirements should be checked and linked to other ESIA chapters, e.g. soil and water.
- Periods during which species can be readily observed and/or identified can be seasonally restricted, and can vary widely between species living in the same habitat. Consequently, unless sites are visited at appropriate times, many species may be underestimated or even missed. In many cases in temperate habitats, weather conditions are also important.
- Where possible, quantitative data such as species abundances and dominance should be obtained. The main options for estimating abundances are outlined in Table 6.2.

Care should be taken to ensure that selected sampling methods are appropriate and will provide results that are compatible with proposed data analysis procedures. Data collection methods include the following:

- Plot sampling involves taking observations, usually for plant species, within defined plots, usually using **quadrats**.

- Plotless sampling is any method in which sampling is not conducted within defined areas. Simple methods include transect walking and the use of line-intercept methods to record all plant species and their abundance along a transect line. This can be applied as a habitat-measurement method for estimation of vegetation. Most other plotless methods involve distance measurements from sampling points. These include methods for estimating: tree densities in woodlands (see Kent and Coker 1992); and animal populations (see Buckland et al. 2001). Some plotless techniques, e.g. for bats, involve a combination of habitat and activity surveys.
- Specialised collecting equipment is often needed in faunal sampling.

The choice of spatial sampling patterns is an important aspect, as this often defines the success of the detailed survey. It involves the questions of where to sample and what pattern of sampling locations is appropriate. The main options are illustrated in Figure 6.4. At the planning stage, the survey methodology must specify how data will be collected, either systematically or randomly. Temporal sampling patterns must also be determined, and can be related to aspects such as the selection of sampling intervals during long-term monitoring; but the commonest reason for considering timing in EcIA is seasonal constraints. Sample size is another critical consideration because data obtained from small samples are generally unreliable and cannot be "improved" by the application of sophisticated analytical procedures. For instance, there is little chance that a few randomly or subjectively placed quadrats will provide representative data for a site. There is no completely objective way of determining the minimum requirement, and the number of observations taken is usually a compromise between the need for precision and the cost in terms of labour and time (Krebs 1998). A percentage-of-area target is sometimes applied in vegetation surveys, e.g. to sample 5 per cent of a study area (Mueller-Dombois and Ellenberg 1974). However, this is rarely achieved, especially on large sites. Sparks et al. (1997) and Greig-Smith (1983) emphasise that sample accuracy is more dependent on the number of observations taken.

6.6.2 Plant surveys

In general, **vascular plants** are relatively easy to sample during appropriate sampling seasons, assuming the necessary taxonomic expertise is available. Consequently, it should be possible to record the majority, or all species, and where necessary to conduct a vegetation survey by measuring their location and abundances. It is recommended that other environmental data are also collected such as aspect, soil depth, and soil chemistry, as these may affect species distribution over relatively small areas.

Sampling bryophytes (mosses and liverworts), lichens, and fungi can be more problematical because many species are inconspicuous and/or difficult to identify. However, they are often important components of communities, and bryophytes and lichens should be included in vegetation surveys wherever

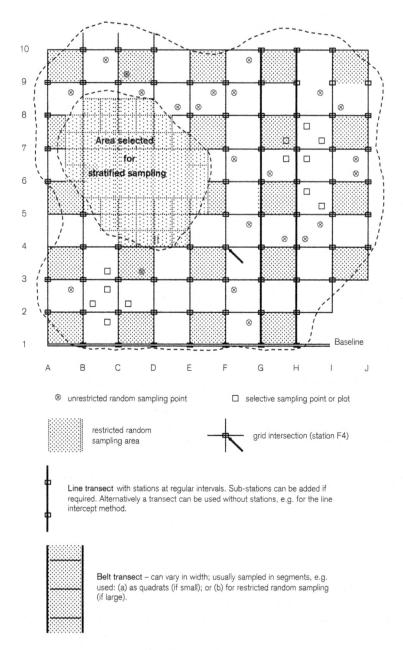

⊗ unrestricted random sampling point   ☐ selective sampling point or plot

restricted random sampling area

grid intersection (station F4)

Line transect with stations at regular intervals. Sub-stations can be added if required. Alternatively a transect can be used without stations, e.g. for the line intercept method.

Belt transect – can vary in width; usually sampled in segments, e.g. used: (a) as quadrats (if small); or (b) for restricted random sampling (if large).

The study area can be a site, a wider area, or a within-site area such as a habitat patch selected for stratified sampling. Similarly, the size of grid square can vary widely, e.g. from 100 m² squares to hectads.

Figure 6.5

Spatial sampling pattern options in relation to hypothetical study areas (Morris and Emberton 2009)

possible. In addition, lichens in particular can be useful indicator species for environmental pollution and the status of a plant community. Freshwater *macrophytes* and *algae* (including phytoplankton) can also pose sampling problems, but should not be ignored and also form the basis for indicator-based methods for assessing quality of freshwater habitats.

6.6.3 Animal surveys

Animal surveys can be difficult, time-consuming, and require careful planning if successful surveying (and monitoring) is to be achieved. Survey methodologies for the same species may vary between countries, especially for migratory species. Animal surveys should use methodologies accepted in the subject country.

Six main issues need to be taken into account when designing detailed animal surveys:

1 While the number of vertebrate species on a site may be small, invertebrate species are usually numerous, especially in ecosystems such as tropical rainforest (current estimates indicate that over half the world's known species, over 1 million species, are invertebrates).

2 Surveys of different taxa often require very different methods and associated expertise in sampling and identification, especially of invertebrates.

3 Many animals are inconspicuous, fugitive, or nocturnal. During parts of the year they may be absent (migratory species). Many hibernate or have inaccessible life cycle stages (e.g. most invertebrates).

4 While some species may only need a particular habitat patch, others may utilise different parts of a site (or wider area) for different purposes such as roosting/shelter, feeding, and breeding.

5 Many animals are very mobile, and periodically move between habitat patches, sites, or wider areas. Consequently, a species present at a given time may be a casual or regular (e.g. seasonal) visitor; but even transitory migrants are dependent on the site, especially if it is on a regular migration route.

6 Determination of distribution and abundance can be difficult, time-consuming and imprecise; and most species vary in abundance from year to year at any site.

These issues inevitably impose limitations on what can be achieved in detailed field surveys, especially when time is limited by developer timescales. Consequently, surveys must be carefully targeted on key and feasible objectives. For example, it is essential to focus on notable species, and on the importance (or potential importance) of a site for these species and for animal taxa in general. It may also be important to consider aspects such as potential indirect impacts that might ensue from impacts on particular animal species, especially keystone species. In any case, animal surveys are likely to be restricted to partial species lists (certainly of invertebrates) and limited quantitative data. Distribution and abundance data may be vital in assessing a species' dependence on a site and the likely viability of the population in the face of impacts; but fully quantitative studies are rarely feasible in EcIA. Similarly, animal species richness/diversity estimates are inevitably limited to partial community data, and often to high-profile and/or easily identified taxa such as butterflies.

Fish

Freshwater fish are often important indicators of ecosystem integrity, are of economic and social value, and can be of great interest to anglers and the public in some cultures. In some countries, fish species can be relatively localised, e.g. to a particular lake or set of lakes (Lévêque 1997). In other countries, native species are more widespread and common, and so can be of lower nature-conservation value (Morris and Emberton 2009). However, even in these situations, some species can be of importance, especially where they are listed as contributory species to habitat protection. In addition, data on migratory species, such as salmonids, may be important for the assessment of potential impacts, and specific surveys and measures may be undertaken to maintain populations of the few rare species where these are present (e.g. Australian Government 2011).

Fish can be surveyed by a variety of methods (Government of Australia 2013, NIWA 2015, Portt et al. 2006, Joy et al. 2013, Natural England 2014, Côté and Perrow 2006, Giles et al. 2005). Active survey techniques include electrofishing from backpack, boat, or bankside; netting, using devices such as seine nets, or scoop and dip nets; fish traps on fish passes; enclosure traps; angling; and visual census using diving, boats or remote control underwater vehicles. Passive techniques include gillnets and fyke nets.

Amphibians

Amphibians utilise both terrestrial and freshwater or brackish habitats, and divide their time between them. Surveying is usually easier when presence within one compartment of a habitat can be predicted, and the individuals come together such as at breeding time. Therefore, in temperate regions, surveys are usually undertaken at breeding sites (ponds, swamps, or slow-flowing rivers) during the breeding season. This varies between species and location. Juveniles and some adults remain in or near water during the summer, so summer surveys of ponds and surrounding areas can provide additional data.

The main methods used are: (a) pond netting for individuals in the water; (b) "torching" at night; (c) bottle trapping; and (d) searches for frog and toad egg masses during the breeding season. Using a combination of survey methods generally proves more effective than one alone, e.g. searches for egg masses in spring, followed by summer netting for juveniles and any remaining adults. However, these methods cannot give more than a crude idea of population numbers, and collecting accurate population data can be time-consuming and expensive. The most frequently used method involves ring-fencing the breeding site to intercept animals moving to or from the surrounding area. Further details about amphibian survey methods can be found in Halliday (2006) and Latham et al. (2005).

In tropical areas, amphibians may use a wide range of habitats from leaf litter to the canopy, and some tropical frogs use small areas of moisture within plant structures for the growth of young. In such circumstances, surveys have to be undertaken using direct observation on a grid or transect approach.

Reptiles

Reptiles should be considered in EcIAs whenever a property is determined likely to contain suitable habitat, which generally is almost always the case (Blomberg and Shine 2006). In temperate areas, surveying is seasonally restricted and concentrates on the use of artificial refuges, which are checked early in the morning over a set period of days. Such refuges are constructed of black mats of material. In warmer climates, indirect methods of trapping, for instance using pitfall traps for smaller species, or direct observation involving transect walking and/or arrays of artificial refuges are most often used. For larger species, direct observation and capture-mark-release techniques can be used to estimate species-presence and population density (Blomberg and Shine 2006, EN 2004b).

Birds

Birds figure to some extent in the majority of EcIAs. One of the main aims of a bird survey should be to evaluate sites and habitats (including small and linear habitats) for birds in general, and for notable species in particular, bearing in mind that habitats may be utilised for various purposes and that most species move between them. They should also aim to record the presence of species and where possible estimate the sizes of populations and their vulnerability to potential impacts. Bird surveys include breeding surveys and transient surveys. The latter can be used to detect the use of sites by birds at different times of the day (for instance for feeding, loafing, or roosting), or for migratory species using the site and its surroundings for part of the year.

Bird surveys require expertise in both visual identification and the recognition of bird calls/song. Surveys are typically time-consuming and require repeated surveys, which may be restricted by season or affected by weather conditions (birds may be less active and conspicuous in wet and windy weather). Guidance on bird census techniques is provided in Sutherland et al. (2004), Bibby et al. (2000), Gibbons and Gregory (2006), and Mustoe et al. (2005).

On land, the most suitable bird survey method is likely to be transect walking, which can be used to estimate breeding territories and densities. However, its value may be limited in small and/or heterogeneous sites when the point-count method (using randomly located observation points) may be more appropriate. Other options include flight-line surveys, radio tracking, and collision-mortality monitoring.

Regional, national, and supranational population data are available for many bird species; so if the local populations can be quantified, both they and the sites that support them can also be evaluated in terms of their representation. Significant overwintering populations of wildfowl and waders are monitored by some national authorities such as the US Fish and Wildlife Service, and by professional organisations bodies on a seasonal or annual basis.

Mammals

Mammal surveys are likely to focus on notable species, some of which may be protected, and for some of which a surveyor must be licensed. Mammals can be divided into three groups – bats, small mammals, and larger mammals – each requiring different survey techniques.

For bats, survey techniques comprise direct observation of individuals and roosts, or the use of ultrasonic detectors for those species using echolocation (Mitchell-Jones and McLeish 2004). Direct observation methods include the observation of roosts (potentially using endoscopes or photographic cameras where roosts are in small cavities), of foraging bats using flight pathways, or of bats emerging from roost structures. Roosts may be found in places such as buildings, trees, caves, mines, and tunnels. They may be detected by the presence of staining, droppings, and insect remains, although it may be necessary to confirm their presence by an ultrasonic or visual search. Flying bats may be observed visually and by ultrasonic detectors, from fixed points or along transects.

Ultrasonic detectors are useful tools for both identifying the species present (using the frequency of ultrasonic emissions by each species) and the activity present. Ultrasonic detectors are usually used in both emergence surveys (when bats emerge or return to their roosts) and along foraging routes/flyways.

Survey methods for small mammals in habitats such as grasslands and hedgerows are relatively easy. However, more complex ecosystems such as tropical rainforest, pose additional issues with species active on both the forest floor and the canopy. The techniques used have to reflect the type of habitat occupied by the species, the trap type and height, and restrictions on access for sample recovery. According to Krebs (2006), possible techniques include:

- Indirect methods such as the use of hair tubes (sections of plastic pipe containing sticky pads to which hairs adhere), and tracks (snow tracks and tracks in wet material, or the use of track plates or sand plates);
- Nest tubes for specific hedgerow and arboreal species;

- Live trapping and mark-recapture using traps such as Longworth traps, Sherman traps, or Elliott traps;
- Snap traps (the use of these is not recommended as the trapped individuals invariably perish);
- Optical and infrared photography using sensor activated equipment.

Larger mammal presence and population figures can be determined by direct observation, although many are fugitive or nocturnal. Radio tracking can be used to define territorial extent and migration routes, and this can be effective at identifying potential impacts from development. Survey extents for such species may extend well outside the development site area. Krebs (2006) suggests direct observation can be enhanced using "field signs" and survey methods such as:

- Hair tubes, live traps, and identification of animal tracks as for small mammals;
- Identification of droppings and latrines;
- Spoor sites/territorial markers;
- Excavations and feeding damage;
- Habitations such as burrows, setts, holts, or dreys.

Invertebrates

Terrestrial invertebrate surveys are seasonally restricted and should ideally involve repeat sampling. Species can be easily missed if they are in a concealed phase when the survey is conducted, e.g. soil-dwelling and stem-boring larvae, and the egg phase of many species. In addition, the activity of many species is restricted to particular times of day or weather conditions.

Consequently, surveys must be carefully targeted, e.g. on notable species, target groups, and indicator species (which can sometimes attest the general suitability of habitats for invertebrates). Target groups are often suggested by national bodies or professional institutes in each country. In the UK, the Institute of Environmental Assessment (IEA 1995) identifies Carabidae (ground beetles), Lepidoptera (butterflies and moths), Orthoptera (crickets and grasshoppers), and Syrphidae (hoverflies) as indicator groups. In USA grassland ecosystems, prairie dogs are used as indicator species (Csanyi 2015), and in India Chital and Indian Hare are used in forest ecosystems (Mathur et al. 2011).

Survey methods for terrestrial invertebrates can be divided into observer-dependent methods, which are carried out by the investigator in the field, and observer-independent methods, which use traps. Table 6.3 outlines commonly used methods. Most of these can also be employed for sampling wetland and semiaquatic species (associated with the margins of water bodies).

Aquatic invertebrates make up a large proportion of the diversity of most freshwater habitats and often contribute significantly to the conservation value

Table 6.3 Methods for sampling terrestrial invertebrates

Observer-dependent methods

Direct searching and recording in selected habitat/vegetation patches. It is not normally quantitative, can lead to misidentification, only records species that are active at the time, and tends to be limited to species that are conspicuous and/or common in the study area.

Transect walking involves the observation, identification, and enumeration of species (usually only of butterflies and day-flying moths) along a set route, within prescribed time and weather conditions.

Sweep netting involves a hand-held net swept through vegetation (that is not woody, thorny, or wet) up to 1 m in height. It collects most species from the vegetation (except the basal parts), but flying insects often escape. It can be quantitative if a standard number of sweeps is taken, but sweeps in different vegetation types are not directly comparable because of differing resistance to the net.

Swish netting is like sweep netting but is restricted to the air boundary immediately above vegetation. It is especially good at collecting Diptera (flies) and Hymenoptera (bees and wasps).

Suction sampling uses a portable vacuum to collect invertebrates from the ground layer and/or basal parts of vegetation. It can be efficient in dry conditions and where there is little vegetation litter, and can provide quantitative data if a set number of samples are obtained.

Soil samples can be taken for identification and enumeration of soil invertebrates. A variety of physical or chemical extraction methods are used to extract the organisms from the soil samples.

Beating uses a stout stick to knock invertebrates off vegetation onto a sheet, from which they are collected. It is usually used to sample the fauna of individual tree species. With care, it can be used to obtain quantitative data, but is not practical in wet conditions.

Subsidiary methods are used by many experts for particular groups. They include observing flower visitors, hand searching vegetation for plant grazers, stone turning, and investigating litter and dead wood.

Observer independent methods

Pitfall traps are placed on a regular grid within selected areas, and provide quantitative data, mainly for ground dwelling beetles, which fall into the traps. They usually contain a killing/preserving fluid.

Malaise traps intercept flying insects by a net, and funnel them into a collection vessel. They can collect large numbers of insects (especially Diptera and Hymenoptera) and obtain quantitative and comparative data, but do not discriminate between insects resident in or flying through the area.

Sticky traps usually consist of a mesh screen on which a viscous oil is applied. They can be used like malaise traps or placed within vegetation. Fragile species may become damaged in trying to escape from the trap, and samples have to be removed using a solvent.

Table 6.3 continued

Water traps rely on the fact that some flying insects (especially flower visitors) are attracted to coloured surfaces. They are simple to use but selective.

Light traps attract night-flying insects, especially if they emit ultraviolet wavelengths. They are useful but require a power source, are not easily transported, and may sample species that are flying over a site rather associated with it.

Emergence traps usually consist of a closed mesh canopy (placed over vegetation) and a collecting vessel, and are designed to collect adult flying insects that were in a developmental stage when the trap was erected. They can be used quantitatively, but must be in place for long periods.

---

of a site (Kerrison et al. 2005). They are also used as biotic indicators of water quality in freshwater ecosystems, and for this purpose the preferred group is widely considered to be benthic macro-invertebrates.

Ideally, surveys should facilitate assessment of the entire value of a property and its different habitats (e.g. mud or submerged plants). Samples collected during surveys from different habitats should be kept separate, and should be replicated to assess whether perceived differences between habitats are likely to be real. Preferably, at least two surveys should be carried out to capture information pertaining to spawning or hatching activity and the emergence of juvenile or adult life stages. Identification to species level should be undertaken, at least in the case of notable species and taxa for which keys are available.

The most commonly used sampling method for aquatic macro-invertebrates is the use of pond nets. However, there are many different techniques, including "kick sampling" and the use of dredges, grabs, and traps. For small standing-water sites, standardised survey methods have been developed which use a three-minute hand-net sample from all significant habitats, and form the basis of the Predictive System for Multimetrics (PSYM), a system for assessing the ecological quality of ponds and small lakes. A three-minute hand-net method can also be used in river surveys. Additional surveys are often conducted for adult emergent insects such as dragonflies, either as they emerge or on the wing.

6.6.4 Habitat surveys

Detailed habitat surveys normally focus on two requirements: (a) to confirm the presence of rare or protected plant species; and (b) to establish the conservation status, protected status, and susceptibility to impacts, of whole "community habitats". Under the IFC Performance Standard 6 (IFC 2012b), the habitat survey is intended to define the status of the habitats present and allow classification of the habitat into one of the following classifications, reflecting habitat "condition" and its importance for maintaining populations of threatened species or other species considered to be a "priority":

a) Modified habitat. Habitat that supports a community which contains a large proportion of non-native or introduced fauna and flora, or where human activities have modified the area's "primary ecological functions". Such areas include agriculture, parks and amenity areas, commercial forestry and plantations, and reclaimed/modified wetlands and rivers.

b) Natural habitat. These are "areas composed of viable assemblages of plant and/or animal species of largely native origin, and/or where human activity has not essentially modified an area's primary ecological functions and species composition" (IFC 2012b).

c) Critical habitat. Critical habitats are the key habitats in maintaining global biodiversity. They include sites which contain high biodiversity values; that provide habitat for critically endangered or endangered species, endemic or restricted-range species, and/or significant concentrations of migratory and or congregatory species for parts of the year; that contain unique and/or highly threatened ecosystems; and areas "associated with key evolutionary processes".

The Performance Standard (IFC 2012a) notes that "Intentional or accidental introduction of alien, or non-native, species of flora and fauna into areas where they are not normally found can be a significant threat to biodiversity, since some alien species can become invasive, spreading rapidly and out-competing native species". It is clearly important, therefore, that the presence of such species on or near the site is identified and recorded in the detailed surveys. The presence of such species is most common in modified habitats, but is not uncommon in natural habitats and less so in critical habitats.

The IFC Performance Standard guidance document (IFC 2012b) criteria for defining the status of critical habitat relate to the percentage of a global population of a particular species that a habitat may support. National systems for classifying habitat types are available in many countries such as the National Vegetation Classification System in the UK. These usually classify vegetative communities based on species composition, species diversity and/or the dominant species present. Such systems are often tied into the definition of valuable habitats types, conservation values and local legislation, and so guide the type of detailed surveys required.

Occasionally, an "extended" detailed vegetation survey may need to be conducted, using a systematic sampling pattern rather than selective sampling procedures. For example, if potential changes in groundwater level or quality may affect the integrity of a high-value wetland site, it may be important to quantify the community-environment patterns and relationships in order to assess the threats and formulate mitigation measures (e.g. Morris 2002). The results from this type of survey can be analysed by GIS and/or multivariate analysis.

6.6.5 Environmental variables and site history

Information on a range of environmental factors affecting the site and its surroundings should also be collected for the EcIA, to understand relationships of species or communities; assist in evaluating a site's conservation value or potential for habitat creation; and facilitate impact prediction. Relevant data to collect may include:

- Topography and aspect of the site;
- Local climate information, including precipitation, temperature and daylight hours;
- Soil conditions such as depth, organic content, composition and pH;
- Water quality, including nutrient status and levels of pollutants;
- River geomorphology;
- Pollution levels;
- Existing and historic land, habitat and site management.

Existing data may be available from sources listed in Table 6.1, and new data may be collected for other disciplines inputting to the ESIA such as landscape, climate, soils and water. Some variables such as pH, conductivity and dissolved oxygen, can be readily measured in the field. However, detailed chemical analysis is time-consuming and if relationships between the results and biological data are required, a sampling pattern may need to be used which allows one sample to be associated with the mean of several biological observations.

An understanding of current ecological systems can often be facilitated by a knowledge of past conditions. Historical and archaeological information may be available; and evidence on ecological conditions may be obtainable using palaeoecological techniques such as dendroclimatology and analysis of sediment (including peat) cores. Such evidence can provide valuable information on changes that have occurred in, e.g. land cover/use, climate, hydrology, and the biota. However, the required techniques are time-consuming and expensive, and existing data for specific receptor sites are rarely available. Past conditions can also be inferred from current ecological features such as floristic richness and indicator species.

6.6.6 Changes without the development

Ecological habitats will tend to "succeed" over time towards the climax community appropriate for the site (unless this is already the case), and so are liable to change over time. Therefore, in order to make valid assessments about the ecological impacts of a project, it is important to consider what changes in biodiversity may occur in its absence (often called the "zero option"). Such changes can result from intrinsic ecosystem processes, or from external factors such as climate change or disease, whether or not these are caused by human activities. The most significant intrinsic changes are likely to be associated with ecological succession. Precise outcomes of this are difficult to foresee, but general predictions can be made. For instance, it can be anticipated that unmanaged temperate grasslands will, over time, succeed to a climax community such as woodland.

The future effects of climate change on community structure and habitat types are important elements to consider. Changes to the temperature and to water balance are already affecting species and habitat ranges. The introduction of disease or disease vectors can lead to devastating effects on a single species. Where this affects a dominant species in a particular habitat such as elm or ash in European temperate woodland, this can lead to significant changes in species composition.

Other external influences that may cause changes in the absence of the project include: large-scale changes in farming practices, biodiversity management programmes, water-resource depletion caused by over-abstraction, and cumulative impacts of land development, including other local projects. These influences may have differing implications in relation to a project, for example:

• Impacts would/would not contribute significantly to cumulative impacts;
• Impacts on a habitat/site would not be important because its current ecological value will decline anyway;
• Although a habitat/site may not be considered worthy of protection in its current condition, its value may increase if impacts from the project are avoided.

6.6.7 Description and evaluation of the baseline conditions

It is critical that the work undertaken on different ecological aspects is coordinated, and that the findings are integrated to produce a clear description and evaluation of the baseline conditions on the site and in surrounding locations. There are no accepted international standards on the layout of an EcIA. Some EcIAs form an appendix in the ESIA, with key descriptions on baseline ecology being drawn forward into the chapter on ecology and biodiversity. Regardless, the key is providing clear information on the ecological baseline of the site, and the ecological context of the property. Therefore, the description should include:

• The aims and scope of the investigations undertaken.
• A description of the ecological surveys undertaken, including clear descriptions of the methods employed in undertaking the baseline assessment; the detailed field surveys undertaken, with information on where and when they were carried out; and weather conditions where this is pertinent.
• The findings of the initial survey, including the desk study/literature review and walkover surveys, indicating sources of any information

obtained. The provenance and age of the data collected should be clearly described.

- A clear presentation of the results of the work, including species lists, tables of quantitative data, clear descriptions and maps of sites and habitats and, where relevant, charts/graphs and GIS presentations.
- An evaluation of all key receptors including: conservation value and status; aspects such as habitat suitability and potential; and susceptibility to impacts.
- An assessment of the environmental factors (including management) controlling the current ecological systems, and of existing trends. Information on current land-use should be presented along with a description of how this is affecting the type and quality of biodiversity present. Where available, information on future pressures on biodiversity in the area should be presented, for instance where investment in conservation projects in a region is anticipated.
- Indications of limitations (e.g. time restrictions and data accuracy) and uncertainties.

6.7 Impact prediction and evaluation

This section describes the methods available for assessing potential ecological impacts arising from new developments. The initial section describes the approach of the IFC Performance Standards (IFC 2012a, b), and subsequent sections provide more detail on assessment methods (Morris and Emberton 2009).

6.7.1 International Finance Corporation

IFC Performance Standard 6 (PS6) states that the assessment should include any "direct, indirect and residual impacts on populations, species and ecosystems and on ecosystem services" which may be affected in the area of influence. It identifies that direct impacts could include a wide range of effects such as

> habitat loss, disturbance, emissions to air, land and water, alterations of surface hydrology and landforms, edge effects and forest gaps, loss of provisioning ecosystem services or access to such services, degradation of regulating, cultural and supporting ecosystem services, etc. Indirect impacts might include the accidental introduction and spread of invasive species, project-induced access by third parties, in-migration and associated impacts on resource use.
>
> (IFC 2012a)

The IFC's guidance (IFC 2012b) on biodiversity conservation and sustainable management of natural resources (GN6) suggests that impact assessment should consider direct, indirect, and cumulative effects on biodiversity and ecosystem services. The latter should address "in combination" effects from other existing, planned, or reasonably foreseeable developments that might affect the same components of biodiversity. Potential impacts should be identified for the area of influence, taking into account the location and scale of project activities, the project's proximity to areas of known biodiversity value or areas known to provide ecosystem services; and the types of technology that will be used (e.g. underground mining versus open pits, directional drilling and multi-well pads versus high-density single-well pads). The area of influence of the project includes not only the areas directly affected (for example areas within the physical footprint of infrastructure) but also areas in ecological connectivity such as habitats associated with rivers some distance downstream.

For large-scale projects, PS6 suggests adopting an "ecosystem approach" as described in the Convention on Biological Diversity. This approach does not imply any specific scale such as a given radius around the development site, but instead relates to the area of influence, however large this may be. The approach also considers each individual component of the ecosystem and the processes operating within it.

PS6 identifies the integration of mitigation and a further round of assessment to identify residual impacts as critical. It highlights the importance of the mitigation hierarchy, and the level of ecological mitigation required in the standard depends upon the classification of the habitats. No net loss of ecology is required in natural habitats, while net gain is required in critical habitats. The incorporation of biodiversity offsetting into project design is encouraged to compensate for residual impacts on natural or critical habitat that are expected to remain despite appropriate avoidance and minimisation measures. Importantly, measures needed to mitigate or offset impacts on natural and critical habitat must be articulated in a biodiversity action plan. This ensures that clear commitments are made as a key output of the EcIA process and that they are documented as part of a project's Environmental and Social Management System. This makes it more likely that they will be implemented and monitored throughout a project's lifetime.

6.7.2 Assessing significance of ecological impacts

Significance of ecological impacts depends on the residual state of biodiversity and ecosystems that remain post-development. Magnitude of impacts needs to be assessed in relative terms. Will the residual animal population still be viable if a proportion of its suitable habitat is lost? Will the ecosystem still function if it is polluted? Impacts lasting for a certain time period might have a long duration when considered from the perspective of a beetle's life cycle, or short when considered from the perspective of a long-lived mammal. The significance of the same environmental change therefore varies depending on the ecological characteristics of the affected feature. The basis for assessing significance needs to be established at the scoping phase to ensure that relevant data are collected, but following the initial and detailed studies, it may be necessary to change the study area, or the scope of assessment. This should be done as the first activity during the impact assessment phase. Factors to consider include:

- Whether there are any protected or designated features that require particular outcomes to be demonstrated by law;
- The role of affected habitats in the overall ecosystem, for example whether they provide important movement corridors;
- Whether the project-affected area provides temporary habitat, for instance in long- or short-distance migration, or for the spread of juveniles from core habitat areas;
- The role of the project-affected area in conferring resilience to climate change on habitat or species.

PS6 requires identification of natural or critical habitats that might be affected and requires impacts on them to be managed according to the mitigation hierarchy so that no net loss or net gain outcomes are achieved respectively. For the first three criteria used to define critical habitats (i.e. critical and endangered species, endemic and restricted-range species, and migratory and congregatory species), PS6 Criteria defines numerical tiers to assist in decision-making by the consenting authorities. The criteria include habitat required to sustain > 10 per cent of a species on the International Union for Conservation of Nature (IUCN 2006) Red List, and habitat known to sustain ≥ 95 per cent of the global population of an endemic or restricted-range species. Development is only considered acceptable within a critical habitat if there will be no measurable adverse impacts on the features it supports. The need to quantify impacts is therefore strongly implied. The performance standard does not include any specific methodologies for assessing impact, as these will be project- and site-specific. However, it identifies the need for assessment to be propor-tionate to "the nature and scale of the proposed project's potential impacts".

6.7.3 Types of ecological impact

Ecological impacts can result from three types of disturbances: pulse (temporary); press (sustained); and catastrophic (highly destructive or irre-versible). Developments can have positive ecological impacts including habitat creation, especially where adequate management is also provided. For example, sensitive and sustainable redevelopment on brownfield sites can, if associated with ecological design or management, improve their ecological value, by incorporating new ecological and landscape features such as "green networks" (e.g. see Angold et al. 2006, Barker 1997, Harrison et al. 1995), or providing new habitats and space for locally indigenous species, e.g. by using locally sourced seed, plants, or turf. The land within the boundaries of long linear projects such as road and rail verges, can be designed to develop into valuable wildlife corridors and habitats, especially for grassland communities. However, a positive impact is only likely to occur where such projects occur in areas of low biodiversity such as within urban areas or areas of intensive agriculture/monoculture. Examples of negative ecological impacts generated by major impact sources are discussed below.

Habitat destruction and fragmentation

These are probably the greatest threats to the biodiversity of a site and surrounding area resulting from development. The amount of natural and semi-natural habitat destruction that is caused directly by a new development will depend largely on how much exists within the project's boundary (permanent and temporary construction and post-construction operational areas). Some loss is usually inevitable. The significance of the habitat destruction will depend upon a range of factors, including:

- The ecological/conservation value of the habitat, and the degree to which notable species depend on it;
- The presence of rare, protected or notable species within the habitat lost;
- The degree to which displaced species can migrate to, and survive in, other suitable sites/habitat patches (which will depend on their availability in the immediate area, and on factors such as the existing density of the species in them, and hence the potential severity of competition);
- The quantity of habitat lost, and the degree to which its loss will affect the fragmentation and integrity of local ecological networks in the area and remaining habitats on-site;
- The importance of the habitat lost in maintaining species composition in ecotones or "edge" communities, or adjacent habitats (some species may use several habitat types for different parts of their life cycle);
- Indirect impacts such as increased flood risk downstream, resulting from development on floodplains (in addition to the loss of valuable wetland habitats by land-take).

New barriers are often created by the removal of interconnecting habitats, or the interposition of a "hostile" feature such as linear infrastructure (i.e. fencing, roads, rails, canals, pipelines, and power lines), buildings, or a series of cultivated fields between habitat patches. A barrier may be physical, preventing wildlife from crossing a site. A barrier may also be behavioural, creating an environmental condition that wildlife is unwilling to enter. Lastly, a barrier may be a hazardous condition, imposing the possibility of injury or death to wildlife that enter the area.

Habitat fragmentation caused by multiple barriers has two primary effects. It divides a habitat patch into smaller patches; and it also introduces barriers between the remaining habitat patches, resulting in increased isolation, and

---

making them unavailable to some species. Habitat fragmentation is one of the indicators for (un)sustainable development and is specified as a particular issue for consideration in the World Bank environmental assessment methods (World Bank 2013, 2002, 2001, 1997a, 1997b).

When they are part of a close-knit mosaic of habitats, small habitat patches can promote biodiversity by introducing habitat diversity and ecotones. However, when small habitat patches are isolated as "islands" in a fragmented landscape, they generally support fewer species and smaller populations than large patches of the same habitat type. Moreover, their biodiversity can decline over time (Joshi et al. 2006), resulting in a longer-term chronic impact on the local ecology. One reason is that a small patch may not contain sufficient resources such as food, water, cover, or habitation sites to support viable populations of some species (Teitje and Berlund 2000). Resident species populations are more susceptible to local catastrophic events such as severe drought or fire, and some species need several habitat patches within a given area. Small patches are also less likely to be colonised, especially by habitat "specialists" (Joshi et al. 2006).

Small patches are susceptible to edge effects and isolation. Edge effects are associated with the increased length of habitat edge relative to its area, i.e. of the boundary/area ratio. A large edge/area ratio generally favours edge-living species at the expense of core species (Bender et al. 1998). Small habitat patches are also susceptible to "external" impacts such as pollution, physical damage and disturbance: these may be more important for some species than reductions in patch size per se (Kirby 1995). They may also be more susceptible to invasion by "foreign" species from neighbouring areas. The boundaries between habitats have different environmental conditions from their interiors. The resulting community often contains species from the bounding habitats, and species specialising in edge habitats.

An important aspect of isolation is the relative ability of species to move between habitat patches within the wider ecological network. The theory of metapopulations states that a "sink" sub-population in a small habitat patch (with insufficient resources to sustain it) may be augmented by immigration from a "source" (reservoir) population living in a larger habitat patch – but only so long as the two populations are not isolated. Consequently, the viability of many species populations in fragmented landscapes will be affected by:

- The species' dispersal capabilities, which vary considerably;
- The nature of the land-use between the habitat patches;
- The degree of isolation between habitat patches – and hence the distances (and likely number and severity of barriers) between them.

In the longer term, lack of genetic exchange may also lead to the decline of isolated sub-populations. Some migrant species such as birds are less affected (Bender et al. 1998), and small isolated patches can still act as *"stepping stone"* **habitats** for these species.

An additional problem that may follow from the isolation of small habitat patches is that conservation management practices such as low-intensity grazing, may be prevented because the remaining areas are too small or are inaccessible. Further, there is increasing evidence that habitat fragmentation inhibits regional dispersal for many species, especially when the remaining habitat patches are small (Collingham and Huntley 2000); and this has serious implications in relation to climate change. Walmsley et al. (2007) discuss the effect of habitat fragmentation on local ecology, and Plantlife International (PI 2005) predicts that the survival of many species will depend on their ability to disperse. They note that many notable species will have difficulty in dispersing effectively unless action is taken.

The significance of barriers varies between species (Eycott et al. 2007). For example:

- Some species have efficient dispersal mechanisms that are not seriously affected by most local barriers;
- Some species have very limited dispersal ability, and such "low-mobility" species are (a) restricted to remaining habitat patches, and (b) unlikely to recolonise isolated habitat patches;
- Animals such as badgers, deer, and otters have wide habitat-area requirements that may exceed the areas of remaining patches (or of patches to which they have been displaced by development) and must risk crossing hazardous barriers;
- Animals such as amphibians may have to cross barriers in order to reach their breeding habitats.

Habitat damage, wildlife disturbance, and direct mortality

Habitat damage may occur from activities such as vegetation trampling, plant damage or removal, and soil compaction or erosion. Many of these activities occur mainly during the construction phase of projects; but it does not follow that this is temporary or reversible, especially when the receptors are long-established semi-natural habitats. Chronic and progressive habitat damage can also result from many forms of increased human activity that a development may generate, e.g. traffic and recreation/visitor pressure, and increased hunting pressure.

Habitat damage can also involve the destruction of microhabitats, and this can affect the whole ecosystem by simplifying habitats. For example, canalisation and river straightening often gives more uniform flow regimes, water depths, and bank profiles, all of which reduce habitat complexity and associated plant and invertebrate diversity. Many species live in (or need) different habitats at various stages of their life cycle, including most migratory species. For example, fish fry benefit from backwaters or bays in which they can develop, and the adults of some aquatic animals need terrestrial habitats. Removing any one of these habitats, or blocking the migration route between them, can therefore eliminate those species from the community.

Wildlife disturbance can result from a variety of sources such as construction, traffic or visitor pressure, and can involve a number of factors, including visual impacts, noise, trampling, and night-time light pollution. The susceptibility of animals to disturbance often varies seasonally, e.g. in relation to breeding periods. This will vary between countries, and the assessor must identify such periods, and where they exist, in the project host country. Vulnerable periods may not be immediately obvious. For instance, invertebrates with mobile adult phases (e.g. butterflies and other flying insects) may be most vulnerable when in developmental stages (eggs, larvae, and pupae) because of damage to their food plants and breeding sites. Most invertebrates in temperate regions are also vulnerable in winter because they are dormant and hence cannot escape. Permanent vertebrate residents are most sensitive to disturbance during the breeding season, but some temperate species may also be vulnerable in overwintering periods, and the risk to migrant birds is during their visit period.

Direct mortality can result from factors such as vegetation destruction, trampling, and fire. Roads present serious long-term threats, especially to animals that need to cross them. As an example, during a 12-month survey of road deaths throughout Britain, 5,675 mammal casualties and 142 bird of prey casualties were recorded (Mammal Society 2002).

Pollution

Water quality varies naturally in relation to local climate and (especially) geology. It makes a major contribution to the diversity of wetland and freshwater systems. Each type of habitat is dependent on a narrow range of water quality, and its integrity is threatened by any deviation beyond this range.

Most developments are likely to release, either accidentally or as part of a permitted discharge, chemicals and materials into the surrounding environment, and this can impact the local ecology. Industrial, urban, and road developments are sources of a wide range of atmospheric pollutants and waterborne pollutants. Water pollution can affect terrestrial ecosystems wherever polluted water periodically inundates the ground surface or soil, e.g. as leachates from landfill sites, mine workings or surface deposits, or runoff from urban and road surfaces. In addition, salt-rich spray regularly falls in the "splash zones" of road verges during winter months, and influences species composition within these zones. However, water pollution is particularly detrimental in wetland and aquatic ecosystems.

The impacts on ecology can occur from the entry of pollutants into, and circulation within, biogeochemical cycles. Those in air or water can affect ecosystems, species and habitats far from their source. This is especially the case with rivers and freshwater habitats because pollutants may bioaccumulate, and bioamplify in food chains. This can have serious consequences, especially for top carnivores. Pollutants can also undergo biotransformation in the environment or within the bodies of individuals (Connell et al. 1999).

The main types of pollutant and pollution affecting freshwater ecosystems are: organic matter; thermal pollution; acidification; eutrophication; sediments; metals, micro-organics and other harmful chemicals; and oils. Contamination of freshwater systems by non-mobile elements (e.g. phosphorus) or non-biodegradable toxins must be minimised since their effect may be permanent and effectively irreversible. Significant impacts are most likely in standing-water systems such as lakes, which act as cumulative sinks for sediment; but water courses can also be affected, and while new pollution events may have little effect in an already polluted reach of a river, they may progressively damage downstream sections.

Most pollution is chronic, putting additional stress on species and reducing their resistance to environmental change, disease, or other factors. However, accidental pollution can be a major threat, especially from heavy industry and transport, and the IFC approach requires the production of a risk assessment (including a worst-case scenario) for this type of impact.

Atmospheric pollution can affect vegetation and animals directly, or indirectly through environmental changes such as those in the chemistry of soils and waters. One significant example is the Black Triangle, an area covering Germany, the Czech Republic, and Poland in central Europe. Here, topography and the burning of brown coal resulted in air pollution which contributed to significant forest deterioration, including 52 per cent loss of forest cover in parts of the Czech Republic (Oulehle et al. 2006).

Water levels

Changes in hydrology brought about by development have the potential to cause significant effects on local terrestrial and aquatic ecology. Changes in ground or surface-water level and stability can affect most ecosystems but are often critical for wetland and freshwater ecosystems (NIWA 2015). The seasonal water regime can be particularly important in marshes and wet grasslands during the breeding season of wetland birds, many of which need a mosaic of varying water levels. Changes in the hydrological regime and climate can also result in significant effects in dry habitats, which depend upon seasonal rainfall, either directly on the area or on local uplands.

Lowered water levels and resulting over-abstraction of water from wetlands and rivers can also increase the incidence of soil drought in terrestrial ecosystems. For example, many wetland habitats in Europe and the USA have been lost or degraded as a result of land drainage or water over-abstraction. A UK report (BCG 2007) identified 101 rivers and 201 wetland protected sites that were in danger of drying out due to groundwater abstraction for agriculture or public water supply. In coastal areas, over-abstraction can result in increased levels of saline intrusion into the groundwater (EEA 2008) and studies in many countries, including the UK,

US, and South Africa have shown the consequent effects on ecology (Sophocleous 2009, CSIR 2007).

Raised water levels, instead, can increase the incidence and duration of inundation and waterlogging of floodplains, which may adversely affect invertebrate diversity. They can be particularly damaging to temporary water habitats such as seasonal ponds and streams, which may host specialised animals and plants that can be of high conservation interest (e.g. Bratton 1990).

The effects of changes in water flow can go far beyond increases in water velocity because this is inevitably accompanied by changes in other variables such as dissolved oxygen concentration, nutrient fluxes, and sediment type and volume. An increase or decrease in flood velocity can indirectly damage communities adapted to the prevailing flow, and may irreversibly modify the physical and biological environment.

Changes in the competitive balance between species

Changes in the species composition of a community often occur because the competitive balance is altered in favour of species that are more tolerant of new conditions. This frequently favours ecological "generalists" at the expense of "specialists", which are often notable species or species confined to high-value habitats.

6.7.4 Methods of impact prediction and evaluation

It is relatively easy to identify primary ecological impacts such as habitat loss or fragmentation, but much more difficult to predict their effects, or to identify the numerous potential secondary impacts that may be generated. Cumulative impacts are particularly difficult to assess, partly because virtually all negative ecological impacts are bound to add to general pressures on biodiversity. The numerous individual activities and impacts associated with development projects rarely affect ecology discretely. Instead, they will have a cumulative overall impact on local biodiversity and ecosystems over time. Research by Didim (2010) in Australia has shown how the different interaction of effects can result in different resulting impacts. This research highlights the need for involvement of professional and experienced ecologists in EcIA, who will be sensitive to the magnification of development project impacts caused by the accumulation of individual activities.

Guidance on the assessment of indirect and cumulative impacts is provided in CEAA (1999), EC (1999) and LUC (2006). Specific sector-based and/or national advice is also often available, for instance by the UK Planning Inspectorate (2015), South Africa DEAT (2004) and the Minerals Council of Australia (2015). Their documents can help to identify methodologies for the collection of information considered important to local stakeholders, non-governmental organisations, and different business sectors. One of the most comprehensive collections of advisory documents is provided by the World Bank's (2013, 2002, 1997a, 1997b) Environmental Assessment Sourcebook and its regular updates.

Impact magnitude (severity) is relatively easy to predict. The use of GIS can assist in quantifying the loss of individual habitat types, so long as mapping on-site has been reasonably accurate (e.g. by the use of satellite or aerial mapping, or by on-site mapping using GIS). Where the location of notable species, habitats and communities are recorded, the proportionate direct losses from these can be measured. Over-reliance on measures such as "percentage-of-site" affected can be misleading. The loss of even a small area of a site may affect the integrity of a site and lead to increased fragmentation. In addition, small impacts on protected sites can be considered of greater impact magnitude than larger loss of less biodiverse sites such as low-diversity agricultural land. However, the majority of impacts are difficult or impossible to quantify and even when some quantification is possible, final assessments of impact magnitude will usually be restricted to qualitative estimates such as slight, moderate, or large.

GIS can be particularly useful, especially if used in conjunction with a system for evaluating quantitative data. Mathematical and statistical models have been used as research tools in ecology for many years, and are increasingly employed in relation to risk assessment, including estimation of minimum critical areas and viable populations (Burgman et al. 1993), ecotoxicology, critical loads, and hydrological processes. However, their use in EcIA can be limited by:

- Current lack of knowledge and understanding of ecosystems' complex interactive processes, and hence of how species and communities will respond to impacts;
- Unsuitability of many models for "off the peg" use;
- Expense and time constraints;
- Difficulty of obtaining sufficient quantitative data on impacts and/or baseline conditions.

These and other challenges suggest that impact predictions often have to be qualitative, despite the emphasis on making precise and, where possible, quantitative, predictions. In the UK, the procedures recommended by IEEM (2006) and the Design manual for roads and bridges (Highways Agency et al. 1993) aim to increase the transparency and rigour of EcIA, and EA (2007) provides guidance on identifying source–pathway–receptor links for impacts on freshwater and wetland sites. However, these methodologies still require a large input of professional judgement and a degree of uncertainty is inevitable and should be acceptable, provided that it is clearly stated in the ESIA report.

Under some assessment methodologies, the likelihood of the impact occurring is also taken into account in assessing the magnitude of impact. This includes risk assessment methodologies such as EPA (2011), IEEM (2006) and IFC (2012b).

The second element required for impact prediction, besides impact magnitude, is the sensitivity of the ecological asset. Receptor sensitivity is a function of the ability of a species or habitat to accept change in the environment and its resilience to any changes which may occur. Some ecosystems are known to be more sensitive or resilient to change than others, but assessing impact significance in relation to these attributes can be one of the most difficult aspects of EcIA. One approach is to consider each potential impact in turn and assess whether any changes are likely to lie within the natural range of perturbation for the ecosystem or for any significant element of it. For example, it may help to consider the most sensitive species (indicator species) and its position in the food chain. Such potential changes should be considered both in the short and long term, and for all phases of the development. The potential of climate change to add further pressure on the habitat/species must also be considered. As a rule, where predicted impacts are within the normal range of the system, or the range present under the "zero option", the level of change is likely to be low. However, where a normal range is exceeded, assessment of impact significance will still have to rely on expert judgement.

Many EcIAs use a matrix (see Figure 1.3), where the magnitude and sensitivity to impact is used to evaluate the significance of impact. To do this, the assessor needs to develop a criteria range for both magnitude and sensitivity, and a matrix for assessing impact. Guidance on determining ecological impact significance is provided in many countries with ESIA regulations in place. This includes, for example, the UK (EN 1999 and IEEM 2006), Australia (Government of Australia 2013), Ireland (NRA 2009), and New Zealand (Government of New Zealand n.d.). In addition, reviews of the identification of significance in ESIA (and EcIA) has been undertaken by a number of reviewers such as Marttunen et al. (2013) and Brigs and Hudson (2013).

In many cases, there will be beneficial effects on some species/habitats and adverse effects on others. Balancing adverse and beneficial effects is especially evident where the project includes some element of ecological design. In these cases it is legitimate to make an overall net assessment. This approach requires careful ecological judgement and detailed descriptions of both the beneficial and adverse impacts, rather than using a simple area or number-of-sites approach. Balancing should err on the side of caution and should be restricted to "slight" or, exceptionally, "moderate" impacts.

6.8 Mitigation

There is a growing opinion that new developments should aim to deliver enhancement (net ecological gain) and not simply strive to limit environmental damage (e.g. Defra 2005, IEEM 2006). Development projects can have positive ecological impacts, especially at previously developed sites; but where a project impacts *semi-natural* habitats, provision of net ecological gain is likely to be both difficult and costly. Mitigation planning provides an opportunity to consider a package of engineering and environmental measures to offset impacts and/or enhance the environmental quality of the project. There are no universal "off the peg" mitigation solutions suitable for all sites. Mitigation can involve several different types of actions, and many developments – especially large developments or those with potential transboundary impacts – use a package of measures. When defining mitigation options suitable for a development, EcIA practitioners are encouraged to adopt a mitigation hierarchy (Treweek 2008):

enhance → avoid → minimise → restore → compensate → offset

This hierarchy can be made operational by the sequential application of mitigation types:

- Mitigation by design: avoidance or minimisation of impacts on species or habitats, enhancement of existing habitats, and introduction of new habitats and ecological features;
- Mitigation by management;
- Remedial measures: creation of alternative habitats and ecological sites;
- Biodiversity offsetting: financial contribution to activities to protect or improve existing habitats off-site.

6.8.1 Mitigation by design

As ESIA practice has developed, it has become much more than a tool to measure and report impacts. Best-practice ESIA is a tool for constructive challenges to the design process, and even identification of site location if applied sufficiently early in a comparative site assessment process. ESIA allows early identification of key ecological features, and an interactive dialogue with the project designers to improve the ecological (and environmental) performance of the project. Because of this iterative "mitigation by design" process, the final project design upon which the ESIA report is based may be significantly less damaging to wildlife than the initial design prepared. The ESIA report should describe this process and the mitigation incorporated into the design. Mitigation by design should comprise an amalgam of three principles: minimise/avoid impacts, enhance habitats, and introduce ecological elements.

The first priority in design should be to avoid or minimise impacts at source, and this will often require modification of a project's location/alignment, design, or construction and operating procedures (IEEM 2006). The location of a project is usually determined largely by socio-economic and technical, rather than environmental, considerations and the choice of sites is often restricted. However, if the proposed siting will clearly cause significant impacts on high-value habitats and/or species, the relocation, rezoning, or no-action options should be considered.

In some plan-led legislative systems, this is done at an earlier stage: important ecological sites are identified during plan development using a SEA, and are consequently not zoned for development. At the project level, EcIA scoping reports can inform such decisions. For example, some developers own or have options on "landbanks" of sites, and ecological information about these can influence decisions on which sites to bring forward and the type of development applied to each site.

The alignment of linear projects such as pipelines, electricity cables, conveyors, roads, and railways can use comparative methods, most of which are based on GIS and score sheets. National mechanisms are available in some countries such as the UK Transport Appraisal Method (DfT 2003, 2004), which converts impacts into a numerical values for comparison, and is designed to identify the least ecologically damaging option. However, this may be precluded by technical and financial constraints, or by conflicting interests, e.g. with other ESIA components.

Retention of existing habitat should be the first option considered in site design as this will have developed over time and is likely to have a more complex ecosystem present than can be created by design and form part of the existing network of habitats in the area. Where possible, areas of semi-natural habitat and connective habitat such as hedgerows, ditches, and lines of trees, should be retained as the basis for the ecological design.

Ecological design is an important aspect of site design and should be incorporated as an integral element of the overall site design, and not as an afterthought. Any habitat creation requires careful planning, which PAA (1994) suggest should involve several basic questions:

- Is it suited to the local geography, climate, soil type and fertility, and geology?
- Is it consistent with (or will it complement) the local ecology and landscape?
- What management will it require and will this be feasible?
- Should it be patches and/or linkages?

Successful habitat creation requires both ecological and engineering design expertise, and the new habitats will need long-term monitoring and management. Appropriate methods of habitat creation vary according to habitat type (Gilbert and Anderson 1998, Parker 1995). For example, new ponds and reed beds are relatively easy to create and maintain (SEPA 2000, Williams et al. 1999), and can be used to create mosaics of habitats on-site, introducing new species to the site, and therefore providing important new sites within the local ecological network. They can also be allied to surface-water and flood management and provide a valuable educational resource. Sensitive design/restoration of river works can often be used as an opportunity to repair damage caused by earlier insensitive schemes and to improve river geomorphology where this is appropriate. Such improvements can be ecologically valuable in the long term for sustainability, as well as valuable for re-introducing features that are important to rare and protected species.

Grasslands may be sown using seed mixes of native species that are tolerant of local climatic and soil conditions (Crofts 1994, Crofts and Jefferson 1999). Species-rich grassland mixes are available from suppliers in most countries. These provide a mix of appropriate species for the location and are often of local provenance. Alternatively, harvesting seed from local species-rich grasslands can provide swards typical of the locale. However, establishment of many species can be problematic and often newly established swards are of lower diversity than those of the donor site. Invasion by pioneer and weed species can also be problematic.

The introduction of blocks of woodland, tree belts, and hedgerows at the development project site can provide additional habitat and shelter for wildlife, and wildlife corridors for wildlife requiring home ranges larger than the project site itself. New tree, shrub, and hedge planting requires the use of a suitable range of native species, and stock of local provenance. The creation of new habitat should be as large as possible, keeping in mind that several small patches, with the same overall area as a single large patch, may provide a mosaic with higher species diversity, but smaller populations of each species.

An alternative strategy is to locate new habitat areas adjacent to existing areas (Buckley and Fraser 1998) as this may promote colonisation of the former and increase the viability of the latter. There is also the opportunity to leave areas free from planting and allow natural recolonisation, permitting the natural development of habitats. A good example of this is the sowing of areas of grassland adjacent to woodland edges and allowing these to succeed to woodland.

Ecological design can also include the provision of artificial features to enhance the attractiveness of a site to new species. This is normally undertaken for particular species or groups of species, and includes:

- Artificial roosts, bird banks and boxes for bird species;
- Artificial roosts and boxes for bats;
- Artificial refuges for small to medium sized mammals, reptiles, and insects;
- Crossing points for mammals over or under roads and long-distance pipelines. This includes sections of underground pipeline across migration routes, and wide "green bridges" which are vegetated;
- Use of vegetation to affect the behaviours of mobile species. An example of this would be the use of "hop over" vegetation to alter bird and bat flight paths.

6.8.2 Mitigation by management

Unlike mitigation by design, which relates to changes in the physical form of the proposed development, mitigation by management relates to the application of non-physical activities and controls to a development, whether during the construction or post-operational phase, to minimise impacts.

Construction-phase impacts are often considered "temporary", but much of the on-site ecological damage caused by developments occurs during the construction phase, especially during site clearance, demolition, or earthworks, and full recovery may take many years. Damage and disruption can also occur to nearby sites of ecological importance during construction. Mitigation by management can be effective at reducing or avoiding these impacts. As with design, a hierarchy of mitigation measures should be used, and it is usual to employ a range of mitigation measures during construction. When considering the potential management options available during construction, the assessor should consider the following:

- Can the activity be timed to avoid the impact? Timing of activities to avoid disturbance during breeding activities, nesting, or migration is often used.
- Is there an alternative method of construction that avoids or reduces the impact? An example is to consider the effects of different methods of piling on noise generation.
- Could controls be put in place that would avoid or reduce the impact of a chosen method of work? This could include the provision of dust or surface-water management controls during construction.

Because mitigation measures can themselves have unexpected consequences, the assessment of the effectiveness of mitigation applied to the whole construction phase has to be considered in a multidisciplinary context. It is only after the "package" of mitigation measures is defined that its overall effectiveness can be assessed. Ecological mitigation measures that have been accepted by regulatory authorities should be incorporated in a Construction Environmental Management Plan that integrates any proposed mitigation and the overall effectiveness of the management measures to be assessed.

Mitigation measures could include:

- Minimisation of storage of construction materials and excavated soils by, e.g. using off-site pre-assembly, just-in-time deliveries, and phasing of works;
- Restriction on the number of access roads, access and service areas, construction compounds, and temporary buildings and materials stores; and care in their routing/siting;
- Reduction of dust emissions during project activities;
- Use of noise mitigation for particular activities such as piling, batching, and other heavy equipment work;
- Institution of surface-water management measures such as particulate control and the use of balancing ponds, to minimise water quality degradation;
- Appropriate storage, handling, and management of soils;
- Avoidance of pollution by correct storage of materials, runoff control and use of bunds, silt traps, and leak nappies;

- Minimisation of waste and use of pollution-prevention measures;
- Minimisation of damage to vegetation and soils during project activities by, e.g. using wide tyres on heavy vehicles, and restricting the size and use of vehicles on soft ground and in ecologically sensitive areas;
- Creation of seed banks before vegetation is damaged by project work, and/or use a local seed initiative for re-sowing after construction;
- Avoiding major construction-phase operations during periods when wildlife and vegetation are particularly vulnerable to disturbance;
- Designating ecological protection zones at the project site by, e.g. defining tree and plant-root protection zones during soil excavations; restricting work adjacent to streams, river banks and other semi-natural habitats; and retaining vegetation, trees, badger setts, and bat roosts (see Mitchell-Jones 2004, Arboricultural Association 2015);
- Protecting adjacent habitats by applying a buffer zone and erecting boundary fences that constrain project work activities.

While mitigation during the operational phase will depend largely on the project design, post-construction management is an important consideration as well. It may involve aspects of operational control such as management/maintenance procedures and the management of open-space areas provided on-site. This is particularly true where areas for biodiversity are provided as part of the project design. In the case of industrial and manufacturing plant, mines, water and waste management, the operational control of a plant will be defined under a set of operational permits with regulators, often backed up by an Environmental Management System adopted by the operator. These ensure agreed levels of performance, which can be taken into account in the EcIA.

However, many development types such as residential and commercial development, and transport infrastructure, are not controlled under long-term environmental permits, and so the post-construction effects on ecology can be more difficult to predict in EcIA. To overcome this, longer-term management proposals have been adopted for some sites such as provision of financial instruments (e.g. bonds or management funds) or long-term management teams; restricting access to valuable wildlife areas; and providing other focuses of attention (see §6.8.4 on offsetting).

The choice of mitigation options, or set of options, to optimise long-term effectiveness in any situation can be difficult, and can also be influenced by the planning/operating/environmental-permitting regulations of the relevant country. As a general rule, point-source pollutants should be dealt with at source, while diffuse source pollution is best dealt with by a combination of measures including biological techniques such as buffer zones, and natural and artificial/constructed wetlands (especially reed beds or ponds). However, these techniques may not be a panacea in the long term. For example, a reed bed used to intercept road runoff may effectively deal with degradable pollutants such as nitrates, but have only a limited capacity to store non-degradable pollutants such as phosphates and heavy metals. Thus, unless suitable maintenance is undertaken, the reed bed may eventually become saturated and then export most of the non-degradable pollutants received. Therefore, a "package" of design and maintenance measures is required.

While mitigation by management can be very effective, and there is significant experience and information on the successes of such approaches, the actual effectiveness of proposed mitigation by management on any particular development will depend on the ecology present, its interactions with the local biodiversity network, the activities occurring on-site and their timing. Therefore, the key element in mitigation by management is monitoring and review. A robust monitoring programme is necessary to demonstrate the effectiveness of the mitigation by management and the attainment of levels of impact-reduction predicted in the EcIA.

Care must also be taken not to over-emphasise the ecological benefits of mitigation to justify other forms of ecological loss or damage. For example, ponds or swales created to intercept urban runoff will have ecological benefits, but are unlikely to be optimal wildlife habitats. Further information on the use and value of natural habitats can be found in COST (2014), CEDR (2011) CIRIA (2007), Cooper and Findlater (1991), Crites et al. (2006), and Nuttall et al. (1998).

6.8.3 Remedial measures

The main remedial measure for habitats and species is translocation. Translocation involves "rescuing" a species or habitat from a donor site (that will be destroyed) and moving it to a receptor site that already contains a suitable semi-natural habitat or (in the case of habitat translocation) is environmentally suitable, e.g. it has similar soil type, hydrology, and climate. This technique has been used across the world for many different species. Restoration is the repair of a habitat that has been damaged or has declined in "value" in the absence of appropriate management. To have any chance of success, both of these methods require a thorough understanding of the ecology of the species or habitat in question.

Species translocations are usually attempted when a project threatens a site hosting a protected, rare, or priority species, or where a site provides an important transit site for migrating species. These programmes have included a wide range of species. They should only be attempted as a last resort because:

- The loss of the habitat will reduce the overall availability of habitat within the area – the threatened habitat may be valuable even if no other notable species are present;
- There may be adverse impacts on the recipient habitat;
- Factors affecting the recipient site may affect its long-term success in hosting the species which are unknown at the time of translocation (e.g. excess predation);

- The chances of success are low because adding to the existing population increases the severity of competition. For example, attempts to transfer the great crested newt into existing ponds have proved unsuccessful because the recipient ponds were unsuitable or were already at their maximum carrying-capacity for the species.

Habitat translocation involves moving vegetation (and some substratum) together with incidentally associated animals. It is usually attempted for high-value habitats that are threatened with destruction, or linear habitats where connectivity must be maintained. The techniques used and their success depend upon the habitat type and the environmental conditions in which they occur. In temperate zones, most habitats can be translocated to some degree, although the success of translocating mature trees has been mixed. In more tropical areas, Emberton (2001) states that "it is not likely that tropical rainforest ecosystems, extensive enough to be viable, could be created as an adjunct to a construction project, not that there would be value in extending less biologically profuse taiga forest in areas where such forest are ubiquitous". Translocation techniques applicable to vegetation are described in Table 6.4.

Translocation work can be successful but a review by Gault (1997) concluded that success could not be certain because there had been insufficient post-translocation time for adequate assessment; most were poorly documented; or there had been a general lack of monitoring. Anderson and Groutage (2003) provide an extensive review of habitat translocation projects, and a best-practice guide that sets out minimum standards. However, it does not promote habitat translocation, and concludes that it "should be regarded a last resort for all sites of high nature conservation value".

Habitat restoration should be undertaken whenever damage by a project to semi-natural habitats is unavoidable. For example, where a project necessitates rerouting a watercourse, steps should be taken to ensure that the channel and river corridor of the new section has high environmental quality and a variety of geomorphological areas typical of the river and its location. Similarly, mines, quarries, and gravel pits can be successfully restored for wildlife (Andrews and Kinsman 1990). Guidance on the restoration of wetland habitats is provided in Acreman et al. (2007), Crofts et al. (2005), EPA (2007), and Treweek et al. (1997).

Severely damaged habitats may be difficult to restore, especially if they are long-established complex ecosystems. In these cases, one option is to effect some basic restoration and rely on time and management to "do the rest" (the more complex the "design" of a restoration scheme, the more likely it is to fail or at least not meet expectations). If this is not considered feasible, the last resort may have to be compensation.

Table 6.4 Techniques for translocating vegetation from a development project site to a protected location as part of mitigation activities

Habitat	Habitat translocation technique
Hedgerows	• Translocation of vegetation root bowls and substrate following coppicing. Sections of hedgerow are raised by a mechanical excavator and placed in a temporary or new permanent location at the site with sufficient watering and maintenance until re-established.
Grasslands	• Turf cutting and replacement; • Mowing to collect seed at several times of the year, temporary seed storage, and resowing the seed until new grasses are established; • Removal of individual plants and placement as plugs in replaced or re-sown grassland areas.
Heathland	• Coppice and burning of heather and larger plants followed by turf cutting and similar activities to those for translocation of grasslands.
Woodland/scrub	• Recovery of woodland soils and replacement under translocated wood or under new wood; • Coppicing of small trees and shrubs followed by transfer of root bowls; • Translocation of larger trees where the site is on sandy soils.
Reed beds	• Recovery of reeds and root systems with some underlying soil matrix followed by replanting, or splitting and replanting.
Freshwater plants	• Removal of plugs of plants with substrate and transfer to new wetland; • Maintenance and protection of plants in temporary or new permanent locations until root growth occurs.
Freshwater invertebrates	• Transfer on transferred plants; • Translocation of recovered silts/bed material into new channel/wetland.

6.8.4 Compensation and biodiversity offsetting

Compensation and offsetting usually involve either the use of financial instruments or some form of habitat creation, and relates to sites remote from the development site. For many projects, these are appropriate components of the mitigation package (Conway et al. 2013, Mavel 2013). Compensation differs from offsetting in that the work is most often delivered directly as part of the project and relates to like-for-like replacement of habitats lost (where "no net loss" is the aim), or a combination of like-for-like replacement and enhancement (where "net gain" is the aim). Compensation can include a number of approaches which replace land of low ecological diversity/interest with sites specifically designed, and maintained, for the development of ecological sites:

- Restoration of agricultural land to ecological habitat;
- Enhancement of derelict or previously developed sites to ecological habitat;
- Enhancement of sites containing disturbed habitat by selective insertion of planting;
- Restoration of river banks and ditches;
- Enhancement of low-diversity woods and hedgerows;
- Enhancement of public open spaces such as parks, school playing areas.

"Set aside" sites, where land is allowed to naturally regenerate to ecological habitat, are excluded as this involves no active intervention by the developer. These compensation activities must be designed by professional ecologists who understand population dynamics, community succession, and habitat maintenance.

Biodiversity offsetting is becoming increasingly attractive to governments and developers, and is used in over 56 countries with over 97 offsetting programmes active in 2014 (OECD 2014). The technique is primarily a financial instrument used to compensate for losses in biodiversity on a development site by funding enhancement activities on one or more different sites. The programmes are set up on the "polluter pays" principle, namely that ecological losses and effects from a development can be effectively compensated for where sufficient suitable habitat can be protected, enhanced, or established elsewhere.

Biodiversity offsetting programmes define the costs of biodiversity loss from development projects by "imposing a cost on the activities that cause adverse impacts to biodiversity" (OECD 2014). Programmers usually use a formula which defines the amount of "investment" required to effectively compensate for the loss of on-site ecology. There are a number of biodiversity offsetting calculators, e.g. Global Footprint Network (2015) and the Environment Bank (2015). Most biodiversity offset programmes aim to deliver "no net loss" (e.g. of a habitat, species, ecological status, ecosystem services). These include programmes operated by the Asian Development Bank, Canada, USA, and China. However, some programmes such as the African Development Bank, Australia, and International Finance Corporation have adopted a "net gain" approach.

Care must be taken when implementing biodiversity offsetting programmes because the use of this approach alone often results in the loss of the original ecology of the site, and the role it provided in the local ecological networks. Because of this the OECD (2014) state that biodiversity offsets "are intended to be implemented only after reasonable steps have been taken to avoid and minimise biodiversity loss at a development site".

While most offsetting programmes seek to replace or compensate for habitat loss, there are examples where the financial incentive is used to reduce external pressures affecting existing habitats of ecological value. This is intended to stabilise, or reverse, the decline in an ecological habitat, thus effectively compensating for ecological effects on the development site. One such programme is currently being developed by the UK Borough of Poole (2015), where investment is being undertaken in techniques to reduce nitrate entering Poole Harbour, where it currently affects the designated habitats and protected species present as well as leisure and commercial activities undertaken within it.

6.9 Monitoring

Post-construction monitoring is a weak point of most ESIA regimes (Glasson 1994, Goldberg 2008). However, it is an important aspect of EcIA, as "hard and fast" rules for complex temporal ecosystems and habitats can be difficult to define. It is important, therefore, for post-construction data to be collected and made available, such that impact-mitigation techniques can be improved, and the effectiveness of different mitigation measures or packages of mitigation can be identified. This is especially true where there are uncertainties concerning the significance of impacts or the effectiveness of proposed mitigation measures, including ongoing management procedures associated with these measures.

Requirements for post-construction EcIA monitoring vary across the world, particularly between international finance organisations and national requirements, and even between different consenting organisations within a single national boundary (e.g. in the USA). EcIA practitioners must understand the particular requirements for post-construction monitoring, and agree these with the consenting bodies and their ecological advisors prior to completion of the ESIA. Although numerous ecological, geographical, and cultural factors influence the scope of EcIA work, the requirements for post-construction ecological monitoring are generally similar in most countries, as shown in Table 6.5.

When planning post-construction monitoring, it is important to understand the aim of the exercise as this will affect the periodicity, location, and type of monitoring undertaken. Where post-construction monitoring aims to confirm the level of predicted ecological impact, it should be targeted at specific habitats or species where clear impacts were identified. This can be achieved by periodic repeat sampling, using the same or similar methods as those employed during completion of the baseline surveys (Spellerberg 2005).

Where monitoring is to identify the success or otherwise of a compensatory, offsetting, or mitigation measure, then clear objectives and criteria must be first set to define the success of the physical design or management activity. This requires the involvement of expert ecologists from both the developer and the

Table 6.5 Post-construction ecological monitoring requirements of selected countries and non-governmental organisations (adapted from Goldberg (2008) with supplemental information)

Country/ institution	Post-construction ecological monitoring requirements
Australia	The Environment Protection and Biodiversity Conservation Act of 1999 (Australian Government 2010) states that the consenting Minister of State may include conditions that require independent environmental auditing or environmental monitoring.
Brazil	ESIA regulations include a requirement to provide a monitoring plan which includes a full specification of the monitoring to be undertaken such as timing, location, and type of monitoring – see CONAMA Resolution 237/97, Article 6(IV).
China	ESIA regulations established in 2009 only include a requirement to provide proposals for construction monitoring (Article 17(6)).
Egypt	ESIA regulations require a fully specified mitigation and management plan that must include monitoring activities before, during, and after development work.
Slovak Republic	Government regulation requires monitoring and specifies the need to compare baseline conditions to post-construction conditions to understand the nature and extent of project-related impacts.
South Africa	The requirements for post-construction monitoring of residual impacts are defined in ESIA regulations established in 2014, and include the requirement for ecological assessment (Appendices 2 and 3). In addition, post-construction guidelines on ecological monitoring are available, such as Aronson et al. (2014) and GGS (2014).
United Kingdom	To date (prior to implementation of the 2014 EU Directive) there are no national requirements for post-construction monitoring of impacts. However, central government road projects have five-year aftercare periods during which ecological monitoring is required where European protected species and ecological design were considered in the development work.
USA	The federal National Environmental Policy Act requires a monitoring and enforcement program as a pre-requisite for obtaining final permits and certifications at the conclusion of a development project, particularly where mitigation work is included in the project. Both state and local governments reserve the right to impose additional monitoring requirements to protect sensitive cultural, economic, and environmental resources.
EBRD	Monitoring is required to demonstrate compliance with bank procedures during the execution and administration of development project work; however, ecological monitoring is not specifically required at present (EBRD, 2015).

Table 6.5 continued

Country/ institution	Post-construction ecological monitoring requirements
EC Directive	The 2014 EIA Directive (EU 2014) includes a new Article 8a.1(b) which requires ESIAs to provide information on "environmental conditions attached to the decision, a description of any features of the project and/or measures envisaged to avoid, prevent or reduce and, if possible, offset significant adverse effects on the environment as well as, where appropriate, monitoring measures". This is reinforced in Annex IV which provides for "a description of the measures envisaged … of any proposed monitoring arrangements (for example the provision of a post project analysis)".
IFC	Monitoring plan is required as part of ESIA.
World Bank	The preparation and implementation of a post-construction monitoring plan is required as part of the ESIA process. World Bank lending and oversight rules require a budget supporting a fully specified plan that includes data collection and knowledge transfer, as well as local training of key stakeholders. OD 4.01 states that borrowing for development projects is conditional on completion of the monitoring plan by the borrower.

consenting organisation. In such cases, methods used in the definition of the baseline may be enhanced with additional methods. A difficult aspect is allowing for changes that may occur without the development, e.g. as a result of natural trends or other developments and activities. The only way to assess this with reasonable certainty is to compare changes in receptor or compensatory sites with those in control/reference sites where these may be available (Bisset and Tomlinson 1990), or against national trends reported for the same habitat or species. Monitoring of representative species that would be expected to exhibit early and obvious signs of change can be used. On identification of agreed levels of change, more extensive monitoring can then be identified.

Post-construction monitoring can be time-consuming and expensive, so the monitoring programme should be carefully targeted, e.g. "on a few key environmental receptors rather than the whole range of effects from a project" (Frost 1997). For EcIA this may include species diversity, establishment or retention of particular species, establishment of defined habitat types (demonstrating defined dominance/species composition classifications), or use of the new ecological area by specific species such as those that are rare or protected in the target country, region, or area. A completion date will probably be imposed, at which time a final audit and report should be produced.

6.10 Conclusions

Ecologically focused impact assessment is an essential element in most ESIAs. While national and international ESIA regulations and guidance vary, and experts may differ in their investigation and assessment approaches, an EcIA can be a tremendous value to local stakeholders, and to organisations and agencies involved in oversight.

Ecological scales rarely, if ever, correspond with jurisdictional scales, which will challenge EcIA proponents to think broadly when identifying direct and indirect impacts and choosing appropriate spatial and temporal scales. The appropriate definition of spatial and temporal scales for ecological analysis is critical for predicting impacts that are based on the interactions of numerous concurrent project activities. The success of EcIA work is enhanced by reliance on quantitative information to the extent practical, because environmental quality and related project-performance metrics are often vital for understanding the significance of any deviations from baseline ecological conditions and ecosystem services, structure, and function. Baseline ecological studies should make maximum use of existing information, but new field surveys (at suitable times of year) will almost always be required. Consequently, it is important to identify significant impacts and ecological receptors (during scoping, if possible) and to target investigation work, mitigation plans, and post-construction monitoring efforts on those aspects of the ecosystem that are most likely to be impacted, either temporarily or in the long term, by the development activity. Impact predictions and mitigation proposals should be as precise and quantitative as possible, although a degree of uncertainty must be accepted.

The EcIA report should include clear explanations of survey methods, results, limitations, uncertainties, and relationships with other ESIA study components. A description of recommended mitigation measures and monitoring activities should be given wherever necessary and appropriate. EcIA should best serve the needs of project proponents, local communities, oversight regulatory agencies, and other stakeholders who collectively wish to preserve and protect natural ecological processes and habitats which do not adhere to anthropocentric boundaries.

References

Acreman MC, J Fisher, CJ Stratford, DJ Mould and JO Mountford 2007. Hydrological science and wetland restoration: some case studies from Europe. *Hydrology and Earth System Sciences* 11(1), 158–169.

Adeyemi A 2014. Changing the face of sustainable development in developing countries: the role of the International Finance Corporation. *Environmental Law Review* 16(2), 91–106.

Anderson P and P Groutage 2003. *Habitat translocation – a best practice guide*. London: CIRIA.

Andrews J and D Kinsman 1990. *Gravel pit restoration for wildlife. A practical manual.* Sandy, Beds: RSPB.

Angold PG, JP Sadler, MO Hill, A Pullin, S Rushton, K Austin, E Small, B Wood, R Wadsworth, R Sanderson and K Thompson 2006. Biodiversity in urban habitat patches. *Science of the Total Environment* 360, 196–204. www.nora.nerc.ac.uk.

Arboricultural Association 2015. http//trees.org.uk/faqs/distance to buildings.

Aronson J, E Richardson, K MacEwan, D Jacobs, W Marais, S Aitken, P Taylor, S Swales and C Hein 2014. South African good practice guidelines for operational monitoring for bats at wind energy facilities. South African Bat Assessment Advisory Panel.

Arts J, HA Runhaar, TB Fischer, U Jha-Thakur, F Van Laerhoven, PP Driessen, and V Onyango 2012. The effectiveness of EIA as an instrument for environmental governance: reflecting on 25 years of EIA practice in the Netherlands and the UK. *Journal of Environmental Assessment Policy and Management* 14(4), 1250025.

Australian Government 2010. *EPBC Act – Environmental Assessment Process.* www.environment.gov.au/system/files.

Australian Government 2011. *Survey guidelines for Australia's threatened fish. Guidelines for detecting fish listed as threatened under the Environmental Protection and Biodiversity Act 1990.* Department of Sustainability, Environment, Water, Population and Communities.

Barker G 1997. A framework for the future: green networks with multiple uses in and around towns and cities. *English Nature Research Report No. 256.* Peterborough, Cambs: English Nature.

BCG (Biodiversity Challenge Group) 2007. *High and dry.* Sandy, Bedfordhire, UK: RSPB.

Beanlands GE and PN Duinker 1983. *An ecological framework for environmental impact assessment in Canada.* Quebec: Federal Environmental Assessment Review Office

Begon M, C Townsend and JL Harper 2005. *Ecology: from individuals to ecosystems,* 4th Edn. Oxford: Blackwell Publishing.

Bender DJ, TA Contreras and L Fahrig 1998. Habitat loss and population decline: a meta-analysis of the patch size effect. *Ecology* 79(2), 517–533.

Bibby CJ, ND Burgess, DA Hill and SH Mustoe 2000. *Bird census techniques,* 2nd Edn. London: Academic Press.

Bisset R and P Tomlinson 1990. Monitoring and auditing of impacts. In *Environmental impact assessment: theory and practice,* P Wathern (ed.), 117–128. London: Routledge.

Blomberg S and R Shine 2006. *Reptiles.* In *Ecological census techniques: a handbook,* 2nd Edn. WJ Sutherland (ed.), 297–307. Cambridge: Cambridge University Press.

Borough of Poole 2015. *Nitrogen reduction in Poole Harbour.* www.poole.gov.uk/planning-and-buildings/nitrogen.

Bratton JH 1990. Seasonal pools – an overlooked invertebrate habitat. *British Wildlife 2,* 22–31.

Brigs S and MP Hudson 2013. Determination of significance in ecological impact assessment: past change, current practice and future improvements, *Environmental Impact Assessment Review 33,* 16–25.

Buckland ST, DR Anderson, KP Burnham, JL Laake, DL Borchers and L Thomas 2001. *Distance sampling: estimating abundance of biological populations.* Oxford: Oxford University Press.

Buckley GP and S Fraser 1998. Locating new lowland woods. *English Nature Research Report No. 283.* Peterborough, Cambs: English Nature. www.naturalengland.communisis.com/NaturalEnglandShop.

Burgman MA, S Ferson and HR Akçakaya 1993. *Risk assessment in conservation biology.* London: Chapman & Hall.

CEAA (Canadian Environmental Assessment Agency) 1999. *Cumulative effects assessment practitioners guide*, Hull, Quebec: CEAA. www.ceaa-acee.gc.ca/013/0001/0004/index_e.htm.

CEDR (Conference of European Directors of Roads) 2011. *Mobility for humans and wildlife – cost effective ways forward*. Task Force 7. CEDR/Danish Roads Authority.

CEQ (Council on Environmental Quality) 1993. *Incorporating biodiversity considerations into environmental impact analysis under the National Environmental Policy Act*. Washington DC: Executive Office of the President.

CFG (California Department of Fish and Game) 2009. Protocols for surveying and evaluating impacts to special status plant populations and natural communities. State of California, California Natural Resources Agency. www.dfg.ca.gov/biogeodata/cnddb/pdfs/Protocols_for_Surveying_and_Evaluating_Impacts.pdf.

CIRIA (Construction Industry Research and Information Association) 2007. *The Sustainable Urban Drainage Systems (SUDS) manual (C697)*. Birmingham: CIRIA, www.ciria.org.uk/suds/publications.htm.

Collingham YC and B Huntley 2000. Impacts of habitat fragmentation and patch size upon migration rates. *Ecological Applications* 10(1), 131–144.

Connell DW, P Lam, B Richardson and R Wu 1999. *Introduction to ecotoxicology*. Oxford: Blackwell Science.

Conway M, M Rayment, A White and S Berman 2013. *Explaining potential demand for and supply of habitat banking in the EU and appropriate design elements for a habitat banking scheme*. London: GHK Consulting Ltd in association with BIO Intelligence Service.

Cooper PF and BC Findlater (eds) 1991. *Constructed wetlands in water pollution*. Oxford: Pergamon.

COST (Cooperation on Science and Technology) 2014. *Handbook: Wildlife and traffic – a European handbook for identifying conflicts and design solutions*. COST 341.

Côté IM and MR Perrow 2006. *Fish*. In *Ecological census techniques: a handbook*, WJ Sutherland (ed.), 2nd Edn, 250–277. Cambridge: Cambridge University Press.

Cowley M and B Vivian 2007. *The business of biodiversity: a guide to its management in organisations*. Lincoln: Institute of Environmental Management and Assessment (IEMA). www.defra.gov.uk/wildlife-countryside/pdfs/biodiversity/bbpg2007.pdf.

Crites RW, EJ Middlebrooks and SC Reed 2006. *Natural wastewater treatment systems*. Boca Raton, FL: CRC Press.

Crofts A 1994. *How to create and care for wildflower meadows*. Lincoln: The Wildlife Trusts.

Crofts A and RG Jefferson (eds) 1999. *The lowland grassland management handbook*, 2nd Edn., Ch. 11: *Grassland creation*. English Nature and The Wildlife Trusts. www.naturalengland.communisis.com/naturalenglandshop/docs/low11.pdf.

Crofts A, L Bardsley and N Giles 2005. *The wetland restoration manual*. Newark: The Wildlife Trusts.

Crossman ND, B Burkhard, S Nedkov, L Willeman, K Petz, P Ignacio, EG Drakon, B Matin-Lopez, T McPhearson, K Boyanova, R Aklemade, B Egoh, MB Dunbar and J Maes 2013. A blueprint for mapping and modelling ecosystem services. *Ecosystem Services* 4, 4–14.

Csanyi 2015. *Examples of indicator species*. Hearst, Seattle Pi, download media.

CSIR (Council for Scientific and Industrial Research) 2007. *Ecological impacts of groundwater abstraction investigated*. Natural Resources and the Environment. CSIR News. www.csir.co.za/news/2007_dec/nre_04.htm.

DEAT (Department of Environment Affairs and Tourism) 2004. *Cumulative Effects*

Assessment. Integrated Environmental Management Information series 7. Pretoria: DEAT.

Defra (Department for Environment Food and Rural Affairs) 2005. *Key facts about: wildlife: scarce and threatened native species (2005).* London: Defra. www.defra.gov.uk/environment/statistics/wildlife/kf/wdkf02.htm.

DfT (Department for Transport) 2003. *Transport Analysis Guidance (TAG): The environment capital approach,* TAG unit 3.3.6. London: DfT. www.webtag.org.uk/webdocuments/3_Expert/3_Environment_Objective/3.3.6.htm.

DfT 2004. *Transport Analysis Guidance (TAG): The biodiversity sub-objective,* TAG unit 3.3.10. London: DfT. www.webtag.org.uk/webdocuments/3_Expert/3_Environment_Objective/3.3.10.htm.

Didim RK 2010. *Ecological consequences of habitat fragmentation.* The University of Western Australia and CSIRO. www.els.net/wileyCDA/ElsArticle/refd=a002.

Dixon JE 1993. The integration of ESIA and planning in New Zealand: changing process and practice. *Journal of Environmental Planning and Management* 36(2), 239–251.

EA (Environment Agency) 2007. *Understanding water for wildlife. Water resources and conservation: assessing the eco-hydrological requirements of habitats and species.* Bristol: Environment Agency. http://publications.environment-agency.gov.uk/pdf/GEHO0407BMNB-e-e.pdf?lang=_e.

EBRD (European Bank for Reconstruction and Development) 2015. *Environmental and social policy. Performance requirement PR1. Environmental and social appraisal and management.* www.ebrd.com.

EC (European Commission) 1999. *Guidelines for the assessment of indirect and cumulative impacts as well as impact interactions.* Brussels: EC, http://ec.europa.eu/environment/ESIA/ESIA-support.htm.

Eccleston C and JP Doub 2012. *Preparing NEPA environmental assessments: a user's guide to best professional practices.* Boca Raton: CRC Press.

EEA (European Environment Agency) 2008. *Impacts due to over-abstraction.* Copenhagen: EEA. www.eea.europa.eu.

EEA 2011. *Environmental risk assessment: approaches, experiences, and information sources.* Copenhagen: EEA. www.eea.europa.eu/publications/GH-07-97-595-EN-C2.

Elliott M and I Thomas 2009. *Environmental impact assessment in Australia: theory and Practice,* 5th Edn. Sydney: Federation Press

Emberton JR 2001. Green engineering. In Carpenter TG. *Environment, construction and sustainable development.* J Wiley & Sons.

EN (English Nature) 1999. *Habitats regulations guidance note 3: The determination of likely significant effect under The Conservation (Natural Habitats &c.) Regulations 1994.* Peterborough, Cambs: English Nature. www.mceu.gov.uk/MCEU_LOCAL/Ref-Docs/EN-HabsRegs-SigEffect.pdf.

EN 2004a. *State of nature. Lowlands – future landscapes for wildlife.* Peterborough, Cambs: English Nature. http://naturalengland.communisis.com/NaturalEnglandShop/browse.aspx.

EN 2004b. *Reptiles: guidelines for developers.* Peterborough, Cambs: English Nature, http://naturalengland.communisis.com/naturalenglandshop/docs/IN15.1.pdf.

Environment Bank 2015. *Biodiversity accounting and offsetting.* www.environmentbank.com/biodiversity offetting.php.

EPA (US Environmental Protection Agency) 2007. *River corridor and wetland restoration.* Washington, DC: Environmental Protection Agency. www.epa.gov/owow/wetlands/restore.

EPA (US Environmental Protection Agency) 2011. *Risk assessment*. Washington DC: EPA. www.epa.gov/risk.

EU (European Union) 2014. *Directive 2014/52/EU on the assessment of the effects of certain public and private projects on the environment*.

Eycott A, K Watts, D Mosely and D Ray 2007. *Evaluating biodiversity in fragmented landscapes: the use of focal species*. Edinburgh: Forestry Commission, www.forestry.gov.uk/PDF/FCIN089.pdf/$FILE/FCIN089.pdf.

Frost R 1997 Chapter 1 in Weston J 1997. *Planning and environmental assessment in practice*. Abingdon: Routledge.

Gault C 1997. *A moving story: species and community translocation in the UK – a review of policy, principle, planning and practice*. Godalming, Surrey: WWF-UK.

GGS (Government Gazette Staatskoerant) 2014. *Environmental impact regulations*. Republic of South Africa, Vol. 594. Pretoria: Department of Environmental Affairs.

Gibbons DW and RD Gregory 2006. *Birds*. In *Ecological census techniques: a handbook*, 2nd Edn, WJ Sutherland (ed.), 308–350. Cambridge: Cambridge University Press.

Gilbert OL and P Anderson 1998. *Habitat creation and repair*. Oxford: Oxford University Press.

Giles N, R Sands and M Fasham 2005. *Fish*. In *Handbook of biodiversity methods: survey, evaluation and monitoring*, D Hill, M Fasham, G Tucker, M Shewry and P Shaw (eds), 368–386. Cambridge: Cambridge University Press.

Glasson J 1994. Life after the decision: the importance of monitoring in ESIA. *Built Environment (1978–)* 20(4), 309–320.

Global Footprint Network 2015. *Ecological footprint calculator*. www.footprintnetwork.org/en/index.php/GF.

Goldberg DM 2008. *A comparison of six Environmental Impact Assessment regimes*. Centre for International Environmental Law, University of California.

Gorman M and D Raffaelli 1993. The Ythan estuary. *Biologist* 40(1) 10–13.

Government of Australia 2013. *Matters of national environmental significance. Significant impact guidelines 1.1 Environmental protection and biodiversity conservation act 1999*. Department of the Environment.

Government of New Zealand (undated). *A guide to preparing your Environmental Impact Assessment (ESIA) For concession applications*. Department of Conservation/Te Papa Atawhai.

Greig-Smith P 1983. *Quantitative plant ecology*, 3rd Edn. Oxford: Blackwell Science.

Halliday TR 2006. *Amphibians*. In *Ecological census techniques: a handbook*, 2nd Edn. WJ Sutherland (ed.), 277–296. Cambridge: Cambridge University Press.

Harrison C, J Burgess, A Millward and G Dawe 1995. *Accessible natural greenspace in towns and cities. A review of appropriate size and distance criteria*. English Nature Research Report No. 153. http://naturalengland.communisis.com/NaturalEnglandShop.

Heuvelhof ET and C Nauta 1997. The effects of environmental impact assessment in the Netherlands. *Project Appraisal* 12(1), 25–30.

Highways Agency, Scottish Office, Welsh Office & Department of the Environment Northern Ireland 1993. *Design manual for roads and bridges (DMRB)*, Vol. 11: *Environmental assessment*, Section 3, Part 4: *Ecology and nature conservation*. London: Highways Agency. www.standardsforhighways.co.uk/dmrb/vol11/section3.htm.

Hill D, M Fasham, G Tucker, M Shewry and P Shaw 2005. *Handbook of biodiversity methods: survey, evaluation and monitoring*. Cambridge: Cambridge University Press.

Hogan C 2012. *Habitat destruction*. www.ecoearth.org/view/article/153224.

Honrado JP, C Vieira, C Soares, MB Monteiro, B Marcos, HM Pereira, and MR Partidário

2013. Can we infer about ecosystem services from ESIA and SEA practice? A framework for analysis and examples from Portugal. *Environmental Impact Assessment Review* 40, 14–24.

IEA (Institute of Environmental Assessment) 1995. *Guidelines for baseline ecological assessment*. London: E & FN Spon.

IEEM (Institute of Ecology and Environmental Management) 2006. *Guidelines for ecological impact assessment in the United Kingdom* (version 7). Winchester: IEEM. www.ieem.net/ecia.

IEEM 2007. *Sources of survey methods (Guidelines for survey methodology)*. Winchester: IEEM, www.ieem.net/survey-sources.

IFC (International Finance Corporation) 2012a. *Performance standards on environmental and social sustainability*.

IFC 2012b. *Guidance note 6. Biodiversity conservation and sustainable management of living natural resources*.

IUCN (World Conservation Union) 2006. *The IUCN Red List of threatened species*, www.iucnredlist.org.

Jennings MD, D Faber-Langendoen, OL Loucks, RK Peet, and D Roberts 2009. Standards for associations and alliances of the US National Vegetation Classification. *Ecological Monographs* 79(2), 173–199.

João E. 2002. How scale affects environmental impact assessment. *Environmental Impact Assessment Review* 22(4), 289–310.

Joshi J, P Stoll, HP Rusterholz, B Schmid, C Dolt and B Baur 2006. Small-scale experimental habitat fragmentation reduces colonization rates in species-rich grasslands. *Oecologia* 148(1), 144–152.

Joy M, B David and M Lake 2013. *New Zealand freshwater fish sampling protocols. Part 1. Wadable rivers and streams*. New Zealand: Massey University.

Juniper T 2007. *Saving planet earth*. London: Collins (by arrangement with the BBC).

Kent M and P Coker 1992. *Vegetation description and analysis: a practical approach*. London: Belhaven Press.

Kerrison P, T Norman and M Fasham 2005. *Aquatic invertebrates*. In *Handbook of biodiversity methods: survey, evaluation and monitoring*. D Hill, M Fasham, G Tucker, M Shewry and P Shaw (eds), 359–367. Cambridge: Cambridge University Press.

Kirby KJ 1995. *Rebuilding the English countryside: habitat fragmentation and wildlife corridors as issues in practical conservation*. English Nature Science No. 10. Peterborough, Cambs: English Nature.

Krebs CJ 1998. *Ecological methodology*, 2nd Edn. New York: Harper & Row.

Krebs CJ 2001. *Ecology: the experimental analysis of distribution and abundance*, 5th Edn. New York: Benjamin Cummings.

Krebs CJ 2006. *Mammals*. In *Ecological census techniques: a handbook*, 2nd Edn. WJ Sutherland (ed.), 351–369. Cambridge: Cambridge University Press.

Latham D, E Jones and M Fasham 2005. *Reptiles*. In *Handbook of biodiversity methods: survey, evaluation and monitoring*, D Hill, M Fasham, G Tucker, M Shewry and P Shaw (eds), 404–411. Cambridge: Cambridge University Press.

Lévêque C 1997. *Biodiversity dynamics and conservation. The freshwater fish of tropical Africa*. Cambridge: Cambridge University Press.

Liebenberg L 1990. *A field guide to the animal tracks of Southern Africa*. Claremont, South Africa: David Philip Publishers.

LUC (Land Use Consultants) 2005. *Going, going, gone? The cumulative impact of land development on biodiversity in England*. English Nature Research Report No. 626.

Peterborough, Cambs: English Nature, http://naturalengland.communisis.com/NaturalEnglandShop/browse.aspx.

LUC 2006. *A practical toolkit for assessing cumulative effects of spatial plans and development projects on biodiversity in England*. English Nature Research Report No. 673. Peterborough, Cambs: English Nature. http://naturalengland.communisis.com/NaturalEnglandShop/browse.aspx.

MA (Millennium Ecosystem Assessment) 2005a. *Living beyond our means: natural assets and human well-being – Statement of the MA Board*. www.maweb.org/en/Reports.aspx.

MA 2005b. *Ecosystems & human well-being: synthesis*. Washington, DC: Island Press. www.maweb.org/en/Reports.aspx.

MA 2005c. *Ecosystems & human well-being; biodiversity synthesis*. Washington, DC: World Resources Institute. www.maweb.org/en/Reports.aspx.

Mammal Society 2002. *National survey of road deaths*. London: The Mammal Society, www.abdn.ac.uk/~nhi775/road_deaths.htm.

Marttunen M, S Viencnen, U Koivisto and E Ikaheimo 2013. *Impact significance determination in environmental impact assessment – literature review*. Imperia and LIFE IIENV/FI/905.

Mathur PK, H Kumar, JF Lehmkuhl, A Tripathi, VB Sawarkar and R De 2011. Mammal indicator species for protected areas and managed forests in a landscape conservation area of northern India. *Biodiversity and Conservation* 20, 1–17.

Mavel J 2013. *Overview of South African framework for biodiversity offsets*. BBOP Community on practice. http://bbopforest-trends.org/documents/file/jman.

Mawdsley JR, R O'Malley and DS Ojima 2009. A review of climate-change adaptation strategies for wildlife management and biodiversity conservation. *Conservation Biology* 23(5), 1080–1089.

Minerals Council of Australia 2015. *Cumulative environmental impact assessment industry guide – adaptive strategies*. www.minerals.org.au.

Ministry of Tourism, Wildlife and Antiquities 2014. *Uganda wildlife policy*. Republic of Uganda.

Mitchell-Jones AJ 2004. *Bat mitigation guidelines*. Peterborough, Cambs: English Nature. http://naturalengland.communisis.com/naturalenglandshop/docs/IN13.6.pdf.

Mitchell-Jones AJ and AP McLeish (eds) 2004. *The bat workers manual*, 3rd Edn. Peterborough, Cambs: JNCC. www.jncc.gov.uk/page-2861.

Morris P 2002. The hydrology and plant communities of Cothill Fen SSSI. *Fritillary* 3, 20–36.

Morris P and JR Emberton 2009. *Ecology – overview and terrestrial systems*. Chapter 11. In *Methods of environmental impact assessment*, 3rd Edn. P Morris and R Therivel (eds). London: Routledge.

Morrison ML, B Marcot, and W Mannan 2012. *Wildlife-habitat relationships: concepts and applications*. Washington, DC: Island Press.

Mueller-Dombois D and H Ellenberg 1974. *Aims and methods of vegetation ecology*. Chichester: John Wiley & Sons.

Mustoe S, D Hill, D Frost and G Tucker 2005. *Birds*. In *Handbook of biodiversity methods: survey, evaluation and monitoring*, D Hill, M Fasham, G Tucker, M Shewry and P Shaw (eds), 412–432. Cambridge: Cambridge University Press.

Nash HL 2014. Defining appropriate spatial and temporal scales for ecological impact analysis. *Environmental Practice* 16(4), 281–286.

Natural England 2014. *Freshwater and migratory fish: surveys and mitigation for development projects*. Wildlife and habitat conservation.

NIWA 2015. *Guidelines for sampling freshwater fisheries.* www/niwa.co.nz/our service/outline services.

NRA (National Roads Authority) 2009. *Guidelines for assessment of ecological impacts of National Road Schemes. Revision 2.* Irish Government.

Nuttall PM, AG Boon and MR Rowell 1998. *Review of the design and management of constructed wetlands.* London: CIRIA.

ODPM (Office of the Deputy Prime Minister) 2005. Circular 06/05. *Biodiversity and geological conservation – statutory obligations and their impact within the planning system.* London: ODPM. www.communities.gov.uk/publications/planningandbuilding/circularbiodiversity.

OECD (Organisation for Economic Co-operation and Development) 2014. *Biodiversity offsets: effective design and implementation.* Policy highlights – preliminary version.

Oulehle F, J Hofmeister, P Cudlín and J Hruška 2006. The effect of reduced atmospheric deposition on soil and soil solution chemistry at a site subjected to long-term acidification, Na etín, Czech Republic. *Science of the Total Environment* 370(2), 532–544.

PAA (Penny Anderson Associates) 1994. *Roads and nature conservation: guidance on impacts, mitigation and enhancement.* Peterborough, Cambs: English Nature.

Parker DM 1995. *Habitat creation – a critical guide.* English Nature Science No. 21. Peterborough, Cambs: English Nature.

PI (Plantlife International) 2000. *Where have all the flowers gone? A study of local extinctions as recorded in the county floras.* Salisbury, Wilts: Plantlife. www.plantlife.org.uk/uk/assets/saving-species/saving-species-publications/where-have-all-the-flowers-gone-2000.pdf.

PI 2005. *Under pressure: climate change and the UK's wild plants.* Salisbury, Wilts: Plantlife. www.plantlife.org.uk/uk/assets/saving-species/saving-species-publications/Under-pressure-climate-change-and-wild-plants.pdf.

Planning Inspectorate 2015. *Advice note 17. Cumulative effects assessment relevant to Nationally Significant Infrastructure Projects.* Bristol: Planning Inspectorate.

Portt CB, GA Coker, DL Ming and RG Randall 2006. A review of fish sampling methods commonly used in Canadian freshwater habitats. *Canadian Technical Report of Fisheries and Aquatic Sciences.* 2604. Fisheries and Oceans Canada.

Richter V and C Freegard C 2009. *Observing animals from secondary signs.* Standard Operating Procedure SOP No: 7.2. Department of Environment and Conservation. Australia: DEC Nature Conservation Service.

SEPA (Scottish Environmental Protection Agency) 2000. *Ponds, pools and lochans: guidance on good practice in the management and creation of small waterbodies in Scotland.* Edinburgh: SEPA. www.sepa.org.uk/pdf/guidance/hei/ponds.pdf.

Sophocleous MA 2009. The impacts of groundwater over-abstraction on the environment. *American Geophysical Union,* Fall Meeting.

Sparks TH, JO Maratfud, ST Manchester, P Rothery and JR Treweek 1997. Sample size for estimating species lists in vegetation surveys. *The Statistician* 46(2), 253–260.

Spellerberg IF 2005. *Monitoring ecological change,* 2nd Edn. Cambridge: Cambridge University Press.

Sutherland WJ (ed.) 2006. *Ecological census techniques: a handbook,* 2nd Edn. Cambridge: Cambridge University Press.

Sutherland WJ, I Newton, and RE Green 2004. *Bird ecology and conservation. A handbook of techniques.* Oxford: Oxford University Press.

Teitje W and T Berlund T 2000. *Land-use planning in oak woodland: applying the concepts of landscape ecology using GIS technology and the CDF oak woodland maps.* IHRMP Oak Fact Sheets No. 52. http://danr.ucop.edu/ihrmp/oak52.htm.

Thorne JH, PR Huber, E O'Donoghue and MJ Santos 2014. The use of regional advance mitigation planning (RAMP) to integrate transportation infrastructure impacts with sustainability; a perspective from the USA. *Environmental Research Letters* 9(6), 065001.

Townsend CR, M Begon and JL Harper 2007. *Essentials of ecology*, 3rd Edn. Oxford: Blackwell Publishing.

Treweek J 1999. *Ecological impact assessment.* Oxford: Blackwell Science.

Treweek J 2008. *Biodiversity that matters.* UKOTCF Conference. Jersey.

Treweek J, M Drake, O Mountford, C Newbold, C Hawke, P José, M Self and P Benstead (eds) 1997. *The wet grassland guide: managing floodplain and coastal wet grasslands for wildlife.* Sandy, Beds: RSPB.

Tucker G 2005. *Biodiversity evaluation methods.* In *Handbook of biodiversity methods: survey, evaluation and monitoring,* D Hill, M Fasham, G Tucker, M Shewry and P Shaw (eds), 65–104. Cambridge: Cambridge University Press.

Tucker JV, SE Kane, CC Twyford, SC Critchfield, D Blatchford, and B Hayes 2012. Practical realities of project financing through equator principal financial institutions. In: *International Conference on Health, Safety and Environment in Oil and Gas Exploration and Production.* Society of Petroleum Engineers

Walmsley CA, RJ Smithers, PM Berry, M Harley, MJ Stephenson and R Catchpole (eds) 2007. MONARCH – *Modelling Natural Resource Responses to Climate Change; a synthesis for biodiversity conservation.* Oxford: UKCIP (UK Climate Impacts Programme). www.ukcip.org.uk/resources/publications.

Warner RM 2013a. Implementing the rule of law for nature in the global marine commons: developing environmental assessment frameworks. In C. Voigt (ed.), *Rule of law for nature. New dimensions and ideas in environmental law* 347–364. Cambridge: Cambridge University Press.

Warner RM. 2013b. Environmental assessments in the marine areas of the polar regions. In E. Molenaar A, GO Elferink and DR Rothwell (eds.), *Law of the sea and polar regions: interactions between global and regional regimes,* 139–162. Leiden, The Netherlands: Martinus Nijhoff Publishers.

Warner RM 2014. Conserving marine biodiversity in areas beyond national jurisdiction: co-evolution and interaction with the law of the sea. *Frontiers in Marine Science,* 1 (Article 6), 1–11.

Wathern P 1999. *Ecological impact assessment.* In *Handbook of environmental impact assessment,* Vol. 1, J Petts (ed.), 327–346. Oxford: Blackwell Science.

Wikramanayake E, E Dinerstein, CL Loucks, DM Olson, J Morrison, J Lancreux, M McKnight and P Heclao 2002. *Terrestrial ecosystems of the Indo-Pacific. A conservation assessment.* WWF US. Washington DC: Island Press.

Williams P, J Biggs, M Whitfield, A Thorne, S Bryant, G Fox and P Nicolet 1999. *The pond book: a guide to the management and creation of ponds.* Oxford: Ponds Conservation Trust.

World Bank 1997a. *Environmental hazard and risk assessment. environmental assessment sourcebook update 21,* 1–10. http://siteresources.worldbank.org/INTSAFEPOL/1142947-1116493361427/20507357/Update21EnvironmentalHazardAndRisk AssessmentDecember1997.

World Bank 1997b. *Biodiversity and environmental assessment. Update No. 20 to the Environmental sourcebook.* Washington DC: World Bank.

World Bank 2001. *Environmental assessment sourcebook and updates.* Washington DC: World Bank. http://go.worldbank.org/D10M0X2V10.

World Bank 2002. *Wetlands and environmental assessment. Update No: 28 to the environmental sourcebook.* Washington DC: World Bank.

World Bank. 2013. *OP 4.01 – Environmental assessment.* http://web.worldbank.org/WBSITE/EXTERNAL/PROJECTS/EXTPOLICIES/EXTOPMANUAL/0,,contentMDK:20064724~menuPK:64701637~pagePK:64709096~piPK:64709108~theSitePK:502184,00.html.

ZSL (Zoological Society of London) 2007. *UK native species conservation.* www.zsl.org/field-conservation/uk-native-species.

7 Coastal ecology and geomorphology

Sian John, David S. Brew and Richard Cottle[1]

7.1 Introduction

The variety of processes, landforms, geology and substrates that characterise the coastal zones around the world, along with the influence of humans, has given rise to a wide range of complex ecosystems that provide highly valued environments. This value has led to the coast being subjected to many different pressures. Consequently its management is often complex due to the need to balance potentially conflicting requirements, such as meeting the demands of recreation and economic development, protecting vulnerable assets from flooding and erosion, and protecting important scenic, geomorphological and ecological systems.

These pressures are well illustrated in Europe. For example, in 2000 the proportion of land area covered by artificial surfaces was 25 per cent higher at the coast than inland, and the annual rate of urban sprawl at the coast (0.66 per cent) was higher than the European average overall (0.52 per cent) (EEA 2006). This land-take was largely driven by residential sprawl. The more recent drivers for urban development are commercial/industrial sites and sports and leisure (EEA 2013). In several coastal regions of Italy, France and Spain the coverage of built-up areas in the first kilometre of the coastal strip exceeds 45 per cent. Population densities in Europe are also higher on the coast by an average of 10 per cent compared to inland (EEA 2006). One of the consequences of this coastal development and population pressure is the loss of semi-natural and natural habitats, with an estimated loss of between 1 and 4 per cent of the area of wetlands, grasslands, pasture and mixed farmland having occurred in the EU between 1990 and 2000 (EEA 2006) and continued losses of wetland habitats since, resulting largely from afforestation, infrastructure development and agricultural conversion (EEA 2013). Between 2006 and 2009 there was a 4.9 per cent increase in impervious areas on the coast. Similar pressures face coastal areas worldwide.

An additional and pressing issue facing coastal areas is the predicted sea-level rise associated with climate change. This is particularly so in countries such as Bangladesh and the Netherlands, where wide areas are situated just above sea level. Over the period 1901 to 2010, global mean sea level rose by an average of 0.19 m, and a further likely rise of 0.26 to 0.55 m is predicted by 2100 (IPCC 2014). In England, the Environment Agency (2013) has produced sea-level rise contingencies for up to 2115 for use in flood risk planning. Similarly, the US Ocean Protection Council (2013) have provided guidance for incorporating sea-level rise projections into project planning and decision-making in California up to 2100, which range from 0.1 to 1.67 m. In addition to risks to human life, settlements and agricultural land, rising sea levels threaten the integrity of significant areas of coastal habitat. For instance, Jones et al. (2011) estimated that coastal margin habitats in the UK will decline in area by approximately 8 per cent by 2060, due to coastal erosion, sea-level rise and reduced sediment supply. Moreover, salt marshes that are 'trapped' between rising sea levels and fixed sea defences are being lost to coastal squeeze. Loss of salt marsh also threatens the integrity of seawalls (defending low-lying areas) that rely on their wave absorbing power (Möller et al. 2014).

This chapter provides a description of key coastal landforms, processes and habitats, and summarises relevant legislation and policy. In the context of the pressures faced in the coastal zone, it sets out proposed approaches to scoping environmental investigations, and coastal ecology and geomorphological surveys and studies. It considers typical impacts that arise in the coastal zone, their sources and nature, and methods of impact prediction, as well as options for impact mitigation and the purpose of and approaches to monitoring.

7.2 Definitions and concepts

7.2.1 The coastal zone

Coasts are among the most dynamic parts of the earth's surface. The land and the sea rarely meet at a constant boundary. The shoreline migrates daily with the tide, changes seasonally, and varies over longer timescales as the coast erodes or accretes, or as sea level changes. Coastal landforms are shaped and reshaped by winds, waves and currents, which in turn vary through time.

The coast everywhere is an interface between the energy of the sea and the resistance of the land. While there are bare rocky cliffs that will take the full force of a storm without apparently suffering much damage, in most places the energy is absorbed by a beach. This is sometimes a wide beach backed by dunes, or a narrow beach in front of cliffs, or a barrier beach behind which salt marshes can accumulate in a sheltered environment (see Figure 7.1). Left alone, coasts will change over time, either building out where sediment accumulates, or moving landward as areas are eroded and the sediment is removed offshore or along the shore.

For consistency across the geomorphological and ecological disciplines, definition of the seaward and landward limits of the coast uses the concepts of the **supralittoral**, **littoral** and **sublittoral** (nearshore) zones. The supralittoral zone lies above the limits of extreme high-water spring tides and, ecologically,

Figure 7.1

Sand spits, sand dunes, tidal inlet and backing lagoon of the Keurbooms Estuary at Plettenberg Bay, South Africa. The position of the spits and tidal inlet fluctuate over periods of decades demonstrating the dynamic behaviour of coasts. Courtesy of Dr. David Brew

can be defined approximately as the limit of influence by salt spray. The littoral zone is the part of the coast that is exposed at low water and submerged at high water (intertidal zone). The nearshore sublittoral zone is the coast seaward of the littoral zone to water depths where sediment is not disturbed during fair weather conditions.

The habitats and communities within all three zones are profoundly influenced by the substratum type, i.e. the rock or sediment type. In any given location, the substrata and other habitat features present depend on the geology and geomorphological processes in the area.

7.2.2 Coastal geomorphology

Coastal geomorphology is the study of coastal landforms, and in particular their nature, origin, development processes and material composition. This includes the study of the shallow marine environment that is influenced by terrestrial factors and the land where the influence of the sea is felt. Coastal geomorphology is important in Environmental and Social Impact Assessment (ESIA) for two main reasons: (i) geomorphological landforms and processes are integral components of coastal ecosystems, and (ii) it has direct relevance to the risk of coastal erosion and flooding. Hence a change in, for example, the hydraulic and sedimentary regimes of an estuarine system is likely to result in indirect effects on intertidal habitats throughout the system (such as changes in intertidal exposure due to modifications of the tidal range, and effects on the rate of erosion/accretion).

The requirements for geomorphological information that underpin the ESIA process can be broadly divided into two categories; process and material. Essential process information includes knowledge of the marine forcing factors present, such as wind, waves, tides, tidal currents and sea-level change, and their strengths, directions and variability with time. Essential material information comprises knowledge of the geology of the coastal zone (offshore as well as onshore, including rock and softer sediments), its topography and bathymetry (underwater topography), and the type (mud, sand, gravel) and distribution of mobile and non-mobile sediments. Of key importance is the interaction of material and processes at the land-sea interface. This includes the nature of erosion, sediment transport and deposition, and spatial and temporal changes in these processes.

In simple terms, therefore, coastal geomorphology is defined by interactions between physical processes (wind, waves, tides and tidal currents) which erode, transport and deposit sediment to create landforms. It is a complex subject area, and only a brief outline of those aspects of most relevance to ESIA is provided here. Further information can be found in many texts, including Pethick (1984), Woodroffe (2002), Bird (2008), Davidson-Arnott (2009) and Masselink et al. (2011).

Physical processes

Waves play a dominant role at the coast where they may form beaches, erode cliffs and dunes, transport sediments onshore, offshore and alongshore, and exert forces on coastal structures. Most waves are wind-generated, and their size is governed by wind speed, the length of time the wind blows at that speed, and the unobstructed distance of sea (known as the fetch) over which the wind blows. Wave action is greatest on coasts exposed to strong onshore winds over extensive areas of sea.

The two main components of tides are rhythm and range. In most locations around Europe, the tidal rhythm consists of approximately two tidal cycles per day. Notable exceptions include the double high water experienced in the Solent, UK. In Karumba, Australia and elsewhere there is just one high and low tide each day. The tidal range (difference in height between high- and low-water levels at a point) varies daily, with a two-weekly cycle of larger spring tides (tides that occur when the tide-generating forces of the sun and moon are acting in the same directions, so the tidal range is higher than average) that advance and retreat much further than the smaller neap tides (tides that occur when the tide-generating forces of the sun and moon are acting at right angles to each other, so the tidal range is lower than average). In addition, there are larger seasonal cycles, with the largest spring tides near the spring and autumn equinoxes (in March and September). These are the extreme high-water spring-tide levels (EHWS) and extreme low-water spring-tide (ELWS) levels. The EHWS level can be extended by waves, especially on exposed shores and during storms.

The mean tidal range (the height difference between mean high-water spring tide (MHWS) and mean low-water spring tide (MLWS) levels) varies considerably between different locations, from 1–2 m (*microtidal*) to as much as 12 m (*macrotidal*). The tidal range affects coastal geomorphology by controlling the vertical distance over which waves and tidal currents are effective. Interaction between waves and tides is dependent on the tidal phase. Storm waves arriving at low tide are unlikely to cause flooding while storm waves at high water may have catastrophic consequences, such as those due to the coastal floods of 1953. In 1953, strong winds combined with very low atmospheric pressure and high spring tides to produce an abnormal rise in local sea level (nearly 3 m above normal high water). This caused defences along the coasts of the Netherlands, Belgium, England and Scotland to be damaged and overtopped by huge waves.

Tides are controlled by the action of the sun and moon, and their height can be predicted accurately for any location and time. However, astronomical tidal predictions seldom agree with those actually observed. This is because tidal movements across an ocean are influenced by global weather systems, with low- or high-pressure systems acting directly on the sea surface over large areas. This causes the tidal elevation to rise or fall, respectively, creating a positive or negative storm surge. In the open sea, the effects of storm surges are minimal, but as the storm event passes into relatively shallow water the effects become more pronounced and it is not uncommon to experience storm surge heights of 0.5 to 1.0 m. The timing of storm events with respect to the phase of the astronomical tide is important. When a storm surge coincides with a spring high tide, the resulting total surge can be many times more devastating than the surge alone; there will be a high risk of flooding unless the defences (or indeed natural geomorphological features) are able to cope with the increased elevation.

Tidal currents are the primary driving force in the formation and maintenance of large-scale sublittoral bedforms, such as sand banks and sand waves. They are capable of importing sediment into estuaries on the flood tide and exporting sediment on the ebb tide, and they have the potential to scour the sea bed if their velocities are high enough.

Sedimentary processes and landforms

Sediment is the product of the weathering and erosion of rocks. Depositional landforms would not exist if they were not supplied with sediment that is redistributed by water and wind. Sediment therefore provides a critical link between form and function, and most physical changes (erosion, transport and deposition) at the coast are associated with its movement.

Erosive processes are widespread, e.g. it has been estimated that 70 per cent of the world's sandy coastline is being eroded (Bird 1985), with waves being the most important agent. The predominant erosional landform types are coastal cliffs and shore platforms. Coastal cliffs are generally steep-faced and consist of consolidated and unconsolidated materials that may range from granites to softer glacial till. Coastal or maritime cliffs can be particularly valuable habitats, often supporting internationally important populations of breeding seabirds. In areas of low sediment supply, a shore platform commonly fronts the cliff (Figure 7.2). Shore platforms are near-horizontal and similar in composition to the lower layers of the cliff. They are essentially the eroded profile of the sea bed over which the cliff has retreated in the past.

Figure 7.2

Cliffs and shore platform composed of London Clay on the Isle of Sheppey, UK. As the platform is lowered the toe of the cliff erodes; toe erosion is in dynamic equilibrium with platform lowering unless modified by humans. Courtesy of Dr. David Brew

Cliffs may suffer severe erosional processes at their base when water is sufficiently shallow for waves to break. The stability of cliffs may also be affected by groundwater seepage and frost action above sea level, and by cliff geology. Soft cliffs of glacial till are often subject to quite rapid erosion, as is the case along many stretches of the east coast of England (Quinn et al. 2009).

Sediment transport depends to a large extent on the particle size and the stress applied to the sediment by the physical process. For example, breaking waves and surf have the potential to transport large volumes of sediment along the coast. Most important to the alongshore transport of sediment are waves approaching and breaking obliquely to the coast, causing a zig-zag movement of sediment particles. The ability to assess the direction and magnitude of alongshore sediment transport is central to the ESIA process. The net transport direction can be established from several field criteria, including the build-up of sediment against cross-shore structures (jetties, groynes; Figure 7.3) and the alongshore growth of spits (Figure 7.1). Alternatively, the potential for sediment transport can be numerically modelled using one of a wide range of models available in the market.

Processes of sediment deposition lead to a variety of coastal landforms, the most important of which are:

- *Beaches* – accumulations of loose sand or shingle that change shape in response to changes in wave energy. The movement of beach sediment, in turn, dissipates some of the energy of a wave breaking on the shore. A beach is therefore able to maintain itself in a state of dynamic equilibrium with its environment due to the mobility of its sediments. Beaches provide a natural barrier that protects a shoreline during storm events and habitats for numerous invertebrates and shorebirds. They are also desirable places to live near (Figure 7.4).

Figure 7.3

Aerial photograph showing the build-up of beach sediment on one side of a set of groynes, indicating longshore sediment transport from left to right. Courtesy of the Environment Agency

Figure 7.4

Wide sandy beach at Filey Bay, UK. Courtesy of Dr. David Brew

- *Spits* – when a coastline turns abruptly landward (such as where a bay or valley opens onto a coast) a finger-like extension of the beach, a spit, may project out across the indentation (Figure 7.1). Generally spits are connected to the end of the attached beach by a narrow neck and are fed by sediment eroded from further up the alongshore transport system.
- *Dunes* – form where a supply of dry sand and the wind to move it exist together. Dune systems are usually fronted by sand beaches (and/or sand bars exposed at low tide), with which they have a close association. Under natural conditions, dunes undergo periods of growth and erosion, with each process contributing to their dynamic evolution. Sand dune systems can provide an important natural coastal flood defence and are also of great importance for nature conservation, recreation and tourism (Figure 7.5). Their development and maintenance depends, to a large extent, on vegetation, which facilitates accretion and stabilisation of the sand (Carter et al. 1992, Nordstrom and Carter 1991).
- *Muddy shorelines* – are common along the upper levels of the intertidal zone of estuaries and sheltered tidal embayments (Figure 7.6). They form where tidal-current velocities are too weak to re-suspend completely the mud that settles out at high-water slack, thus permitting the net accretion necessary to form mudflats and salt marshes. In estuaries, accretion is enhanced by the mixing of freshwater and seawater, which causes

Figure 7.5

Sand dunes fronted by a beach at the mouth of the Tees Estuary, UK. Hartlepool nuclear power station is in the background behind the dunes. Courtesy of Dr. David Brew

flocculation and hence settling of waterborne sediments (Dyer 1998). The tide is the central feature around which these habitats function, through accretion and the development and maintenance of tidal channel and creek networks. The tide also sets the maximum height to which saltmarshes can vertically accrete. Saltmarshes and mudflats are important

Figures 7.6a and 7.6b

Examples of mature marsh (left) and mudflat (right) in the Severn Estuary, UK. Courtesy of Dr. David Brew

for flood defence, as they dissipate wave and tidal energy, reducing their erosive power. They are also of high natural value, supporting large populations of invertebrates and birds.

All of these systems depend on a continuous supply of sediment (from river discharge, coastal erosion etc.), which can be interrupted by activities such as coastal protection works and dredging (§7.5.2). Changes in sediment supply are the commonest cause of downdrift effects (impacts on the lee side of coastal activities), including downdrift erosion.

At a landscape scale, sediment erosion, transport and deposition can be defined within the context of coastal sediment cells. These are lengths of shoreline within which the cycle of bedload sediment erosion, transportation and deposition is essentially self-contained. The up-coast boundary of a littoral cell is typically a headland, littoral barrier or a sink such that the transport of sediment into the cell from the adjacent up-coast compartment is restricted. Sediment enters a coastal cell primarily from cliff or dune erosion, and rivers draining to the shoreline, and is then transported alongshore and cross-shore within the cell. Ultimately, sediment is permanently lost from the cell offshore or onshore into coastal dunes. The littoral cell boundary concept does not apply to fine-grained sediment.

A coastal sediment cell includes the beach above the highest tides, windblown sand, and any sediment within the surf zone and out to the depth where sediment is not disturbed during fair weather conditions. Ideally, each coastal cell exists as a distinct entity with little or no transport of sediment between adjacent cells. Although many headlands form nearly total barriers to alongshore transport, it may be that under certain conditions, such as large storms, significant sediment is carried into adjacent cells.

The upper sections of estuaries and coastal inlets may form coastal floodplains, which can provide protection to more inland areas against flooding resulting from storm surges and sea-level rise. They will flood due to raised sea levels, river floodwaters, or a combination of both (Environment Agency 1997). However, in many countries the seaward sections of former coastal floodplains are now protected by flood defences and support a variety of land uses ranging from grazing marsh to urban and industrial development. These areas are susceptible to flooding due to poor drainage, or breaching or overtopping of coastal defences.

7.2.3 Supralittoral habitats

Supralittoral habitats lie above the limits of the extreme high-water spring tides, and support terrestrial vegetation. In near-littoral locations which are affected by wave splash and spray, salt-tolerant plant species often dominate the vegetation communities. The supralittoral zone includes several important habitats, including coastal cliffs and vegetated shingle. As in littoral habitats, an important controlling factor is substratum type (see Table 7.1).

Coastal cliffs and slopes vary widely in character, reflecting local geology and land forms, and may possess faces ranging from vertical to gently sloping. These features support a variety of habitats, such as rock crevices and ledges, seepages, coastal grasslands, heathlands and scrub. These habitats are considered to extend inland to at least the limit of salt-spray deposition, and hence sometimes encompass whole headlands or islands (UKBG 1999).

Vegetated shingle sometimes supports scrubby vegetation or a grass sward, but more exposed areas have open vegetation with scattered vascular plants and lichens (Packham and Willis 1997, Packham et al. 2001) (see Figure 7.7). This habitat is particularly sensitive to human disturbance and once damaged, recovery can be slow (Covey and Laffoley 2002).

Sand dunes typically consist of several dunes aligned approximately parallel to the coastline and increasing in age inland, interspersed with depressions (Figure 7.5). They are complex systems which include a range of habitats that are sensitive to disturbance. Information on sand dune ecology is provided in Gimmingham et al. (1989), Packham and Willis (1997), Crawford (1998), Martinez and Psuty (2004) and McLachlan and Brown (2006).

Figure 7.7

Example of vegetated shingle on the Dungeness foreland, UK. Courtesy of Dr. David Brew

Supralittoral habitats are usually backed by agricultural land or developed land (e.g. residential properties, industry, recreation and communication infrastructure). In the absence of these human land uses, these habitats would grade inland into fully terrestrial habitats such as woodland. Consequently, from an ecological perspective the landward extension of the supralittoral coastal zone may not be clearly definable.

7.2.4 Littoral habitats

The littoral zone can be very narrow, e.g. steep shingle beaches, or quite extensive, e.g. mudflats. The approximate upper and lower boundaries of this zone are the extreme high-water and low-water spring-tide levels. In most littoral habitats the resident species are essentially marine, but adapted to a regime of cyclical immersion. The communities usually exhibit clear zonation along the land-sea axis, although this is controlled by elevation rather than distance from high- and low-water levels.

The substrate characteristics of littoral habitats vary depending on local geology, landform type, and exposure to wave action, tides and tidal currents. For instance, mud accumulates in sheltered locations, and rocky shores occur where exposure prevents any sediment deposition (see Table 7.1).

Rocky shores provide a generally impenetrable substratum that precludes burrowing or penetration by all but a few organisms, but supports seaweeds and animals that adhere to rock surfaces. The wide variation in wave exposure and rock types leads to a wide variety of associated community types. Sheltered

Table 7.1 Typical locations of, and relationships between, littoral and supralittoral habitats in relation to wave exposure

Location	Littoral zone	Supralittoral zone
Exposed coastlines and headlands	Rocky shores	Coastal cliffs and slopes
Fairly exposed coastlines, typically where oblique waves transport shingle alongshore to deposition locations	Shingle beaches	Vegetated shingle
Exposed or fairly exposed coastlines, often in bays or at the mouths of estuaries	Sandy shores	Sand dunes and machairs
Estuaries and sheltered inlets (sometimes behind sand dunes or barriers/spits)	Mudflats and saltmarshes	
Depressions partially cut off from seawater, usually by barriers of sand or shingle	Coastal lagoons	
Tropical intertidal zone	Mangrove forest	

shores are normally dominated by seaweeds and support a diverse fauna, including gastropods (e.g. topshells, whelks and winkles). By contrast, most seaweeds and animal species are excluded from very exposed rocky shores, which are usually dominated by barnacles, mussels and limpets. Further information on rocky-shore ecology can be obtained from Little and Kitching (1996), Denny and Gaines (2007), and Raffaelli and Hawkins (2011).

Shingle beaches are a hostile environment, as a result of the constant movement and grinding action of sediment, in which few resident species can survive. However, supralittoral shingle is a valuable habitat (Figure 7.7).

Sandy and muddy shorelines are soft, unstable substrates that are not generally suitable for species requiring anchorage, but can host burrowing molluscs and marine worms that live in the substratum. Sandy shores are a relatively hostile environment (see Brown and McLachlan 2006) and most are too unstable for plant growth. However, seagrass beds occur on some muddy sands. Mudflats are usually coated with a film of microscopic **algae** (Coles 1979) and normally contain a diverse and abundant invertebrate infauna, principally of bivalves (e.g. cockles), and marine worms, especially in estuaries.

Saltmarshes develop on mudflats where there is sufficient shelter and elevation, and the mud is sufficiently stable, to permit colonisation by salt-tolerant **vascular plants**, the growth of which facilitates further sediment accretion and stabilisation of the substrate. Saltmarshes are largely restricted to the zone between mean high-water neap tides and mean high-water spring tides, so only the lower fringes are submerged by the daily tidal cycle throughout the year, and the upper levels are only subject to inundation at EHWS tides. Although dominated by essentially terrestrial vegetation, the plants are adapted to live in saline conditions. The communities usually exhibit zonation, along the land-sea axis, in relation to the frequency and duration of inundation.

Saltmarshes are often located in *estuaries*, which are unique ecosystems in which the mixing of freshwater and saltwater is a fundamental component of their ecology. Information on saltmarsh and estuarine systems is provided in Adam (1993), Packham and Willis (1997), Little (2000), McLusky and Elliott (2004) and Saintilan and Rogers (2013). Because factors such as the relatively warm seas, mild winters, and nutrient inputs from the land provide suitable conditions for the development of abundant and species-rich invertebrate communities in intertidal areas, estuaries are among the most biologically productive ecosystems in the world (Rothwell and Housden 1990). They provide rich feeding grounds for birds and, in particular, form vital links between the breeding and overwintering grounds of migratory waders and wildfowl.

Coastal lagoons are bodies of saline or brackish water that are partially separated from the sea, but retain some seawater at low tide (see UKBG 1999). They often support unusual communities that include algae, vascular plants, and invertebrates that rarely occur elsewhere (see Barnes 1994 and Anthony et al. 2009).

Mangrove forests are found in tidal areas in tropical and subtropical regions within 30 degrees of the equator (see Figure 7.8). They are characterised by a rich biodiversity and create sheltered nursery grounds. They also protect the coastline from extreme wave action by absorbing and dispersing tidal surges and facilitating sedimentation (Barnes and Hughes 1988). Mangroves are pioneered by the hardiest salt and wind tolerant species which trap sediment and raise the level of the shore, decreasing the period of tidal inundation. Other less tolerant but competitively stronger species can then establish, creating a succession of species.

7.2.5 Sublittoral habitats

The upper limit of the sublittoral zone is the extreme low-water spring-tide level. The seaward limit is less clear, but can be taken to include the shallow seas to the depth where sediment is not disturbed during fair weather conditions. The environment fluctuates less widely than that of the littoral zone, and the sea bed is usually dominated by soft sediments, with rocky substrates restricted to narrow zones adjacent to coastlines.

The sublittoral zone includes **benthic** (seabed) habitats and communities ranging from nearshore algal or seagrass-dominated communities (where light levels permit photosynthesis) to deeper water animal-dominated communities on a variety of sediment types. Sublittoral sediment communities generally

Figure 7.8

Example of a mangrove forest in Western Port, Australia. Courtesy of Dr. David Brew

have high proportions of polychaetes, bivalves and echinoderms, and can include biogenic reefs built by polychaetes, bivalves, coral reefs or cold-water corals (which can extend down to c.2,000 m). Benthic communities exhibit appreciable variation due to:

- Substrate type – which largely controls the range and types of organisms present in an area (e.g. soft sediment, burrowing fauna or cobble/boulder-attached fauna) and, to a lesser degree, those in the water above (which include organisms that depend on the seafloor for food, shelter or reproduction).
- The considerable variation in water movement, including turbulence, currents and tidal movement.
- Differences in salinity and *turbidity*. Near the mouths of estuaries, salinity may be reduced by freshwater inputs, and turbidity increased by suspended sediments, especially in wet weather. Turbidity also tends to be high in areas with a muddy or sandy sea bed, especially when sediments are disturbed during storms.
- Temperature and water clarity – with coral reef structures forming in clear tropical and semi-tropical waters, creating a highly diverse ecosystem and providing coastal protection.
- Topography and currents – with cold-water corals forming in deep cold water on hard-sloping ground in strong currents, creating a rich oasis of food and shelter, providing breeding grounds and refuge for fish and other species.
- Light – allowing photosynthesising kelp forest (large brown seaweeds) to establish in cool clear waters on hard substrata, and seagrass beds to form on sand and mudflats.
- Geothermal activity – causing hydrothermal vents at continental plates where seawater meets hot magma through gaps in the earth's crust. These support extremophiles which work with bacteria to create energy from the chemical compounds in the water (Barnes and Hughes 1988).

As in all ecosystems, sublittoral and open-sea communities depend on flows of energy and nutrients along food chains that are based on photoautotrophs: organisms, typically plants, that convert inorganic materials into organic materials using sunlight as their energy source. The exception is deep-ocean hydrothermal vents, where the food chains rely on chemoautotrophs that utilise chemical energy. Apart from algal and seagrass beds, the primary producers are *phytoplankton* which form the basis of food chains in both free-floating/swimming and benthic communities. A clear difference between sublittoral and terrestrial communities is that a much larger proportion (up to 80 per cent) of the 'primary pasturage' (the phytoplankton) is consumed by *zooplankton*, and passes along grazing food chains. The ecology of the sublittoral zone is discussed in Earle and Erwin (1983), Hiscock (1998), Gray and Elliot (2008) and Speight and Henderson (2010).

7.3 Key legislation, policies and guidance

7.3.1 Legislation

Legislation in the coastal zone worldwide is complex, typically with no overall authority responsible for its management. Responsibility is often shared between several central and local agencies with varying responsibilities and limits of jurisdiction. However, there have been, and continue to be, moves towards greater unification of responsibility. For instance, in the UK the Marine Management Organisation (MMO) was established under the Marine and Coastal Access Act 2009, as an umbrella organisation with overall responsibility to license, regulate and plan marine activities in the seas around England and Wales.

International legislation/agreements and EU Directives relevant to the coast are set out in Table 7.2. Much of this legislation refers to both inland and coastal waters. While most legislation on nature conservation is not specific to the coast, it is often highly relevant to the coastal zone. To this end the general policies on nature conservation outlined in §6.3 also apply to the coastal zone.

International nature conservation agreements often set out obligations, which are then incorporated into EU Directives which, in turn, are transposed into country-specific laws. For example, the EU meets its obligations for bird species under the Bonn Convention and more generally the Bern Convention by means of the Birds Directive 2009/147/EC, which is then transposed into the legislation of individual Member States. Similarly, the Espoo Convention on Environmental Impact Assessment in a Transboundary Context sets out the obligations of parties (of which there are 56 globally) to assess the environmental impact of certain activities at an early stage of the planning process. Directive 97/11/EC brought EIA practice in line with the Convention and increased the types of projects covered, and the number of projects requiring mandatory EIA (Annex I). It also provided for new screening arrangements, including new screening criteria (at Annex III) for Annex II projects, and established minimum information requirements. The EIA Directive and its subsequent amendments (most recent of which is 2014/52/EU) are transposed into the UK through a raft of legislation including: the Town and Country Planning (EIA) Regulations 2011, the Marine Works (EIA) Regulations 2007 (as amended) and the Infrastructure Planning (EIA) Regulations 2009.

The Water Framework Directive (WFD) is one of the most substantial pieces of European water legislation to date. The requirements of the WFD need to be considered during all stages of the planning and development process. Requiring all EU Member States to prevent deterioration and to protect and enhance the status of aquatic ecosystems, it ensures all new schemes do not adversely impact upon the status of aquatic ecosystems. The WFD requires all inland, estuarine and coastal waters to reach 'good status' by 2021 or, in some cases, 2026. Good status is achieved by establishing a river basin district structure within which demanding environmental objectives will be set, including ecological targets for surface waters (see also Chapter 2). In

Table 7.2 International agreements and EU legislation relevant to the coastal zone

Agreement for the Conservation of Cetaceans (whales and dolphins) in the Black Sea, Mediterranean Sea and Contiguous in the North and Baltic Seas (ACCOBAMS) 1996.

Agreement on the Conservation of Small Cetaceans of the Baltic and North Seas (ASCOBANS) 1991.

Bathing Water Directive (BWD) (76/160/EEC) revised in 06/7/EC. Integrated into all other EU measures that protect the quality of rivers, lakes, groundwaters and coastal waters through the Water Framework Directive.

Convention on Biological Diversity (CBD) 1992.

Convention on the Conservation of European Wildlife and Natural Habitats (the Bern Convention) 1982.

Convention on the Conservation of Migratory Species of Wild Animals (Bonn Convention or CMS) 1985.

Convention on the Prevention of Marine Pollution by Dumping of Wastes and Other Matter 1972, updated by the London Convention 1992 – to protect the marine environment from human activities.

EIA Directive 85/337/EEC revised in 2014/52/EU – includes in Schedule 1: large ports and piers, and in Schedule 2: coast protection works (other than maintenance or reconstruction); large fish farms; reclamation; shipyards; marinas > 0.5 ha; and construction of harbours and ports > 1 ha. Other particularly relevant projects include oil or gas extraction plants and pipelines, and extraction of minerals by fluvial dredging (but not marine dredging or disposal).

ESPOO Convention 1991 implemented by the EIA Directive.

Floods Directive 2007/60/EC requires Member States to assess if all water courses and coast lines are at risk from flooding, to map the flood extent and assets and humans at risk in these areas and to take adequate and coordinated measures to reduce this flood risk.

Habitats Directive 1992/43/EEC.

Marine Strategy Framework Directive 2008/56/EC aims to establish a framework within which Member States will take measures to maintain or achieve 'good environmental status' in the marine environment by 2020.

North Sea Conference 2006 to protect and improve the marine environment of the North Sea.

Oslo and Paris Convention for the Protection of the Marine Environment of the North-East Atlantic (OSPAR) 1992 accepted in EU Council Decision 98/249/EC – signatories agreed to continually reduce emissions of hazardous substances, with the aim of achieving near background levels of naturally occurring substances and near zero concentrations of synthetic substances by 2020.

Urban Waste Water Treatment Directive 91/271/EEC.

Waste Framework Directive 2008/98/EC requires that waste is managed without endangering human health and harming the environment.

Water Framework Directive 2000/60/EC amended by 2013/39/EU primarily concerned with the ecological status of aquatic ecosystems (i.e. biological elements, and the chemical, physico-chemical and hydromorphological elements supporting the biological elements) and preventing deterioration/enhancement of the status of aquatic ecosystems.

Wild Birds Directive 79/409/EEC amended in 92/43/EEC (commonly referred to as the Birds Directive).

addition, EU Directives such as Directive 2008/98/EC, the Waste Framework Directive, aim to control pollution of all surface waters, including coastal waters, and cover some port and harbour operations.

The Marine Strategy Framework Directive (see Table 7.2) aims to protect more effectively the marine environment across Europe by achieving 'good environmental status' in the EU's marine waters by 2020 and the resource base upon which marine-related economic and social activities depend. It is the first EU legislative instrument related to the protection of marine biodiversity, as it contains the explicit regulatory objective that 'biodiversity is maintained by 2020' as the cornerstone for achieving good environmental status.

Internationally, marine protected areas (MPAs) are protected areas of seas and oceans that typically restrict human activity to protect natural or cultural resources. They are protected by local, state, territorial, native, regional or national authorities and differ substantially among nations. This variation includes different limitations on development, fishing practices, catch limits, moorings and bans on the removal or disruption of marine life. Milford Sound and Goat Island in New Zealand are strict marine reserves, and 'no take' zones, with legal protection against fishing and development. In contrast, in the Phoenix Island Protected Area in the Republic of Kiribati MPAs provide revenue. The largest MPAs are in the Indian and Pacific Oceans. In 2014, more than 6,500 MPAs encompassed just over 2 per cent of the world's oceans. Marine sites designated in the UK include: Marine Nature Reserves (MNRs), Marine Conservation Zones (MCZs) and MPAs.

Elsewhere a number of voluntary marine nature reserves have been established by agreement between non-governmental organisations, stake-holders and user groups, and coastal sites may have national designations and non-statutory designations (e.g. Heritage Coasts).

The greatest protection, of course, is afforded to 'international' sites such as Special Areas of Conservation (SACs), Special Protection Areas (SPAs) or Ramsar sites. The Ramsar Convention (formally, the Convention on Wetlands of International Importance, especially as Waterfowl Habitat) is an international treaty for the conservation and sustainable utilisation of wetlands, recognising the fundamental ecological functions of wetlands and their economic, cultural, scientific and recreational value. Presently there are 169 contracting parties. The Ramsar List of Wetlands of International Importance included 2,208 sites covering over 210,734,269 ha at the end of 2015, but this number is constantly increasing, with China having designated three new sites in early 2016. The highest number of sites is in the UK at 170, and the country with the greatest area of listed wetlands is Bolivia, with over 140,000 km². The Ramsar definition of wetlands is fairly wide, including 'areas of marine water the depth of which at low tide does not exceed six meters' as well as fish ponds, rice paddies and salt pans. In the EU, many Ramsar sites are also SPAs classified under the Birds Directive. Further data and information on Ramsar sites elsewhere in the world can be located via the Ramsar Wetland Data Gateway as well as on the Ramsar sites database.

In spite of the stringent obligations often imposed by designation, many estuaries and sections of coastline worldwide, recognised as internationally important wildlife or geological sites, continue to be subject to significant development pressures.

7.3.2 Policies and guidance

The EC Fifth Environmental Action Programme (EC 1993) called for *sustainable development* of coastal zones in accordance with the *carrying capacity* of the coastal environments. Integrated Coastal Zone Management (ICZM) has been called for by several UN and international conferences, including the Rio Earth Summit and the World Coastal Conference 1993.

This has led to a focus on improved and integrated coastal management in order to deal with historically complex, sectoral management arrangements and also to reflect increased awareness of the pressures facing the coastal zone. In the UK, Shoreline Management Plans, Estuary Management Plans and Coastal Habitat Management Plans (English Nature et al. 2003) were developed to promote the good and prudent management of the coastline. In addition to providing a framework for coastal management (including ESIA), these identify (among other things) the flood and coastal defence works likely to be needed to conserve key coastal assets, including the nature conservation interest of SACs, SPAs and Ramsar sites. This is particularly important where the current defence line may be unsustainable or could cause substantial losses, either by preventing intertidal habitats from migrating inland (*coastal squeeze*) or, as a result of retreat, threatening freshwater habitats located behind the current defence line.

In March 2013 the European Commission adopted a proposal for a directive establishing a framework for Maritime Spatial Planning and Integrated Coastal Management. The proposed instrument will require Member States to establish coastal management strategies that build further on the principles and elements set out in the Council recommendation on ICZM of 2002 and the Protocol to the Barcelona Convention on ICZM, ratified by the EU in 2010. Integrated coastal management covers the full cycle of information collection, planning, decision-making, management and monitoring of implementation; and it is important to involve all stakeholders across the different sectors. The coherent application of ICZM with maritime spatial planning will improve the sea-land planning and management interface (e.g. related to the effects of infrastructure works to protect coastlines against erosion or flooding on activities in coastal waters such as aquaculture or the protection of marine ecosystems).

According to the United Nations Educational, Scientific and Cultural Organization, marine spatial planning is a public process of analysing and allocating the spatial and temporal distribution of human activities in marine areas to achieve ecological, economic and social objectives that usually have been specified through a political process. Essentially, marine spatial planning is a planning tool that enables integrated, forward-looking and consistent decision-making on the use of the sea. Current policy for the coastal zone in the UK is set out at a high level in the UK Marine Policy Statement.

A number of other international conventions and agreements also influence policy, including the London Convention, the Convention for the Protection of the Marine Environment of the North-East Atlantic (OSPAR), and Agreement on the Conservation of Small Cetaceans in the Baltic, North East Atlantic, Irish and North Seas (ASCOBANS) (see Table 7.2). The London Convention was developed by the International Management Organisation to promote the effective control of all sources of marine pollution. Contracting parties must take effective measures to prevent pollution of the marine environment caused by dumping at sea. The OSPAR Convention is the mechanism by which 15 Governments and the EU cooperate to protect the marine environment of the North-East Atlantic; and ASCOBANS is an agreement between 18 countries on the conservation of small marine mammals of the Baltic, North-East Atlantic, Irish and North Seas.

7.4 Scoping and baseline studies

7.4.1 Introduction

Much of the ecological interest of the coastal zone is linked to its geomorphology, and ecological studies must take this into account. Moreover, geomorphological processes and changes can have important implications for coastal defence policy and practice. Given this, establishing a refined probable impact area (zone of influence) is an important starting point, but may be difficult because of the challenge associated with determining both the potential extent of influence associated with a change (e.g. in the tidal range or current velocities) and boundaries in the coastal zone, especially of the sublittoral zone. It is, therefore, sensible for initial geomorphological predictions to be completed early in the assessment in order to inform the potential impact area of the proposed activities on ecology. Initial estimates may also have to be revised in the light of information that emerges during the study. Consideration of the lateral extent of most geomorphological processes should be confined to coastal sediment cells (§7.2.2).

The coastal zone is also affected by developments in associated freshwater *catchments* and so another important aspect for the definition of the scope of a coastal ESIA may be catchment hydrology (Chapter 2). Furthermore, since many of a project's impacts may only be realised in combination with other activities (i.e. they are cumulative), it is important to seek information on other predicted developments and trends (e.g. in recreational use).

7.4.2 Scoping

Scoping is a (if not *the*) key early stage of the pre-application ESIA process. The key issues that should be considered by stakeholders with respect to coastal studies include:

- the extent of the impact area in relation to marine ecology (habitats and species);
- potentially significant impacts requiring further assessment and insignificant impacts that do not;
- the periods during which 'scoped in' species may be present (in turn determining the required timing of surveys);
- the adequacy of existing baseline data and scope for further studies, should they be deemed necessary;
- the approach to be taken for coastal process modelling studies, should any be required;
- the methods for assessing the magnitude and significance of effects/impacts (including cumulative and residual effects).

The main focus of this stage should be on 'scoping-in' those issues that could be potentially significant in this context and 'scoping-out' those that will not. For those issues scoped in, it is also important to closely define to what extent the subject needs to be examined (rather than simply examining all elements of it) and to design the survey and assessment work to be undertaken based on this. This should then be verified and agreed with the relevant stakeholders and, as far as possible, this should be signed up to. For those issues to be scoped out, again, confirmation of the Regulators' agreement should be obtained (as far as possible) and recorded. Adopting this approach will result in better-informed Regulators, better-informed stakeholders, (ideally) a more precise scoping opinion, a more efficient ESIA process and a more successful consent application.

In general, coastal ecological impact assessments should employ the scoping procedures outlined in §6.4, including adopting a phased approach for baseline surveys. As for all ecological surveys, planning coastal surveys should accommodate the seasonal windows of key species (see §7.4.5). An ecological impact assessment (EcIA) may be deemed to be insufficient by the Regulators if surveys have not been undertaken during key periods of use (or potential use) by notable taxa.

A central tenet of good scoping is that a formulaic approach is not simply applied. Effective ESIA scoping must be undertaken in relation to the specific context and requirements of the project in question, taking into account the considered views of all stakeholders. Suitably qualified and experienced persons should exercise judgement. This can save significant time and costs in data collection, report preparation and design/construction. For example, for some (but not all) marine licence applications for works in the marine environment, all landside issues (transport, air quality, noise, etc.) can be scoped out, because these are not directly relevant to the marine works and this requirement may be fulfilled by reference to separate terrestrial consents.

For major infrastructure projects with potentially multiple consenting routes, scoping and the subsequent ESIA process can benefit significantly from 'route-mapping' the consenting and associated ESIA requirements. This allows the subtle but important differences between consenting legislation and ESIA requirements to be accommodated in the process and this can be resolved at the scoping phase; with the marine ecology inputs tailored to meet all relevant needs.

7.4.3 Use of existing information

Much of the information required for a coastal assessment can be compiled through a desk study. Aerial photographs, satellite data, topographic maps and bathymetric charts can provide information on the current and historic morphology of the coast, and may reveal patterns of coastal erosion or accretion and/or change.

Information is also available in various inventories and databases (Table 7.3 includes some European examples). Although many of the data will not refer to the immediate vicinity of a project, they can still be useful. For example, tidal regimes can often be calculated from data for the nearest ports. In addition, some of the websites provide links to other sites, often worldwide, and the organisations involved may hold, and be willing to supply, information other than that available on the websites. For example, the environmental protection authorities have a duty to supply relevant information, on request, for EcIAs (although there is likely to be a charge for the provision of this information).

In spite of the increasing range and extent of existing information, much of it may be out of date or inadequate in terms of quality or resolution required for the purposes of a specific EcIA, and new surveys should be conducted where necessary. For example, the statutory nature conservation organisations will often require data to be relatively recent (e.g. valid within the last three years) and may ask for more than two years of data in a series (e.g. for overwintering waterbird data) for projects that may influence the intertidal zone. However, the time and effort to be invested in data collection must be reasonable in the context of the certainty that the data will provide. For example, if a year of estuary-wide bird data will only provide a snapshot of bird use for that period of time, and a second year will provide the same level of evidence, then the value of requiring a second year of bird data is questionable. Similar arguments can be made for fish and marine mammal data.

This form of data collection challenge may also be relevant for sediment samples. It is important that surface sediment samples are current, given the potential for the quality of surface sediment to change relatively rapidly (unless

Table 7.3 Inventories and databases relating to the European coastal zone

---

British Oceanographic Data Centre (BODC) European Directory for Marine Environmental Data (EDMED). An inventory of marine data and Data Holding Centres, currently describing over 3800 data sets from a wide range of disciplines, held at over 600 Data Holding Centres across Europe. The directories for the data sets are searchable online.

BODC Marine Environment Monitoring and Assessment National database (MERMAN) (www.bodc.ac.uk/projects/uk/merman). Database which holds and provides access to data collected under the Clean Safe Seas Environmental Monitoring Programme (CSEMP). Database holds information on significant contaminants, benthic biology and biological effects in estuarine and coastal waters for about 1,000 monitoring sites around the UK.

British Geological Survey (BGS) National Geoscience Data Centre (NGDC) (www.bgs.ac.uk/services/ngdc). Online spatial index of BGS data holdings (e.g. sea bed datasets including sediment particle size and geochemistry (including contaminants)). Data is held in a GIS format and costings can be provided for the supply of more specific information.

European Union for Coastal Conservation (EUCC) (www.coastalguide.org/projects/index.html). Database containing worldwide applied coastal (and marine) projects and regional case studies. Data is searchable using key words (data, name or country etc.).

Mapping European Seabed Habitats (MESH) (www.emodnet-seabedhabitats.eu/default.aspx?page=2003). A range of facilities including guidance on mapping, interactive GIS maps and a metadata catalogue of survey data.

Marine Life Information Network (MarLIN) (www.marlin.ac.uk). Includes data on (a) species listed in Conventions and EU/UK legislation: (b) species' identification, biology, habitat preferences, distributions, sensitivity (to a range of factors), recoverability, and importance; (c) information on MHCBI biotopes; (d) links to other UK datasets.

Surge Watch (www.surgewatch.org). Contains information about 96 large storms taken from tide gauge records, which record sea levels back to 1915. It shows the highest sea levels the storms produced and a description of the coastal flooding that occurred during each event.

---

the system is closed), but samples at depth may be valid beyond a standard three-year data life period. That is, this material (and especially geological material) is highly unlikely to have changed in the short term.

7.4.4 Geomorphological surveys

Geomorphological parameters can be measured by a variety of methods, using *in situ* recording instruments and remote sensing techniques: see Cooper et al. (2000), Andrews et al. (2002), Woodroffe (2002), Miller et al. (2005) and Klemas (2013). The methods are generally time-consuming and expensive and, although coastal geomorphology can be very dynamic, changes tend to occur relatively slowly. This means that many methods require repeat measurements over extended periods. Consequently (i) the assessment of trends will normally have to rely on existing information, and (ii) new surveys for ESIA are likely to be restricted to large projects and post-development monitoring programmes (in which case it may be beneficial to initiate appropriate studies at the baseline-survey stage). In making decisions about the need for new data, and the selection of appropriate methods, advice should be sought from nature conservation and environmental agencies.

If geomorphological monitoring is deemed necessary at the baseline stage, then a wide range of techniques of varying sophistication can be employed, including tidal-current meter and wave stations, sediment sampling, repeat bathymetric and topographic surveys, Laser Induced Direction and Range (**LiDAR**: measurement of distance using a laser) and aerial photography.

Data on waves, tides and currents are ideally derived by direct measurement using wave recorders, tide gauges and current metres. The main difficulty with deployment of any equipment is how representative the site is of the wider study area. This is not a problem if the results are not extrapolated or applied beyond the set of conditions that the site represents. The most important factor is how representative the period of measurement is. Often intensive measurements can only be carried out over a limited time period and, therefore, only a limited range of conditions are measured, which are only characteristic of that time period and cannot be directly applied to other time periods (such as different times in the spring-neap tidal cycle or seasons of the year).

Sediment composition and distribution along the coast can be evaluated through a campaign of surface and subsurface sampling, followed by laboratory analysis. Particle size and other textural parameters, such as sorting, can then be interpreted in the context of sedimentary processes and temporal change. More specialised analysis of sediment may be required for samples from muddy shorelines. Clay mineralogy, organic content and geotechnical properties may be investigated in addition to particle size, to characterise the sediment type. The geotechnical parameters give an indication of the sediment stability, shear strength and its susceptibility to erosion. Organic content is critical as it influences the infaunal communities (organisms that live below the surface of the sea bed) and can cause deoxygenation, which can in turn cause major losses of biota.

A common form of field morphological monitoring is beach profiling. Beach morphology can be monitored using cross-shore profile data to assess changes in width, slope and volume, and to describe beach behaviour and its variability. These data can be used to identify trends and areas of high net change and variability. Several techniques are available for collecting beach survey data. The least sophisticated method (although not necessarily the least accurate) is a survey using a quick set-level, staff and chain. More advanced methods include using a total station with electronic distance measurement to a survey reflector prism, and computer logging of data points. Current best practice involves the use of the Real Time Kinematic Global Positioning System.

In view of the costs and practical difficulties of regular *in situ* monitoring of large areas of coast, there is an increasing role for **remote sensing** techniques from aircraft, satellite or merely from a high point on the coast. Vertical aerial surveys of a shoreline can provide quantitative data on large-scale changes to the coast, such as the retreat of cliffs, dune morphology or movement of the salt marsh edge (Figure 7.9). The process of reviewing and assessing geomorphology from aerial photography generally requires time-series analysis. This is most easily achieved through the registration of the data into digital systems such as GIS that allow the data to be correctly spatially located. This process is called georectification and allows the image to be fixed in the horizontal plane in relation to a standard spatial georeferencing framework, such as WGS84 latitude longitude, or a national projection system. A vertical aerial photograph image can be displayed in the system, relative to other data sets and the features of interest can be digitised to permit time-series analysis of change/variation.

Airborne Laser Induced Direction and Range (LiDAR) is a remote sensing technique for the collection of topographic data. It uses laser technology to 'scan' the ground surface, taking up to 10,000 observations per square kilometre with a very rapid speed of data capture (up to 50 km² per hour). The system can

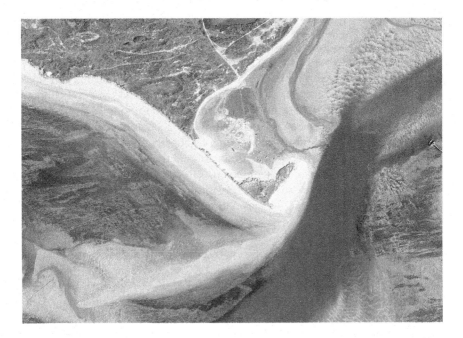

Figure 7.9

Example of an aerial photograph (2013) showing the dunes, spit and rock platform at Crow Point, Taw-Torridge Estuary, UK. Courtesy of the Environment Agency

operate both day and night, and with light cloud cover, although it is affected by rain. It can operate on beaches and mudflats but care needs to be taken in areas of standing water, as with the normal settings the laser beam is absorbed by water rather than reflected.

Two main techniques are available for the collection of **bathymetric** data: single beam echo sounding and multi-beam echo sounding. Single beam echo sounding typically involves using a transducer attached to the hull of a vessel. The echo sounder calculates the water depth beneath the transducer, by transmitting a sound pulse that is returned to the vessel via reflection off the sea bed. Standard single beam echo-sounders collect data for a narrow zone along the track of the vessel and hence the system is limited in its seabed coverage. A multi-beam echo sounder, instead produces a number of beams at right angles to the vessel track, forming a swathe of sound pulses or acoustic energy. This can provide a greater density of soundings allowing faster coverage of a site. The main advantage of multi-beam systems is that they can provide 100 per cent coverage of the sea bed without the need to interpolate between lines.

7.4.5 Ecological surveys

The coastal zone presents special problems for ecological sampling, especially of the sublittoral zone. However, if information does not exist, baseline surveys should be undertaken to cover all habitat types and specific taxa that occur within the potential impact area and/or are identified as particularly sensitive to potential changes in coastal processes. A range of different scales of survey can be applied using defined habitat classification systems and specific survey methods for individual species groups. As with all EcIAs, sampling and identification of many taxa in the coastal zone can be difficult, time-consuming and expensive, so surveys may need to be targeted, e.g. to focus on **notable** or representative species. Experts in both sampling methods and identification will usually need to be involved.

The timing of field surveys and (where possible) repeat sampling is particularly important in coastal zone assessment because many of the ecosystems present have a high degree of seasonality. While some animals may be present all year round, the presence and abundance of many fish and bird populations vary in relation to breeding and overwintering strategies. In particular, many waders and other migratory waterbirds and seabirds are resident on coasts and in estuaries only during the winter months, and some species undertake shorter 'stopovers' on spring and autumn migrations (see Figure 7.10). Saltmarsh vegetation grows and flowers relatively late in the summer, and sand dune fauna (and some annual plants) should generally be sampled earlier than the most suitable period for a general vegetation survey. Resident shore communities can be sampled at most times of year, but neap tides do not expose the lower shore, and sampling is best conducted during the large spring-tide periods in March or September. The sublittoral environment, by contrast, does not exhibit significant seasonal differences.

Figure 7.10

Colour-ringed curlew. Courtesy of Lucy Wright, BTO and Tidal Lagoon Power

Habitat surveys

Ecological surveys of sublittoral and littoral habitats typically comprise some form of mapping. Various techniques may be used, depending on the scale of the area being studied. Remote sensing and aerial imagery, together with ground-truthing and/or substrate sampling, can provide large-scale information on broad habitat types. At the smaller scale, or where more detail is required on a complex habitat, walkover surveys for coastal (littoral) habitats and intensive sampling (e.g. grab samples, video drop-down surveys) in marine (sublittoral) habitats can be employed. At all scales, a classification system that provides a means to define the habitats encountered should be utilised.

For instance, in Europe the European Nature Information System (EUNIS) provides a unified system for the description of all habitats across terrestrial, freshwater and marine environments. EUNIS brings together information on species, habitats and sites across Europe, and in particular on *Natura 2000* sites and the work of the European Environment Agency. In the UK, surveys of maritime and littoral habitats often employ the JNCC phase 1 and phase 2 habitat survey methods: desk studies and relatively high-level phase 1 (dominant and conspicuous species) surveys are carried out initially, with more detailed phase 2 (detailed species identification) surveys carried out as required. However, the JNCC classification does not cover the sublittoral zone and surveys of both sublittoral and littoral habitats should follow the Marine Habitat Classification for Britain and Ireland (Connor et al. 2004).

In the USA, the Coastal and Marine Ecological Classification Standard (CMECS) provides a comprehensive national framework for organising information about coasts and oceans, and their living systems; the physical, biological and chemical data collectively define coastal and marine ecosystems. CMECS classifies the environment into biogeographic and aquatic settings that are differentiated by features influencing the distribution of organisms and by salinity, tidal zone and proximity to the coast. Within these systems there are four underlying components that describe different aspects of the seascape, namely: the water column, geoform, substrate and biotic. The components can be mapped independently or combined as needed. CMECS uses the setting and component units to identify **biotopes**, which can then be mapped to provide detailed information for each relevant study area.

Surveys of maritime and benthic species and communities

In general, detailed fauna and flora surveys of maritime (supralittoral) habitats can follow the procedures described in §6.6 for terrestrial or freshwater systems. Remote sensing techniques can also be employed at the wider scale, e.g. see Brown et al. (2003).

Coastal birds are included here because they are most frequently surveyed from the land. In addition to the general census techniques referred to in §6.6.3, a number of specific methods have been developed for seabirds: e.g. see Tasker et al. (1984) and Camphuysen et al. (2004) for seabirds at sea; Lloyd et al. (1991) and Walsh (1995) for seabird breeding colonies; Bibby et al. (2000) for breeding waders; and the British Trust for Ornithology for estuarine/coastal high-tide and low-tide counts (www.bto.org/volunteer-surveys/webs). Information on seabird distributions and numbers is available in a number of publications, including Kober et al. (2010) and Stone et al. (1995). Consequently, data will already exist for many sites and species, enabling the importance of an area for coastal birds to be determined. If it is suspected that information is out of date or that more specific data on assemblages of local or regional importance is required, then surveys should be undertaken.

In the UK, *vegetation surveys* can employ National Vegetation Classification (Rodwell 2000, JNCC 2004a, Hill et al. 2005). The system for measuring **species abundances** is SACFOR ratings (superabundant, abundant, common, frequent, occasional, rare) which are based on ranges of per cent cover or density (see Table 6.2) depending on the species being sampled. This is because some species, such as seaweeds and encrusting animals, are best sampled by per cent cover, while most animals are best sampled as density. Details of the system, and guidance on survey and mapping methods, is given in Connor et al. (2004). A National Vegetation Classification standard has also been developed in the USA (http://usnvc.org) and covers coastal vegetation types. The classification allows users to produce uniform statistics about vegetation resources at a range of spatial scales and for vegetation from broad-scale formations (like forests) to fine-scale plant communities. Similar classification systems are also utilised in other countries (e.g. the National Vegetation Information System for Australia and the Canadian National Vegetation Classification).

On *rocky shores*, **quadrat** sampling can be employed because seaweeds and most animal residents are immobile and easily visible at low tide (see Baker and Crothers 1987, and Davies et al. 2001). Rocky-shore communities normally show clear zonations along the land-sea axis, so the use of transects along this axis is usually a suitable sampling pattern.

A similar sampling pattern may be suitable for *sandy and muddy shores and mudflats*, but sampling their fauna requires different techniques. Subsurface **macro-invertebrates** are an important group in these habitats because they are at the base of the food chain. They can be surveyed by a number of methods ranging from a simple inspection of the sediment (e.g. to estimate the densities of lugworms from their casts) to methods which employ the use of corers and grabs to estimate densities and **biomass** (see New 1998, Wolff 1987 and Davies et al. 2001).

A problem affecting *sublittoral benthic* surveys is the need for specialist equipment and personnel (e.g. boats and/or divers). Baseline studies may have to rely on existing information unless these habitats are of particular interest and may be adversely affected by a proposed development or activity (e.g. development of offshore renewable energy projects and marine aggregate extraction). Techniques for survey and monitoring of sublittoral habitats are described in Eleftheriou (2013), JNCC (2004b), Davies et al. (2001) and Ware and Kenny (2011). Guidance on seabed-mapping techniques is available from the European Marine Observation and Data Network (EMODNet) and the Marine Environmental Data and Information Network (MEDIN).

Surveys of pelagic species and communities

Like those of sublittoral benthic habitats, pelagic (free-swimming or floating) species and communities are relatively inaccessible and can be difficult to survey.

The survey of **plankton** presents problems because (a) they are very small and diverse, (b) they are widespread over large areas of sea, and (c) abundance may undergo rapid fluctuation in time and space (e.g. in relation to currents). Satellite and airborne sensors that respond to chlorophyll-a fluorescence may provide detailed distribution maps for phytoplankton, and are used in *eutrophication* studies. However, the method is very expensive and therefore generally unrealistic for EcIA, and most plankton sampling employs nets and samplers that can be filled with seawater at prescribed depths. These methods, and techniques for analysing the samples, are explained in Newell and Newell (2006) and Tett (1987), who suggests that for survey purposes it is convenient to adopt categories based mainly on ecological rather than taxonomic criteria. Baseline data may be available from the Sir Alister Hardy Foundation for Ocean Science (www.sahfos.ac.uk), which operates the Continuous Plankton Recorder survey in the North Sea and North Atlantic.

Fish survey techniques are numerous and variable in their level of complexity. They are influenced by various characteristics of the fish populations and communities, including: distribution (vertical and horizontal); size and mobility; and population and community dynamics, e.g. single or mixed-species shoals, and seasonal migration and breeding patterns. Reviews of methods are provided in Blower et al. (1981), Côté and Perrow (2006), Davies et al. (2001), Jennings et al. (2001), Pitcher and Hart (1982), and Potts and Reay (1987). They can be grouped under two broad headings:

- observation, e.g. aerial, direct underwater, underwater photography and acoustic surveys;
- capture, e.g. by traps, hook and line, hand nets, set nets, seines, trawls, lift, drop and push nets – most of which can provide specimens for mark-recapture programmes (Table 6.2).

The samples obtained by these methods can be analysed to provide information on species abundance, age structure, fish health, dietary requirements and site productivity (see Potts and Reay 1987). This information can indicate the relative worth of a site to fish stocks and hence the significance of a development's potential impact.

Marine mammal data from existing data sources provide highly useful information for EcIAs. In the UK, the Inter-Agency Marine Mammal Working Group provide guidance on appropriate management units and the associated population estimates for key cetacean (whales, dolphins and porpoises) and seal species. The US National Marine Mammal Laboratory conducts research (including census surveys and telemetry studies) on marine mammals important to the mission of the National Marine Fisheries Service and the National Oceanic and Atmospheric Administration, and focuses on marine mammals off the coasts of Alaska, Washington, Oregon and California. The Australian Marine Mammal Centre coordinates Australia's marine mammal research expertise and provides database applications to support marine mammal conservation and policy initiatives.

Site-specific data for at-sea marine mammals can be collected visually by aerial survey, boat-based survey or land-based vantage-point surveys (Hammond 1987, Hammond and Thompson 1991, Hiby and Hammond 1989 and Diederichs et al. 2008). These are often combined with surveys of seabirds.

Visual data are often supplemented with passive acoustic data to detect cetaceans which are vocalising in the area.

Where quantitative impact assessments are appropriate, robust site-specific datasets with high levels of sightings may be required to allow statistical analysis of species-density estimates. Site-specific surveys may also help to determine what a site is being used for, depending on the behaviour of the marine mammals recorded (e.g. foraging or transiting) and any seasonal trends. Telemetry surveys of seals can provide useful information of transiting routes and diving behaviour.

Where a project is located close to haul-out sites for seals, surveys to count the numbers of adults and juvenile seals during the breeding or moulting season may be required to understand the numbers and demographic of seals using the site (Thompson and Harwood 1990, and Ward et al. 1988).

7.4.6 Evaluation of baseline conditions

When evaluating baseline conditions, particular attention should be paid to sensitive geomorphological systems and high-value species, habitats and sites. European guidelines tend to focus on SACs, SPAs and Ramsar sites, but this should not preclude the thorough evaluation of 'less important' and small sites, especially if these support notable species or habitats. In evaluating habitats, consideration should also be given to 'secondary' attributes (e.g. ecological function). For example, in addition to their ecological value, sand dunes and salt marshes act as natural defence systems. Attention should be paid to the sensitivity of species and habitats to any effects likely to arise, along with their mobility (behaviour), longevity and rate of recovery.

7.5 Impact prediction and evaluation

7.5.1 Introduction

The difficulty associated with accurately predicting impacts and ecosystem complexity particularly applies to the coastal zone because of its dynamic nature and the diversity of the habitats and species that occur there. In addition, each type of development brings with it a suite of potential effects and issues and there can be significant variability between and within individual development types. For example, the potential impacts of a salmon farm on an inshore sea loch are likely to be very different from those of a nuclear power station or barrage scheme on the open coast. However, while activities and development types may differ, some effects may impact in similar ways upon sensitive *receptors*. Thus, both a barrage scheme and a salmon farm may have similar implications for water quality and productivity, but via different impact pathways. It is therefore important in impact prediction to determine and understand in detail both the nature of the effects of the proposed development and the likely response of the receptors to the potential effects. Because of the value and fragility of many coastal ecosystems, any development which has the potential to disrupt the fine balance of interacting processes and the habitats and species that these processes influence should be viewed with a degree of concern.

Coastal ecosystems are dynamic, and so a combination of natural trends and human influences will inevitably lead to changes, even in the absence of development or coastal management. For example, some soft cliffs are currently experiencing rapid erosion, some estuaries are changing through progressive erosion or sedimentation, and systems such as sand dunes are intrinsically unstable and can be significantly altered by storms. A new project must be considered within the context of dynamic natural change and also in the knowledge that many of its potential impacts may be *cumulative*. However, although a coastal system may be changing, that will not necessarily make an anthropogenic contribution to the change acceptable (or unacceptable).

Within this potential complexity, adopting a structured approach to EcIA at the coast is essential. The assessment of hydrodynamic effects and ecological impacts should involve four distinct phases:

1   identification of the effect and potential impact;
2   description of the effect and/or impact;
3   assessment of the effect and/or impact;
4   the derivation of impact significance.

The distinction between hydrodynamic effects and ecological impacts is made here because a predicted 'change' to or 'effect' on the hydrodynamic and sedimentary regime will not necessarily equate to an 'impact'; that is an effect that is predicted to arise due to the proposed scheme (e.g. an increase in current speeds) may or may not lead to a consequential impact. Coastal processes themselves are not considered to be receptors sensitive to change, and while a change to a process can be predicted and described with respect to the known baseline in terms of its magnitude, it is not appropriate to predict the significance of an impact on the process. Given this, changes in hydrodynamic processes should not be defined, in the first instance, as impacts in ESIA; rather they should be described in a chapter dealing with the hydrodynamic and sedimentary regime, and any impacts associated with the changes (e.g. on intertidal benthic communities or waterbird populations) should be addressed in subsequent chapters. An exception to this rule is when the hydrodynamic and sedimentary processes themselves have an intrinsic value (e.g. a bore or shifting sandbanks) that could be impacted.

The potential effects of a proposed scheme on the hydrodynamic and sedimentary regime will typically be fully realised on completion of the construction phase and therefore, for the purpose of the ESIA process, should be assumed to arise in the operational phase only.

7.5.2 Sources and types of impact

The identification of potential project-related impacts is a key stage within any ESIA and should be based on:

- inputs from consultation with the public, key marine users and regulatory authorities;
- a review of existing survey data;
- a review of impacts associated with similar schemes;
- the findings of specialist studies undertaken in relation to the ESIA (e.g. modelling);
- the expertise and judgement of the assessors.

Following the identification of potential site-specific impacts, each impact should be described fully. Ideally, all potential impacts should be identified, quantified and expressed as testable hypotheses, based on the results of earlier studies, where these are available. These hypotheses should then be tested through the use of well-designed data collection and/or monitoring programmes.

Major causes and associated types of impact in the coastal zone are shown in Figure 7.11 and discussed below. However, this is not a complete list and all of the possible development/impact interrelationships are not covered. As in all ecosystems, in the coastal zone primary impacts can lead readily to secondary, tertiary and, potentially, cumulative impacts.

Urban, industrial and commercial development is prolific, and is the greatest source of potential impacts in the coastal zone. According to the Australian Bureau of Statistics, 83 per cent of Australians live within 50 km of the coastline. In England and Wales around 31 per cent of the coastline is developed; and around 33 per cent of the total population is located in the 10 km zone (Defra/EA 2002, National Trust 2015).

More recently developed, the coastal plain of the Al Batinah coast of Oman has experienced planned and unplanned development over the past four decades, including housing, roads, ports and fishing harbours, with supporting infrastructure such as seawalls and groynes. This coast is comprised of up to 90 per cent sandy beaches, with longshore sand transport driven by low-wave energy (Al-Hatrushi et al. 2014). These developments have interrupted sand transport processes along the coast, leading to localised erosion and an increased risk of coastal flooding. The construction of dams on the major wadis of the Al Batinah coast has exacerbated the potential for erosion because the supply of sediment to the coast has been reduced; almost 50 per cent of sediment transported down the wadis is now trapped in recharge dams. Although the overall volume of sediment supplied by the wadis is still sufficient to meet the demand of the longshore sand transport rate, local erosion is taking place close to the wadi mouths where major dams occur upstream.

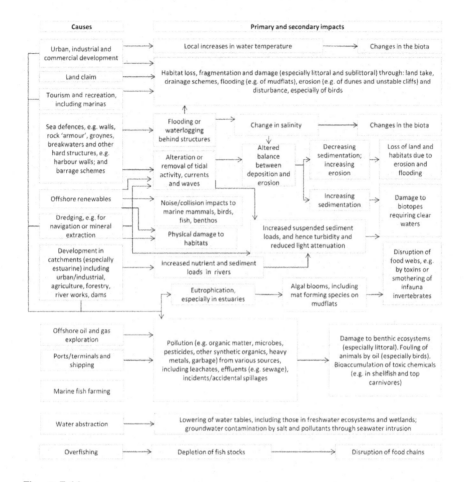

Figure 7.11

Causes and types of impact in the coastal zone (modified from Thompson and Lee 2001)

Around 40 per cent of UK industry, much of which is heavy (including chemical) industry, is situated at or near the coast (see Figure 7.12). As indicated in Figure 7.11, major impacts due to development include habitat loss and fragmentation, and pollution. However, as elsewhere, much of the land that has remained undeveloped is now protected by landscape or nature conservation designations.

A principal reason for industry siting itself at the coast is for the transport of goods via ports, but historically another reason has been the need for the use of large quantities of water in industrial processes and the ease of disposal of

Figure 7.12

Port of Felixstowe, Orwell Estuary, UK. Courtesy of Hutchison Ports UK Ltd

unwanted by-products through discharge into coastal and estuarine waters. Industrial discharges are now under tighter control through the implementation of various national and international legislative requirements.

Pollution ignores administrative boundaries, and effects across entire water bodies demonstrate that land-based developments can have significant impacts on the whole continental shelf. *Point-source pollution* from urban and industrial developments (particularly via sewage effluent) is a major source of coastal and estuarine water pollution, along with *diffuse pollution* from urban areas and land under agricultural production. *Bioaccumulation* of toxic pollutants by coastal and marine organisms can have significant adverse physiological and ecological impacts, especially in shellfish and top carnivores, and may also have serious implications for human health (see Brouwer et al. 1990, Walker et al. 2012 and Schäfer et al. 2015). Eutrophication (particularly from sewage effluent, urban areas and farmland) can also have various consequences (especially in estuarine and nearshore coastal waters), including contamination of shellfish by toxins from algal blooms.

Nutrient inputs above normal levels can lead to a variety of deleterious effects, including oxygen depletion and mortalities of benthos and fish. Changes in the ratio of nitrogen or phosphorus to silicate in nutrient inputs can also affect the marine food web by altering the balance between diatom and other taxa in the phytoplankton community (Heath 2010). Most run-off is transported to the coast by rivers, so estuaries are particularly affected by upstream land management. Eutrophication can be particularly problematic in small estuaries or those estuaries in which tidal flushing has been reduced by other activities. An assessment of 141 US estuaries found that two-thirds of the estuaries evaluated exhibited moderate to high levels of eutrophication (Bricker et al. 2007). Nutrient additions (e.g. via run-off from agricultural land or poorly treated sewage) can cause a progression of eutrophic symptoms that most often begin with increased phytoplankton productions and/or macroalgal blooms. Excessive algal blooms may lead to other more serious impacts, including loss of submerged aquatic vegetation, a shift from benthic to **pelagic** dominated system productivity, low dissolved oxygen and occurrences of nuisance/toxic blooms (Bricker et al. 2007, 2008). Algal blooms can also have a negative effect on bird populations as they reduce the quantity and availability of infaunal invertebrates. Studies during the 1990s indicated a decrease in the number of birds in the Ythan Estuary related to an increase in macroalgal biomass (Gorman and Raffaelli 1993).

Coastal waters also suffer from pollution due to rubbish from land-based sources, ships and pleasure craft. The stomachs of fulmars in the North Sea, storm petrels in the Antarctic and albatrosses in Hawaii have all been found to contain plastic discarded by consumers or industry. Some birds may consume hundreds of plastic fragments and many have died as a result. Globally, up to 90 per cent of all individual seabirds may contain plastic in their gut (Wilcox et al. 2015). A study of fulmar carcasses from the North Sea found that 19 out of every 20 had plastic in them (Save the North Sea 2005).

Tourism and recreation pressures are continually increasing on the coast and in estuaries. Direct impacts include visitor pressure on sensitive coastal ecosystems, particularly sand dunes, very few of which are not impacted by development, leisure facilities or artificial sea defences (Covey and Laffoley 2002). Once dune vegetation cover becomes seriously damaged, it no longer acts to stabilise the sand and wind action will rapidly remove the exposed loose sand, forming a blow-out. Once initiated, a blow-out can spread rapidly and large areas of dune can be affected, even in mature dunes. In addition, marinas and other tourist infrastructure can increase disturbance pressure on wildlife, especially birds, which are reliant upon undisturbed feeding sites.

Reclamation of intertidal areas has a very long history of re-configuring the morphology of the coastline, and consequently altering sedimentation and erosion patterns. About one-third of all UK intertidal estuarine habitat and half the salt marsh area has been reclaimed since Roman times, largely for agricultural use (Thornton and Kite 1990). As an example, the intertidal area of the Tees estuary was reduced by around 90 per cent in the 100 years between 1890 and 1990 (Rothwell and Housden 1990). Over the past century most of San Francisco South Bay's saltmarshes (over 150 km^2) were diked to create 'ponds' for salt production through the evaporation of seawater (Goals Project, 1999).

Today, the large-scale reclamation of intertidal areas for agricultural purposes is less widespread. However, land-claim for maritime infrastructure projects (e.g. 'coastal cities', ports, marinas and associated developments) is still ongoing, although such schemes are now typically subject to stringent regulatory controls. The key impacts associated with reclamation are the obvious direct intertidal and, in some cases, subtidal habitat loss, with implications for bird and fish feeding and nursery grounds, as well as the indirect consequences of changes in hydrodynamic process and patterns of sediment transport, erosion and accretion. This can include effects on the tidal range. In addition, impacts on some fish species and marine mammals may occur due to noise associated with piling.

Barrage schemes fall into two basic categories: permeable and impermeable. Some *impermeable barrages*, such as the Thames Barrier, provide flood defence against high tides and tidal surges, and are only used when conditions dictate, but many are intended for total exclusion of the tide, primarily for amenity purposes such as water sports or to provide consistent views for waterside developments. The immediate impacts of the latter include the loss of marine habitats, such as mudflats. The implications of such changes on marine/coastal fauna and flora can be profound, as demonstrated by the numerous studies of ecological impacts associated with the Cardiff Bay Barrage which was completed in 1999. Here, monitoring revealed significant declines in the populations of waders and waterfowl associated with (lost) intertidal mudflats and a rise in the populations of species associated with terrestrial open water habitat such as mute swan and coot, as well as several species of gull (Burton et al. 2003). Barrages also have the potential to significantly affect water quality (YSI environmental 2011) and sedimentary regimes both locally (Phillips 2007) and for many kilometres along the coastline, often enhancing erosion at susceptible sites.

Permeable barrages normally harness tidal power in order to generate electricity. These barrages may also change sedimentation patterns and reduce tidal activity upstream, potentially increasing eutrophication. Further information on the design and environmental impacts of tidal barriers is provided in Burt and Watts (1996). A large-scale tidal power installation has been in operation on the Rance Estuary in France since 1966, and in 2011 a new facility in South Korea, the Sihwa Lake Tidal Power Station, became the world's largest tidal power installation.

In the UK the potential use of the tidal range (and stream) for electricity generation has come to the fore again in recent years as a result of government commitments to cut CO_2 emissions. Discussions about harnessing the Severn tide have been ongoing for nearly a century. Options for tidal power development in the Severn were investigated again in 2010, including one proposal to construct a 16-km barrage across the Severn Estuary (DECC 2010). This ambitious scheme would have the potential to generate 5 per cent of UK electricity needs, but the impacts of a barrage on the existing environment were predicted to be immense, with the potential loss of up to 16,000 ha of intertidal area and significant implications for migratory fish populations (DECC 2010). By contrast, in 2015, consent was given for the construction of a relatively small (11.5 km²) *tidal lagoon* in Swansea Bay, Wales. This project was deemed to have an acceptably low level of environmental impact and has the capability, through turbines installed into the walls of the lagoon structure, to generate 495 GWh per annum of electricity (around 11 per cent of Wales' domestic electricity) by harnessing changes in tidal levels. Proposals are currently being considered for a tidal lagoon near Cardiff which would generate enough low carbon electricity to power every home in Wales.

Coastal and flood defences protect many settlements, transport routes and agricultural land from flooding and coastal erosion. However, hard defences and structures (such as sea walls and groynes) can disrupt physical processes, leading to geomorphological change. Such structures fail to dissipate wave energy and, by deflecting waves and currents, affect deposition and erosion processes. For example, although sea walls do not directly control longshore sediment transport processes they can have a major impact on beach elevation due to changes in cross-shore processes. Reflection of wave energy from the face of a vertical sea wall can cause increased scour at the toe of the wall, resulting in lowering of the beach profile and associated loss of beach width. This reduction in beach size can deprive a coastal system of sediment which may be vital in the replenishment of beaches further along the coast, thus causing downdrift erosion (Komar 1983).

Historically, problems have arisen as a result of the management of coastal defences in relation to administrative rather than geomorphological boundaries, when, for example, erosion along one frontage led to coastal defences being installed that limited deposition along the frontage of another district (Clayton 1993). Although the legacy of this management approach is still apparent, the current management of coastal defences is typically at the level of the coastal sediment cell. This, in conjunction with coastal strategies that consider: whether to hold the line (of defence); advance the line or retreat the line (via managed realignment or no active intervention) in the short term (0–20 years), medium term (20 to 50 years) and long term (50 to 100 years); will help to minimise such impacts in the future. An additional problem associated with hard sea defences is *coastal squeeze*, which is the loss of intertidal habitats due to rising sea levels where defences limit the ability of the frontage to roll back. Retreat can also threaten freshwater habitats that have developed behind seawalls.

Dredging is carried out for various purposes. Globally, 57 per cent of the total dredging market is for projects associated with the growth in world or seaborne trade: harbour extensions, new ports, navigation channels and maintenance dredging (www.dredging.org). The boom in LNG (liquefied natural gas) has led to several large dredging projects, such as the construction of ports to accommodate LNG vessels in Australia and Qatar. Dredging can also involve trenching for the laying of oil and gas pipelines or work related to offshore wind farms; and it is undertaken for coastal protection, urban development (such as land reclamation for city expansion), and leisure (beach replenishment).

Impacts associated with dredging and the disposal of dredged material include (Posford Haskoning 2004, Cooper and Brew 2013):

- an increase in turbidity during the works, which may reduce light-penetration, and hence primary production and visibility, and cause acute physiological responses in some organisms;
- possible release of toxins and nutrients which normally remain locked up in the sediment, thus creating toxicity or eutrophication problems;
- physical damage to the dredge/disposal site, and associated habitat loss or disruption;
- the settlement of material released into suspension by the dredging or disposal process, potentially affecting habitats and species outside the immediate dredge/disposal area;
- changes in estuarine bathymetry which may in turn affect local hydrodynamic conditions and sedimentary processes;
- deepening of inshore waters, potentially increasing shoreface slopes and allowing larger waves to break closer to the shore, thus increasing the risk of shoreline erosion.

These factors are believed to be having ecological impacts on the Great Barrier Reef in Australia. According to McCook et al. (2015) dredging and sediment disposal could change the physical and chemical environment and affect the biological values of this World Heritage site by damaging and/or smothering coral reefs and potentially killing seagrass meadows. However, many of the impacts driven by these phenomena will be context-dependent and will differ between locations, types and the extent of dredging and sediment disposal activities.

Land use within river catchments is important because most sections of the coastline form an integral component of river catchments; so river flow, groundwater levels and water quality (including nutrient, sediment and toxic pollutant loadings) at the coast can be affected by land-use practices and infrastructure development anywhere in the catchment, often many kilometres inland. Water abstraction, irrigation schemes and developments such as dams can reduce (i) groundwater levels, and (ii) river flows and sediment loads, potentially leading to lower sediment accretion rates in estuaries, for example. Conversely, urban development, river works and agro-forestry (including deforestation) can increase:

- *run-off* (including flash floods during storm periods);
- soil erosion and consequent suspended sediment loads, which can lead to increased sedimentation in estuary systems and result in the need for dredging to maintain navigation;
- nutrient and toxic pollutant loadings in estuarine and marine waters.

Climate change can affect the magnitude of natural nutrient inputs due to changes in ocean upwelling and currents, and changing patterns of rainfall

---

over land catchments. It may also affect the patterns of **anthropogenic** inputs, primarily through rainfall patterns and the effect on river flows. Disentangling trends in nutrient concentrations due to changing climate, human populations and industrialisation, and relating these to eutrophication status, is a major scientific challenge (Heath 2010). Continued high inputs of nutrients through human activity combined with the effects of climate change is anticipated to lead to increased eutrophication in estuaries and shallow coastal waters in the future (Statham 2012).

The potential geomorphological effects and ecological impacts of *offshore wind farms* and *wave and tidal arrays* relate to the construction, operation and decommissioning of a development (including all associated infrastructure), the export cables and the coastal landfall site (ABPmer 2002, Lambkin et al. 2009). During the construction period discrete short-term disturbances of the sea bed are likely as the device foundations are installed and the export and inter-array cables are laid sequentially across the development site. Seabed sediments have the potential to be released into the water column, resulting in the formation of sediment plumes. At the landfall site, construction activities may result in short-term changes to the sediment budget, as infrastructure on the coast can cause temporary blockages to alongshore sediment transport. The decommissioning phase is generally considered to have a similar or lesser effect than the construction phase.

The greatest potential for changes to the wave- and tidal-current regimes occurs during the operational stage of the development (due to the offshore infrastructure). No significant effects will arise due to the export and inter-array cables because, during operation, they will be buried or surface-laid and protrude (with armouring) only a small distance above the sea bed.

From an ecological perspective, the impacts of offshore renewable installations are largely a function of location, size, construction methods and their operational attributes. Construction invariably leads to local disturbance and small-scale loss of benthic habitat as a result of the installation of the devices and associated offshore infrastructure, and the laying of export and inter-array cables. Piling works for device foundations may significantly increase underwater noise levels, leading to disturbance effects on fish and marine mammals (Bailey et al. 2014). Conversely, if a gravity-based structure is used (avoiding the need for piling), a larger footprint is required on the seabed, causing greater habitat loss. During operation, bird mortality as a result of collision with rotating turbine blades is a key concern for wind farms, although research has shown that over 99 per cent of seabirds will alter their flight paths to avoid collision (BTO 2014). Seabird colony level effects as a result of collision mortality are unlikely to arise due to individual wind farms, but may occur where multiple wind farms are sited, particularly if these are located close to or within the foraging ranges of seabirds from individual colonies (Bradbury et al. 2014). Collision risk of marine mammals, birds and fish is a key concern for offshore tidal developments. Small-scale changes in fish populations have been observed in operational wind farms, largely as a result of changes in habitat complexity (Bergström et al. 2013). For safety reasons, commercial fishing activities may be excluded from some wind farms, which may have localised beneficial effects for some pelagic and benthic fish populations.

Oil and gas exploitation involves exploration (including seismic surveys), laying of pipelines, construction of offshore rigs and onshore terminals, and eventual decommissioning. All of these activities cause at least local disturbance to marine species and ecosystems (e.g. increased sound levels during seismic surveys may cause injury and disturbance to marine mammals). However, the greatest hazard related to this activity is probably the accidental release of oil and the consequent chronic and acute effects that this may have on the marine environment. The transport of oil by ocean-going tankers is probably the most prominent and visible source of oil pollution events, largely through accidental spillage. However, the shipping sector as a whole inputs a significant amount of oil and oily waste products to the sea as a result of routine operational discharges, illegal discharges and accidental spills (GESAMP 2007).

Ballast water and *biofouling* represent other significant risks associated with shipping. Ballast water and sediments represent one of the top four current threats to global biodiversity, due to their role in the transfer of invasive species, harmful aquatic organisms and pathogens. The International Maritime Organization's International Convention for the Control and Management of Ships' Ballast Water and Sediments was adopted in 2004 and requires ballast-water management which largely refrains from unmanaged discharge. Guidance is provided in the Convention and recent marine biosecurity planning guidance is available in Scotland (SNH 2016). In the US and Russia, ballast-water risk assessments (and in some instances ship testing) are commonplace, and they are increasingly being used elsewhere.

Marine farming (e.g. salmon and prawn farms) is typically undertaken in an enclosed section of the sea where sea conditions are calm, e.g. the sheltered waters of sea lochs or in Kerala, India where the tide is used to help harvest shrimps (around 75 per cent of farmed shrimp is produced in Asia and particularly China and Thailand). Mariculture has expanded rapidly over the last two decades due to new technology and increasing demand for seafood products, but commonly identified risks from marine farms include waste from cage cultures, the introduction of invasive species, genetic pollution and disease, parasite transfer and habitat modification. Marine farms have a high potential to lower the water quality in and around the rearing cages as there is often a heavy reliance upon chemicals to control pest outbreaks. Further pollution of the water column and sea bed can also result from the high loadings of organic and nitrogenous compounds present in faecal material and uneaten food (Thompson et al. 1995). The noted rise of parasitic lice infections associated with salmon farming has led to declines of wild salmonid stocks (Krkošek et al. 2005, Vollset et al. 2015). Additional concerns include disturbance caused by marine farm operational activities, the excessive use of wild stocks of fish to feed captive fish, and the genetic decline of wild fish stocks as a result of interbreeding with escaped captive stock. However, since the late 1990s generally stronger regulation of the industry by governments worldwide, aimed at promoting more sustainable farming practices, has been introduced.

Water abstraction within the coastal zone is an important issue because many coastal communities rely on groundwater for their drinking water supply. By depleting the groundwater in **aquifers**, abstraction can lead to intrusion by seawater. The main result is saline intrusion but the groundwater can also be contaminated by pollutants present in the seawater. Removal or alteration of certain habitat types, such as sand dunes, can have a similar effect, as dune formations act as small-scale aquifers and may maintain the water table at an elevated level in relation to surrounding areas. Saline intrusion can also affect the biota of maritime fresh or brackish water habitats. In some areas, the combination of abstraction and infrastructure development has caused the land to sink relative to sea level (IAH 2006).

Over-fishing represents a significant problem for marine ecosystems around the world. Although there are some well-known instances of the collapse of fish stocks, such as the loss of the once-prolific cod fishery of the Grand Banks off Newfoundland and the demise of the North Sea cod fishery, the scale of the loss of ocean and coastal fish stocks has only relatively recently been determined. Catch records from the open sea give a picture of declining fish stocks. The Food and Agriculture Organisation (FAO) estimate that 53 per cent of the world's fisheries are fully exploited, and 32 per cent are overexploited, depleted or recovering from depletion (FAO 2011). Bigger vessels, better nets, and new technology for spotting fish are not bringing the world's fleets bigger returns – in fact the global fishing fleet is two to three times larger than the size the oceans can sustainably support, and it is estimated that stocks of all species currently fished for food could collapse by 2048 (Worm et al. 2006). The situation is actually likely to be much worse than reported, as the data used by the FAO does not fully take account of the volume of fish taken by small-scale fisheries, illegal fishing and discards (Pauly and Zeller 2016). Although many factors are likely to have contributed to these recorded collapses, over-fishing is considered to be the prime driver behind the observed declines. In particular, the continued capture of small species of fish for use in the animal feed or fertiliser industries can disrupt food chains, with particularly serious consequences for top carnivores, including large fish, birds and marine mammals.

7.5.3 Methods of impact prediction and assessment

The 'impact assessment' phase follows the identification and description of potential impacts. An *impact* arises when a particular *effect* interacts with a *receptor* to cause a *change* to the environment (adverse or beneficial). Therefore, during the impact-assessment stage for a coastal or marine EcIA, the nature/magnitude of the effect and the nature/sensitivity of the receptor should be assessed, since the derivation of significance is a function of both of these.

The following aspects of the magnitude of the effect should be defined, as far as possible:

- *Scale* – The scale of change that the effect may cause compared to the baseline.
- *Spatial extent* – The spatial extent over which the predicted effect will arise (e.g. regional, local (say within 5 km of a dredge zone) or at the site of impact).
- *Duration* – The length of time over which the effect will occur (prior to recovery or replacement of the resource or receptor).
- *Reversibility* – An irreversible (permanent) impact is one from which recovery is not possible within a reasonable timescale or for which there is no reasonable chance of action being taken to reverse it.
- *Frequency* – The number of times that the effect will occur within the duration of the activity.
- *Likelihood* – The likelihood that an effect will actually occur.

The key aspects of coastal and marine receptors that require consideration are (Posford Haskoning 2004):

- *Vulnerability* – The likelihood (or risk) of an effect interacting with (or affecting) the receptor (e.g. could a fish species avoid the effect by moving away?).
- *Sensitivity/intolerance* – The sensitivity (level of intolerance) of the receptor to the effect under consideration, i.e. whether the species/population or some of the species/population likely to be damaged/destroyed (e.g. fish with swim bladders in close proximity to the inner ear are far more sensitive to noise than fish species lacking a swim bladder) and how its viability will be affected. Sensitivity is a function of a receptor's capacity to accommodate changes in baseline conditions resulting from the activity and/or as a result of ongoing natural processes, and reflects its capacity to recover if it is affected. Marine Life Information Network (MarLIN) (Table 7.3) provides detailed information on the sensitivity of many key features of the marine environment, although the data should be critically appraised, as they may be extrapolated from other similar species.
- *Recoverability* – How quickly the receptor can recover to its pre-impact state following exposure to an effect (distinguishing between partial and full recovery).
- *Value/importance* – Whether the receptor is 'important' based on a number of criteria, including its occurrence (e.g. is it rare or unique?); its biodiversity/social/community/economic value on a local, regional, national and international scale; and its ecosystem function.

Tables 7.4 and 7.5 present examples of the type of criteria that can be used to determine and classify (define) effect magnitude and receptor sensitivity and value for marine ecology.

Provided with this context, potential changes to geomorphological processes and coastal ecological interests should be assessed in relation to baseline conditions, known linkages and interactions, and information on past, current and predicted trends. Due to the highly dynamic nature of the coastal

Table 7.4 Criteria that can be used to define the magnitude of effects on marine ecology

Magnitude of impact	Criteria
High	The quality and availability of habitats and species are degraded to the extent that locally rare populations and habitats are destroyed and protected species and habitats experience widespread change, such that the integrity of the ecosystem and the conservation status of a designation may be compromised. Activities predicted to occur and affect receptors continuously over the long term, and during sensitive life stages. Recovery, if it occurs, would be expected to be long-term i.e. ten years following the cessation of activity. Impacts not limited to areas within and adjacent to the development.
Medium	The quality and availability of habitats and species are degraded to the extent that the population or habitat experiences reduction in number or range. Activities predicted to occur and affect receptors regularly and intermittently, over the medium to short term and during sensitive life stages. Recovery expected to be medium-term timescales, i.e. five years following cessation of activity. Impacts largely limited to the areas within and adjacent to the development.
Low	The quality and availability of habitats and species experience some limited degradation. Disturbance to population size and occupied area within the range of natural variability. Activities predicted to occur intermittently and irregularly over the medium to short term. Recovery expected to be short term, i.e. one year following cessation of activity. Impacts limited to the area within the development.
Very Low	Although there may be some impacts on individuals, it is considered that the quality and availability of habitats and species would experience little or no degradation. Any disturbance would be in the range of natural variability. Activities predicted to occur occasionally and for a short period. Recovery expected to be relatively rapid, i.e. less than approximately six months following cessation of activity. Impacts limited to the area within the development.

Table 7.5 Criteria that can be used to define sensitivity and value for marine ecology

Definition	Value and sensitivity guidelines
High	**Value** Feature/receptor possesses key characteristics which contribute considerably to the distinctiveness, rarity and character of the site/receptor, e.g. designated features of international/national importance (Ramsar, SAC and SPA). Feature/receptor possesses important biodiversity, social/community value and/or economic value. Feature/receptor is rarely sighted. **Sensitivity** Receptor populations have very low capacity to adapt to, or recover from, proposed form of change, i.e. population is highly sensitive to change.
Medium	**Value** Feature/receptor possesses key characteristics which contribute considerably to the distinctiveness, rarity and character of the site/receptor, e.g. designated features of regional/county importance. Feature/receptor possesses moderate biodiversity, social/community value and/or economic value. Feature/receptor is occasionally sighted. **Sensitivity** Receptor has low capacity to accommodate proposed form of change, i.e. is moderately sensitive.
Low	**Value** Feature/receptor only possesses characteristics which are of district or local importance. Feature/receptor not designated or only designated at the district or local level. Feature/receptor possesses some biodiversity, social/community value and/or economic value. Feature/receptor is relatively common. **Sensitivity** Feature/receptor is tolerant to changes within the range of natural variation, i.e. is only slightly sensitive.
Very Low	**Value** Feature/receptor characteristics do not make a contribution to the character or distinctiveness locally. Feature/receptor not designated. Feature/receptor possesses low biodiversity, social/community value and/or economic value. Feature/receptor is abundant. **Sensitivity** Feature/receptor is generally tolerant of the proposed change, i.e. of low sensitivity.

environment and the complexity of relationships between geomorphological processes and the habitats that they support and modify, prediction of outcomes as a result of human activity at the coast is uncertain and often difficult to define. Assessment needs to consider the likely changes in processes that an action may cause and then the result of any process change on coastal and marine habitats, species and other environmental parameters (e.g. navigation). Assessing impacts may, therefore, involve several steps or iterations and needs to consider both the primary impacts of an action (e.g. land-take) and also secondary impacts that result from the effects of any process change (e.g. modification of a sediment transport pathway and the effect that this may have on a habitat).

Where good historical data exists, predicting change in coastal and estuarine processes and morphological features can be undertaken using observed trends. This historical trend analysis method involves the interrogation of time-series data to identify directional trends and rates of processes and morphological change over time. However, because coastal sedimentation and erosion are very dynamic processes, even where good historical records are available it cannot be assumed that the same conditions will continue to apply in the future (MAFF 2000). Hence, expert geomorphological assessment should ideally incorporate not only historical trend data but also information about current physical processes, geological constraints and sediment properties, and general relationships between processes and morphological responses (HR Wallingford et al. 2007).

In addition to the more conceptual methods, a large number of numerical models have been developed for predicting changes in coastal and estuarine systems and/or for coastal management. But, as discussed above, due to the complexity of the systems and interactions involved, accurately modelling these processes and systems and determining responses is difficult. The results from such work, therefore, should be considered in light of limitations of knowledge about the processes, the ability of models to accurately replicate 'natural' dynamics and other relevant contextual information (i.e. modelling results should not be used in isolation for predictive purposes). However, where changes to coastal processes are likely to result from works in the coastal zone, the use of predictive models in determining potential changes is highly advisable. Advice on available models, and on the feasibility of utilising them in an ESIA, can be sought from organisations such as Royal HaskoningDHV, HR Wallingford, ABPmer, Delft Hydraulics, Moffat Nichol, GHD and Proudman Oceanographic Laboratory. A range of relevant predictive approaches were assessed in detail as part of the EMPHASYS (2000) programme (see also Brew and Pye 2002). Background information on assessing flood and coastal erosion risks is provided in Thorne et al. (2007).

Where knowledge or resources are limited, predictions may have to rely on relatively simple methods, such as those outlined in §6.7.4. The implications of some geomorphological changes can be predicted, for example, using standard risk-assessment methods (Chapter 18) such as the calculation of *return periods* (MAFF 2000; Penning-Rowsell et al. 1988). However, many regulators are increasingly requiring more complex modelling approaches as part of consent applications.

The derivation of significance for individual impacts is typically based on the relationship between the magnitude of an effect and the value and/or sensitivity of the affected resource or receptor. Following assessment of these factors, it is then possible to assign a level of significance to the impact, perhaps using a simple matrix (see Figure 1.3). However, although this method enables the significance of an impact to be described, it does not take into account a range of other important factors related to both the effect and the receptor which will influence the overall significance of the impact. Therefore, in order to ensure that the impact is fully described, it is recommended that all of the information related to the nature of the effect and the receptor is utilised (as described above and in Table 7.6). It may, for example, be useful to differentiate between pulse, press and catastrophic disturbance types in assessing impact significance (Glasby and Underwood 1996). A framework for risk assessment has been developed by Defra and Cranfield University (2011).

A *pulse disturbance* is a short-term disturbance of potential high intensity, which may result in a temporary response in a population or process. Examples might be the short-term impacts associated with the construction of a building near a coastal waterway which results in disposal of spoil to that waterway, or the temporary changes in beach profile and extent associated with a beach recharge scheme. Many coastal and shallow marine habitats and species groups are adapted to natural pulse disturbances, such as fluctuations in sediment transport or high-energy storm events.

A *press disturbance* is a sustained or chronic disturbance to the environment which may cause a long-term response. For example, any permanent development, such as a coastal defence scheme, may cause long-term changes to the sediment balance, perhaps enhancing erosion or sediment accretion (which may have positive or negative consequences). Other examples could be the long-term discharge of a thermal plume from a nuclear power station, causing changes in the distribution of littoral biota, or the increased presence of fish near the intake screens of a water-cooling system, and their subsequent entrapment on the sieve system.

A *catastrophic disturbance* is a major habitat destruction from which populations are unlikely to recover, or that may lead to a complete change in habitat type. An example is the permanent flooding of intertidal mudflats by a static barrage scheme. Similarly, cliff-collapse caused by the construction of buildings on unstable cliffs might result in the permanent loss of valuable geological or geomorphological features (Baird 1994).

Although these definitions are clear, in practice a project may generate combinations of disturbance types and responses to them may vary between organisms (Glasby and Underwood 1996). For example, a pulse disturbance to a population of very long-lived organisms may be a press disturbance to a population of organisms with a short lifespan. Similarly, a local geomorpholog-

Table 7.6 Key aspects of the nature of the 'effect' and the 'receptor' – an example

Nature of effect	
Description	**Deposition of sediment in dredging overflow (discharged water).**
Spatial extent	Judged to be **local** (within 5 km of the dredge zone).
Magnitude	**High** – based on the high level of deposition compared to baseline deposition levels.
Duration	**Short term** – six months to a year.
Frequency	**Infrequent** – no maintenance dredging is envisaged.

Nature of receptor	
Description	**Herring spawning ground (evidence based).**
Is the receptor vulnerable to the effect?	**Yes** – spatial analysis using GIS indicates that the deposition footprint and the herring spawning ground overlap.
Sensitivity (intolerance) of receptor to the effect	**High** – the predicted depth of deposition is higher than smothering values that would have adverse effects on herring spawning. It is predicted that the deposition will change the substrate composition to a degree that makes the ground unsuitable for herring spawning.
Recoverability of receptor to the effect	**High** – the receptor has a rapid recovery rate. The impact would be temporary, as following the cessation of dredging (and deposition), excess sediment would be removed by natural processes and depths of sediment would return to baseline levels within five years. Therefore, the ground would once again be suitable for herring spawning.
Value (importance) of the receptor	**High** – it is the only known spawning ground for this species within the wider study area and the pre-spawning aggregation of herring in this area represents an important component of the local commercial fishery (i.e. it has economic value).

Note: Sensitivity and recoverability will vary based on differing spatial extents, magnitude, duration and frequency of effects.

ical pulse disturbance, such as dredging a channel, may upset the sediment balance and lead to catastrophic disturbances elsewhere in the coastal sediment cell.

A potential framework for integrating information on vulnerability, sensitivity, recoverability and the value of receptors in the assignment of significance-criteria to impacts is provided in Figure 7.13. This framework is relevant to the assessment of impacts on marine ecology (including designated sites), fish and shellfish resources, commercial fisheries and archaeology. It is not

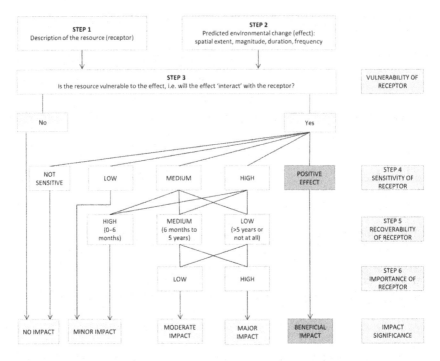

Figure 7.13

Potential decision framework for assigning significance to impacts on marine ecology, fish and shellfish, and archaeology (modified from Posford Haskoning 2004)

relevant to the assessment of impacts related to changes in physical processes (waves, tidal currents and sediment transport) or navigation (where navigational risk assessments are more useful).

With respect to the hydrodynamic regime, extensive information is often available but may be difficult to categorise by means of the framework described above. This is because the relevant components (e.g. tidal currents) represent forcing parameters that, if altered, can result in a change which may or may not directly translate into an impact (e.g. a change in current velocity). However, its indirect influence on other parameters (such as the benthic resource) could cause an impact to arise, but this would be considered as part of the assessment of impacts on marine ecology. Therefore, the assessment of the hydrodynamic regime tends to focus on describing change rather than defining impacts.

Clearly, the assessment of potential impacts and derivation of significance is heavily reliant on our level of understanding, and on the quantity and quality of data on potential effects and the sensitivity and recoverability of certain species/habitats to those effects (e.g. smothering and habitat removal). The complexities inherent in coastal ecosystems illustrate how inadequate data and/or understanding of the coastal system hamper impact prediction, and explain why scientists are often loath to make concrete statements regarding changes or losses that a project may generate. Given that data gaps often exist and that knowledge is incomplete with respect to baseline conditions and certain effects or impacts then, as a matter of good practice, the **precautionary principle** should be applied when such situations are faced in the assessment process.

The 'Rochdale Envelope' is an approach to ESIA, named after a UK planning-law case, which allows a project description to be broadly defined, within a number of agreed parameters, for the purposes of a consent application. This allows for a certain level of flexibility while a project is in development. The ESIA should set out, to the best of the applicant's knowledge, what the maximum extent of the proposed development may be and assess, on that basis, the effects that the project could have. This approach is commonly adopted in ESIAs for offshore wind farms and tidal lagoons, for example, where the technology is still developing.

7.6 Mitigation

Ecologists and geomorphologists involved in the EcIA of coastal developments should have the formulation of appropriate mitigation measures as one of their main objectives. Descriptions of proposed measures should be detailed where impacts are predicted to arise, along with implementation mechanisms and proposals for monitoring their effectiveness. Wherever possible, proposed mitigation measures should avoid or minimise potentially harmful geomorphological changes; pollution, including eutrophication; habitat loss or fragmentation; and disturbance of species and communities.

Means of avoiding or mitigating the potential geomorphological effects that may be associated with coastal development or offshore works, and the consequent adverse impacts on ecological interests, are largely based on coastal engineering techniques and good site management (ENCORA 2007, Webb et al. 2012, USACE 2008, John et al. 2015). For example, it is now widely recognised that if a project requires the construction or modification of sea defences, it is desirable that these are soft rather than hard (§7.5.2). Options include:

- Replenish shallow sloping beaches, which are more effective at dissipating wave energy and maintaining the erosion/deposition regime (Rogers et al. 2010, Bird and Lewis 2015).
- Use groynes to stabilise beaches where replenishment is not an option, e.g. due to a lack of suitable sediment. Groynes are usually effective in the short term, but by their nature they disrupt deposition patterns. This can be reduced by minimising their encroachment into the littoral zone or, for

many beaches, by placing them at intervals along the coast (French 2000, Rogers et al. 2010).

- Encourage the maintenance and development of natural barriers such as saltmarshes, mangroves, forests and sand dunes, which also have positive ecological impacts (Royal Haskoning 2007, FAO 2007).
- Use managed realignment to replace the loss of intertidal habitat (salt marsh and mudflats) that may be associated with coastal development directly, through land-take, or indirectly, through the disruption of sediment supply (Defra/EA 2002, Leggett et al. 2004).

Mitigation measures can also involve the use of sensitive construction methods. For example, during the construction phase of projects such as quay walls, impacts on sediment balance can be minimised by conducting the work on the leeward side of existing structures and/or by the use of floating platforms for construction machinery. The use of a 'soft start' for piling or shrouded piles can also reduce noise effects. Further guidance on good construction practice in the coastal and marine environment is provided by John et al. (2015).

The impacts of dredging in estuaries and on the open coast can be reduced by carefully planned extraction programmes and controlled techniques (Simpson and John 2005). In estuaries, operations can be confined to defined tidal periods in order to facilitate the transport and distribution of sediment disturbed during the dredging process to specific parts of the estuary. Through this approach sediment can either be maintained in the system, so as to promote accretion, or sediment can be taken out of the system in order to minimise potential increases in nutrient loadings, high turbidity and sedimentation rates at sites where biota may be sensitive to these changes. For the dredging of fine sediments, appropriate hydrodynamic techniques can be utilised to minimise impacts (see ABP 1999). Such techniques enable sediment to be redistributed in nearshore and estuarine systems and significantly reduce or avoid the need for the disposal of dredged sediments. In situations where dredged sediments require disposal, this is undertaken at a licensed (normally offshore) site, where environmental characterisation has demonstrated that the likely impacts of disposal will be minimal. It is also a requirement of any dredging programme to consider the beneficial use of any material arising from the dredging process. If undertaken and planned with regard to environmental sensitivities, the beneficial use of dredged material can provide the opportunity to offset adverse impacts associated with the dredging activity. Guidance on the beneficial use of dredged material is provided in Burt (1996) and USACE (2013).

In addition to maintaining and enhancing natural features, such as sand dunes, that maintain water table levels, mitigation against groundwater contamination by seawater intrusion can be achieved by methods such as artificial recharge of the aquifer, e.g. by importing freshwater from outside the catchment or by re-routing streams or storm run-off into infiltration pits, which reduce *evapotranspiration*. However, care is needed to ensure that such measures do not generate other impacts on the freshwater systems involved.

It is important to avoid or minimise habitat loss or fragmentation on both the landward and seaward sides of a project. Together with disturbance of wildlife, these impacts depend largely on project location and design, including the siting of associated infrastructure such as new roads or construction compounds; so mitigation measures must focus on sensitive siting and design. If loss of valuable habitat is unavoidable, compensation may be considered as an alternative (and is required under the European Habitats Directive where the activity is predicted to have an adverse effect on the integrity of a European designated site), but this should be a last resort since it can be difficult and expensive to recreate many ecological processes and habitat types (see §6.8.4). John et al. (2007) provide an example of the creation of compensatory habitat associated with port development. Other good examples of intertidal habitat creation in the UK include Tollesbury Wick, Wallsea Island, Trimley Marshes, Streat Marshes and Medmerry.

In the US, a good example of large-scale wetland restoration is the South Bay Salt Pond Restoration Project in San Francisco Bay, which aims to restore up to 54 km^2 of reclaimed salt ponds to tidal action (EDAW et al. 2007). Potential geomorphological impacts of such a large change in the bay coastline were discussed by Brew and Williams (2010). Less precedent exists for compensation of the loss of subtidal habitats, but Voordelta at the Port of Rotterdam (see Box 7.1) and the San Francisco Bay Subtidal Habitat Goals Project are good examples.

Apart from protection in reserves that are closed to the public, damage to fragile habitats (such as sand dunes) by visitor pressure can be limited by measures such as the exclusion of vehicles, provision of boardwalks and management to control access and control or repair wind erosion. These may include the use of netting or brushwood fencing or, more effectively, replanting and protecting vegetation, especially marram grass (see Doody 1985, Pye et al. 2007).

7.7 Monitoring

Monitoring should be undertaken by experts and in consultation with the statutory/regulatory authorities and relevant non-governmental organisations (NGOs). It forms an integral component of ESIA in that it can:

- test impact hypotheses and thus improve understanding and predictive capabilities for the future;
- verify the effectiveness of mitigation measures, and allow their modification if there are unpredicted harmful effects on the environment;
- assess performance and monitor compliance with any agreed conditions;
- provide early warning of undesirable change so that corrective measures can be implemented;
- provide evidence to refute or support claims for damage compensation;

Box 7.1 Example of compensation for subtidal habitats

Development of the Port of Rotterdam took place within the Maasflakte 2 (Voordelta) site. The Voordelta is designated at an international level for a range of habitats and species, including submerged sandbanks and birds. It is also a commercial fishery and heavily used for recreation. An appropriate assessment under the Habitats Directive identified that the proposed development would have an adverse effect on the subtidal sandbanks and on velvet scoter, common and sandwich tern. It was decided (under Article 6(4) of the Habitats Directive) that there were no alternatives and that the project was necessary for reasons of overriding public interest. Compensation was therefore required.

Creation of new subtidal sandbanks outside of the Voordelta site was not considered viable, and therefore research was undertaken into other ways to compensate for the loss of this habitat. It was found that, while the subtidal sandbanks were in favourable condition, removing the use of beam trawlers could result in an increase in habitat quality of 10 per cent. Therefore, beam trawling was removed from an area ten times that of the area to be lost. This is not 'like for like' compensation but it was acknowledged that 'like for like' compensation for losses of some estuarine habitats, particularly those which are subtidal, can be very difficult if not impossible to achieve through the creation of new habitats outside of existing European sites. Extensive monitoring is now in place to assess the effectiveness of this measure, and there is agreement about further measures to be taken if the required degree of habitat improvement is not achieved (Schouten et al. 2008).

• further the knowledge base relating to the actual effects of a particular activity.

The first step in devising a monitoring plan is to define the relevant objectives, which must be realistic and measurable. It is important to ensure that the scale of monitoring relates to the scheme and that the results will be meaningful and provide effective guidance within the context of the assessment process.

An effective baseline against which future monitoring can be compared must be available, which should take account of other activities that occur or could occur within the ESIA study area. The natural variability within a system will need to be determined, as far as possible, in order to predict possible changes in factors such as species abundance and composition in benthic communities, seabed mobility and changes in suspended sediment concentrations (e.g. due to storm activity). Control sites selected for monitoring will need to take account of these factors to ensure that results are not biased.

In the coastal environment, dealing with variability in the natural system and therefore developing a baseline against which to compare monitoring data may be difficult, particularly for small schemes or projects. To combat this issue, particularly in relation to coastal defence schemes, large-scale regional monitoring programmes may be set up. The co-ordination of data gathering at this scale enables resources to be combined, better characterisation of wider coastal dynamics to be achieved, and the provision of data to enable sound and sustainable management decisions to be made.

Monitoring criteria should effectively be defined by the predicted impacts and proposed mitigation measures. Requirements for monitoring should be site-specific and based on the findings of the baseline surveys and subsequent interpretation. The methodology used for monitoring environmental effects should be the same as that used for determining the characteristics of the relevant parameter during the baseline survey. The frequency of sampling during monitoring will need to be based on the objectives and the criteria for the monitoring, as changes in different parameters may occur over a variety of timescales and the frequency of monitoring will need to take this into account (i.e. it cannot be standardised for all parameters).

Depending on the objective of and criteria for monitoring, it is important to define a level above or below which an effect is considered to be unacceptable, i.e. an 'environmental threshold' (e.g. for the maintenance of the affected habitat or species), and an appropriate response. Without this, monitoring for many parameters can only be justified on the basis of improving the knowledge base of the particular effect. *Adaptive monitoring and management* plans associated with development in the coastal environment are now commonplace.

Defining thresholds can be a problem due to the difficulty of accurately determining the level above or below which an effect becomes sufficiently adverse to warrant action being taken. It often requires detailed knowledge of the sensitivity of various receptors to environmental change. Where knowledge is lacking, the precautionary principle may need to be applied. However, its application should be appropriate to the specific situation under investigation and be based on a realistic scenario and the latest information. The results of monitoring should be analysed and interpreted using the same techniques as applied to the baseline data in order to provide valid comparison over time.

Some geomorphological parameters can be monitored using fairly simple techniques. For example:

- On rocky coasts, cliff recession can be measured with pegs driven into the rock, and beach profiles can be measured using conventional field surveying techniques. Other methods of measuring processes are reviewed in Sutherland (2007).
- On a defended coast, rates of deposition can be monitored by indicators such as accumulation/erosion at breakwaters and groynes, dilution rates of particles in sediment of known source, or the use of sediment traps or tracers such as dyes.
- Rates of mud accretion (e.g. on saltmarsh) can be measured using standard levelling techniques or sediment traps (Thayer et al. 2005).

- Sediment transport can be monitored directly by sampling water, or indirectly by beach profile and groyne-height exposure measurements, benthic sampling or remote sensing.
- Photographic or video records can be made, e.g. of beach profiles and sand dune erosion or recovery (SNH 2000; Thayer et al. 2005).

Beyond such simple approaches, significant advances have been made in the use of remotely sensed data in monitoring coastal parameters (e.g. LiDAR data) and, for example, sediment tracer tests (Karstens et al. 2015). Advice on the use of other methods, e.g. for sublittoral monitoring and the use of models, is available from a variety of published sources (e.g. Davies et al. 2001) and relevant organisations (see Table 7.3).

7.8 Conclusions

Over the past two decades there have been significant improvements in the methodologies used to undertake ESIA in the coastal zone. In particular, greater emphasis has been given to the assessment of cumulative impacts; ecosystem function and services; and changes in interactions between geomorphological processes and ecological interests. Much of this has been driven by the requirements of legislation and the designation of numerous coastal and marine habitats. However, there are still issues that development proposals and the associated ESIA process need to consider further if ESIA is to act as a tool that drives sustainability.

The coastal zone is often very dynamic, with significant fluctuations in physical and biological processes occurring over a variety of timescales. To adequately assess potential impacts in relation to this dynamic behaviour requires that data are collected over timescales that reflect the range of environmental conditions, and that sufficient effort is then expended in the monitoring and documentation of impacts and the success, or otherwise, of mitigation measures. This issue also requires an element of risk to be accepted in the development of mitigation measures, since it will be almost impossible to know with certainty how a coastal system will respond to an effect. Best estimates and predictions must therefore be used in making management decisions. Activities must then be fully monitored and a system put in place to allow appropriate response to the findings (i.e. active and adaptive management).

The wider aspects and content that ESIA needs to consider is now being provided by planning and policy objectives set at the regional, national and strategic level, e.g. in relation to the exploitation of marine and coastal areas for the production of offshore energy (and associated Strategic Environmental Assessment). In the UK, National Policy Statements for energy and ports are examples of this.

Increasingly, the strategic planning principles inherent in Integrated Coastal Zone Management are being adopted and implemented; a process that will hopefully lead to the removal of piecemeal development and provide an arena that brings together conflicting and overlapping interests. ICZM also offers the opportunity to plan for a more sustainable response to the pressures that climate-change-linked sea level rise and increased storminess will generate (e.g. TCPA 2015, EPA 2015). Strategic assessment, planning and management will not remove the need for project-based ESIAs, but it should facilitate their execution and effectiveness.

The coastal zone is an outstanding area for wildlife and ESIA provides a means of providing checks on development activities which could undermine its ecological worth. Both ecological and geomorphological science have an obvious role in the process. While there has been significant progress in describing and determining ecological and geomorphological processes on the coast, there is still much work to do in understanding the linkages between the physical and biological environment and how, in particular, habitats and species respond to changes in coastal processes. If coastal ESIA is to develop as a tool for environmental management which helps to realise the goals of conservation and sustainability, it is important that ecologists and geomorphologists have a greater input to the process.

Note

1   With input from David Tarrant on policy, guidance and case studies, Gemma Keenan on marine mammals and Jen McMillan on habitat types and marine impacts; and support from Royal HaskoningDHV.

References

ABP (Associated British Ports) Research and Consultancy Ltd 1999. *Good practice guidelines for ports and harbours operating within or near UK European marine sites.* English Nature, UK Marine SACs Project Report.

ABPmer (ABP marine environmental research) 2002. *Potential effects of offshore wind developments on coastal processes.* Energy Technology Support Unit.

Adam P 1993. *Saltmarsh ecology.* Cambridge: Cambridge University Press.

Al-Hatrushi SM, AY Kwarteng, A Sana, AS Al-Buloushi, A MacLachlan, KH Hamed 2014. *Coastal erosion in Al Batinah, Sultanate of Oman.* Academic Publication Board, Sultan Qaboos University, 261.

Andrews BD, PA Gare and JD Colby 2002. Techniques for GIS modelling of coastal dunes. *Geomorphology* **48**, 289–308.

Anthony A, J Atwood, PV August, C Byron and S Cobb 2009. *Coastal lagoons and climate change: Ecological and social ramifications in U.S. Atlantic and Gulf Coast ecosystems.* The University of Rhode Island, Biological Sciences Faculty Publications.

Bailey H, KL Brookes and PM Thompson 2014. Assessing environmental impacts of offshore wind farms: lessons learned and recommendations for the future. *Aquatic Biosystems* **10**(8). www.aquaticbiosystems.org/content/10/1/8.

Baird WJ 1994. Naked rock and the fear of exposure. *Geological and landscape conservation.*

D O'Halloran, C Green, M Harley, M Stanley and J Knill (eds), 335–336. London: The Geological Society.

Baker JM and JH Crothers 1987. Intertidal rock. *Biological surveys of estuaries and coasts*, JM Baker and WJ Wolff (eds), 157–197. Cambridge: Cambridge University Press.

Barnes RSK 1994. *The brackish-water fauna of Northwest Europe. An identification guide to brackish-water habitats, ecology and macrofauna for field workers, naturalists and students.* Cambridge: Cambridge University Press.

Barnes, RSK and RH Hughes 1988. *An introduction to marine ecology*, 2nd Edn. Oxford: Blackwell Scientific Publications.

Bergström L, F Sundqvist and U Bergström 2013. Effects of an offshore wind farm on temporal and spatial patterns in the demersal fish community. *Marine Ecology Progress Series* **485**, 199–210.

Bibby CJ, ND Burgess, DA Hill and S Mustoe 2000. *Bird census techniques*, 2nd Edn. London: Academic Press.

Bird ECF 1985. *Coastal changes: a global review*. Chichester: Wiley.

Bird ECF 2008. *Coastal geomorphology: an introduction*, 2nd Edn. Chichester: John Wiley & Sons.

Bird ECF and N Lewis 2015. *Beach renourishment*. New York: Springer.

Blower JG, LM Cook and JA Bishop 1981. *Estimating the size of animal populations*. London: Allen & Unwin.

Bradbury G, M Trinder, B Furness, AN Banks, RWG Caldow and D Hume 2014. Mapping seabird sensitivity to offshore wind farms. *PLoS ONE* **9**(9): e106366.doi:10.1371/journal.pone.0106366.

Brew DS and K Pye 2002. *Guidance notes for assessing morphological change in estuaries*. Defra/Environment Agency R&D Technical Report FD2110/TR1.

Brew DS and PB Williams 2010. Predicting the impact of large-scale tidal wetland restoration on morphodynamics and habitat evolution in south San Francisco Bay, California. *Journal of Coastal Research* **26**, 912–924.

Bricker, SB, B Longstaff, W Dennison, A Jones, K Boicourt, C Wicks and I Woerner 2007. *Effects of nutrient enrichment in the nation's estuaries: A decade of change, National Estuarine Eutrophication Assessment Update*. NOAA Coastal Ocean Program Decision Analysis Series No. 26. Silver Spring, MD (USA): National Centers for Coastal Ocean Science.

Bricker SB, B Longstaff, W Dennison, A Jones, K Boicourt, C Wicks and J Woerner 2008. Effects of nutrient enrichment in the nation's estuaries: A decade of change. *Harmful Algae* **8**, 21–32.

Brouwer A, AJ Murk and JH Koeman 1990. Biochemical and physiological approaches in ecotoxicology. *Functional Ecology* **4**, 275–281.

Brown AC and A McLachlan 2006. *Ecology of sandy shores*, 2nd Edn. London: Academic Press.

Brown K, C Hambidge and A Matthews 2003. *The development of remote sensing techniques for marine SAC monitoring*. English Nature Research Report 552. Peterborough, Cambs.

British Trust for Ornithology 2014. *The avoidance rates of collision between birds and offshore turbines*. British Trust for Ornithology (BTO) Report No. 656. Authors: Cook ASCP, EM Humphreys, EA Masden, W Band and NHK Burton.

Burt N 1996. *Guidelines for the beneficial use of dredged material*. Report SR 488. HR Wallingford. London: DfT (Department for Transport).

Burt N and J Watts 1996. *Barrages: engineering, designs and environmental impacts*. Chichester: John Wiley & Sons.

Burton HHK, MM Rehfisch and NA Clark 2003. *The effect of the Cardiff Bay Barrage on waterbird populations. Final report.* British Trust for Ornithology Research Report No. 343. Thetford, Cambs: British Trust for Ornithology.

Camphuysen CJ, AD Fox, MF Leopold and LBK Petersen 2004. *Towards standardised seabirds at sea census techniques in connection with environmental impact assessments for offshore wind farms in the U.K. A comparison of ship and aerial sampling methods for marine birds, and their applicability to offshore wind farm assessments.* Published for the Collaborative Offshore Wind Research Into the Environment (COWRIE). www.offshorewindfarms.co.uk.

Carter RWG, TGF Curtis and MJ Sheehy-Skeffington (eds) 1992. *Coastal dunes: geomorphology, ecology and management for conservation.* Proceedings of the 3rd European Dune Congress. Galway June 1992. Rotterdam: Balkema.

Clayton KM 1993. *Coastal processes and coastal management.* Cheltenham, Glos: Countryside Commission.

Coles SM 1979. Benthic microalgal populations on intertidal sediments and their role as precursors to salt marsh development. *Ecological processes in coastal environments,* RL Jefferies and AJ Davy (eds), 25–42. Oxford: Blackwell Science.

Connor DW, JH Allen, N Golding, KL Howell, LM Leiberknecht, KO Northen and JB Reker 2004. The Marine Habitat Classification for Britain and Ireland, Version 04.05. Peterborough, Cambs: JNCC.

Cooper NJ and Brew DS 2013. Chapter 5. Impacts on the physical environment. *Aggregate dredging and the marine environment: An overview of recent research and current industry practice,* RC Newell and TA Woodcock (eds), 68–87. The Crown Estate.

Cooper NJ, DJ Leggett and JP Lowe 2000. Beach-profile measurement, theory and analysis: practical guidance and applied case studies. *Water and Environment Journal* **14**(2), 79–88.

Côté IM and MR Perrow 2006. Fish. *Ecological census techniques: a handbook,* 2nd Edn., WJ Sutherland (ed.), 250–277. Cambridge: Cambridge University Press.

Covey R and D Laffoley 2002. *Maritime state of nature report for England: getting onto an even keel.* Peterborough, Cambs: English Nature.

Crawford RMM 1998. Shifting sands: plant survival in the dunes. *Biologist* **45**(1), 27–32.

Davidson-Arnott R 2009. *Introduction to coastal processes and geomorphology.* Cambridge: Cambridge University Press.

Davies J, J Baxter, M Bradley, D Connor, J Khan, E Murray, W Sanderson, C Turnbull, and M Vincent 2001. *Marine monitoring handbook.* Peterborough, Cambs: JNCC.

DECC (Department of Energy and Climate Change) 2010. Severn tidal power feasibility study: conclusions and summary report.

Defra/EA (Department for Environment, Food and Rural Affairs / Environment Agency) 2002. *Managed realignment review: project report.* Flood and Coastal Defence R&D Programme, Policy Research Project FD2008.

Defra and Cranfield University 2011. *Guidelines for environmental risk assessment and management. Green Leaves III.* London: Defra.

Denny MW and SD Gaines (eds) 2007. *Encyclopedia of Tidepools and Rocky Shores.* Berkeley, CA: University of California Press.

Diederichs AG, M Nehls, S Dahne, S Adler and U Koschinski Verfuß 2008. *Methodologies for measuring and assessing potential changes in marine mammal behaviour, abundance or distribution arising from the construction, operation and decommissioning of offshore windfarms.* BioConsult SH report to COWRIE Ltd.

Doody JP (ed.) 1985. *Sand dunes and their management.* Focus on Nature Conservation No. 13. Peterborough, Cambs: JNCC.

Dyer KR 1998. *Estuaries: a physical introduction,* 2nd Edn. Chichester: John Wiley & Sons.

Earle R and DG Erwin (eds) 1983. *Sublittoral ecology, the ecology of the shallow sublittoral benthos.* Oxford: Oxford University Press.

EC (European Commission) 1993. *Towards sustainability: the fifth EC Environmental Action Programme.* Brussels: EC, http://ec.europa.eu/environment/actionpr.htm.

EDAW, Philip Williams and Associates, HT Harvey and Associates, Brown and Caldwell and Geomatrix 2007. *South Bay salt pond restoration project final environmental impact statement/report.* US Fish and Wildlife Service California Department of Fish and Game.

EEA (European Environment Agency) 2006. *The changing faces of Europe's coastal areas.* EEA Report 6/2006, http://reports.eea.europa.eu/eea_report_2006_6/en.

EEA 2013. *Balancing the future of Europe's coasts – knowledge base for integrated management.* EEA Report 12/2013.

Eleftheriou E (ed.) 2013. *Methods for the study of marine benthos,* 4th Edn. Oxford: Wiley Blackwell (UK).

EMPHASYS 2000. *A guide to prediction of morphological change within estuarine systems,* Version 1B. EMPHASYS Consortium for MAFF Project FD1401, http://books. hrwallingford.co.uk/acatalog/free_downloads/tr114.pdf.

ENCORA 2007. *European platform for sharing knowledge and experience in coastal science, policy and practice,* http://encora.eu.

English Nature, Environment Agency, Defra, LIFE and NERC 2003. *Living with the sea: good practice guide.* Peterborough, Cambs: English Nature.

Environment Agency 1997. *Our policy and practice for the protection of floodplains.* Bristol: Environment Agency.

Environment Agency 2013. *Climate change allowances for planners. Guidance to support the National Planning Policy Framework,* September 2013. www.gov.uk.government/ uploads/system/uploads/attachment_data/file/296964/LIT_8496_5306da.pdf.

EPA (Environment Protection Agency) 2015. *Coastal adaptation toolkit.* www.epa.gov/ cre/coastal-adaptation-toolkit.

European Marine Observation and Data Network (EMODNet) (n.d.) www.emodnet-seabedhabitats.eu/default.aspx.

FAO (Food and Agriculture Organization) 2007. *Coastal protection in the aftermath of the Indian Ocean tsunami: What role for forests and trees?* Chapter 4: Protection from coastal erosion, www.fao.org/ docrep/010/ag127e/AG127E00.htm#Contents.

FAO 2011. *The State of World Fisheries and Aquaculture* (SOFIA) 2010, 197.

French PW 2000. *Coastal protection: processes, problems and solutions.* London: Routledge.

GESAMP (Group of Experts on the Scientific Aspects of Marine Environmental Protection) 2007. *Estimates of oil entering the marine environment from sea-based activities.* International Marine Organisation. Report No. 75. London: GESCAMP, http://gesamp. net/page.php?page-3.

Gimmingham CH, W Ritchie, BB Wiletts and AJ Willis (eds) 1989. *Coastal sand dunes.* Edinburgh: Royal Society of Edinburgh.

Glasby TM and AJ Underwood 1996. Sampling to differentiate between press and pulse disturbances. *Environmental Monitoring and Assessment* **42**, 241–252.

Goals Project 1999. *Baylands ecosystem habitat goals: A report of habitat recommendations.* San Francisco (CA). US Environmental Protection Agency.

Gorman M and D Raffaelli 1993. The Ythan estuary. *Biologist* **40**(1), 10–13.

Gray JS and M Elliot 2008. *The ecology of marine sediments*, 2nd Edn. Oxford: Oxford University Press.

Hammond PS 1987. Techniques for estimating the size of whale populations. *Symposium of the Zoological Society of London* **58**, 225–245.

Hammond PS and PM Thompson 1991. Minimum estimation of the number of bottlenose dolphins *Tursiops truncatus* in the Moray Firth, N.E. Scotland. *Biological Conservation* **56**, 79–87.

Heath, M 2010. Nutrient enrichment in MCCIP *Annual Report Card 2010–11*, MCCIP Science Review. www.mccip.org.uk/arc.

Hiby AR and PS Hammond 1989. Survey techniques for estimating abundance of cetaceans. *Report to the International Whaling Commission (Special Issue No. 11)*, 47–80.

Hill D, M Fasham, G Tucker, M Shewry and P Shaw 2005. *Handbook of biodiversity methods: survey, evaluation and monitoring*. Cambridge: Cambridge University Press.

Hiscock K (ed.) 1998. *Benthic marine ecosystems of Great Britain and the north-east Atlantic*. Peterborough, Cambs: JNCC.

HR Wallingford, ABPmer and J Pethick 2007. *Review and formalisation of geomorphological concepts and approaches for estuaries*. Defra/Environment Agency R&D Technical Report FD2116/TR2.

IAH (International Association of Hydrogeologists) 2006. *4th World Water Forum. Groundwater for life and livelihoods – the framework for sustainable use*. Kenilworth, UK: International Association of Hydrogeologists.

IPCC 2014. *Climate change 2014: Synthesis report*. Contribution of Working Groups I, II and III to the Fifth Assessment Report of the Intergovernmental Panel on Climate Change [Core Writing Team, Pachauri RK and Meyer LA (eds)]. Geneva, Switzerland: IPCC.

Jennings S, MJ Kaiser and JD Reynolds 2001. *Marine fisheries ecology*. Oxford: Blackwell Science.

JNCC 2004a. *Common Standards Monitoring guidance for saltmarsh habitats*. Peterborough: JNCC.

JNCC 2004b. *Common Standards Monitoring guidance for inshore sublittoral sediment habitats*. Peterborough: JNCC.

John SA, M Simpson and T Gray 2007. Compensatory habitats: defining needs and demonstrating success. *International conference on coastal management*, Institution of Civil Engineers, 183–192. London: Thomas Telford Publishing.

John S, N Meakins, K Basford, H Craven and P Charles (eds) 2015. *Coastal and marine environmental site guide (second edition) (C744)*. London: CIRIA.

Jones L, S Angus, A Cooper, P Doody, M Everard, A Garbutt, P Gilchrist, J Hansom, R Nicholls, K Pye, N Ravenscroft, S Rees, P Rhind and A Whitehouse 2011. Coastal margins. In *The UK National Ecosystem Assessment Technical Report*. UK National Ecosystem Assessment. Cambridge: UNEP-WCMC.

Karstens S, F Schwark, S Forster, S Glatzel and U Buczko 2015. Sediment tracer tests to explore patterns of sediment transport in coastal reed beds – a case study from the Darss-Zingst Bodden Chain. *Rostock. Meeresbiolog. Beitr* **25**, 41–57.

Klemas V 2013 Airborne remote sensing of coastal features and processes: an overview. *Journal of Coastal Research* **29**, 239–255.

Kober K, A Webb, I Win, S O'Brien, LJ Wilson and JB Reid 2010. An analysis of the numbers and distribution of seabirds within the British Fishery Limit aimed at identifying areas that qualify as possible marine SPAs. JNCC Report No. 431.

Komar PD 1983. Coastal erosion in response to the construction of jetties and breakwaters. *Handbook of coastal process and erosion*. PD Komar (ed.), 191–204. Boco Raton, FL: CRC Press.

Krkošek M, MA Lewis and JP Volpe 2005. Transmission dynamics of parasitic sea lice from farm to wild salmon. *Proceedings of the Royal Society B* **272**: 689–696.

Lambkin DO, JM Harris, WS Cooper and T Coates 2009. *Coastal process modelling for offshore wind farm environmental impact assessment: best practice guide*. London: COWRIE Limited.

Leggett DJ, N Cooper and R Harvey 2004. *Coastal and estuarine managed realignment – design issues* (CIRIA 628). London: CIRIA. www.ciria.org/acatalog/c628.pdf.

Little C 2000. *The biology of soft shores and estuaries*. Oxford: Oxford University Press.

Little C and JA Kitching 1996. *The biology of rocky shores*. Oxford: Oxford University Press.

Lloyd CS, ML Tasker and KE Partridge 1991. *The status of seabirds in Britain and Ireland*. London: T & A Poyser.

MAFF (Ministry of Agriculture, Fisheries and Food) 2000. *Flood and coastal defence project appraisal guidance: approaches to risk*, FCD-PAG4. London: MAFF.

Martinez ML and NP Psuty (eds) 2004. *Coastal dunes: ecology and conservation*. Berlin and Heidelberg: Springer-Verlag.

Masselink G, MG Hughes and J Knight 2011. *Introduction to coastal processes and geomorphology*, 2nd Edn. London: Routledge.

McCook LJ, B Schaffelke, SC Apte, R Brinkman, J Brodie, P Erftemeijer, B Eyre, F Hoogerwerf, I Irvine, R Jones, B King, H Marsh, R Masini, R Morton, R Pitcher, M Rasheed, M Sheaves, A Symonds and MStJ Warne 2015. Synthesis of current knowledge of the biophysical impacts of dredging and disposal on the Great Barrier Reef: report of an independent panel of experts. Townsville: Great Barrier Reef Marine Park Authority.

McLachlan A and AC Brown 2006. *The ecology of sandy shores*, 2nd Edn. London: Elsevier Inc.

McLusky DS and M Elliott 2004. *The estuarine ecosystem: ecology, threats and management*, 3rd Edn. Oxford: Oxford University Press.

Miller RL, CE Del Castillo and BA McKee 2005. *Remote sensing of coastal aquatic environments: technologies, techniques and applications*. Berlin: Springer-Verlag.

Möller I, M Kudella, F Rupprecht, T Spencer, M Paul, BK van Wesenbeeck, G Wolters, K Jensen, TJ Bouma, M Miranda-Lange and S Schimmels 2014. Wave attenuation over coastal salt marshes under storm surge conditions. *Nature Geoscience* **7**, 727–731.

National Trust 2015. *Mapping our shores: 50 years of land use change at the coast*. Swindon, UK: National Trust.

New TR 1998. *Invertebrate surveys for conservation*. Oxford: Oxford University Press.

Newell GE and RC Newell 2006. *Marine plankton: a practical guide* (CD or paperback). Lymington, Hants: Pisces Conservation.

Nordstrom K and W Carter 1991. *Coastal dunes: form and process*. Chichester: John Wiley & Sons.

Packham JR and AJ Willis 1997. *Ecology of dunes, saltmarsh and shingle*. London: Chapman & Hall (Kluwer Academic).

Packham JR, RE Randall, RSK Barnes and A Neal 2001. *Ecology and geomorphology of coastal shingle*. Otley, W. Yorks: Westbury Academic and Scientific Publishing.

Pauly D and D Zeller 2016. Catch reconstructions reveal that global marine fisheries catches are higher than reported and declining. *Nat. Commun.* 7, 10244.

Penning-Rowsell E, P Thompson and D Parker 1988. Coastal erosion and flood control: changing institutions, policies and research needs. *Geomorphology in environmental planning*, JM Hooke (ed.), 211–230. Chichester: John Wiley & Sons.

Pethick J 1984. *An introduction to coastal geomorphology*. London: Edward Arnold.

Phillips MR 2007. Beach response to a total exclusion barrage: Cardiff Bay, South Wales, UK. *Journal of Coastal Research* **23**, 794–805.

Pitcher TJ and PJB Hart 1982. *Fisheries ecology*. London: Croom Helm.

Posford Haskoning 2004. *Marine aggregate environmental impact assessment: approaching good practice* (SAMP1.03). London: DCLG. www.dclgaggregatefund.co.uk/docs/final_reports/samp_1_031.pdf.

Potts GW and PJ Reay 1987. Fish. *Biological surveys of estuaries and coastal habitats*, JM Baker and WJ Wolff (eds), 342–373. Cambridge: Cambridge University Press.

Pye K, S Saye and S Blott 2007. *Sand dune processes and management for flood and coastal defence. Part 4: Techniques for sand dune management*. Defra/Environment Agency R&D Technical Report FD1392/TR, May 2007.

Quinn JD, LK Philip and W Murphy 2009. Understanding the recession of the Holderness coast, east Yorkshire, UK: a new presentation of temporal and spatial patterns. *Quarterly Journal of Engineering Geology and Hydrogeology* **42**, 165–178.

Raffaelli D and S Hawkins 2011. *Intertidal ecology*. London: Chapman & Hall.

Rodwell JS (ed.) 2000. *British plant communities*, Vol. 5: *Maritime communities and vegetation of open habitats*. Cambridge: Cambridge University Press.

Rogers J, B Hamer and A Brampton 2010. *Beach management manual*, 2nd Edn (C685B). London: CIRIA.

Rothwell PI and SD Housden 1990. *Turning the tide: a future for estuaries*. Sandy, Beds: RSPB.

Royal Haskoning 2007. *Saltmarsh management manual*. Joint Defra/Environment Agency Flood and Coastal Erosion Risk Management R&D Programme Technical Report SC030220.

Saintilan N and K Rogers 2013. The significance and vulnerability of Australian saltmarshes: implications for management in a changing climate. *Marine and Freshwater Research* **64**, 66–79.

Save the North Sea 2005. *Save the North Sea – final report. A project targeting change of attitudes and behaviour towards marine litter in the North Sea*. Interreg IIIb North Sea Programme.

Schäfer S, G Buchmeier, E Claus, L Duester, P Heininger, A Körner, P Mayer, A Paschke, C Rauert, G Reifferscheid, H Rüdel, C Schlechtriem, C Schröter-Kermani, D Schudoma, F Smedes, D Steffen and F Vietoris 2015. Bioaccumulation in aquatic systems: methodological approaches, monitoring and assessment. *Environmental Sciences Europe* **27**(1), 1–10.

Schouten P, K Jennings, C McMullon, B Paterak, C Smit and H Verbeek 2008. *Natura 2000 and development in estuarine areas*. Mitigation and Compensation issues: Results of the first EU peer-exchange Natura 2000 estuaries group.

Simpson M and SA John 2005. *Port development and estuarine stewardship*. Proceedings of the 3rd International Conference on Marine Science and Technology for Environmental Sustainability. Newcastle-upon-Tyne.

SNH (Scottish Natural Heritage) 2000. *A guide to managing coastal erosion in beach/dune systems*. Wallingford: HR Wallingford, www.snh.org.uk/publications/on-line/heritage management/erosion/1.shtml.

SNH 2016. Marine biosecurity planning guidance, www.snh.gov.uk/policy-and-guidance/guidance-documents/document/?category_code=Guidance&topic_id=1628.

Speight MR and Henderson PA 2010. *Marine ecology: concepts and applications*. Oxford: Wiley Blackwell. www.bookdepository.com/book/9781444335453.

Statham PJ 2012. Nutrients in estuaries – an overview and the potential impacts of climate change. *Science of the Total Environment* **434**, 213–227.

Stone CJ, A Webb, C Barton, N Ratcliffe, TC Reed, ML Tasker and MW Pienkowski 1995. *An atlas of seabird distribution in north-west European waters*. Peterborough, Cambs: JNCC.

Sutherland J 2007. *Inventory of coastal monitoring methods and overview of predictive models for coastal evolution*. European Commission.

Tasker ML, PH Jones, TJ Dixon and BF Blake 1984. Counting seabirds from ships: a review of methods employed and a suggestion for a standardised approach. *AUK* **101**, 567–577.

TCPA (Town and Country Planning Association) 2015. *No regrets: planning for sea-level rise and climate change and investing in adaptation.* www.tcpa.org.uk/data/files/4305 _No_Regrets_August2015_final_LR.pdf.

Tett PB 1987. Plankton. *Biological surveys of estuaries and coastal habitats*, JM Baker and WJ Wolff (eds), 280–341. Cambridge: Cambridge University Press.

Thayer GW, TA McTigue, RJ Salz, DH Merkey, FM Burrows and PF Gayaldo 2005. *Science-based restoration monitoring of coastal habitats. Vol. 2: Tools for monitoring coastal habitats*. NOAA Coastal Ocean Program. Decision Analysis Series No. 23, Vol. 2. Silver Spring, MD: NOAA/CCMA. http://coastalscience.noaa.gov/ecosystems/ estuaries/restoration_monitoring.html.

Thompson PM and J Harwood 1990. Methods for estimating the population size of common seals, *Phoca vitulina*. *Journal of Applied Ecology* **27**, 924–938.

Thompson S and J Lee 2001. Coastal ecology and geomorphology. Ch. 13 in *Methods of environmental impact assessment*, 2nd Edn, P Morris and R Therivel (eds). London: Spon Press.

Thompson S, JR Treweek and DJ Thurling 1995. The potential application of Strategic Environmental Assessment (SEA) to the farming of Atlantic salmon (*Salmo salar* L.) in Scotland. *Journal of Environmental Management* **45**, 219–229.

Thorne CR, EP Evans and EC Penning-Rowsell 2007. *Future flooding and coastal erosion risks*. London: Thomas Telford.

Thornton D and DJ Kite 1990. *Changes in the extent of the Thames estuary grazing marshes*. Peterborough, Cambs: Nature Conservancy Council.

UKBG (Biodiversity Group) 1999. *Tranche 2 Action Plans. Vol. V: Maritime species and habitats*. Peterborough, Cambs: English Nature (the plans are available at www.ukbap.org.uk).

US Ocean Protection Council 2013. *State of California sea-level rise guidance document,* March 2013 update.

USACE (US Army Corps of Engineers) 2008. *Coastal engineering manual*, Coastal and Hydraulics Laboratory. http://chl.erdc.usace.army.mil/chl.aspx?p=s&a=ARTICLES; 101.

USACE 2013. *Beneficial use of dredged materials*. Building Strong. www.mvn.usace.army.mil/About/Offices/Operations/BeneficialUseofDredged Material.aspx.

Vollset KW, RI Krontveit, PA Jansen, B Finstad, BT Barlaup, OT Skilbrei, M Krkošek, P Romunstad, A Aunsmo, AJ Jensen and I Dohoo 2015. Impacts of parasites on marine survival of Atlantic salmon: a meta-analysis. *Fish and Fisheries* doi: 10.1111/faf.12141.

Walker CH, SP Hopkin, RM Sibly and DB Peakall 2012. *Principles of ecotoxicology*, 4th Edn. London: CRC Press, Taylor & Francis.

Walsh PM 1995. *Seabird monitoring handbook for Britain and Ireland: a compendium of methods for survey and monitoring of breeding seabirds*. Peterborough, Cambs: JNCC.

Ward AJ, D Thompson and AR Hiby 1988. Census techniques for grey seal populations. *Symposia of the Zoological Society of London* **58**, 181–191.

Ware S and A Kenny 2011. *Guidelines for the conduct of benthic studies at marine aggregate extraction site*, 2nd Edn. Cefas, Lowestoft (UK). Project Code: MEPF 08/P75.

Webb T, B Miller, R Tucker, P von Baumgarten and D Lord 2012. *Coastal engineering guidelines for working with the Australian coast in an ecologically sustainable way*, 2nd Edn. National Committee on Coastal and Ocean Engineering; Engineers Australia.

Wilcox C, E Van Sebille and BD Hardesty 2015. Threat of plastic pollution to seabirds is global, pervasive and increasing. *Proceedings of the National Academy of Sciences of the United States of America* **112**(38), 11899–11904.

Wolff WJ 1987. Identification. *Biological surveys of estuaries and coastal habitats*, JM Baker and WJ Wolff (eds), 404–423. Cambridge: Cambridge University Press.

Woodroffe CD 2002. *Coasts: form, process and evolution*. Cambridge: Cambridge University Press.

Worm B, EB Barbier, N Beaumont, J Emmett Duffy, C Folke, BS Halpern, JBC Jackson, HK Lotze, F Micheli, SR Palumbi, N Sala, KA Selkoe, JJ Stachowicz and R Watson 2006. Impacts of biodiversity loss on ocean ecosystem services. *Science* **314**(5800), 787–790.

YSI Environmental 2011. *Water quality is key to the success of Cardiff Bay restoration*. YSI, A592 0811.

8 Ecosystem services

• •

Jo Treweek and Florence Landsberg

8.1 Introduction

The Millennium Ecosystem Assessment (MA 2005) highlighted global decline in the extent, condition and functionality of ecosystems and their capacity to provide services, with potentially serious consequences for human well-being. Over the previous 50 years, 60 per cent of all ecosystem services were concluded to have declined significantly as a direct result of global growth in agriculture, the exploitation of forestry and fisheries, industrial development and urbanization. Subsequent assessments from the global to sub-national scales have revealed a similar picture (see IPBES 2016 for a database of assessments of biodiversity and ecosystem services).

Although natural ecosystems generate economically and socially important services for society, current economic markets and established approaches to environmental assessment largely fail to ensure that these services are adequately safeguarded for the future (Kinzig et al. 2011). For example, the UK's National Ecosystem Assessment estimated that the country's ecosystem services were worth billions of pounds (UK NEA 2011), but also found these values to be poorly recognised in development planning at national or local levels, and rarely prioritised for protection or restoration. Furthermore, the costs of ecosystem degradation are rarely accounted or mitigated for when specific development projects are planned, even when they are significant. For example, Tardieu et al. (2015) estimated losses of ecosystem services as a result of a high speed rail project in France at approximately €228,000 every year, even for the least damaging option.

Among many benefits, ecosystems control soil erosion, sustain food production and regulate air and water quality. Such benefits and the costs of restoring or replacing them are largely overlooked and unpriced in conventional economic terms, and tend to be omitted from "trade-off" decisions, despite potentially far-reaching and long-term social and economic ramifications. Ecosystems also provide a wide range of less tangible aesthetic and cultural benefits such as the "unique sense of place" associated with a particular landscape or the cultural importance of sacred trees or forests. Such benefits are challenging to measure but may nevertheless be highly valued, their loss having significant impacts on the well-being of project-affected communities. Indeed, one of the fundamental challenges of mainstreaming ecosystem services into decision-making is that of linking changes in ecosystems with changes in human well-being, particularly where less tangible values and benefits are involved (Guerry et al. 2015).

Incorporating ecosystem services at the beginning of decision-making processes can help development planners to consider the full range of benefits and costs associated with their actions. It follows that Environmental and Social Impact Assessments (ESIAs) should make impacts on ecosystem services explicit so that they can be appropriately considered when development is planned.

Although the values and benefits of ecosystem services have not been well-considered in EIA/ESIA in the past, this is changing. The International Finance Corporation (IFC) requires ecosystem services to be addressed in ESIAs for developments seeking its financial support, with references to this in several of its Performance Standards (IFC 2012). Banks that have adopted the Equator Principles generally follow these standards. The Inter-American Development Bank also takes a strategic approach through a Biodiversity and Ecosystem Service Program that supports countries in Latin America and the Caribbean to integrate the value of ecosystem services into key economic sectors and protect priority regional ecosystems. These examples reflect growing recognition within the business community that the unsustainable "draw down" of natural capital and associated decline in ecosystem services can be a "material risk" to successful development. A number of businesses and companies have recently agreed a "Natural Capital Protocol" for incorporating natural capital in their business planning and decision-making, based on standardised principles (Natural Capital Coalition 2015). Importantly, ESIA is seen as the expected means of measuring project impacts and dependencies on natural capital and ecosystem services, and there is a growing body of guidance on how to assign costs and values and use these to evaluate predicted changes (e.g. de Groot et al. 2010).

Ecosystem services are a multidimensional concept and assessing impacts and dependencies on them demands a holistic view of environmental and social changes. This can be challenging in practice, as ESIA involves many specialist inputs, which are often addressed in isolation from one another. Advocates of including ecosystem services in ESIA emphasise the need for explicit, systematic consideration of ecosystem services because they embody material risks and impacts that have not been addressed well in the past, creating legacies of ecological and social damage that can be costly to rectify. Landsberg et al. (2013) argued that, by focusing attention on the socio-economic dimensions of a project's ecological impacts, consideration of ecosystem services in ESIA can "capture the unanticipated costs and benefits of projects more fully than a standard EIA, and identify stakeholders who might otherwise be missed".

In this chapter, some important definitions and concepts are explained to provide a basis for considering impacts on ecosystem services and human well-being (for more information on addressing project dependencies on ecosystem services, see Landsberg et al. 2013). These have a bearing on preferred approaches and methods, with some divergence between those preferring to interpret impact significance in quantitative terms and those taking a more qualitative approach. Available guidance on incorporating ecosystem services in ESIA is also summarised. Most available guidance has suggested that the "ecosystem services perspective" should be used to complement, rather than replace existing approaches, being seen as a means of capturing issues that would not otherwise be given due consideration. Appropriate consideration of ecosystem services in ESIA therefore requires a more collaborative mindset, with active engagement between "specialist" dimensions. This, together with in-depth and systematic review of impacts on ecosystem services, can bring valuable insights and improve outcomes for developers as well as people affected by development.

8.2 Definitions and concepts

8.2.1 Ecosystems, natural capital and ecosystem services

No standard terminology for ecosystem services assessment exists and the lack of a common vocabulary can generate confusion (Olander et al. 2015). This section describes some key concepts and provides brief definitions of terminology relevant to the assessment of impacts on ecosystem services. In particular, it is important to differentiate clearly between ecosystem service supply, demand, use and benefit, as these have to be described or quantified in order to predict impacts in ESIA.

An *ecosystem* is a dynamic complex of plant, animal, and micro-organism communities and their non-living environment interacting as a functional unit (UN 1992). Ecosystems are sometimes referred to as natural capital, mirroring economic semantics of capital stock and flows. **Natural capital** is the stock of living *and* non-living environmental resources potentially available to generate value, while ecosystem services are the flow of benefits that this stock provides. "Essentially, natural capital is about nature's assets, whilst ecosystem services relate to the goods and services derived from those assets" by particular users or beneficiaries (British Ecological Society 2016).

Ecosystem services are the direct and indirect contributions made by ecosystems to human well-being and also to project performance. They are generally classified into four types (adapted from MA 2003):

1 *Provisioning services* – goods or products such as food, timber and freshwater.

2   *Regulating services* – contributions to human well-being arising from an ecosystem's control of natural processes, such as climate regulation, disease control, erosion prevention, water flow regulation, and protection from natural hazards.

3   *Cultural services* – the non-material contributions of ecosystems to human well-being, such as recreation, spiritual values, and aesthetic enjoyment.

4   *Supporting services* – the natural processes needed to maintain the other services, such as primary production, or nutrient and water cycles. Usually, these processes are only considered from an ecological perspective, failing to understand their importance to people's well-being. While they might be appropriately omitted from valuation exercises (to avoid double counting their benefits) it is important to understand their role in maintaining people's well-being and to understand how they will be affected by a proposed project, particularly where cumulative impacts on ecosystems are likely to occur.

Ecosystem services have also been defined as the conditions and relationships through which natural ecosystems sustain and fulfil requirements for human life (Daily 1997). This definition emphasises the fact that human well-being is based on benefits derived from people's actual use of ecosystem services.

While some ecosystem services contribute directly to human well-being, others do so indirectly by supporting other services. For example, livestock farming provides direct value to human well-being through income or by providing meat or milk for subsistence, whereas hay production contributes indirectly by supporting livestock production. The former are referred to as "final services" and the latter as "intermediate services".

In other words, ecosystem services are "final" at "point of use", in which case it should be possible to identify a specific beneficiary, whereas for "intermediate" services it can be less straightforward to identify a particular "end-user". This doesn't make them any less necessary or important, but introduces a challenge from an ESIA perspective, particularly where multiple ecosystem services have multiple beneficiaries and complex "supply chains". Boyd and Banzhaf (2007) explain intermediate and final ecosystem goods and services in the context of identifying indicators for use in environmental accounting and provide advice on how to identify indicators that are relevant and meaningful in terms of key interdependencies.

8.2.2 Ecosystem service benefits and beneficiaries

Ecosystem service benefit is the gain in human well-being derived from the use of an ecosystem service, often in combination with other inputs (e.g. labour and capital) (adapted from van Oudenhoven et al. 2012).

Beneficiaries are the people who derive benefit from an ecosystem service. In contrast to natural capital, ecosystem services cannot exist without a user or a beneficiary.

As emphasised by the US National Ecosystem Services Partnership (2016):

> Ecological features and processes are essential for the provision of ecosystem services but are not the same as services. Until there is some person somewhere who benefits from a given element or process of an ecosystem, that element or process is not a service.

Beneficiaries may be individuals, whole communities or particular groups within a community, for example children unable to attend school, women, or members of an indigenous community. One of the key tasks of an ecosystem services assessment is to establish which services are being used by whom in a pre-development context, and thereby identify the constituency of people who may experience significant changes in the benefits they derive from ecosystems when a project is developed, such that their livelihood or well-being might suffer. ESIAs may need to identify beneficiaries at very different spatial scales, for example local communities might benefit from services provided by a National Park, but so may international visitors, or even people who never visit the park, but derive benefit from knowledge that it exists to conserve endangered species.

A change in the capacity of an ecosystem to supply services can have very different implications for beneficiaries, depending on their level of dependence on a service and their ability to access or use alternatives. Therefore impact assessment needs to be quite precise and consider how different beneficiaries might be affected by the same changes in an ecosystem. This may require information on levels or frequency of use by different beneficiaries. One of the main challenges in ESIA is to develop defensible methods for detecting key dependencies.

For example, an ecosystem services review for oil development in Uganda (TEC 2015) revealed high levels of dependence of hired cattle herders on livestock production from open-access grazing, an ecosystem service-dependence that did not emerge as being particularly significant in social baseline surveys carried out for the ESIA. The proposed development was expected to catalyse increased levels of land enclosure and settled agriculture, with potential benefits in terms of reducing pressure on forests from existing livestock grazing regimes and boosting local food production. However, an integrated approach through ecosystem services review revealed a specific and significant risk from land enclosure to hired herders. This group are particularly vulnerable because they don't own land, depend absolutely on employment for income and food, and have highly restricted alternatives. Mitigation recommendations therefore included efforts to improve land-use zoning so that open-access grazing could persist at a sustainable level. Herders already living in poverty were also identified as a vulnerable group, potentially requiring targeted interventions in future to accommodate to changing agricultural practices.

Projects are often beneficiaries of ecosystem services themselves and can fail if their own dependence on ecosystem services is not fully recognised. Many developments need a sustainable supply of ecosystem services for their planned operational performance, but it is not unusual for these dependencies to be poorly understood. For example, a development might depend on cooling water derived from a lake ecosystem, but the proponent may have limited understanding of how reliable the water supply will be in future. There may be plans for the development of hydro-power on the river systems supplying the lake, which need to be taken into account when estimating water availability; or poor land management around the lake might cause sedimentation at rates that will compromise water levels in future or block abstraction points. When considering new projects, accurate, upfront assessment of both impacts on, and dependencies on, ecosystem services is therefore essential.

8.2.3 Valuation of ecosystem services

Although many people benefit and profit from ecosystem services, they do not necessarily pay for them in any conventional sense and therefore many ecosystem services have no market price. Nevertheless, ecosystems and the services they supply have value because people derive utility from their actual or potential use and also value them for reasons not connected with use (i.e. non-use values). Two ways in which ecosystem services contribute to economic welfare are by contributing to the generation of income or well-being and by preventing damages and associated costs.

Economic valuation techniques attempt to elicit preferences for changes in ecosystems (impacts) in monetary terms. It is not necessary to use economic valuation techniques to describe or quantify benefits from ecosystem services but there are some situations in which they may be useful or appropriate. As a general rule, they come into play when changes in well-being have been identified and their significance is being assessed. However they can also be used to compare the costs and benefits of development alternatives, or prioritise which services to consider at the scoping stage. Many efforts have been made to develop robust valuation techniques, and research in this area merits a whole book in itself. There are many well-established "preference evaluation methods" (including market and non-market economic valuation), as well as non-monetary methods that can be used to estimate values for ecosystem services. The challenges are how to apply them and how to obtain the data needed for them to be reliable and effective (Bateman et al. 2011).

One problem with using economic valuation techniques routinely in ESIA is the shortage of experienced practitioners and the time and resources that may be needed to obtain reliable data and results, particularly when new data need to be collected. Other, more qualitative indicators of ecosystem service benefit may, therefore, often be needed.

Payments for Ecosystem Services (PES) are a mechanism for people who manage ecosystems (whether they are government conservation agencies, private or community landholders, or non-governmental organizations (NGOs)) to benefit from generating ecosystem services that are used or enjoyed by others. PES schemes are market-like mechanisms that allow ecosystem services to be valued and traded. PES can also be a proxy for costs of damage, making them relevant in terms of the "polluter pays" principle, whereby those who benefit from ecosystem services must pay for those services and whoever damages the environment must compensate for the damage (OECD 2010). Carbon markets are an example of this, allowing users to buy and sell carbon emission rights in a "cap and trade" system. PES schemes exist for carbon sequestration in China and the United Kingdom; watershed protection in South Africa and Mexico; and biodiversity conservation in the United States, Costa Rica, and Nicaragua (Kinzig et al. 2011). In the context of ESIA, PES probably offer most potential as a mechanism for investing in and delivering mitigation or offsets.

8.2.4 Ecosystem service demand and use

Ecosystem service demand is the level or amount of ecosystem service desired by beneficiaries to achieve a target level of human well-being. If demand is realised or met, it equates to "use".

Ecosystem service use is the level of a particular ecosystem service actually consumed or enjoyed by beneficiaries (adapted from Boyd and Banzhaf 2007). It can be consumptive (e.g. agriculture crops for food, water for drinking) or non-consumptive (e.g. recreational and spiritual appreciation of a landscape or wildlife, pollination of crops by bees).

The concept of use is particularly important in ESIA as a means of linking ecosystems (supply side) with benefit (user side). It is also important to assess the sustainability of use in the baseline (i.e. in a future without the project) and with the project.

The area people use to obtain a particular service is their "use-shed" for that service, for example, the area around a village used to collect firewood or the extent of a river catchment people benefit from to obtain drinking water. Not only can a project affect the relation between supply and use for an ecosystem service within a specific use-shed, development projects can also alter use-sheds by introducing barriers or conversely by introducing new access routes or even new technologies. Development of paved roads through a forest, in conjunction with increased economic opportunity and ability to purchase vehicles, for example, could expand a use-shed from a few kilometres around a village, to hundreds of kilometres. Use-sheds for non-use values can be very large and, for services such as climate amelioration, might be global.

8.2.5 Ecosystem service supply and access

Ecosystem service supply is the maximum amount of "service" that can be drawn from an ecosystem without damaging its future productive capacity, for example the number or weight of fish that can be taken from a fishery without causing its yield to decline, or the amount of timber that can be harvested from a forest without affecting its capacity to produce timber at the same level in future (this might be measured in terms of maximum sustainable yield). This definition of supply is adapted from UNEP-WCMC (2011) and Kareiva et al. (2011).

If use exceeds supply, there is a risk of significant impact on the benefits that people are able to derive from that service over the long term. However, whether benefits will be affected within a project timeframe depends on how close a service is to its sustainability threshold.

Supply depends primarily on ecosystem type, and condition or health. The ecosystem type determines which services are provided; ecosystem condition determines the capacity of the ecosystem to supply them and can be affected by management (negatively or positively). The geographic extent of an ecosystem, its biodiversity, structure and processes can also affect the quantity and quality of services the ecosystem supplies and need to be considered when assessing the condition of an ecosystem or its capacity to supply a service.

Whether a beneficiary will be able to use an ecosystem service and "enjoy" the resulting benefits is generally determined by whether or not there is an "*accessible*" supply. This is particularly true of provisioning services and not necessarily all other types. A proposed development might have no effect on supply of firewood from a woodland, for example, but might prevent its use by introducing an impermeable barrier (such as a very busy road) that makes it impossible for people to harvest firewood any more. This means supply and access need to be considered when identifying impacts on ecosystem service supply.

Related to accessibility, the "supply-shed" is the area people need access to for an ecosystem service to be sustainable in terms of the level of benefit they need or depend on. For example, the area of woodland people would need to obtain a sufficient, sustainable amount of firewood.

Figure 8.1 illustrates how ecosystem services interface between the biophysical environment (ecosystem) and human well-being (social), high-lighting the requirement for assessment through different "specialist" dimensions in ESIA.

8.3 Key policy, legislation, guidance and standards

Government positions on ecosystem services vary, but policies encouraging planning within environmental limits, better consideration of the benefits and values associated with ecosystem services and efforts to avoid costs of ecosystem damage are becoming more common. For example, in the USA the White House released a memorandum requiring federal agencies to incorporate ecosystem services into federal planning and decision-making in 2015 (Dickinson et al. 2015) and China has even announced plans to track natural capital and ecosystem services through a new metric, "gross eco-system product", to be reported alongside GDP (Zhiyun 2013).

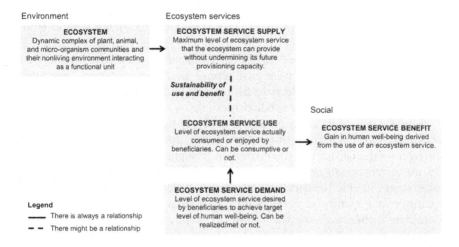

Environment

ECOSYSTEM
Dynamic complex of plant, animal,
and micro-organism communities and
their nonliving environment interacting
as a functional unit

Ecosystem services

ECOSYSTEM SERVICE SUPPLY
Maximum level of ecosystem service
that the ecosystem can provide
without undermining its future
provisioning capacity.

*Sustainability of
use and benefit*

Social

ECOSYSTEM SERVICE USE
Level of ecosystem service actually
consumed or enjoyed by
beneficiaries. Can be consumptive or
not.

ECOSYSTEM SERVICE BENEFIT
Gain in human well-being derived
from the use of an ecosystem service.

ECOSYSTEM SERVICE DEMAND
Level of ecosystem service desired
by beneficiaries to achieve target
level of human well-being. Can be
realized/met or not.

Legend
——— There is always a relationship
– – There might be a relationship

Figure 8.1

Key interrelationships between ecosystem, and ecosystem service supply, use, demand, and benefit

Some government legislation requires consideration of ecosystem services in EIA/ESIA and it is also an expectation of some international agreements and standards.

8.3.1 International agreements and policy objectives

At the Nagoya UN Biodiversity Summit in October 2010, 192 countries (and the European Union) agreed to a new "Strategic Plan" of action on biodiversity and ecosystems, with the following vision: "By 2050, biodiversity is valued, conserved, restored and wisely used, maintaining ecosystem services, sustaining a healthy planet and delivering benefits essential for all people." Parties to the Convention on Biological Diversity (CBD 2010) also agreed a shorter-term ambition to:

> take effective and urgent action to halt the loss of biodiversity in order to ensure that by 2020 ecosystems are resilient and continue to provide essential services, thereby securing the planet's variety of life, and contributing to human well-being, and poverty eradication.

The need to safeguard essential ecosystem services was therefore central to the Strategic Plan. Parties agreed on a set of strategic goals and targets referred to as the "Aichi" targets to drive action on biodiversity and implement the plan.

Responding to these targets, in 2011 the EU published a vision and 2020 mission for biodiversity (EC 2011):

- By 2050, European Union biodiversity and the ecosystem services it provides – its natural capital – are protected, valued and appropriately restored for biodiversity's intrinsic value and for their essential contribution to human well-being and economic prosperity, and so that catastrophic changes caused by the loss of biodiversity are avoided.
- Halt the loss of biodiversity and the degradation of ecosystem services in the EU by 2020, and restore them insofar as is feasible, while stepping up the EU contribution to averting global biodiversity loss.

At a global and EU-level, this established clear policy and objectives for both protecting and restoring ecosystem services, recognizing their essential contribution to well-being and economic prosperity. Despite the explicit inclusion of ecosystem services in the European Commission's proposal to amend EIA Directive 2011/92/EU (EC 2012), however, they were not incorporated in the final text of the adopted revised EIA Directive (2014/52/EU) (EC 2014).

8.3.2 National ecosystem services policy, legislation and outcomes: examples from the UK and Vietnam

The UK was one of the first countries to carry out a national-level Ecosystem Assessment, in line with the Millennium Ecosystem Assessment's approach (UK NEA 2011). This demonstrated clearly the value of the UK's ecosystems in providing essential services such as flood protection, concluding that coastal wetlands could be providing benefits worth £1.5 billion annually by buffering the effects of storms and managing flooding. It also highlighted that a third of the ecosystem services provided from the UK's ecosystems are in decline and that this trend can be expected to worsen as a result of population growth and climate change. This assessment prompted the explicit goal of England's Biodiversity 2020 strategy to "support healthy well-functioning ecosystems and establish coherent ecological networks, with more and better places for nature for the benefit of wildlife and people" (Defra 2011).

In terms of outcomes, England's Biodiversity 2020 Strategy established targets to conserve 17 per cent of areas of particular importance for biodiversity and ecosystem services through management and the establishment of "nature improvement areas" and to restore 15 per cent of degraded ecosystems as a contribution to climate change mitigation and adaptation.

While the UK's National Planning Policy Framework states that the planning system should recognise the wider benefits of ecosystem services, national policies are not reinforced by strong legislation to enforce effective avoidance or mitigation of impacts in practice. The UK's position is therefore limited to recognizing the values that ecosystems may provide, giving an indication of policy, and providing some advice and guidance on plan-making or decision-making. This position is weak in terms of requiring specific outcomes for ecosystem services that take into account the needs or dependencies of particular beneficiaries.

A similar situation prevails in many other countries. Hanh et al. (2010) provide a well-documented example in their Review of Laws and Policies Related to Payment for Ecosystem Services in Vietnam, published by the IUCN (International Union for Conservation of Nature). This concluded that many policies and laws imply the need to protect ecosystem services, but none explicitly require it. The national Law on Water Resources 1998, Land Law 2003, Law on Forest Protection and Development 2004, and Law on Environmental Protection 2005, all recognise certain elements of the ecosystem services provided by ecosystems (biodiversity protection; landscape beauty; watershed protection; and carbon sequestration), but none of these laws actually regulate the valuation or use of ecosystem services. While the Law on Forest Protection and Development (2004) states that climate regulation is one of the purposes of the protection of forests (Article 4), the Law on Environmental Protection (2005) does not specifically enable or require carbon sequestration or establish any outcome indicators.

8.3.3 International finance institutions

Some of the international finance institutions place a stronger emphasis on the consideration of ecosystem services in their standards and requirements. IFC Performance Standards 1, 4, 5, 6, 7 and 8 (IFC 2012) refer to ecosystem services and PS 6 in particular requires clients wishing to procure IFC finance to develop a project to demonstrate how the benefits from ecosystem services will be maintained and to indicate what mitigation measures will be used to "maintain the value and functionality of priority services". The goal is to ensure that the benefits people derive from these priority services are maintained when projects are developed, operated and then closed. Some of the benefits people derive from ecosystem services relate directly to their livelihoods in the sense of IFC PS5, and may be affected primarily in the event of land acquisition for development, and dealt with through standardised land-acquisition processes. Others may be non-financial: these are typically harder to account for in ESIA and are often omitted as a result.

Similarly, for services used and depended on by a project, IFC PS6 suggests that developers should assess their own dependence on ecosystem services with a view to ensuring that they will have access to a sustainable supply throughout the project's planned operational life. The IFC refers to assessment of impacts and dependence on ecosystem services as "Ecosystem Services Review" and requires it in cases where "priority" ecosystem services are likely to be affected. Identifying priority services is largely based on the judgement of the practi-tioner, who is expected to consider levels of dependence on services and the extent to which beneficiaries have viable alternatives. The European Bank for Reconstruction and Development has a similar requirement in Performance Requirement (PR) 6 relating to the assessment of "use of, and dependence on, ecosystems by potentially affected communities" (EBRD 2008).

8.3.4 Guidance on integrating ecosystem services in ESIA

As for many other specialist aspects of ESIA, there is no "one size fits all" process for integrating ecosystem services. The information needed to define, measure or model ecosystem services may be gained at different rates and at different stages in the ESIA process.

Nevertheless, guidance on assessing impacts on ecosystem services is available from various sources: Waage (2014) provides an overview of ecosystem services methods and tools. Overall guiding principles and steps are provided by Baker and Scott (2013) and by the IPIECA and OGP (2011) for the oil and gas sector. More detailed instructional guidance is provided by Everard and Waters (2013) and Landsberg et al. (2013). In addition to project-level guidance, policy-level or strategic assessment guidance is also available (e.g. Geneletti 2011; Partidário and Gomes 2013; GIZ 2012).

The following steps for assessing project impacts on people who use and depend on ecosystem services are in line with most current guidance:

1   Given predicted changes in ecosystems as a result of the planned development, identify the ecosystem services for which supply might change. Also identify ecosystem services for which the ability of users to access them might be affected by the planned development.
2   Identify the users and beneficiaries of these services.
3   If intending to comply with IFC Performance Standards, select "priority ecosystem services" (those on which beneficiaries have high levels of dependence, with limited or no available alternatives).
4   Establish the baseline for these priority ecosystem services, assuming current levels of use: will there be a sustainable relationship between supply and use in the absence of the planned development?
5   Predict project impacts on priority ecosystem services (their likely future supply, use and benefits as appropriate), using current and projected levels/relationships as the baseline.
6   Mitigate project impacts on priority ecosystem services to ensure that benefits are maintained.

Ultimately, the goal is to identify the specific benefits that people derive from ecosystem services, understand their level of dependence on them and determine their ability to maintain the levels of benefit that they need through alternative means, if they lose access to the services that underpin them or if supply goes down. This makes it very important for ESIA, and any social assessments that are carried out as part of it, to include beneficiaries who depend strongly on priority services. There may be people who are highly vulnerable to changes in ecosystem service supply who might not be identified through typical social assessments, which sometimes focus more strongly on economic aspects of livelihood.

In the US, guidance on ecosystem services and their inclusion in impact assessment has recently been issued for federal agencies, as they are responsible for a variety of actions and projects that may influence ecosystem conditions and change the provision of ecosystem services that are valued by the public (Olander et al. 2015). Previously decisions affecting ecosystems had generally relied on ecological assessments with little or no consideration of the benefits and value of ecosystem services to their users. The guidance emphasises the need to consider both an ecosystem's capacity to provide services and the extent to which those services are needed for social benefit. It suggests three guiding principles for ecosystem services assessment or review:

1 Extend assessments beyond purely ecological measures to measures of ecosystem services that are directly relevant to people.
2 Assess ecosystem services using a structured and systematic approach, based on well-defined measures that go beyond narrative description.
3 Include all services that may be important, even those that are difficult to quantify.

The guidance also suggests the use of "Benefit Relevant Indicators" to structure assessments and link ecosystem capacity with use by different beneficiaries. They may be monetary values, but can also reflect less tangible non-use values that can be difficult to quantify and are often excluded from ESIA.

In the UK, the Department for Environment, Food and Rural Affairs published useful introductory guidance on valuing ecosystem services (Defra 2007), which explains the strengths and limitations of different economic valuation techniques and gives examples of their application in policy and decision-making. Some advice on applying these techniques to environmental assessment is also given, EIA being seen as a means of providing evidence or inputs to economic valuation exercises. This guidance also emphasises the need to use a range of techniques for quantifying and valuing benefits to ensure that the full range of relevant costs and benefits are considered when decisions are made.

8.4 Scoping and baseline studies

The scoping stage is particularly important for ESIAs setting out to incorporate ecosystem services, because of the need to integrate across specialist disciplines and to allow time and resources for an appropriate level of stakeholder engagement. It is given particular emphasis in this chapter because of this need for a "front-loaded" process to address impacts on ecosystem services thoroughly.

The main purpose of scoping is to establish the ecosystem services context for a planned project based on:

- likely impact sources;
- ecosystems likely to be exposed to change;
- beneficiaries who could be affected.

Based on initial findings, the scoping phase should then confirm:

- proposed spatial scope or scale of analysis;
- proposed temporal scope;
- ecosystem services and their beneficiaries to be addressed in further steps of the ESIA.

8.4.1 "Entry points" for scoping an ecosystem services review (ESR)

Because ecosystem services are a multidimensional concept, there are different possible "entry points" for assessing impacts on them. The practitioner may choose to start by mapping the proposed project activities and associated infrastructure against ecosystems to identify those for which significant changes might be expected as a result of the construction, operation or closure of the project. Ecosystems might be affected because they are within a project's infrastructure footprint, or because the project will induce biophysical or social changes that might alter ecosystems and ecosystem service supply, or the ability of people to access particular services. Potential users of these ecosystems can then be identified and the implications of a project considered in more detail when it is known which ecosystems are likely to change. It is important to be aware of potential indirect effects and the need to consider all project-related infrastructure, as well as related developments, otherwise cumulative impacts on ecosystem services will not be captured very well.

On the other hand, the practitioner may choose to start by identifying people who use and depend on ecosystem services within a project's area of influence, identify the services they use and then consider how their ability to access them might be affected by a project's impacts. This approach can be less straightforward because some beneficiaries may depend on ecosystems within a project's area of influence but be located well outside it. So long as all relevant aspects are considered, it doesn't necessarily matter which entry point is used and it is generally desirable for ecologists and social practitioners to share and integrate their initial findings at an early stage so that they can influence each other's baseline data collection.

8.4.2 Establishing the spatial scope

Defining the spatial scope of ecosystem services review can be challenging, as services "produced" in the study area may be used by people considerable distances away (e.g. upland land use affecting exposure to flood risk downstream) and conversely people within the study area may benefit from services (such as air-quality regulation or the recruitment of fish stocks) which are largely or wholly "produced" by distant ecosystems. To understand an ecosystem's ability to supply services to a particular human population or beneficiary it is generally necessary to consider its entire extent or distribution, so that its functioning and overall viability is known. This is in line with accepted good practice for ecological impact assessment, but may not coincide well with the spatial scale for other environmental impacts and typically requires a broader perspective than might be needed for other aspects of ESIA.

Establishing a suitable spatial scope is likely to require information on:

1   The locations of planned development, infrastructure and activities and their associated zones of impact or "effect distances".
2   The distributions and extents of ecosystems that might be affected, including those for which direct land-use changes are expected, as well any ecosystems in functional connectivity (e.g. if part of a watershed or ecological unit would be affected, implications for the whole should be considered). This might include ecosystems that are within the proposed physical footprint of planned development and infrastructure as well as those potentially affected by land-use change as a result of demographic shifts or improved infrastructure.
3   Distributions or locations of beneficiaries of ecosystem services affected by planned developments and infrastructure.

8.4.3 Sources of information and tools for scoping and baseline studies

Scoping (and subsequent impact assessment) is likely to be an iterative process. A pragmatic starting point might be a map of broad ecosystem types and settlements/infrastructure. There are also some generic checklists of ecosystem services that can be used to identify the types of ecosystem service that different ecosystems can be expected to supply (see Table 8.1). These are generally based on the Millennium Ecosystem Assessment categorisation but can be adapted for any particular development country or context.

Key sources of information are likely to include:

•   Satellite imagery, aerial photos or other sources showing distributions and locations of ecosystems and communities.
•   Ecological assessments and reports focusing on ecosystem type, distribution and condition, threats and pressures affecting ecosystems and how they has changed over time. Information is needed on sustainable production, harvest or yield, if possible.
•   Reports and information on planned development and the design and locations of planned infrastructure and activities.
•   Social reports and results of initial stakeholder engagement or consultation, focusing on demographic, cultural, economic characteristics and uses of/dependence on natural resources and ecosystems.

It is not possible to conduct a good ecosystem services review or assessment without active communication and collaboration between ESIA specialists. Input from ecological and social practitioners is necessary as well as that from the development proponent. Engagement with other specialists may also be needed, for example to understand changes in water supply or quality, or to obtain information about possible changes in infrastructure or transport that may affect access to ecosystem services. A participatory approach is essential to ensure that interested and affected parties are identified and that the ESR is framed to address key issues of concern to stakeholders. The earlier this takes place in the ESIA process the better, or opportunities to obtain essential information in social and ecological surveys will be lost and considerable duplication of effort likely.

As well as identifying possible links between ecosystems and users, it is also necessary to identify the particular benefits that users derive. It may be possible to gain this knowledge from existing reports or information, or specific baseline investigations may be needed. Figure 8.2 provides an example of the many intra- and interrelationships that may exist between ecosystems, services and

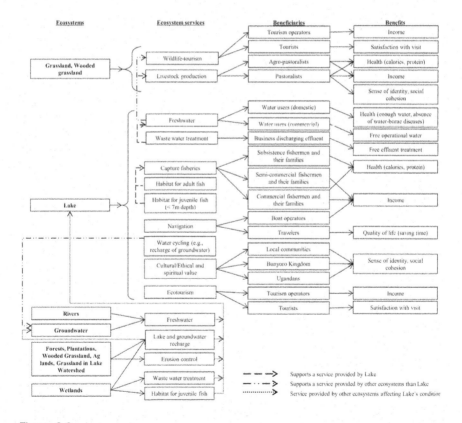

Figure 8.2

Linking ecosystems with benefits; oil development project in Uganda (TEC 2015)

Table 8.1 Scoping checklist for ecosystem services potentially affected by a land-based project (adapted from TEC 2015)

Ecosystem services	Natural forest	Plantation forest	Riverine forest	Woodland	Grassland	Wetland/ swamp	Open water (Lake rivers)
Provisioning							
Crops			○	○		○	○
Livestock products (meat and milk)				○	×	×	×
Capture fisheries			○		○	×	×
Wild food (mushrooms, nuts, fruit)	×		○	○	○	×	○
Biological raw materials		○		×	×	×	
Timber	×	×	○	×	○	×	○
Freshwater	×		×	×	○	×	×
Biochemicals, natural medicines	×		○	×	×	○	
Regulating							
Air quality regulation	×	○	○	○	○	○	○
Global climate regulation	×	×	×	×	○	○	○
Water regulation	×	○	○	○	○	×	×
Erosion regulation	×	×	×	×	×	×	×
Water purification	×	×	×	×	×	×	×
Disease regulation	○	○	○	○	○	○	○
Soil quality regulation	×	○	×	×	○	○	
Pest/invasive species regulation	×		○	○			○
Pollination	×		×	×	×	×	
Natural hazard regulation	×	○	×	×	○	×	×
Cultural							
Recreation and eco-tourism	×	○	○	○	×	×	×
Spiritual values	×	×	×	×	○	×	×
Ethical/non-use values, e.g. existence value of wildlife	×	×	×	×	×	×	×

Note: Importance of ecosystem service based on conclusions from literature: *High* × *Medium/low* ○ *Not relevant/Negligible [blank]* benefits. Each ecosystem might provide many benefits to many different users. Equally, one group of people may use multiple services to derive all the benefits they need. This means there can be complex interrelationships to assess and manage. Constructing flow charts or network diagrams can be an effective, systematic way to tease these out. During the impact assessment, a causal chain – also known as a path model or means-end diagram – could build on this diagram and track how a project action or impact can be expected to propagate through the ecosystem to effect changes in the provision of ecosystem services and benefits to particular beneficiaries. Causal chains commonly used in ecological assessments often end with expected environmental changes and omit impacts on benefits to society. In contrast, a causal chain in an ecosystem services assessment leads to effects on human well-being or benefit (Olander et al. 2015).

During scoping and baseline studies, it is also necessary to understand patterns of supply and use without the planned development so that it is possible to predict how these can be expected to change as a result of development in the impact assessment. Sometimes sufficient baseline understanding is available at the scoping stage but often there are important gaps. Risks from unplanned events or accidents should also be considered.

8.4.4 Ecosystem services baseline assessment

Developing a reliable baseline for ecosystem services can be particularly challenging because of the need to obtain information on:

- current levels of ecosystem service supply, use and benefit;
- interrelationships between supply, use and benefit;
- future supply and use in the absence of the planned project, taking account of existing threats and pressures and their relationship with people's livelihoods, health, safety, and culture (benefits).

Availability of relevant information will vary between projects. For ecosystem services, transparency about assumptions and data limitations is particularly important, because these affect conclusions about significance of any loss of benefit from ecosystem services and have implications for livelihoods and well-being. If existing baseline information is limited it will be necessary to design studies to obtain information needed to assess impacts on ecosystem services. This information is typically obtained from a variety of social and ecological surveys and assessments, including, among others:

- land cover and land-use mapping;
- vegetation survey and classification;
- surveys of biodiversity and ecosystems;
- ethno-botanical surveys;
- agricultural surveys;

- Social Impact Assessments and results of stakeholder engagement;
- Health Impact Assessments;
- livelihood surveys;
- water studies.

When designing baseline studies, consideration should be given to how environmental impact significance will be interpreted in terms of human well-being. Some guidance suggests the use of indicators, in which case these should be identified as early as possible so that they can be used to structure information collection. This also means the results can readily be linked with significance evaluations at the end of the impact assessment, as well as potentially providing a framework for monitoring. For example, suitable indicators to establish the significance of a wetland's contribution to the health of the people depending on it for accessing clean water might include:

- volume of water the wetland can treat;
- quantity of water people abstract for drinking and/or cooking;
- the incidence of water-borne diseases.

On the other hand, it may not be possible to select good indicators until a certain amount of baseline-information collection has already taken place. If indicators are used, they should reflect the key aspects of ecosystem supply, use and benefit (Landsberg et al. 2013, Olander et al. 2015, Geneletti 2011). It is important to focus on ecological changes that are relevant to the provision of specific services and also to ensure that any indicators will be relevant and meaningful for potentially affected people. An indicator becomes benefit-relevant when it is expressed in a way that resonates with beneficiaries because it affects their welfare directly. For example, "numbers of fish consumed/sold that can be caught without undermining future catch" is likely to be more relevant to people who depend on fishing for their livelihood than other measures, such as total fish stock and fish stock diversity – even though these parameters might reflect the state of fish populations very well (Olander et al. 2015).

It will also become clear in the section on mitigation that good practice requires a strong emphasis on the avoidance of impacts on ecosystems, so that ecosystem service supply is not jeopardised by a project. To determine what scope exists for avoidance, it may be necessary to model alternative scenarios. These scenarios can be developed using tools such as InVEST – "Integrated Valuation of Environmental Services and Tradeoffs" (Natural Capital Project 2016). These largely focus on mapping ecosystems and estimating ecosystem service supply, but have also been making significant progress in mapping demand. The applicability of such tools in ESIA depends critically on access to scale-relevant spatial data, but if they are used from the outset they can support a robust, spatially explicit approach.

Modelling alternatives for specific beneficiaries can help in identifying options that minimise impacts on vulnerable groups, who depend heavily on particular services and have limited alternatives. Who these are may have been identified during the scoping phase, or it may be necessary to go back and engage further with stakeholders when there is better understanding of how ecosystem service supply or use can be expected to change. In order to avoid impacts on ecosystem services, early consideration of alternatives is necessary, meaning a "front-loaded" process is likely to be needed as well as one which allows for timely stakeholder engagement.

8.4.5 Priority ecosystem services

It can be challenging to address project impacts on all ecosystem services and beneficiaries within typical ESIA time frames, therefore it is often necessary to focus ESIA on "priority" ecosystem services: those for which changes due to a project are most likely to affect the well-being of beneficiaries, because they depend on them to a great extent and have limited or no alternatives. This is the approach suggested by IFC Performance Standard 6, for example (IFC 2012) and some guidance suggests an explicit prioritization step, for example (Geneletti 2011; Landsberg et al. 2013). However, it is important to recognise the fact that some people may depend on a variety of ecosystem services to meet their needs, and this may make it inappropriate to screen some services out too early. It is also important to recognise that decline in one ecosystem service may increase levels of dependence on others. This emphasises the need to consider overall resilience to change among beneficiaries, given the full range of services they depend on. Prioritization may therefore have to be conducted iteratively as knowledge of dependence on ecosystem services improves throughout the ESIA process.

The decision tree illustrated in Figure 8.3 reflects the criteria that might be used to identify priority ecosystem services:

1 The supply or use of the ecosystem service is affected by the project, whether it is supplied from ecosystems which are located in areas exposed to land-use change as a direct or indirect result of the project, or because the presence of the project will affect the ability of users to access it.
2 Project impacts on the ecosystem service might lead to a change in the benefits it provides to people.
3 The benefits derived from the service are important to the overall well-being of its beneficiaries.
4 The beneficiaries have no or limited viable alternatives to the service to maintain their well-being.

Prioritizing ecosystem services generally requires stakeholder engagement to:

1 Identify ecosystem services that contribute directly to livelihood or well-being.
2 Establish how ecosystems and ecosystem services are used, and determine whether this varies between different stakeholder groups.

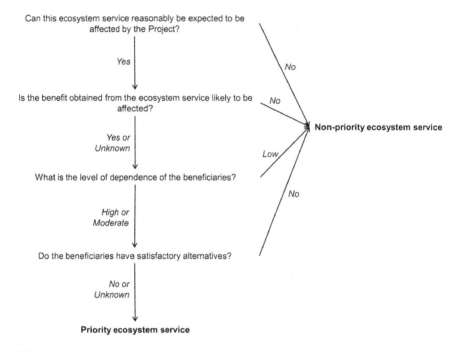

Figure 8.3

Decision tree for identifying priority ecosystem services affected by the project

3   Identify the use-shed, i.e. where people access the ecosystem services they depend on.
4   Establishing the extent of use, dependence and benefits derived.
5   Discussing the acceptability of alternatives.

The stakeholder engagement needed to prioritise services may be undertaken as part of social studies already planned for the ESIA or established purely for purposes of ecosystem services assessment. The appropriate timing will be entirely context-specific but should start early, either in the scoping stage or to elaborate the baseline. Stakeholder engagement to identify typical uses or users of ecosystems across the landscape might be possible through discussion with representatives of organisations and departments involved in natural resource allocation. However planning and management, engagement to establish levels of benefit derived from a particular service, or the alternatives available to people, will require the engagement of affected communities.

Focus-group discussions (FGD) can be a good way to gather together people from similar backgrounds or specific "user communities" to discuss levels of use and dependence on ecosystem services, and their likely trends in the absence of the project. Women in fishing communities might have different perceptions about benefits associated with, and dependence of their families on, fishing than men, and men who fish might have a better understanding of changes in the supply of fish over time than do people who purchase fish for consumption. FGDs are a practical way to conduct dialogue with different beneficiary groups and to ensure that such differences in vulnerabilities and dependencies are recognised and taken into consideration.

For each of the priority ecosystem services, the current socio-economic benefits derived by affected stakeholders need to be established and linked to current or baseline levels of use. During impact prediction and evaluation, changes in ecosystem service benefits resulting from project-related changes in ecosystem service supply and/or use can then be estimated and compared against this baseline.

8.4.6 Scoping and baseline assessment outcomes

At the end of the scoping and baseline studies, the ESIA should present:

- A list of ecosystem services to be considered, and stakeholders to be engaged in further stages of the ESIA process (including results of any explicit prioritisation) to confirm which ecosystem services are most important in terms of users' dependence and availability of alternatives. In other words, services for which decline in access or use would mean significant or unacceptable loss of benefit.
- Shared understanding of ecosystems, their types and condition, and potential changes in supply in the absence of the project.
- Shared understanding of level of use and type, and level of benefits derived from ecosystem services, and potential changes in use and benefits in the absence of the project.

It is important to revisit scoping and the baseline as more information is known about the project, its potential impacts, the affected people and their dependence on ecosystems for their well-being.

8.5 Impact prediction and evaluation

Impact prediction needs to consider the effects of the planned development on the supply and use of ecosystem services and the implications of any changes for people's livelihoods, health, safety, and culture. This establishes the situation in the presence of planned development as compared with the baseline and generally builds on information from other specialist topic areas in the impact assessment.

A project may affect beneficiaries' well-being in two ways (see Figure 8.4):

1   *Impact on supply* (dash arrows): It might contribute to existing direct drivers of ecosystem-change or introduce new ones, for example, by polluting waterways, overharvesting water, or draining a wetland, thereby reducing the capacity of an ecosystem to supply services. Information on the state of ecosystems and their capacity to supply is likely to build on ecological studies. A project may also accelerate or decelerate ecosystem-change by affecting indirect drivers of change, for example by increasing the local population and the rate at which services are used, potentially in excess of their sustainable supply.
2   *Impact on use* (dash-dot-dot arrows): It might affect beneficiaries' well-being and therefore their demand for ecosystem services. Information on beneficiaries is likely to build on results of social studies. For example, a project may provide an alternative source of income that makes people less dependent on ecosystem services for meeting that aspect of their well-being and therefore diminishes their use of that service. It might also affect indirect drivers of ecosystem-change such as governance and cultural norms, which can affect the legal and social enforcement of ecosystem protection and use.

Predicting how human well-being will change as a result of project impacts on ecosystem services requires information on current levels of use and benefit in relation to supply, and on future trends with a project in place. It is important to recognise that levels of supply and use might alter without affecting level of benefit. It is therefore necessary to consider use relative to "accessible" supply. If supply is comfortably exceeding use of a service before a development is implemented, it may be possible for users or beneficiaries to tolerate a reduction in supply without experiencing significant loss of well-being. On the other hand, if supply is already trending towards unsustainability, an increased local workforce might increase levels of use, causing an ecosystem service to crash. If use is close to being unsustainable (i.e. very close to level of supply), then even a very slight decrease in supply might cause the relationship between use and supply to become unsustainable, and in an extreme case might lead to the collapse of the ecosystem service. Impact significance depends on the extent to which predicted changes in well-being can be accommodated by beneficiaries. If they are living in poverty and are already suffering from malnutrition, any loss of food supply will be significant. On the other hand, if a project provides income or builds local capacity to improve food production, small changes in access to wild food may not be significant.

Impacts may therefore occur because a development affects the capacity of ecosystems to supply services, because the ability of people to access services is altered, or because the relationship between supply and use changes. ESIA must distinguish clearly between these scenarios so that appropriate mitigation interventions can be identified.

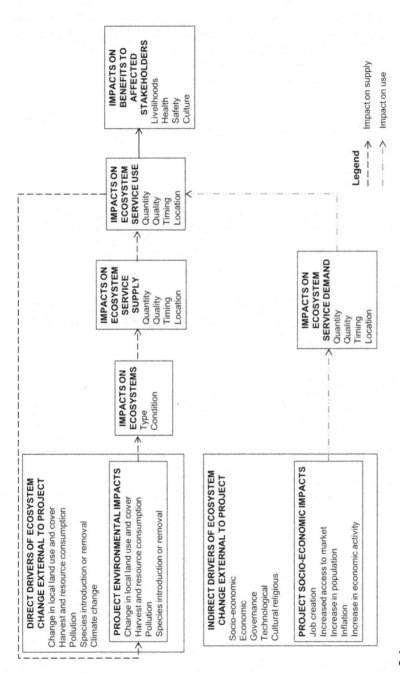

Figure 8.4

Impact pathway: impacts on ecosystem services supply, use and benefits (based on Landsberg et al. 2014)

The significance of project impacts will be a function of beneficiaries' sensitivity to predicted impacts on ecosystem service benefits and the magnitude of loss in these benefits (Landsberg et al. 2013). These in turn depend on beneficiaries' human, social, physical, financial and natural capital, and the predicted duration, frequency and reversibility of impacts on the benefits they need and, therefore, on their well-being. For example, referring back to the example of Ugandan hired cattle herders, those for whom herding is their sole source of income are likely to endure a significant impact from enclosure of open-access grazing whereas herders who have organised into collectives to purchase communal land will be less affected.

8.6 Mitigation and enhancement

The overall goal of mitigation is to achieve an outcome where there is "no unacceptable loss of ecosystem service benefits" (adapted from Landsberg et al. 2014, IFC 2012). To achieve this, Landsberg et al. (2013) suggest following a mitigation hierarchy that focuses first on supply and then on benefits (Figure 8.5). This approach also underpins IFC Performance Standards for biodiversity and ecosystem services.

Mitigation strategies should follow the mitigation hierarchy, with avoidance, minimisation or restoration measures being used as appropriate to ensure that ecosystem service supply and long-term benefits are maintained. Offsets can then

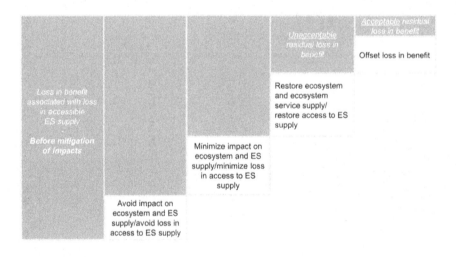

Figure 8.5

Mitigation hierarchy for ecosystem services (adapted from BBOP 2013)

be used to compensate for residual impacts on benefit. The best outcomes are obtained if the capacity of ecosystems to continue supplying ecosystem services in future is sustained by avoiding impacts in the first place. For this reason, IFC (2012) and other guidance emphasise the need for efforts to avoid impacts before moving to subsequent steps in the mitigation hierarchy, which focus on limiting impacts, and then possibly taking measures to restore ecosystem capacity post-impact, such that the supply of services remains sustainable.

Outcomes from restoration of ecosystems are unpredictable and have not been monitored for many ecosystem types, though there are some examples of relatively successful restoration of ecosystem service supply, particularly from wetlands. Meli et al. (2014) carried out a meta-analysis of 70 wetland restoration projects in 14 countries to evaluate their effectiveness in restoring biodiversity and ecosystem services. They found overall supply of ecosystem services to be 43 per cent higher in restored wetlands than in degraded ones, but 13 per cent lower than in natural wetlands. This review suggests that it is possible to restore ecosystem service supply to some extent through mitigation measures involving ecosystem restoration, at least for some wetlands. Similar follow-up studies are not available for other ecosystem types, however, so it is not known how common it is for restoration to succeed. ESIAs therefore need to be designed to take restoration-effectiveness into account.

If there will be residual impacts on well-being despite effective avoidance, minimization, and restoration measures, the focus for mitigation then has to shift to offset measures, focusing on the beneficiary and the benefits needed to maintain their well-being. These offset measures may be other ecosystem services but could also be non-ecosystem-based alternatives, for example access to modern health facilities might be provided to compensate for loss of access to traditional medicines that were previously provided by a forest. This is the stage in ESIA where payments for ecosystem services can be considered as one option for generating income from offsets, while also maintaining levels of benefit for ecosystem service users. Implementing PES schemes in practice can be challenging but the Katoomba Group has developed a PES Contracts Toolkit (Katoomba-CARE 2016) that provides advice on how to draw up legal contracts for PES Schemes and OECD (2010) provides a comprehensive overview of approaches and examples.

If it proves necessary to trade-off different types of benefit, it will not be possible to conclude the ecosystem services assessment without further stakeholder engagement. The key test is the ability to achieve an outcome that maintains benefits for affected parties in a manner that they consider to be acceptable. It is good practice to discuss the implications of different options so that affected parties are fully informed about the risks and opportunities. They might need advice on the pros and cons of "land for land" vs financial compensation, for example.

8.7 Monitoring

Monitoring is important as the basis for adaptive management, particularly where an ESIA has had to deal with high levels of uncertainty, and impact predictions have therefore involved assumptions about either levels of supply, use or dependence on specific ecosystem services. Follow-up may be needed to ensure that project-affected people do have access to the benefits they need, with a project in place. Monitoring or follow-up may also be needed to check that mitigation has been effective. Indicators can be particularly useful at this stage, as they support systematic follow-up in a manner relevant to review conclusions from the ESIA.

Follow-up studies in which ESIA outcomes for ecosystem services have been monitored systematically are rare because of the relative lack of reports documenting practical experiences of incorporating ecosystem services in project-level ESIA. However, some mining companies have conducted long-term monitoring of their projects in partnership with NGOs and have been independently audited for compliance with IFC and other standards. This should mean that availability of cases and examples improves over time.

Monitoring of impacts on ecosystem services needs to monitor aspects of supply and use, the relationships between them and the benefits derived. TEC (2015) designed indicators of supply, use, and benefit for use in ESIA and follow-up monitoring of oil development activities for Tullow Oil, Uganda. Some examples are given in Table 8.2, although monitoring results are not yet available for this case.

Table 8.2 Selected indicators of supply, use and benefit for monitoring impacts of oil development in Uganda on two priority ecosystem services

Ecosystem service	Indicator		
	Supply	Use	Benefit
Woody biomass for fuel for cooking and heating water	Annual yield of woody biomass	Amount of firewood collected per year	Number of cooked meals as a proxy for nutritious food; alleviation of wood fuel burden (time spent not collecting wood fuel)
Fresh water	Water yield from catchment	Amount of water collected; amount of water above WHO drinking water standards	Incidence of waterborne disease; proportion of households reporting insufficient availability of water to drink; alleviation of water burden (time spent not collecting water)

8.8 Conclusions

Ecosystem Services Review (ESR) for purposes of ESIA is based primarily on integrated social and ecological perspectives, with specific benefits providing the "currency" for assessing impacts and trade-offs. It may be led by social or ecological practitioners, but requires involvement of both and close interaction between them. Interaction with other environmental specialists is also likely to be needed, for example with hydrologists, to understand project impacts on water supply, or with engineers to understand the amounts of water likely to be abstracted from an aquifer used by people. If benefits are to be valued in financial terms, inputs from economists may be necessary. The interactive and integrated nature of ESR means that the process should generally be "front-loaded" with active engagement and discussion taking place during the scoping and baseline stages.

While ESIAs have a history of being conducted by specialists working in sequence or in parallel, incorporation of ecosystem services has prompted increased collaboration, particularly between social and ecological specialists, and has improved consideration of issues that are typically excluded if they don't have obvious economic consequences. All in all, incorporating ecosystem services in ESIAs requires a degree of integration and stakeholder engagement that is healthy in terms of managing the multidimensional issues and challenges that typify most ESIAs.

To incorporate assessment of impacts on ecosystem services, ESIA needs to forecast:

- How development might alter ecosystem service-**supply** and the extent to which, in combination with other drivers (which may be external to the planned development), it might alter the **benefits** people are able to derive from ecosystem services in the long term.
- How development might alter the **accessibility** of ecosystem services, for example by introducing physical barriers that prevent access to an ecosystem, or by improving access to new markets. Changing the accessibility of ecosystem services to beneficiaries can also affect the extent to which they **benefit**.
- How development might induce demographic change or population growth, altering **demand** for ecosystem services and therefore levels of **use** and available benefit.

By distinguishing carefully between these aspects of supply, demand, use, and benefit (TEC 2015), ESIA can deliver more effective mitigation recommendations for impacts on ecosystem services, allowing a range of interventions to be identified that will maintain the livelihoods or well-being of particular beneficiaries or users.

Even in ESIA for individual projects, however, effective ESR may require landscape-scale perspectives on ecosystem (and social) patterns and processes (e.g. to identify an ecosystem for which sustainability thresholds are close to being reached, or identify beneficiaries who are already experiencing loss of well-being because of declining access to essential services). This can be challenging within the restricted spatial and temporal confines of many ESIAs. Hence ecosystem services are not always considered to be tractable at a project level, but there is increasing availability of computerised models that facilitate "system" approaches and support assessment of scenarios with multiple impact sources, pathways, and receptors. Even without such techniques, simply considering the potential consequences of ecosystem-change for different groups of people who use associated ecosystem services, and identifying the particular benefits they depend on, can be very beneficial in assessing the social consequences of environmental impacts. It is also beneficial for businesses or developers to improve their understanding of the ecosystem services on which they depend to support their planned operations and ensure a sustainable supply. ESIA can play an important role in generating the evidence on levels of use and dependence that is needed for developments to be compatible with the sustainable supply of ecosystem services to both businesses and society.

ESIAs have generally been poor at linking specific biophysical changes from project infrastructure or activities to subsequent changes in well-being for particular ecosystem service users or beneficiaries. However, consideration of ecosystem services in ESIA is becoming more common and changing this situation (Honrado et al. 2013).

Increased awareness of the importance of ecosystem services to sustain human well-being, improved policy and regulation frameworks, and improved access to practical guidance and tools are all helping to ensure that ESIAs support decision-making by considering the many ways ecosystems contribute to human well-being and are clarifying the types of intervention that are most likely to result in acceptable outcomes for affected people.

References

Baker J and A Scott 2013. *Support for incorporating ecosystem services into environmental impact assessment*. National Ecosystem Approach Toolkit. http://neat.ecosystem-sknowledge.net/pdfs/environmental_impact_assessment_ecosystem_proofed_tool.pdf.

Bateman IJ, GM Mace, C Fezzi, G Atkinson and K Turner 2011. Economic analysis for ecosystem service assessments. *Environmental and Resource Economics* **48**, 177–218.

BBOP (Business and Biodiversity Offsets Programme) 2013. *Mitigation hierarchy*. http://bbop.forest-trends.org/pages/mitigation_hierarchy.

Boyd JW and S Banzhaf 2007. What are ecosystem services? The need for standardised environmental accounting units. *Ecological Economics* **63**, 616–626.

British Ecological Society 2016. *Ecosystem services and valuing natural capital*. www.britishecologicalsociety.org/public-policy/policy-priorities/ecosystem-services-and-valuing-natural-capital.

Convention on Biological Diversity 2010. COP Decision X/2, *Strategic Plan for Biodiversity 2011–2020*. Montréal: CBD.

Daily GC (ed.) 1997. *Nature's services: societal dependence on natural ecosystems.* Washington DC: Island Press.

Defra (Department for Environment Food and Rural Affairs) 2007. *An introductory guide to valuing ecosystem services.* London: Defra.

Defra 2011. *Biodiversity 2020: A strategy for England's wildlife and ecosystem services.* London: Defra. www.gov.uk/government/uploads/system/uploads/attachment_data/file/69446/pb13583-biodiversity-strategy-2020-111111.pdf.

de Groot R, B Fisher, M Christie, J Aronson, L Braat, J Gowdy, R Haines-Young, E Maltby, A Neuville, S Polasky, R Portela and I Ring 2010. Integrating the ecological and economic dimensions in biodiversity and ecosystem service valuation. Ch. 1 in *The economics of ecosystems and biodiversity: ecological and economic foundations,* P Kumar (ed.). London: Earthscan.

Dickinson T, T Male and A Zaidi 2015. *Incorporating natural infrastructure and ecosystem services in federal decision-making.* www.whitehouse.gov/blog/2015/10/07/incorporating-natural-infrastructure-and-ecosystem-services-federal-decision-making.

EBRD (European Bank for Reconstruction and Development) 2008. *Performance Requirement 6 on the Conservation and Sustainable Use of Biodiversity.* London: EBRD.

EC (European Commission) 2011. *Our life insurance, our natural capital: an EU biodiversity strategy to 2020.* COM(2011) 244. Brussels: EC.

EC 2012. *Proposal for a Directive of the European Parliament and of the Council amending Directive 2011/92/EU on the assessment of the effects of certain public and private projects on the environment.* http://eur-lex.europa.eu/legal-content/EN/TXT/?uri=CELEX:52012PC0628.

EC 2014. *Directive 2014/52/EU of the European Parliament and of the Council amending Directive 2011/92/EU on the assessment of the effects of certain public and private projects on the environment.* http://eur-lex.europa.eu/legal-content/EN/TXT/PDF/?uri=CELEX:32014L0052&from=EN.

Everard M and R Waters 2013. *Ecosystem services assessment: how to do one in practice.* London: Institution of Environmental Sciences. www.the-ies.org/resources/ecosystem-services-assessment.

Geneletti D 2011. Reasons and options for integrating ecosystem services in strategic environmental assessment of spatial planning. *International Journal of Biodiversity Science, Ecosystem Services and Management* **7**(3), 143–149.

GIZ 2012. *Integrating ecosystem services into development planning: A stepwise approach for practitioners based on the TEEB approach.* Eschborn: GIZ. www.cbd.int/doc/case-studies/inc/giz-2012-en-integr-ecosys-serv-in-dev-planning.pdf

Guerry AD, S Polasky, J Lubchenco, R Chaplin-Kramer, GC Daily, R Griffin, M Ruckelshaus, IJ Bateman, A Duraiappah et al. 2015. Natural capital and ecosystem services informing decisions: From promise to practice. *Proceedings of the National Academy of Sciences* **112**(24), 7348–7355.

Hanh VT, P Moore and L Emerton 2010. *Review of laws and policies related to payment for ecosystem services in Vietnam.* IUCN. http://cmsdata.iucn.org/downloads/080310_pes_vn_legal_review_only_legal_sections_final.pdf.

Honrado JP, C Vieira, C Soares, MB Monteiro, B Marcos, HM Pereira and MR Partidário 2013. Can we infer about ecosystem services from EIA and SEA practice? A framework for analysis and examples from Portugal. *Environmental Impact Assessment Review* **40**, 14–24.

IFC (International Finance Corporation) 2012. *IFC performance standards on environmental and social sustainability.* Washington DC: IFC. www.ifc.org/wps/wcm/connect/

c8f524004a73daeca09afdf998895a12/IFC_Performance_Standards.pdf?MOD=AJPERES.

IPBES (Intergovernmental Science-policy Platform on Biodiversity and Ecosystem Services) 2016. *Catalogue of assessments on biodiversity and ecosystem services.* http://catalog.ipbes.net.

IPIECA (Global oil and gas industry association for environmental and social issues) and OGP (International Association of Oil and Gas Producers) 2011. *Ecosystem services guidance: biodiversity and ecosystem services guide and checklists.* London: IPIECA and OGP. www.ipieca.org/sites/default/files/publications/ecosystem_services_guidance_8.pdf.

Kareiva P, H Tallis, TH Ricketts, GC Daily and S Polansky 2011. *Natural capital: theory and practice of mapping ecosystem services.* New York: Oxford University Press.

Katoomba-CARE 2016. *Payments for ecosystem services (PES) contract toolkit.* www.katoombagroup.org/regions/international/legal_contracts.php.

Kinzig AP, C Perrings, FS Chapin, S Polasky, VK Smith, D Tilman and BL Turner 2011. Paying for ecosystem services – promise and peril. *Science* **334**(6056), 603–604.

Landsberg F, M Ozment, M Stickler, J Henninger, J Treweek, O Venn and G Mock 2011. *Ecosystem services review for impact assessment: introduction and guide to scoping.* WRI Working Paper. Washington DC: World Resources Institute. www.wri.org/sites/default/files/ecosystem_services_review_for_impact_assessment_introduction_and_guide_to_scoping.pdf.

Landsberg F, J Treweek, N Henninger, M Stickler and O Venn 2013. *Weaving ecosystem services into impact assessment: a step-by-step method.* Washington DC: World Resources Institute. www.wri.org/sites/default/files/weaving_ecosystem_services_into_impact_assessment.pdf.

Landsberg F, J Treweek, M Stickler, N Henninger and O Venn 2014. *Weaving ecosystem services into impact assessment technical appendix (Version 1.0).* Washington DC: World Resources Institute. www.wri.org/sites/default/files/weaving_ecosystem_services_into_impact_assessment_technical_appendix.pdf.

MA (Millennium Ecosystem Assessment) 2003. *Ecosystems and human well-being: a framework for assessment.* www.millenniumassessment.org/en/Framework.html.

MA 2005. *Ecosystems and human well-being: biodiversity synthesis.* Washington DC: World Resources Institute

Meli P, JMR Benayas, P Balvanera and MM Ramos 2014. Restoration enhances wetland biodiversity and ecosystem service supply, but results are context-dependent: a meta-analysis. *PloS ONE* **9**(4), p.e93507.

Natural Capital Coalition 2015. *The draft natural capital protocol and sector guides for business: An updated overview.* www.naturalcapitalcoalition.org/js/plugins/filemanager/files/Natural_Capital_Protocol_An_Overview_WorkingDoc-16Nov2015.pdf.

Natural Capital Project 2016. *InVEST: integrated valuation of environmental services and tradeoffs.* www.naturalcapitalproject.org/invest.

National Ecosystem Services Partnership 2016. *Federal resource management and ecosystem services guidebook,* 2nd Edn. Durham, USA: National Ecosystem Services Partnership, Duke University, https://nespguidebook.com/assessment-framework/what-are-benefit-relevant-indicators/

OECD (Organisation for Economic Co-operation and Development) 2010. *Paying for biodiversity: enhancing the cost-effectiveness of payments for ecosystem services.* Paris: OECD.

Olander, L, RJ Johnston, H Tallis, J Kagan, L Maguire, S Polasky, D Urban, J Boyd, L

Wainger, and M Palmer 2015. *Best practices for integrating ecosystem services into federal decision-making.* Durham, USA: National Ecosystem Services Partnership, Duke University.

Partidário MR and RC Gomes 2013. Ecosystem services inclusive strategic environmental assessment. *Environmental Impact Assessment Review* **40**, 36–46.

Tardieu L, S Roussel, JD Thompson, D Labarraque, and J Salles 2015. Combining direct and indirect impacts to assess ecosystem service loss due to infrastructure construction. *Journal of Environmental Management* **152**, 145–157.

TEC 2015. *A method for ecosystem services review at the landscape level.* Report produced for Tullow Oil, Uganda.

UK NEA 2011. *The UK national ecosystem assessment.* Cambridge: Cambridge University Press. http://uknea.unep-wcmc.org.

UN (United Nations) 1992. Convention on Biological Diversity. www.cbd.int/doc/legal/cbd-en.pdf.

UNEP-WCMC (United Nations Environment Programme-World Conservation Monitoring Centre) 2011. *Developing ecosystem service indicators: Experiences and lessons learned from sub-global assessments and other initiatives.* Montréal, Canada: Secretariat of the Convention on Biological Diversity. www.cbd.int/doc/publications/cbd-ts-58-en.pdf.

van Oudenhoven AP, K Petz, R Alkemade, L Hein and RS de Groot 2012. Framework for systematic indicator selection to assess effects of land management on ecosystem services. *Ecological Indicators* **21**, 110–122.

Waage S 2014. *Making sense of new approaches to business risk and opportunity assessment: Integrating ecosystem services into investor due diligence and corporate management.* BSR. www.bsr.org/en/our-insights/report-view/making-sense-of-new-approaches-to-business-risk-and-opportunity-assessment.

Zhiyun O 2013. Gross ecosystem product: concept, accounting framework and case study. *Acta Ecol Sin* **33**, 6747–6761.

9 Noise

••

Graham Wood and Riki Therivel

9.1 Introduction

Virtually all development projects have noise impacts. Noise during construction may be due to activities such as land clearance, piling, and the transport of materials to and from the site. During operation, noise levels may decrease for some forms of developments such as science parks or new towns, but may remain high or even increase for developments such as new roads or industrial processes. Decommissioning and demolition is a further cause of noise. Consequently, Environmental Social and Impact Assessments (ESIAs) for most development projects are likely to consider noise.

Noise is a major and growing form of **pollution**. It can interfere with communication, increase stress and annoyance, cause anger at the intrusion of privacy and disturb sleep, leading to lack of concentration, irritability, and reduced efficiency (WHO 1999). It can contribute to stress-related health problems such as high blood pressure and increased risk of heart attacks (WHO 2009; Basner et al. 2014). Prolonged exposure to high noise levels can cause deafness or partial hearing loss. Noise can also change the character of a landscape or historical setting, affect property values, community atmosphere, and overall quality of life.

The effects of noise pollution are widespread and significant. For example, in Europe the World Health Organisation identifies noise as the second largest environmental risk to public health, estimating that between 1–1.6 million healthy life years are lost every year from traffic-related noise (WHO 2011). In the US, Hammer et al. (2014) estimate that over 140 million people may be at risk of noise-related health effects, with noise consistently rated as the number one 'quality of life' issue in New York City.

Although most ESIAs focus on the impact of noise on people, noise may also affect wildlife (Table 9.1) and domestic animals, and in certain cases ESIAs will need to include specialist studies on these impacts. This chapter focuses on airborne environmental noise, although some of the principles considered are also relevant to ground-borne noise and vibration (see Box 9.1 for a basic overview of principles of vibration). Occupational-noise assessments are not considered in this chapter, neither are underwater acoustics relevant to offshore marine projects such as oil and gas development (see Farcas et al. 2016 and Williams et al. 2015 for useful reviews).

Table 9.1 Effects of noise on wildlife (based on Wardell Armstrong 2011)

Type of effect	Primary	Secondary
Auditory	Hearing loss Auditory threshold shift	Change in predator–prey relationships Mating interference/success Reduction in functioning
Physiological	Stress Metabolic change Hormonal change	Reduced reproductive capacity Weakened immune system Reduction in functioning
Behavioural	Signal masking Avoidance behaviour	Change in predator–prey relationships Population reduction Migration and loss of habitat Mating interference/success

Box 9.1 Basic principles of vibration

Humans distinguish vibration from audible sound in terms of the way it is sensed or 'experienced' by the recipient. Thus, whilst sound is detected by hearing, vibration is felt as it is transmitted through solid structures directly to the human body (Environment Agency 2004).

Vibration refers to the oscillation of an object about a reference point, with the number of oscillations per second giving the frequency of vibration (Hz). Vibration can occasionally occur at a single frequency, but normally several different frequency components occur simultaneously (e.g. different components of a car engine vibrate at different frequencies). Vibration frequencies between 1–80 Hz are perceptible to humans.

A particle may vibrate vertically, longitudinally, and horizontally (often involving a simultaneous combination of all three). Vibration levels are commonly quantified in ESIA reports in terms of Peak Particle Velocity (PPV), measured in millimetres per second (mm/s). The threshold of human perception of vibration is typically within the range 0.14 mm/s to 0.3 mm/s PPV. Road traffic and railways are common sources of vibration relevant to ESIA, along with blasting associated with mineral extraction and quarrying. Complaints about vibration from quarries are often not caused by ground-borne vibration, but are linked to the air pressure wave that results from blasting (e.g. windows rattling or ornaments shaking on the mantelpiece). BS 5228-2 (BSI 2009) indicates that construction activities generally only cause vibration impacts when they are located less than 20 metres from sensitive locations.

9.2 Definitions and concepts

9.2.1 Fundamentals of acoustics: amplitude, frequency and the decibel scale

Noise is essentially unwanted sound. This definition holds within it one of the core aspects of noise impact assessment: namely it deals with people's subjective responses ('unwanted') to an objective reality ('sound'). The physical level of noise does not clearly correspond to the level of annoyance it causes – contrast the perspective of revellers in a nightclub with that of nearby residents – yet it is the annoyance caused by noise that is critical in ESIA. Noise impact assessment revolves around the notion of quantifying and 'objectifying' people's personal and subjective responses. The following definitions and concepts all relate to this issue.

Sound consists of pressure variations that are detectable by the healthy human ear. These pressure variations have two key characteristics: frequency and amplitude. Sound **frequency** refers to how quickly the air vibrates, or how close the sound waves are to each other (in cycles per second, or Hertz (Hz)). For example, the sound from a transformer has a wavelength of about 3.5 m, and hums at a frequency of 100 Hz; a television line emits waves of about 0.03 m, and whistles at about 10,000 Hz or 10 kHz. Frequency is subjectively felt as the **pitch** of the sound. Broadly, the lowest frequency audible to humans is approximately 20 Hz, and the highest is 20,000 Hz.

A pure tone is one in which the sound wave is characterised by a single frequency, e.g. in music, the sound emitted by a 'tuning fork' closely approximates a pure tone. Environmental noise can be dominated by a particular frequency or it may be described as 'broad band', i.e. comprising a range of different frequencies. Most sounds in an environmental context consist of many different frequencies, although equipment such as turbines, compressors, cooling fans, or transformers can generate noise with discrete frequencies that produce distinct tonal characteristics. This **tonal noise** is typically more noticeable and has the potential to cause greater annoyance than non-tonal noise of the same level (Environment Agency 2004). For convenience of analysis, the audible frequency spectrum is often divided into standard octave bands of 32, 63, 125, 250, 500, 1 k, 2 k, 4 k and 8 kHz. More sophisticated sound level meters are capable of disaggregating the frequency characteristics of environmental noise.

Sound **amplitude** refers to the amount of pressure exerted by the air, which is often visualised as the height of the sound waves. Amplitude is described in units of pressure per unit area, microPascals (μPa). The amplitude is sometimes converted to sound **power**, in picowatts (10^{-12} watts), or sound intensity (in 10^{-12} watts/m^2). Sound intensity is subjectively felt as the **loudness** of sound. However, none of these measures are easy to use because of the vast range which they cover (see Table 9.2). As a result, a logarithmic scale of **decibels** (dB) is used. The sound pressure level (L_p) in decibels is given by

$$L_p = 10 \log_{10} (P/p)^2 \, dB$$

where P is the amplitude of pressure fluctuations, and p is $20\mu Pa$, which is considered to be the lowest sound audible to the healthy human ear. The sound intensity level can also be described as:

$$L_i = 10 \log_{10} (I/i) \, dB$$

where I is the sound intensity and i is 10^{-12} watts/m^2, or by

$$L_w = 10 \log_{10} (W/w) \, dB$$

where W is the sound power, and w is 10^{-12} watts. The sound power level (L_w) is the energy output of a source, calculated from measurements made around the equipment, under carefully controlled conditions. Manufacturers of equipment (e.g. construction plant, industrial fans, etc.) will often provide L_w data for their products, which can then be used within a noise impact assessment to calculate the sound pressure level (L_p) at specific locations (see §9.5.2) The range of audible sound is generally from 0dB to 140dB, as is shown in Table 9.2.

Because of the logarithmic nature of the decibel scale, a doubling of the power or intensity of a sound (for instance combining two identical sound levels), leads to an increase of 3 dB, not a doubling of the decibel rating. For example, two lorries, each at 75 dB, together produce a total sound level of 78

Table 9.2 Sound pressure, intensity and level

Sound pressure (µPa)	Sound power (10^{-12} watt) or intensity level (10^{-12} watt/m^2)	Sound level (dB)	Example
200,000,000	100,000,000,000,000	140	threshold of pain
	10,000,000,000,000	130	riveting on steel plate
20,000,000	1,000,000,000,000	120	pneumatic drill
	100,000,000,000	110	loud car horn at 1 m
2,000,000	10,000,000,000	100	alarm clock at 1 m
	1,000,000,000	90	inside underground train
200,000	100,000,000	80	inside bus
	10,000,000	70	street-corner traffic
20,000	1,000,000	60	conversational speech
	100,000	50	business office
2,000	10,000	40	living room
	1,000	30	bedroom at night
200	100	20	broadcasting studio
	10	10	normal breathing
20	1	0	threshold of hearing

dB. Multiplying the sound power by ten (e.g. ten lorries) leads to an increase of 10 dB. Table 9.3 shows an approximate method that can be used to add or subtract sound levels in dB. Box 9.2 provides an equation that can be used for combining several noise sources where a more precise method is required.

Under laboratory conditions, a change in noise level of 1 dB is just perceptible. Subjectively, a change of 3 dB is generally perceptible by the human ear under normal conditions, providing that the change in sound pressure level is not accompanied by some change in the character of the sound. A change of 6 dB will be obvious and a change of 10 dB is broadly perceived as a doubling/halving of loudness. Consequently, the logarithmic

Table 9.3 Adding and subtracting noise levels (Environment Agency 2004)

Difference between the two sound levels	Quantity to be added (or subtracted) from the higher sound level
0	3.0
1	2.5
2	2.1
3	1.8
4	1.5
5	1.2
6	1
7	0.8
8	0.6
9	0.5
10 or more	0

Example: 57 dB + 60 dB = 61.8 dB

Box 9.2 Method for combining several noise levels

$Lp_{Total} = 10Log [10^{(Lp1/10)} + 10^{(Lp2/10)} + 10^{(Lpn/10)}+......]$
Lp_{Total} is the total sound level
Lp_1 = sound level 1
Lp_2 = sound level 2
Lp_n = sound level n

Example – Combine the following four noise levels: 65, 76, 68, 69 dB
$Lp_{Total} = 10 * log (10^{65/10} + 10^{76/10} + 10^{68/10} + 10^{69/10})$
$= 10 * log (10^{6.5} + 10^{7.6} + 10^{6.8} + 10^{6.9})$
$= 10 * log (3162277 + 39810717 + 6309573 + 7943282)$
$= 10 * log (57225849)$
$= 77.6dB$

decibel scale, in addition to simplifying the necessary manipulation of a very large range of sound pressures/intensities, is conveniently related to the human perception of loudness.

The human ear is more sensitive to some frequencies than to others. It is most sensitive to the 1 kHz, 2 kHz, and 4 kHz octaves and much less sensitive at the lower audible frequencies. For instance, tests of human perception of noise have shown that a 70 dB sound at 4 kHz sounds as loud as a 1 kHz sound of about 75 dB, and a 70 dB sound at 63 Hz sounds as loud as a 1 kHz sound of about 45 dB. Since most sound analyses, including those in ESIA, are concerned with the loudness experienced by people (rather than the actual physical magnitude of the sound), an **A-weighting curve** is used to give a single-figure index which takes account of the varying sensitivity of the human ear; this is shown in Figure 9.1. Most sound level meters incorporate circuits which carry out this weighting automatically, and all ESIA results should be A-weighted (dB(A)). Other weightings exist, but are rarely used.

9.2.2 Metrics for characterising noise over time

Noise levels are rarely steady: they rise and fall in accordance with the activity taking place in the area over a given period. Time-varying noise levels can be described in a number of ways. The principal measurement index for environmental noise is the **equivalent continuous noise level**, LA_{eq}. The LA_{eq} is a notional steady-noise level which, over a given time, would provide the same energy as the time-varying noise: it is calculated by averaging the sound pressure/power/intensity measurements, with a bias towards the louder noise

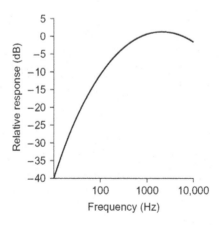

Figure 9.1

A-weighting curve

events, and converting that average into the dB scale. Most environmental noise meters read this index directly. LA_{eq} has the dual advantages that it: takes into account both the energy and duration of noise events; and it is a reasonable indicator of likely subjective response to noise from a wide range of different noise sources. In circumstances where the noise level is reasonably constant, a relatively short measurement period will be sufficient to characterise LA_{eq}, but if the noise fluctuates or has cyclical elements, then a longer period of measurement is required in order to obtain a representative sample. The time of day and the measurement period used must always be stated when reporting values for LA_{eq} and other statistical parameters.

In addition to LA_{eq}, statistical indices are used as the basis of some types of noise assessment for ESIA, particularly when characterising the baseline conditions (see §9.4). LA_{90}, the dB(A) level which is exceeded for 90 per cent of the time, is used to indicate the noise levels during quieter periods, also referred to as the **background noise**. LA_{10} is the dB(A) level which is exceeded for 10 per cent of the time and is representative of the louder sounds, and is often used as the basis of road traffic noise assessment in the UK. Note that in all cases, $L_{10} > L_{eq} > L_{90}$, as shown in Figure 9.2. In addition to LA_{eq} and the statistical indices it can be useful to consider the maximum noise level, the LA_{max}. The LA_{max} can be important when night-time noise and the potential for sleep disturbance is considered.

Many noise standards specify the length of time over which noise should be measured. For instance UK Planning Practice Guidance for mineral extraction projects refers to dB LA_{eq} (one hour), the equivalent continuous noise level, in

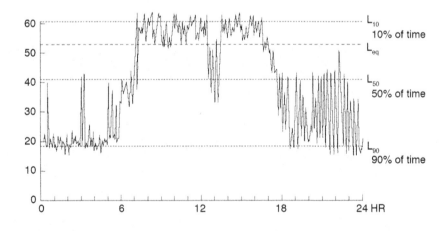

Figure 9.2

Sound levels exceeded for stated percentage of the measurement period dB(A), during one hour of a working day (DCLG 2014a). When considering noise criteria which are expressed in terms of LA_{eq}, the measurement period can be particularly important. The slow passage of a heavy goods vehicle (HGV) at a distance of 10 m, for instance, may give rise to a 12-second LA_{eq} of 75 dB(A), a five-minute LA_{eq} of 61 dB(A) and a one-hour LA_{eq} of 50 dB(A).

Beyond the metrics for summarising noise levels, it may be necessary to consider other physical characteristics of the sound being assessed, e.g. whether the sound is characterised as constant and **continuous**, constant but **fluctuating** in level, or **intermittent** in character. It may be important to consider whether the sound is **impulsive** (it contains distinct clatters and thumps, e.g. pile driving), **tonal** (whine, scream, hum) or whether it contains **information content** (such as speech or music).

9.2.3 Factors influencing noise impacts

The principal physical factors which influence how much effect a sound will have upon a potentially affected receptor are the **level of the sound** being assessed and the **level of other sounds** which also affect the receptor. For instance, people in rural environments would expect lower sound levels than those in a busy city centre. This interplay of location and noise is not often seen in noise standards, though the OECD (1996) recommends different noise levels for urban, suburban and rural areas. The level of sound being assessed is determined by several factors that influence the way sound propagates over space.

First, with increasing distance from a source of sound, the level of noise from the source decreases. The principal factor contributing to this attenuation is typically **geometric dispersion of energy**. With increasing distance away from a sound source, the sound power from the source is spread over a larger and larger area (think of the way that ripples diminish from a stone thrown into a pond). For a point noise source, the sound wave expands over distance in the form of a sphere with increasing radius, leading to a reduction of 6 dB in the sound pressure level with doubling of distance. For line sources of noise (such as major roads), the rate at which this happens is 3 dB per doubling of distance. It is because of this principle that noise fades rapidly near a noise source, but more slowly at greater distances from it (it is why, for instance, motorways can be heard over long distances).

The next most important factor in governing noise levels at a distance from a source is **whether the propagation path from the noise source to the receiver is obstructed**. If there is a large building, a substantial wall or fence, or a topographic feature which obscures the line of sight, this can reduce noise levels by, typically, a further 5–15 dB(A). The amount of attenuation (reduction) depends upon the geometry of the situation and the frequency characteristics of the noise source. Trees, unfortunately, do not generally act as effective barriers, with UK Environment Agency guidance (2004) suggesting that a tree belt must be greater than 100 m thick and very densely planted to achieve any significant excess attenuation. Nonetheless, through visual screening, i.e. blocking the noise source from sight, trees may have a significant psychological effect (when in leaf).

If the sound is travelling over a reasonable distance (generally hundreds rather than tens of metres), the **type of ground** over which it is passing can have a substantial influence on the noise level at the receiver. Acoustically hard surfaces (e.g. concrete, asphalt, or water) tend to reflect sound, but if the sound is passing at a reasonably low physical level over soft ground (grassland, crops, trees, etc.) there will be an additional attenuation beyond that caused by geometric dispersion. Ground effects will be particularly important when predicting impacts over relatively long distances, as the potential for absorption increases in line with distance. Thus additional attenuation of around 3 dB occurs over distances of 100 m, increasing up to about 9 dB at 1,000 m. The degree of attenuation is greater with higher frequency sound and is influenced by the height of propagation above the ground. It should be noted, however, that where barrier attenuation is included in calculations, soft-ground attenuation is generally ignored in sound propagation modelling.

Other physical effects which may have to be addressed include reflection and meteorological effects. Probably the most important aspect of reflection that needs to be considered is whether the propagation relates to **free-field** conditions (i.e. at least 3.5 m from reflective surfaces other than the ground) or at the **façade** of a building (i.e. 1 m from the façade of the potentially affected receptor). Measurements that are made close to the façade of a building will include both direct noise as well as a reflected component, which adds approximately 3 dB to the level of the noise. In reality, façade effects vary from source to source, depending on whether the sound field is directional or diffuse. Whether calculation or measurement results are free-field or façade is critical, however, as the differentials that have to be assumed are considerable. Other reflection effects occur where hard surfaces act as acoustic mirrors, increasing the sound pressure level or intensity (not the power) of a source. This may require consideration where detailed calculations are being carried out.

Meteorological effects generally only need to be considered where calculations are being made over large distances (upwards of 100 m or so). Wind speed and direction can affect noise levels. A gentle positive wind (the wind blowing from the noise source to the receptor) slightly increases noise levels compared with calm conditions, but a negative wind has a larger effect (i.e. it reduces noise levels more than a positive wind increases them).

Temperature gradients in the atmosphere also influence the behaviour of sound waves. Where the rate of cooling in the atmosphere is close to the environmental lapse rate (i.e. approximately 0.6°C with every 100 m increase in height, during the daytime), sound rays are **refracted** upwards, leading to a 'shadow zone' of reduced sound levels around the source. When combined with wind, the shadow zone upwind of the source will be further enhanced. Under calm conditions, when temperature increases with height (a temperature inversion), sound waves are refracted downwards towards the ground leading to elevated noise levels. In both cases, the nature of these effects depends upon the frequency of the sound (ISO 1996).

Sound attenuation also occurs due to **atmospheric absorption**, as sound energy dissipates due to friction and the vibration/rotation of nitrogen and oxygen molecules in the atmosphere. Over small distances (up to a few hundred metres), atmospheric absorption can generally be ignored, as the effect is minor compared with that of geometrical dispersion. However, the effect is more notable over longer distances and is frequency-dependent, with higher frequencies (e.g. 4 kHz) attenuated far more than lower frequencies. These effects depend upon temperature and humidity, and a method for calculating atmospheric absorption in relation to meteorological conditions is provided in ISO 9613 (ISO 1996).

Clearly, as distances increase from a noise source, the degree of certainty to which noise levels can be estimated rapidly diminishes. Where large distances are involved, and noise level estimates are critical (as they can be for power stations or large petrochemical plants, for instance) it is essential that the conditions for which any noise predictions are expected to hold are clearly defined.

9.3 Key legislation, guidance and standards

Legislation relevant to the control of noise varies around the globe and hence it is important to consult with the relevant environmental authority early in the ESIA process to seek advice on the appropriate regulations and guidance for appraising a given development project.

Policy and legislation to control noise can involve setting standards that place limits on the emission of noise from particular sources, e.g. European Directive 70/157/EEC relating to the permissible sound level from vehicle exhaust systems, or the US Federal emission standards set up by the Noise Control Act of 1972. Policy approaches can also seek to manage overall environmental noise levels and hence the impact of noise on individuals, particularly in terms of health. For example, in Europe, Directive 2002/49/EC (EC 2002) – the Environmental Noise Directive – requires Member States to map noise in densely populated areas and from major transport projects, and to introduce plans to manage noise where necessary, and also to prevent specified quiet areas from getting noisier. However, European legislation does not specify fixed quantitative noise or vibration limits, leaving this decision to individual Member States. Table 9.4 summarises other international examples, including those used for the measurement and assessment of vibration.

World Health Organisation (WHO)

The *Guidelines for Community Noise* (WHO 1999) are health-based guidelines that incorporate influential noise standards as part of a framework for noise

Table 9.4 Examples of key standards and guidelines for the measurement and assessment of noise and vibration

Standard/Guidelines	Title	Description
International Standard ISO 1996-2:2007	*Description, Measurement and Assessment of Environmental Noise Part 2: Determination of Environmental Noise Levels*	Describes procedures for determining sound pressure levels by direct measurement, extrapolation of measurement through calculation, or exclusively by calculation. Recommends preferable conditions for measurement or calculation to be used in cases where other regulations do not apply. Provides guidance on evaluating the uncertainty of noise assessment results.
International Standard: ISO 9613-2:1996	*Acoustics – Attenuation of Sound During Propagation Outdoors – Part 2: General Method of Calculation*	Provides algorithms for the prediction of noise levels in the community from sound emission sources. Key mechanisms of sound attenuation include geometric divergence, atmospheric absorption, ground effect, reflection from surfaces and screening. Algorithms are widely adopted in commercially available software.
International Institute of Noise Control Engineering 2011	*Guidelines for Community Noise Impact Assessment and Mitigation*	Non-technical guidance aimed at policy makers involved with noise regulation and control through EIA. Focuses primarily on the broad approach to undertaking a noise impact assessment. Also provides information on dose-response relationships and land use planning to control exposure to environmental noise.
World Health Organisation 2009	*Night Noise Guidelines for Europe*	Provides additional information on the health effects of night noise, including health-based guideline values. Proposes an Interim Target level of 55 dB L_{night} outside.
Association of Noise Consultants 2012	*Measurement and Assessment of Groundborne Noise and Vibration* (2nd edn.)	Practical guidance with particular attention paid to railway vibration and ground-borne noise. Provides advice on overcoming problems associated with widely different procedures, criteria and equipment adopted across the industry.

Table 9.4 continued

Standard/Guidelines	Title	Description
British Standard 5228-1:2009+A1:2014, 5228-2:2009	Code of Practice for Noise and Vibration Control on Construction and Open Sites. Part 1: Noise and Part 2: Vibration	Part 1 provides a methodology for calculating noise levels generated for a range of common construction plant (both fixed and mobile). Includes a database of equivalent continuous noise source levels (LA$_{eq}$ dB) for various plant, for use in the absence of measured data. Outlines a simple noise propagation model that incorporates allowances for source-receiver distances, ground effects, surface reflection, barriers, and duration and timing of activities. The standard also outlines methods for determining the significance of noise effects and provides guidance on minimising potential effects through mitigation. Part 2 provides guidance on measuring vibration, including procedures for estimating vibration attributable to vibratory rolling and piling activities. Annex B provides guidance on the effects of vibration levels on human receptors, as well as guide values for cosmetic damage to building structures. The standard provides information on vibration control, including a review of relevant vibration criteria.
Environmental Protection Authority (EPA) Western Australia 2014	Environmental Assessment Guideline: Consideration of Environmental Impacts from Noise	Assists project proponents in determining whether noise emissions may cause significant impacts; explains how potential noise impacts are considered by the EPA and assessed within the EIA process; and directs proponents to appropriate regulatory standards and technical guidance.
Environmental Protection Department Hong Kong	EIA Ordinance. Technical Memorandum Annex 13: Guidelines for Noise Assessment	Basic guidance on commonly adopted approaches and methodologies for assessment of noise impacts arising from designated EIA projects.

Table 9.4 continued

Standard/Guidelines	Title	Description
British Standard 4142:2014	*Methods for Rating and Assessing Industrial and Commercial Sound*	Guidance on the monitoring and assessment of industrial and commercial sound sources. Provides a methodology and criteria for assessing the impacts of new or existing sound sources by comparing the operational sound level (the 'rating level') with the background level (i.e. the baseline without the development). The rating level can incorporate a 'rating penalty' based on a subjective or objective assessment of its characteristics (e.g. tonal, impulsive).
British Standard 8233:2014	*Guidance on Sound Insulation and Noise Reduction for Buildings*	Provides guidance for the control of noise in and around buildings. Applicable to new buildings, or refurbished buildings undergoing a change of use. Provides indicative internal and external guideline noise values. Provides design guidelines for internal acoustic environments within buildings dependent upon their function.
British Standard 6472 Part 1: 2008	*Guide to Evaluation of Human Exposure to Vibration in Buildings. Part 1: Vibration Sources Other Than Blasting*	Guidance on the magnitude of vibration (expressed as a Vibration Dose Value) at which adverse comments may arise. Is also referred to within BS5228-2:2009.
British Standard 7385 Part 2: 1993	*Evaluation and Measurement for Vibration in Buildings – Part 2: Guide to Damage Levels from Ground-borne Vibration*	Guidance on the levels of vibration (expressed as peak particle velocity) at which cosmetic damage is likely to occur within buildings.
Environment Agency 2004	*Horizontal Guidance for Noise Part 2 – 'Noise Assessment and Control'*	Provides basic noise theory and describes the principles of noise measurements and prediction. Considers the control of noise through design, operational and management techniques and abatement technologies.

Table 9.4 continued

Standard/Guidelines	Title	Description
Department for Transport and Welsh Office 1988	*Calculation of Road Traffic Noise (CRTN)*	Gives procedures for calculating road traffic noise levels, incorporating factors including: traffic volume, composition, and speed; road gradient; road surface; distance between the road and receptor; ground cover; barrier screening; and reflection from façades. The noise parameter calculated is the LA_{10}, 18hr and is based on the 18-hour Annual Average Weekday Traffic.
Highways Agency 2011	*Design Manual for Roads and Bridges (Volume 11 Section 3 Part 7 HD213/11 Noise and Vibration)*	Describes procedures for assessing the impacts/effects of road traffic using noise descriptors based upon statistical noise level LA_{10}, 18hr i.e. over an 18-hour period between 06:00 and 24:00 (the traffic noise index).

management. The guidelines recommend internal and external noise levels that will prevent detrimental effects on a community, including rest, sleep, and work that requires concentration, amongst others:

- To protect the majority of people from serious annoyance during the daytime, the noise level on balconies, terraces and outdoor living areas should not exceed 55 dB LA_{eq} for a steady, continuous noise. To protect the majority of people from being moderately annoyed during the daytime, the outdoor noise level should not exceed 50 dB LA_{eq}.
- At night, noise levels at the outside façades of the living spaces should not exceed 45 dB LA_{eq} and 60 dB LA_{max}, so that people can sleep with bedroom windows open.

These guideline levels are derived from research and relate to continuous noise sources (e.g. road traffic), and so are not applicable to construction noise. The values are based on the assumption that noise levels experienced inside with the window partly open are reduced by approximately 15 dB. Guideline values for specific environments with varying degrees of sensitivity to noise are also provided (see Table 9.5).

International Finance Corporation (IFC)

The IFC Performance Standards provide an international benchmark for environmental and social risk management and reference to noise is contained within PS 1, 2, 3 and 6. Section 1.7 of the IFC Environmental, Health, and Safety (EHS) General Guidelines (IFC 2007) provides further information on noise assessment and management, including absolute noise level limits. These relate to noise experienced beyond the property boundaries of the proposed development project (Table 9.6) and in contrast to the WHO guidelines, the levels distinguish between just two main categories of receptor (residential and industrial). Where the existing ambient baseline noise level is already above the prescribed level, the guidance suggests that the noise source should not raise the ambient levels at the nearest off-site receptor by more than 3 dB.

The guidance relates to noise from stationary noise sources and it is commonly applied as design standards for industrial facilities. It is not directly applicable to transport or mobile noise sources, and hence is most relevant for the control of operational noise impacts rather than the assessment of temporary construction effects.

IEMA Guidelines for Environmental Noise Impact Assessment 2014

The Institute of Environmental Management and Assessment (IEMA) Guidelines outline the key principles of noise impact assessment within the context of ESIA. They provide advice on scoping a noise assessment, assessing the baseline noise environment, predicting changes in noise levels due to

Table 9.5 World Health Organisation guidelines for community noise

Specific environment	Critical health effect(s)	LA$_{eq}$ (dB)	Time base (hours)	LA$_{max}$ fast (dB)#5
Outdoor living area	Serious annoyance, daytime and evening	55	16	–
	Moderate annoyance, daytime and evening	50	16	–
Dwelling, indoors	Speech intelligibility and moderate annoyance, daytime and evening	35	16	–
Inside bedrooms	Sleep disturbance, night-time	30	8	45
Outside bedrooms	Sleep disturbance, window open (outdoor values)	45	8	60
School classrooms and pre-schools, indoors	Speech intelligibility, disturbance of information extraction, message communication	35	during class	–
Pre-school bedrooms, indoors	Sleep disturbance	30	sleeping time	45
School, playground outdoor	Annoyance (external source)	55	during play	–
Hospital, ward rooms, indoors	Sleep disturbance, night-time	30	8	40
	Sleep disturbance, daytime and evenings	30	16	–
Hospitals, treatment rooms, indoors	Interference with rest and recovery	#1		
Industrial, commercial, shopping and traffic areas, indoors and outdoors	Hearing impairment	70	24	110
Ceremonies, festivals and entertainment events	Hearing impairment (patrons: < 5 times/year)	100	4	110
Public addresses, indoors and outdoors	Hearing impairment	85	1	110
Music through headphones/earphones	Hearing impairment (free-field value)	85 (#4)	1	110
Impulse sounds from toys, fireworks, firearms	Hearing impairment (adults)	–	–	140 #2
	Hearing impairment (children)	–	–	120 #2
Outdoors in parkland and conservation areas		#3		

Notes:
#1: as low as possible
#2: peak sound pressure (not LA$_{max}$, fast), measured 100 mm from the ear
#3: existing quiet outdoor areas should be preserved and the ratio of intruding noise to natural background sound should be kept low
#4: under headphones, adapted to free-field values
#5: see §9.4 for an explanation of fast vs slow measurements

Table 9.6 IFC EHS noise level guidelines

Receptor	Maximum allowable ambient noise levels LA$_{eq}$,1hr, dB(A) Free field	
	Daytime 07:00–22:00	Night-time 22:00–07:00
Residential, institutional, educational	55	45
Industrial, commercial	70	70

development, and on defining and evaluating the significance of effects. The guidance recognises that the significance of the noise-effect associated with a development will be dependent on factors including (but not limited to) the sensitivity of the receptor, frequency and duration of the noise source, and time of day.

National noise policy, land-use planning and noise guidance: examples from England

Box 9.3 illustrates how national noise policy in England links to land-use planning and related noise guidance. The approach taken is unusual in that it avoids prescribing specific fixed quantitative noise levels, instead promoting a *noise exposure hierarchy* approach (see Table 9.7), inspired by concepts used in toxicology, to aid the interpretation of 'significant adverse impacts'.

Box 9.3 National noise policy and guidance for England

In 2012 the National Planning Policy Framework (NPPF) set out the Government's land-use planning policies for England and how these are expected to be applied to decision-making (DCLG 2012). Regarding noise, NPPF paragraph 123 states that planning policies and decisions should aim to:

- Avoid noise giving rise to significant adverse impacts on health and quality of life as a result of new development;
- Mitigate and reduce to a minimum other adverse impacts on health and quality of life arising from noise from new development, including through the use of conditions;
- Recognise that development will often create some noise and existing businesses wanting to develop in continuance of their business should not have unreasonable restrictions put on them because of changes in nearby land uses since they were established;
- Identify and protect areas of tranquillity which have remained relatively undisturbed by noise and are prized for their recreational and amenity value for this reason.

The NPPF refers to the 2010 Noise Policy Statement for England (NPSE) for guidance regarding the term 'significant adverse impact'. The NPSE (Defra 2010) defines the following terms:

- 'No Observed Effect Level' (NOEL) – the level below which there is no detectable effect on health and quality of life due to the noise;
- 'Lowest Observed Adverse Effect Level' (LOAEL) – the level above which adverse effects on health and quality of life can be detected;
- 'Significant Observed Adverse Effect Level' (SOAEL) – the level above which significant adverse effects on health and quality of life occur.

The NPSE avoids prescribing specific fixed quantitative noise levels stating, 'It is not possible to have a single objective noise-based measure that defines SOAEL that is applicable to all sources of noise in all situations. Consequently, the SOAEL is likely to be different for different noise sources, for different receptors and at different times'.

Planning Practice Guidance (PPG) (DCLG 2014b) sets out additional guidance on how planning can manage potential noise impacts of new development, indicating that planning authorities should consider:

- whether or not a significant adverse effect is occurring or likely to occur;
- whether or not an adverse effect is occurring or likely to occur;
- whether or not a good standard of amenity can be achieved.

The PPG states that these potential effects should be evaluated by comparison with the SOAEL and LOAEL. To illustrate these thresholds and help identify where noise could be a concern, PPG provides a noise exposure hierarchy (see Table 9.7) which includes a fourth effect level not previously incorporated in the NPSE, namely:

- 'Unacceptable Observed Adverse Effect Level' (UOAEL) – the level above which extensive and regular changes in behaviour and/or an inability to mitigate the effect of noise lead to psychological stress or physical effects.

In line with the Noise Policy Statement for England, the PPG does not provide specific numerical values for the different effect levels, instead indicating that 'the subjective nature of noise means that there is not a simple relationship between noise levels and the impact on those affected. This will depend on how various factors combine in any particular situation'. Local authorities must therefore consider the noise exposure hierarchy and attempt to align it with significance criteria, having regard to guidance provided in various British Standards, the World Health Organisation and other relevant sources of information.

9.4 Scoping and baseline studies

The ESIA scoping stage identifies relevant potential sources of noise, the people and resources likely to be affected by noise from the proposed development (the receivers or 'receptors'), and monitoring locations for baseline data collection. The scoping stage should also set out the noise policy context and identify national/international guidelines and standards applicable to the assessment (see §9.3). The baseline studies involve identifying existing information on noise levels, carrying out additional noise measurements at appropriate locations where necessary, and considering future changes in the baseline conditions. These stages – which are interlinked and do not necessarily happen consecutively – are discussed below.

The project details should be analysed and **each potential source of noise impact identified**. Where possible, the characteristics of these new noise

Table 9.7 Noise exposure hierarchy (based on DCLG 2014b)

Perception	Examples of outcomes	Action
Not noticeable	No effect	No specific measures required
Noticeable and not intrusive	Noise can be heard, but does not cause any change in behaviour or attitude. Can slightly affect the acoustic character of the area but not such that there is a perceived change in the quality of life.	No specific measures required
Noticeable and intrusive	Noise can be heard and causes small changes in behaviour and/or attitude, e.g. turning up volume of television; speaking more loudly; where there is no alternative ventilation, having to close windows for some of the time because of the noise. Potential for some reported sleep disturbance. Affects the acoustic character of the area such that there is a perceived change in the quality of life.	Mitigate and reduce to a minimum
Noticeable and disruptive	The noise causes a material change in behaviour and/or attitude, e.g. avoiding certain activities during periods of intrusion; where there is no alternative ventilation, having to keep windows closed most of the time because of the noise. Potential for sleep disturbance resulting in difficulty in getting to sleep, premature awakening and difficulty in getting back to sleep. Quality of life diminished due to change in acoustic character of the area.	Avoid
Noticeable and very disruptive	Extensive and regular changes in behaviour and/or an inability to mitigate effect of noise leading to psychological stress or physiological effects, e.g. regular sleep deprivation/awakening; loss of appetite; significant, medically definable harm, e.g. auditory and non auditory	Prevent

sources should be identified, e.g. whether it is continuous or intermittent in nature, and the timing, duration, and tonal characteristics of the noise. Both on-site and off-site noise sources should be considered for the project construction, operation and (where appropriate) the decommissioning stage. During scoping, each source of impact should be considered and a judgement made with regard to the need for (a) carrying out further, detailed assessment; (b) carrying out further but less detailed assessment; or (c) discarding the source of impact from the main ESIA stage on the grounds that any resultant effects are highly unlikely to be significant. The reasoning for the ranking of sources of impact should be made explicit and transparent. This process enables the ESIA to concentrated on assessing noise from the sources of impact most likely to give rise to significant effects.

Ultimately, the effects of noise are dictated by the characteristics of the **potentially affected receptors**. Various maps can help to identify noise receptors in the area, but this should be confirmed by a site survey. The people affected by a development are not only local residents but also people working nearby, users of public places such as parks and footpaths, and users of other outdoor areas such as private playing fields and fishing lakes. ESIAs should identify any potentially particularly noise-sensitive receivers such as schools, hospitals, libraries, and places of worship.

Consultation with the decision-making body/relevant environmental agency can help to determine the spatial and temporal scope of the assessment, including the identification of particular local concerns and **sites for baseline monitoring**. Monitoring is normally targeted at the most sensitive receptors, concentrating on locations that are likely to experience adverse effects. However, where there are many receivers, for instance along linear developments such as roads or railway lines, representative receivers will need to be identified. A systematic approach is required, splitting potentially affected receivers into residential; non-residential and noise-sensitive; and non-residential and not noise-sensitive. Often the latter (e.g. factories and other industrial premises) can be scoped out of the assessment. Noise-sensitive non-residential receivers may need a further degree of sub-classification (a major broadcast studio may be potentially more sensitive than a shopping centre for instance).

Because noise is primarily a local impact, typically only **limited existing information** can be obtained from desktop studies, and virtually all ESIAs rely on noise measurements carried out at the site. Information about the wider area may be gleaned from the strategic noise maps, e.g. in Europe those prepared in line with Directive 2002/49/EC on the Assessment and Management of Environmental Noise, or the CPRE (2016) 'tranquillity maps' in England. However, these maps are largely intended to provide an indication of the relative degree of noise/tranquillity at different locations, rather than a definitive and precise assessment of noise levels, and their limitations and assumptions need to be recognised (IEMA 2014). Local government environmental audits may include noise data, but are unlikely to be site-specific.

Measurement of the ambient baseline noise environment is normally achieved by carrying out measurements at the potentially most affected noise-sensitive receptors. Different receptor locations may be important for different phases of a development. Every effort should be made to carry out measurements at times that are most relevant to the phase of the development when the new source will occur. For instance, this should reflect the day of the week and time when the proposed development will be in operation, or when sensitive receptors such as places of worship or schools are being used. In some cases, noise levels measured at the time of the main assessment may not represent the relevant time for the baseline noise levels required, e.g. in circumstances where there is a long lead time before a development becomes operational. In this case it may be necessary to calculate the future baseline levels, e.g. for road schemes in the UK, baseline noise levels are determined for the year of opening and for the 'design year', 15 years later.

Noise measurements should represent typical ambient conditions: normal prevailing winds, wind speed < 5 m/s, dry conditions (unless wet conditions are more normal for the area), dry roads and during normal weekdays and weekends as appropriate. Different receptors may also be most sensitive at different times of the day and days of the week. The noise survey may also record the quietest conditions which typically occur in an area (e.g. on a quiet Sunday morning). This is because the biggest increase in noise caused by a proposed development will be in comparison with these quiet conditions.

If, under particular conditions (e.g. a specific wind direction), higher background levels commonly occur, these are also recorded. For some projects (wind farms for instance) it may be appropriate to carry out assessments for a range of climatic conditions; care should be taken, however, to exclude the effects of atypical climatic conditions, such as temperature inversions. IEMA (2014) provide a summary of research studies that demonstrate the degree of variability that can occur in noise monitoring data.

Sound measuring equipment is portable and battery-powered, and usually consists of (a) a microphone which converts changes in ambient pressure into an electrical quantity (usually voltage), protected by a windshield, (b) a sound level meter which amplifies the voltage signals, averages them, and converts them to dB, (c) an analyser which records noise descriptors (e.g. LA_{eq}, LA_{10}, LA_{90}, LA_{max}) over a period of time, and (d) a reference sound source against which to calibrate the equipment. Several of these will normally be incorporated into the same piece of machinery.

The **sound level meter** will have different types of settings, corresponding to different ways of averaging voltage over time. Slow or 'S' weighting uses an averaging time of approximately one second and is used in circumstances where the noise level is relatively constant, or varies slowly. Fast or 'F' weighting uses an averaging time of approximately 0.125 second, appropriate where noise levels fluctuate. Impulse or 'I' weighting normally requires specialist equipment that responds rapidly to sudden noise increase, e.g. controlled explosions/blasting in mineral extraction and quarrying projects.

The precise procedures for measuring sound, for instance the length of time of measurement, location of equipment, and measurement indicator levels and averaging periods are sometimes specified in the relevant regulations or guidelines (see §9.3). It is generally advisable to agree the noise monitoring regime with the relevant environmental agency, which will have a good understanding of local conditions and any particular 'hot spots'. A typical survey strategy may include a limited number of positions where long-term (24 hours or more) unattended measurement are carried out using automatic data loggers, plus several positions where shorter term (15 minutes or more) attended sample measurements are taken. Where additional information on the frequency characteristics of the baseline noise environment is required, a sound level meter with a frequency analyser will be required. An acoustic windshield should always be used for environmental noise measurements.

Broadly, noise measurements involve:

- taking note of the equipment used, including manufacturer and type;
- taking note of the date, weather conditions, wind speed, and wind direction;
- calibrating the sound meter and microphone;
- setting up the microphone at the appropriate site (check relevant guidelines/legislation for details);
- noting the precise location where measurements are taken (e.g. on a map or using grid references/GPS coordinates);
- taking measurements using the criteria from the relevant guidelines (e.g. continuous for 24 hours, or for one hour; using fast weightings for traffic or slow for construction noise);
- noting start and finish times, identifying the principal influences on the noise environment (particularly the major influences on the LA_{eq}, LA_{90} and LA_{max}) during the measurement period, and any other factors (e.g. whether the equipment was attended or not) which could affect the measurements;
- checking the calibrations.

Table 9.8 gives an example of baseline noise data. Generally an ESIA includes such data, a description of how they were collected, and a map showing the location of the measurement points.

A final stage of scoping and baseline studies is to consider whether baseline noise levels are likely to change in the future in the absence of the proposed development. For instance, if a development is proposed near an industrial complex that is currently under construction, then the future baseline is likely to change. In some cases the future baseline may be established through calculations, particularly intensification of a route corridor where the level of noise from the existing traffic can be readily calculated.

Table 9.8 Example of baseline sound data

Date	Start of period	Sound levels, in dB(A)					Comments
		L_{90}	L_{50}	L_{10}	LA_{max}	L_{eq}	
1 April	15:00	56	57	60	62	58	Mostly traffic noise
	22:00	46	49	53	55	50	Traffic, dog barking
2 April	07:20	55	57	59	61	57	Traffic, birdsong

Notes: The most important things to be noted are generally:
• principal influence on LA_{eq}
• principal influence on LA_{90}
• whether the samples can be considered representative.

9.5 Impact prediction and evaluation

The aim of noise prediction in ESIA is to identify the changes in noise levels which may occur, both in the short and long term, as a result of the development; and to determine the significance of the change. Predicting noise levels is a complex process which takes into consideration a wide range of variables, including:

• existing and likely future baseline noise levels;
• the type of equipment, both mobile and fixed, used at the site;
• the duration of various stages of construction and operation;
• the time of day when the equipment is used;
• the actions of the site operator;
• the location of the receivers and their sensitivity to noise;
• the topography of the area, including the main forms of land-use and any natural sound barriers;
• meteorological conditions in the area.

These factors will affect the amount and type of sound coming from the site (e.g. type of equipment, duration of workings), how that sound travels (e.g. distance between source and receptor, topography, meteorology), and the response of the receptors (e.g. timing of workings, sensitivity to noise). Essentially, noise level prediction involves determining the sound power level at the source; predicting the sound level at each receptor site (which represents certain receivers) using corrections for factors such as distance, screening and ground attenuation (see §9.2.3); and adding the new sound levels to the ambient levels. Table 9.9 shows an illustrative example.

Information on the sound power of the source may be available from the manufacturers of equipment; alternatively the sound pressure level associated with the source for a specific reference distance can be used. For example, BS 5228-1:2009 provides a database of equivalent continuous noise levels (LA_{eq}

Table 9.9 Example of noise predictions

Receiver no.	Noise source	Distance (m)	Sound power level at source (dB(A))	Distance correction (dB(A))	Screening attenuation (dB(A))*	Soft ground attenuation (dB(A))	Predicted LA_{eq} at receiver (dB(A))	Ambient noise levels (LA_{eq})	Increase in noise (dB(A)L_{eq})
1		470	110	−64.4	−5	0	40.6	52.4	0.3
2	Loading operations	335	110	−61.5	0	−8.2	40.3	42.9	1.9
3		135	110	−53.6	0	0	56.4	60.1	1.5

Note: * Either screening/barrier or soft-ground attenuation is valid for a given site, not both.

Table 9.10 Examples of typical sound levels from construction equipment (BS5228)

Type of equipment	Sound level, in L_{eq} dB(A), at 10 m
Bulldozer (41 tonne)	80
Hand-held pneumatic breaker (roads)	95
Tracked concrete/rubble crusher	84
Dump truck (tipping fill)	79
Hydraulic hammer rig	89
Water pump	65
Articulated dump truck	81
Large concrete mixer	76
Tower crane	76
Circular bench saw (concrete blocks)	85
Diesel generator (lighting)	65
Dust suppression unit trailer	78

dB) generated by fixed and mobile construction plant at a reference distance of 10 m from the source (see Table 9.10) and this is often used in international assessments.

9.5.1 Calculating distance attenuation due to geometric dispersion: sound pressure level data

By using sound pressure level data measured at existing emission sources, or sourced from databases such as BS5228, approximate calculations of distance attenuation due to geometric dispersion can be determined as follows.

Point noise source

Most industrial sources and construction plant can be treated as a point source. Due to geometric dispersion, as the sound wave is spreading away from a point source, the 'inverse square law' applies, i.e. doubling the distance from the source produces a reduction in sound level of 6 dB, under free-field conditions. For a point noise source, the equation to determine the reduction in noise due to distance attenuation is:

$$L_{p2} = L_{p1} - 20\log(r2/r1)$$

Where:
L_{p1} = the sound pressure level measured for the noise source in dB at distance r1 (metres).
L_{p2} = the sound pressure level in dB for a receptor located at distance r2 from the source (metres).

For example, using the BS5228 data provided in Table 9.10, a 41-tonne bulldozer has a sound pressure level of 80 L_{eq} dB(A) at 10 m. To determine the distance attenuation and hence the noise level experienced at a receptor located 200 m away the calculation would be:

$$L_{p2} = 80 - 20\log(200/10)$$
$$L_{p2} = 54 \text{ dB(A)}$$

Line noise source

In the case of line sources of noise (e.g. road traffic, railway, a conveyor or a line of roof fans), sound dissipates in the form of a cylinder, rather than a hemisphere. If the distance from the line source doubles, the sound pressure halves resulting in a 3 dB reduction per doubling of distance. For a line noise source, the equation to determine the reduction in noise due to distance attenuation is:

$$L_{p2} = L_{p1} - 10\log(r2/r1)$$

Taking the example of a road, assuming that noise monitoring data reveals a sound pressure level of 75 dB $LA_{10, 18hour}$ at 5 m, to determine the distance attenuation and hence the noise level experienced at a receptor located 315 m away the calculation would be:

$$L_{p2} = 75 - 10\log(315/5)$$
$$L_{p2} = 57 \text{ dB LA}_{10, 18\text{hour}}$$

As a general rule of thumb, a line source must be at least three times as long as the distance between the source and receiver, otherwise it behaves as a point source (Environment Agency 2004).

9.5.2 Calculating distance attenuation due to geometric dispersion using sound power data

An alternative method for calculating the noise level at a receptor uses the sound power level of the source (as opposed to the sound pressure level at a specified distance, as used above in §9.5.1). Sound power level data is often supplied by manufacturers of plant and machinery and can be used to explore the implications of design and layout changes, or the substitution of alternative equipment as a mitigation measure (see §9.6). The following basic equation is used:

$$L_{p1} = L_w - 20\text{Log r} - 11$$

Where:
L_{p1} = the sound pressure level at a distance of r metres from the source
L_w = the sound power of the source

For example, a manufacturer may specify that a compressor turbine has a sound power L_w of 107 dB(A). To determine the distance attenuation and hence the noise level experienced at a receptor located 250 m away, the calculation would be:

$$L_{p1} = 107 - 20\text{Log } 250 - 11$$
$$= 48 \text{ dB(A)}$$

Environment Agency guidance (2004) suggests that if the ground between the source and the receptor is hard (e.g. concrete), the correction of 11 is replaced with a value of 8, since the sound that would have been absorbed is now reflected so that it only 'fills' half the volume, i.e. a hemisphere as opposed to a sphere.

9.5.3 Modelling multiple sources of noise

Where a development project has multiple sound sources that are close together, they will normally be considered together as one source by adding their levels (e.g. using the methods described in Table 9.3 or Box 9.2). Where multiple sound sources are not close together, each source's sound level at each receiver is calculated, and these sound levels are then added together (again using Table 9.3 or Box 9.2) for each receiver. Box 9.4 gives a very basic example to illustrate these principles.

Box 9.4 Noise predictions for dispersed multiple sound sources

Assume that a receiver will be affected by sound from three separate point sources (say industrial pumps):

- source A has a sound pressure level of 95 dB at 1 m, and is 64 m from the receiver;
- source B has a sound pressure level of 97 dB at 1 m, and is 128 m from the receiver;
- source C has a sound pressure level of 109 dB at 1 m, and is 256 m from the receiver.

Assuming geometric dispersion under free field conditions (i.e. doubling the distance from the source produces a reduction in sound level of 6 dB), the additional sound at the receiver will thus be 59 dB from source A, 55 dB from source B, and 61 dB from source C.

The total additional sound at the receiver will be 59 + 55 + 61 dB giving (from Box 9.2) a total of 63.7 dB.

Whilst the equations outlined above describe distance attenuation due to geometric dispersion, they do not take account of other important environmental factors that can affect predictions over distance (see §9.2.3). Detailed procedures for predicting sound levels from different types of development, and different stages of development (construction, operation, decommissioning) are specified in many of the guidelines and standards referred to in §9.3. In particular ISO 9613:1996 is a widely used method for predicting outdoor sound propagation in ESIA (particularly for industrial sources), and provides algorithms for the effects of atmospheric absorption, ground effect, reflections and screening, as well as geometric divergence.

Proprietary software packages (e.g. SoundPLAN and CadnaA) incorporate the algorithms contained in standards such as ISO 9613:1996 and are commonly used in ESIA, particularly in cases involving a large number of noise emission sources over an extensive area. These packages provide a systematic approach to the collection of input data, and have the potential to simulate different conditions (e.g. the influence of different atmospheric conditions upon noise levels at a receptor) and scenarios (e.g. the effect of introducing a noise barrier to mitigate noise). Such models also have the advantage of allowing the user to generate graphical outputs such as *isoline plots* that can be useful for communicating results with an ESIA report. Model outputs may also be used within Geographical Information Systems (GIS) including digital terrain models to facilitate further analysis of noise levels around a development project site, and 3D acoustic modelling packages are becoming increasing popular to analyse noise impacts at various heights, e.g. different floors in a residential tower block.

The Department of Trade and Industry (DTI 2007) provides an accessible guide to predictive modelling for environmental noise, including a summary of advanced techniques from research. More complex models that incorporate a wider range of variables inevitably require more input data that may not always be available for the project location. In practice, the principle of parsimony is often important, and a simplified prediction model based upon geometric attenuation and barrier effects may be fit for purpose.

9.5.4 Assessing significance

The effects of noise are primarily subjective and whilst the assessment process should strive to be as consistent and objective as possible, determining the **significance** of the effects will inevitably involve an element of professional judgement. Significance criteria are widely used in practice, both to ensure consistency of approach and to enhance the transparency of the assessment when communicating the results within an ESIA report (Wood 2008). The significance of noise effects generally depends on (a) the sensitivity of the receptor (Table 9.11) and (b) the magnitude of the predicted change in sound levels (the difference between existing and predicted future levels) experienced at the receivers. Typically, some form of impact-significance matrix is used to guide the evaluation (see Figure 1.3), including quantitative noise criteria linked to reputable guidance such as the WHO thresholds (1999).

When interpreting changes in noise levels, it should be remembered that due to the logarithmic nature of the decibel scale, a change of 3 dB is just detectable whereas a change of 10 dB corresponds subjectively to a doubling (or halving) of loudness. Also, an increase in noise in an area already subjected to high noise levels may be more significant than a similar increase in an area with lower baseline noise levels. The same level of noise at a noise-sensitive location will be more significant than that at a less sensitive location.

In addition to noise magnitude, other important factors to consider in judging significance include the nature of noise (continuous, intermittent, etc.); the frequency and duration of occurrence; the time of day the impact occurs; and the tonal characteristics. For example, a new industrial source could be tonal or impulsive, or a new specialist commercial source (say, perhaps, a cinema complex or a nightclub) may give rise to appreciable levels of low-frequency noise. In these instances, a description of the impact in terms of change or absolute levels of A-weighted sound pressure levels may not be an adequate indicator to allow potential effects to be assessed, and more detailed

Table 9.11 Examples of criteria for determining sensitivity

Sensitivity	Description	Examples of receptors
High	Receptors where occupants or activities are particularly susceptible to noise	Residential accommodation Private gardens Quiet outdoor areas used for recreation Conference facilities Auditoria/studios Schools in daytime Hospitals/residential care homes Religious institutions, e.g. churches or mosques Universities and research facilities Community facilities Public rights of way Designated areas and sites of historical importance
Medium	Receptors moderately sensitive to noise, where it may cause some distraction or disturbance	Offices Bars/cafes/restaurants Shops Temporary holiday accommodation Sports grounds where spectator noise is not a normal part of the event and where quiet conditions are necessary (e.g. golf or tennis)
Low	Receptors where distraction or disturbance from noise is minimal	Residences and other buildings not occupied during working hours Factories and working environments with existing high noise levels Sports grounds where spectator noise is a normal part of the event
Negligible	Receptors where distraction or disturbance from noise is very limited	Warehouses Light industry Car parks Agricultural land Night clubs

descriptions will be necessary. Care is also needed to make sure the appropriate averaging period is used for any noise indicator, as this can serve to mask effects (although this is often set within guidance and standards). For example, the use of an overly long averaging period can serve to mask high-magnitude noise that occurs infrequently, whilst a very short averaging period may not be sufficient to give a realistic and representative measure of prevailing conditions.

Box 9.5 provides a case study of an approach to significance assessment in an Ethiopian ESIA, and IEMA (2014) provides other examples used within the UK EIA process.

9.6 Mitigation

Mitigation will be necessary if the noise from the proposed development, or the change of noise levels caused by the development, are likely to exceed the levels recommended in the relevant standards (see §9.3). However, it may be

Box 9.5 Noise impact significance criteria: Yara Dallol Potash Project ESIA, Ethiopia (ERM 2015)

In this example the magnitude of predicted noise levels (for construction and operational phases) are compared to defined criteria to evaluate the impacts at all receptor locations taking into account the duration/frequency of the impact and the sensitivity of the receptors. The noise criteria are developed using a combination of the Ethiopian and the IFC EHS noise guidelines (IFC 2007) as follows:

- The 'daytime' period as defined in the Ethiopian noise standard is adopted (06:00 to 21:00 hours), as is the 'night-time' period (21:00 to 06:00 hours);
- Disturbance criteria are based on an LA_{eq}, 15 min assessment period (the time interval used with the Ethiopian standards);
- Amenity criteria (expressed as LA_{eq}, period) are determined by adding 3 dB to the existing baseline noise level or Assessment Background Level;
- Project specific noise criteria are determined by the most stringent of the IFC Disturbance and Amenity criteria; and the WHO Community Noise Guidelines (1999) values.

Initially, noise levels are predicted *without* mitigation. The assessment results are reported in Table 9.12 which follows. The meaning of the four impact significance ratings used is defined as follows:

- Negligible – no detectable effects, no need to consider in decision making, no mitigation required;
- Minor – the effect may be detectable, but small enough that noise management practices would ensure impacts are reduced to be Negligible;
- Moderate – a detectable effect, an impact that is significant, noise management practices and/or mitigation should be considered. Mitigation is likely to affect design and cost;
- Major – a detectable effect, an impact that is significant, noise management practices and mitigation must be considered. Mitigation will alter project design and cost. Impacts are undesirable if not addressed.

The ESIA indicated that impacts rated as 'Moderate' or above will be mitigated 'where practicable, feasible and reasonable with proportionately more emphasis as the rating increases. These criteria will provide the basis for developing performance standards and acoustic specifications for the proposed project'. The findings are subsequently used in the development of a noise management plan for the project.

Table 9.12 Project specific noise impact assessment criteria (based on ERM 2015)

Project phase and duration	Receptor type	Period[1]	Noise impact significance scale						
			Negligible	Minor		Moderate		Major	
			$PNL^2 <$	PNL >	PNL <	PNL >	PNL <	PNL <	PNL >
Operation long term/constant	Residential receptors with very low background noise levels (viz. Asabuya village)	Daytime	35	35	40	40	45	45	45
		Night-time	30	30	35	35	40	40	40
	Other residential and tourist (viz. Mount Dallol) receptors	Daytime	40	40	45	45	50	50	50
		Night-time	35	35	40	40	45	45	45
	Lower sensitivity residential receptors (viz. military camp)	Daytime	55	55	60	60	65	65	65
		Night-time	45	45	50	50	55	55	55
Construction medium term/often	Residential receptors with very low background noise levels (viz. Asabuya village)	Daytime	40	40	45	45	50	50	50
		Night-time	35	35	40	40	45	45	45
	Other residential and tourist (viz. Mount Dallol) receptors	Daytime	45	45	50	50	55	55	55
		Night-time	40	40	45	45	50	50	50
	Lower sensitivity residential receptors (viz. military camp)	Daytime	60	60	65	65	70	70	70
		Night-time	50	50	55	55	60	60	60

Notes:
1 Daytime = 6 am to 9 pm and Night-time = 9 pm to 6 am
2 PNL = predicted LA_{eq}, 15 min Project Noise Level

useful to implement noise mitigation measures even if standards are met, to prevent annoyance and complaints and as part of best-practice procedures. The best noise mitigation is that which is integrated into the project design: the siting of machinery and buildings, choice of equipment, and landscaping to reduce noise are all easiest, cheapest and most effective if they are designed-in rather than added towards the end of the process.

For a new project that has potential to produce significant noise impacts, mitigation of noise is best carried out at the source, i.e. before the noise has 'escaped'. Failing this, barriers and the siting of buildings can be used to separate noise sources from potentially affected noise-sensitive locations. As a last resort, noise can be controlled at the receiver's end through the provision of, say, secondary glazing or other noise insulation measures. In some cases, financial compensation for the loss of property value may be required.

Control of noise at the source can take a number of forms. First, the equipment used or the modes of operation can be changed to produce less noise. For instance, rotating or impacting machines can be based on anti-vibration mountings. Internal combustion engines must be fitted with silencers. Airplanes can be throttled back after a certain point at take-off, to reduce their noise. Traffic can be managed to produce a smooth flow instead of a noisier stop-and-start flow, and use of quieter road surfacing materials can significantly reduce tyre noise. Limiting the hours of operation and the duration of noisy activities can help to minimise the impact experienced outside the boundary of a proposal. Well-maintained equipment is generally quieter than poorly maintained equipment.

Second, the source can be sensitively located. It can be located (further) away from the receivers, so that noise is reduced over distance. A buffer zone of undeveloped land can be left between a noisy development and a residential area. The development can be designed so that its noisier components are shielded by quieter components; for instance, housing can be shielded from a factory's noise by less sensitive retail units. Effective use of natural ground contours or artificially constructed topography including earth bunds or landscaping can be used to screen the source.

The source can be enclosed to insulate or absorb the sound. Sound insulation reflects sound back inside an enclosure or barrier, so that sound outside the enclosure is reduced. However, merely enclosing the source is not the optimum solution, since the noise reverberates within the enclosure, and effectively increases the strength of the enclosed sound. Providing sound absorption within the enclosure avoids this happening. Sound absorption occurs where the enclosure or barrier absorbs the sound, converting it into heat. Most enclosures are constructed of both insulating and absorbing materials.

Details of requirements for noise enclosures and their effectiveness are very complex and require specialist knowledge. The reader is referred to the relevant standards and to textbooks on noise control (e.g. Crocker 2007 or Smith et al. 2011). However, some general points can be made here. Methods of measuring sound insulation usually distinguish between airborne sound (noise) and structural sound (vibration), and any reference to insulation should distinguish between them. Broadly, the ability of a panel to resist the transmission of energy from one side of a panel to the other, or its transmission loss, will depend on (a) the mass of the panel (more mass = more transmission loss), (b) whether it is layered or not, with or without discontinuities between the layers, (c) whether it includes sound-absorbing material, and (d) whether it has any holes or apertures.

Acoustic fencing or other screens, either at the source or at the receiver, can also reduce noise by up to 15 dB. The effectiveness of screens depends on (a) their height and width (larger is better), (b) their location with respect to the source or receiver (closer is better), (c) their form (wrapped around the source or receiver is better), (d) their transmission loss, (e) their position with respect to other reflecting surfaces, (f) the area's reflectivity, and (g) whether they have any holes or apertures.

Noise screens can consist of topographical features or tree plantings as well as of artificial materials. For instance, earth mounds (bunds) are often built alongside roads to absorb and reflect traffic noise away from nearby buildings. Thick areas (> 30 m) of very dense trees and underbrush may reduce noise by up to 3–4 dB at low frequencies and 10–12 dB at high frequencies. Although thinner tree belts have little actual effect on noise, the visual barrier they form can make people think that noise levels have been reduced. A mixture of deciduous and coniferous trees will give maximum noise reduction in the summer, and some reduction in the winter when the deciduous trees' leaves have fallen. It must be remembered that saplings take time to mature, and are unlikely to reduce noise for several years after planting.

Control of noise at the receiver's end is often similar to that at the source. Good site planning can minimise the impact of noise; for instance, in a house by a busy road the more noise-sensitive rooms (e.g. bedroom, living room) can be shielded from the road noise by the less noise-sensitive rooms (e.g. kitchen, bathroom). A screen can be erected to reflect sound away from the receiver, for instance an acoustical screen between a highway and house. The equivalent of a noise enclosure can be achieved by soundproofing a house using double-glazed windows.

9.7 Monitoring

Depending on a country's regulatory regime, conditions imposed as part of a project's permission to develop may be enforceable, including conditions related to noise. These can apply not only to noise levels (e.g. during construction, operation; during the day/night), but also to noise monitoring conducted by the developer (e.g. distance from the site boundary, frequency). If no planning conditions are set, local environmental regulatory bodies may still monitor noise from a site, for instance in response to local residents' complaints, to determine whether it is a statutory nuisance.

For major development projects, the ESIA may include a 'Noise Management Plan' as part of a wider Environmental and Social Management Plan for the scheme (Chapter 20). This sets out the actions required to manage the noise effects associated with the construction and operation of the development, including details on when each action should occur and who is responsible for its delivery.

Comparisons of noise monitoring data with the noise predictions made in ESIAs are rare (see Wood 1999) although some countries have mitigation and follow-up monitoring requirements. A best-practice ESIA could propose not only noise-related planning conditions, but also a noise monitoring programme, and relate its findings to the ESIA to improve future noise prediction methodologies. The sites and noise measurement techniques used in carrying out baseline noise surveys should be such that comparable monitoring data can later be collected.

9.8 Conclusion

Noise is of increasing concern internationally, and the assessment of noise from proposed developments is important in order to ensure that health and quality of life are not compromised. The assessment of potential impacts can be technically complex, involving computer models and quantitative metrics to characterise existing and predicted noise levels. Ultimately, noise is experienced by individuals as a subjective response, and so the basis for judgements of impact significance must be robust and transparent so that any potential impacts are clearly communicated to readers of the ESIA report.

Acknowledgements

The authors are grateful to Mike Breslin for his work on a previous edition of this chapter.

References

Basner M, W Babisch, A Davis, M Brink, C Clark, S Janssen and S Stansfeld 2014. Auditory and non-auditory effects of noise on health. *The Lancet*, **383** (9925), 1325–1332.

BSI (British Standards Institute) 2009. *Code of practice for noise and vibration control on construction and open sites. Part 2: vibration*. BS5228-2. London: BSI.

CPRE (Campaign to Protect Rural England) 2016. *Tranquillity*. http://maps.cpre.org.uk/tranquillity_map.html.

Crocker MJ, ed. 2007. *Handbook of noise and vibration control*. New Jersey: John Wiley & Sons.

DCLG (Department for Communities and Local Government) 2012. *National Planning Policy Framework*. London: DCLG.

DCLG 2014a. *Assessing environmental impacts from minerals extraction – noise emissions.* http://planningguidance.communities.gov.uk/blog/guidance/minerals/assessing-environmental-impacts-from-minerals-extraction/noise-emissions.

DCLG 2014b. *Planning practice guidance: Noise.* http://planningguidance. communities.gov.uk/blog/guidance/noise.

Defra (Department for Environment, Food and Rural Affairs) 2010. *Noise policy statement for England (NPSE).* London: Defra.

DTI (Department of Trade and Industry) 2007. *Guide to predictive modelling for environmental noise assessment.* www.npl.co.uk/upload/pdf/guide-to-predictive-modelling-env-noise-assessment.pdf.

EC (European Commission) 2002. Directive 2002/49/EC of the European Parliament and the Council of 25 June 2002 relating to the assessment and management of environmental noise. *Official Journal of the European Communities* L189. http://eur-lex.europa.eu/legal-content/EN/TXT/?uri=celex%3A32002L0049.

Environment Agency 2004. *Horizontal guidance for noise part 2 – noise assessment and control.* Bristol: Environment Agency.

ERM 2015. *Yara Dallol BV Potash Project Environmental and Social Impact Assessment.* www.erm.com/en/public-information-sites/yara-dallol-bv-potash-project.

Farcas A, PM Thompson and ND Merchant 2016. Underwater noise modelling for environmental impact assessment. *Environmental Impact Assessment Review* **57**, 114–122.

Hammer MS, TK Swinburn and RL Neitzel. 2014. Environmental noise pollution in the United States: developing an effective public health response. *Environmental Health Perspectives* **122**, 115–119.

IEMA (Institute of Environmental Management and Assessment) 2014. *Guidelines for environmental noise impact assessment.* Lincoln: IEMA.

IFC (International Finance Corporation) 2007. Environmental, health, and safety (EHS) guideline 1.7 – Noise. www.ifc.org/wps/wcm/connect/06e3b50048865838 b4c6f66a6515bb18/1-7%2BNoise.pdf?MOD=AJPERES.

ISO (International Organization for Standardization) 1996. *ISO 9613-2. Acoustics – Attenuation of sound during propagation outdoors – Part 2: General method of calculation.* Geneva: ISO.

OECD (Organisation for Economic Co-operation and Development) 1996. *Pollution prevention and control environmental criteria for sustainable transport.* Paris: OECD.

Smith BJ, RJ Peters and S Owen 2011. *Acoustics and noise control.* Harlow: Pearson Education Limited.

WHO (World Health Organisation) 1999. *Guidelines for community noise.* Geneva: WHO. www.who.int/docstore/peh/noise/guidelines2.html.

WHO 2009. *Night noise guidelines for Europe.* Copenhagen: WHO Regional Office for Europe.

WHO 2011. *Burden of disease from environmental noise. Quantification of healthy life years lost in Europe.* Copenhagen: WHO Regional Office for Europe.

Williams R, AJ Wright, E Ashe, LK Blight, R Bruintjes, R Canessa, CW Clark, S Cullis-Suzuki, DT Dakin, C Erbe and PS Hammond 2015. Impacts of anthropogenic noise on marine life: publication patterns, new discoveries, and future directions in research and management. *Ocean and Coastal Management* **115**, 17–24.

Wood G 1999. Assessing techniques of assessment: post-development auditing of noise predictive schemas in environmental impact assessment. *Impact Assessment and Project Appraisal* **17**, 217–226.

Wood G 2008. Thresholds and criteria for evaluating and communicating impact significance in environmental statements: 'See no evil, hear no evil, speak no evil?' *Environmental Impact Assessment Review* **28**, 22–38.

10 Transport

•••

Chris Ferrary and Polash Banerjee (based on Fry and Therivel 2009)

10.1 Introduction

Transport is a key factor in the design, approval and success of new development. Development requires good access for residents, employees and customers. It also needs good servicing arrangements. How development relates to the surrounding transport network is critical to the delivery of sustainable planning policies. The interaction of development and the transport network is an important consideration in determining planning consents, as are the environmental effects associated with travel to and from the development proposed. The nature and location of the development, and any proposed transport provision, will determine the nature of trips to and from the site, as well as the potential for achieving a modal shift through increased walking, cycling and use of public transport.

Some ESIAs will be for a transport infrastructure proposal (e.g. road, airport, rail link) as such developments require assessment of their environmental impact as part of consenting processes in most jurisdictions. However, almost all ESIAs (e.g. for non-transport development such as urban development, power stations, afforestation) will include consideration of the environmental impacts of transport due to movement of people and goods generated during the construction, use and decommissioning of that development. Both proposals for new transport infrastructure and the travel generated by other development will cause environmental impacts. These include noise and vibration, air pollution, impacts on biodiversity, community severance, visual intrusion, and accidents. Ensuring good access is also key to economic regeneration.

In most jurisdictions, ESIA legislation does not require the specific assessment of transport impacts. However, it is often a requirement of planning or consenting authorities, and is also necessary to assess a project's environmental impacts properly, particularly in developing countries. The extent to which these issues are considered and analyzed as part of consent processes depends on local regulatory requirements and practice. The resource available to devote to planning and development of projects is also a key consideration. Depending on this context, many methods used by transport planners will be of particular interest to ESIA practitioners, including the prediction and evaluation of demand, and flows and effects of various transport modes (i.e. aviation, highways, rail and road-based public transport, pedestrians, cyclists etc.), as this will often provide source data for ESIA where resources permit.

An important resource often used is the World Bank technical paper *Roads and the Environment: A Handbook* (Tsunokawa and Hoban 1997). This provides practical methods to design and carry out effective ESIA for projects ranging from major new highways to minor maintenance works on existing roads. These techniques are applicable both to in-depth ESIA or modest environmental action plans for small projects. The first part of the handbook provides an overview of the ESIA process and recommends the detailed methodological steps to be followed in carrying out ESIA for highways. The second part provides more detailed information on each of the major categories of environmental effects arising from building and using road projects, describing possible impacts, the nature and scale of the impacts, and common mitigation options. Each chapter provides a checklist of common ways of minimizing the impacts on the specific environmental resource considered, together with a list of information sources, which users can refer to for more details.

10.2 Definitions and concepts

Transport demand is typically modelled using what has been termed the "classic four stage model" (Willumsen 2014). This approach, developed in the 1960s, is illustrated in Figure 10.1 and involves:

- *Stage 1*: Identifying a zoning and network system representing in abstract the transport network and the land uses it serves.
- *Stage 2*: Collecting data on population, land use and traffic flows to populate the model and using this to estimate the total number of trips generated and attracted by each zone identified in the model (this is sometimes termed a "gravity model").
- *Stage 3*: Representing the distribution of origins and destinations of trips in space in a *trip matrix*, and allocating them to specific modes (e.g. cars, buses, lorries, trains, cycles, taxis, etc.).
- *Stage 4*: Assigning trips by mode to the corresponding transport networks in the model, and producing predicted flows for each link in the abstracted network, together with travel times and generalized costs.

The outputs of such models provide estimates of traffic and/or passenger flows on the links in the modelled network, which represent roads or public transport routes in the real world. These estimates in turn provide inputs to the prediction, estimation or modelling of environmental effects that are correlated to these flows, such as noise and air pollution. The application of these is discussed later in this chapter.

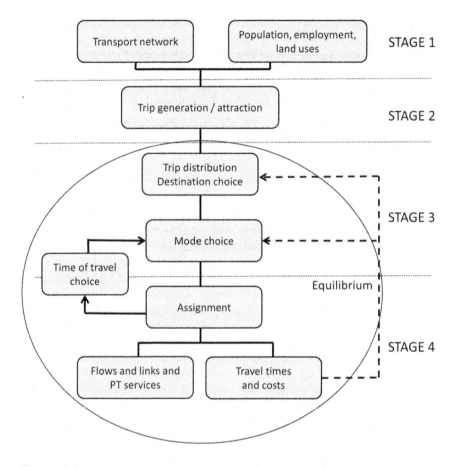

Figure 10.1

Classic transport model (after Willumsen 2014)

Until relatively recently, the use of demand models in transport planning led to the dominant approach being characterized as "predict and provide" (Owens 1995). This essentially meant that the planning of transport focused on the prediction of unrestrained demand for transport, typically based on past trends, and the provision of capacity in the form of infrastructure to provide for this, regardless of the unintended consequences of worsening congestion and environmental damage. This focus on **mobility** – ease of movement – led to a strong emphasis on the provision of road (and to a lesser extent other forms of transport) infrastructure. This can be characterized by features termed the "traditional" approach to transport planning (Grant 1977), that formed the backbone of the training and education of highway engineers and transport planners until the mid-1970s, and remains a key influence today. Essentially, this approach views transport problems as resulting from congestion, caused by growth in travel that is seen as a natural consequence of economic growth. The only restraint on this growth is viewed as the inadequacy of transport networks in coping with it. The role of transport planning, therefore, was seen as ensuring that networks can cope with the amount of travel, or at the very least that delays be kept to a minimum. This, broadly speaking, was the prevalent view of transport professionals until concerns about environmental issues during the mid-1960s and the large increase in energy prices experienced during the 1970s began to bring many professionals to develop new ways of thinking about transport.

In the UK, this culminated in the publication of an influential government-sponsored report of 1994 that concluded that new infrastructure can generate new traffic as well as catering to existing needs, especially where high levels of congestion are preventing people from making as many journeys as they would ideally wish to (SACTRA 1994). In other words, it concluded that we couldn't build our way out of congestion. Similar changes in thinking could be observed in other European countries at the same time. For example, in 1988 the Dutch Government published its Second Transport Structure Plan that set out the policy requirements to achieve a compromise between mobility, accessibility and environmental protection (Haq and Bolhuis 1998). Similarly, in Germany transport policy shifted in a coordinated way to encompass taxes and restrictions on car use, to help limit car use and mitigate its harmful impacts; provide high-quality, cheap, coordinated public transport services; improve infrastructure for non-motorized travel to increase the safety and convenience of walking and cycling; and production of urban development policies and land use planning to encourage compact, mixed-use development (Buehler and Pucher 2009).

This resulted in a shift of emphasis in transport policy aiming to improve **accessibility** – the "ease of reaching". This can include ease of access to the transport system, ease of access to facilities, ease of participation in activities and delivery of goods, and the possibility of having an alternative way of accessing services even if this is used only rarely (like when the car breaks down). Accessibility can be improved by, for instance, careful siting of housing vis-à-vis employment, and other land uses to reduce the need to travel, promotion of walking and cycling, use of information technology to substitute for physical journeys, and efficient use of the existing transport infrastructure. Since the 1990s, highways-building programmes have been dramatically cut in many developed countries. That said, in practice most transport assessments still focus heavily on road infrastructure, although this emphasis is slowly changing. This has in no small part been due to the pressures on public sector budgets following the global economic crisis of 2008, with capital investment being increasingly focused on larger infrastructure projects and a consequent shift of transport policy towards measures that do not involve capital expenditure at the local level.

Nevertheless, governments of countries that are enjoying rapid economic growth, such as the so called "BRIC" grouping (Brazil, Russia, India and

China), typically want to promote consumer spending by providing affordable small cars in the market and encouraging easy loans to buy them. This has led to a rapid rise in private vehicles in these countries, where urban road networks have often not been sufficiently equipped to accommodate this surge in traffic. Consequently, many cities in these countries suffer frequent traffic congestion, rampant traffic-rule violations and deteriorating air quality.

Several **modes of transport** are associated with new developments: vehicular traffic, heavy and light rail, cycling and walking. Vehicular traffic can be further subdivided into private cars and taxis, vans, goods vehicles, buses and motorcycles. The different modes have different impacts, and many governments in developed countries aim to promote a **modal shift** from private vehicles to walking, cycling and public transport.

Each existing road link, road junction, or link on a public transport network has a finite **capacity**: a maximum number of vehicles or passengers that it can nominally accommodate. As traffic levels approach and then exceed capacity, congestion, queuing and overcrowding occur. Public transport services may have licences that do not permit them to exceed their capacity.

Different types of trips (see Figure 10.2) will have different impacts. **New trips** are those that did not occur anywhere else on the transport network prior to the development. **Pass-by trips** are made as part of another journey, such as stopping off on the way home from work. **Diverted trips** are similar to pass-by

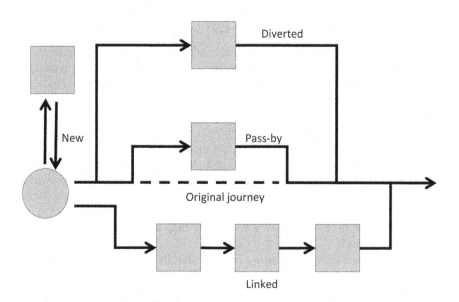

Figure 10.2

Types of journey

trips, but involve a longer diversion from the existing route to the site. **Linked trips** are trips with multiple destinations. **Transferred trips** are trips that are already being made and that would be transferred to the proposed development (UK DCLG/DfT 2007). For example, a new housing development that generates 1,000 new trips per day will have very different transport impacts compared to a new shopping centre that also generates 1,000 trips per day but where most of these trips may be transferred or pass-by.

The most significant traffic problems on a given route are likely to occur at times of **peak traffic flow**. These are typically weekday morning and evenings but could be different, for instance near major distribution centres or airports, or on weekends in tourist areas. It is at times of peak flow that the worst congestion occurs on roads, public transport capacity is most likely breached, and any unusual event (e.g. an accident) has the greatest repercussions. One way of making the best use of existing transport infrastructure is to support projects whose peak traffic flow does not coincide with peak flows on the surrounding transport network.

Finally, new transport developments can cause **community severance**. This describes the phenomenon where introduction of a new road or rail line, faster traffic, access controls, and physical barriers cuts existing lines of travel or communication. They can essentially cut a community in two.

10.3 Key guidance

10.3.1 Policy and guidance on EIA for transport projects

Many international institutions provide guidance on the development of transport policy and the delivery of specific transport interventions aimed at promoting sustainable development, as well as how to examine the environmental effects of these.

The World Bank guide (Tsunokawa and Hoban 1997), for example, introduces the environmental assessment process as a planning tool, emphasizing the importance of understanding the impacts of roads in different land-use settings, such as urban or rural, and new-build or regeneration projects. The guide identifies five types of Environmental Impact Assessments (EIAs) applied to road projects as follows:

- Project-specific EIA, assessing individual new roads (or public transport infrastructure);
- Programmatic EIA, examining a number of individual projects that may be delivered together as a package or programme;
- Summary environmental evaluation, focusing on specific categories of impact and their mitigation;
- Regional EIA, examining all the planned projects within a specified geographical area at the sub-national level;

- Sectoral EIA, strategically assessing all projects of a certain type, such as transport infrastructure in a relatively large geographical area.

The guide makes clear that without proper administrative support, adequate technical capacity or regulatory and monitoring functions, an EIA will be merely a paper exercise. It advocates early and careful planning, acknowledging that, although time-consuming, this will ensure that work is well-thought out and optimized. The guide also provides details for project-specific EIA. It notes that transport agencies should have clearly designated staff with overall responsibility for environmental matters and knowledge of environmental laws and regulations, with access to senior management and their support in coordinating environmental actions throughout the organization. Identification and assessment of environmental impacts should be integral to the project cycle and commence early to enable the full consideration of alternatives and to avoid later delays and complications. Assessments should be followed up with action plans and monitoring and remedial measures to ensure the effectiveness of environmental recommendations and decisions. Community involvement is also identified as an essential element of environmental management of roads.

Similarly, the European Investment Bank (EIB 2013) finances investment in more sustainable forms of transport (e.g. rail, water, and public passenger transport), including intermodal transport systems; promotion of teleworking by accelerating the next generation of information and communications technology; balanced regional development; and support for integrated urban regeneration and development. More generally, the Bank recognizes the global and regional dimension of environmental issues, such as climate change, and the need to reduce the ecological impact of Europe on the rest of the world by promoting improved resource management and reduced pollution. Through international cooperation, the Bank supports a number of regional environmental programmes, particularly in the Mediterranean Basin, the Baltic Sea and the Balkans.

Different countries have national guidance, for instance the Indian Government's *Guidelines for Environmental Impact Assessment for Highway Projects* and *Environmental Impact Assessment Guidance Manual for Highways* (ASCI 2010); Botswana's *Planning and Environmental Impact Assessment of Road Infrastructure* (Republic of Botswana 2001), and the UK's *WebTAG* guidance (UKDfT 2014) and *Design Manual for Roads and Bridges Vol. 11* (Highways England 2015).

The European Union has published guidance on how to carry out EIA for large-scale transboundary projects, including long-distance rail lines and motorways (EU 2013). It focuses on how to carry out EIA consultation in several European countries, and how to consider alternatives that span several countries, etc. Its principles are likely to be useful in other cases of transboundary projects and ESIAs.

10.3.2 Policy to reduce transport impacts and improve accessibility

In relation to major developments that will generate significant levels of new travel and traffic, it has become good practice worldwide for a Transport Statement or Transport Assessment to be prepared. This enables decision-makers to take account of whether:

- Opportunities for sustainable transport modes have been taken up to reduce the need for new major transport infrastructure;
- Safe and suitable access to the site can be achieved for all people;
- Improvements can be undertaken within the transport network that cost-effectively limit the significant impacts of the development.

In most jurisdictions, development plans and the basis for individual planning decisions are increasingly aiming to ensure developments that generate significant movements are located where the need to travel will be minimized and the use of sustainable transport modes can be maximized, although this needs to take account of other planning policies – particularly in rural areas. Plans are now more frequently required to protect and exploit opportunities for the use of sustainable transport modes for the movement of goods or people. This typically would be done by locating and designing developments to:

- Accommodate the efficient delivery of goods and supplies;
- Give priority to pedestrian and cyclist movement, and have access to high-quality public transport facilities;
- Create safe and secure layouts which minimize conflicts between traffic and cyclists or pedestrians, avoid street clutter and where appropriate establish "home zones";[1]
- Incorporate facilities for charging plug-in and other ultra-low emission vehicles;
- Consider the needs of people with disabilities in all modes of transport.

The usual key tool adopted to facilitate this is a travel plan. For larger-scale residential developments in particular, planning policies are often required to promote a mix of land uses in order to provide opportunities to undertake day-to-day activities, including work on-site. Where practical, particularly within large-scale developments, key facilities such as primary schools and local shops should be located within walking distance of most properties.

10.4 Scoping and baseline studies

Typically, the transport issues relating to new development and relevant to an ESIA will include:

- The planning context of the development proposal;
- Highway trip generation and trip distribution;
- Public transport capacity, walking/cycling capacity and road network capacity;
- Measures to promote sustainable travel;
- Assumptions about the development proposal;
- Safety implications of development;
- Mitigation measures (where applicable), including scope and implementation strategy.

10.4.1 Transport baseline

Although Transport Assessments or Statements are often specifically prepared for planning applications, it is unlikely that these will fulfil the specific role required of the transport element of an ESIA. This is particularly the case where environmentally sensitive areas are close to the proposal and/or where the development could have implications for breach of statutory thresholds in relation to noise and air quality because of traffic generated by the development or the impact of existing traffic on the site under consideration.

The scope and level of detail required for the transport baseline in ESIA will vary from site to site but the following should be considered:

- Site layout, transport access and layout across all modes of transport;
- Neighbouring uses, amenity and character, existing functional classification of the nearby road network;
- Existing public transport provision, including provision/frequency of services and proposed public transport changes;
- Data on current traffic flows on links and at junctions (including by different modes of transport and the volume and type of vehicles) within the study area and identification of critical links and junctions on the highways network;
- Analysis of accident records on local roads for the most recent three- or five-year period;
- Qualitative and quantified description of the travel characteristics of the proposed development, including movements across all modes of transport that would result from the development and in the vicinity of the site;
- Parking facilities in the area and the parking strategy of the development.

It is possible to describe the baseline stream of traffic on a length of road at a particular time with reference to:

- Highway width, structural condition and link capacity;
- Junction capacity (which is often more restricted than highway link capacity);
- Driver delay/queuing time at junctions;

- Average speed of travel;
- Turning movements;
- Number of accidents (slight, serious, fatal) or the rate of accidents per vehicle/km;
- Proportion of heavy goods vehicles;
- Number of bus movements;
- Pedestrian and cycle flows;
- Location and type of on-street car parking;
- The nature of frontage land uses (Hughes 1994).

On rail and tramway networks, the pertinent factors are:

- Line capacity (single or dual);
- Station capacity (stairwells, platform width etc.);
- Platform length;
- Rolling stock passenger capacity;
- Frequency of service and station wait time;
- Time delays at any railway crossings;
- Junction capacity and signalling;
- Layover capacity;
- Proportion of freight trains;
- Proportion of stopping and non-stopping services;
- Speed.

Other public transport, such as buses and paratransit,[2] can be characterized by similar factors to rail and tramway systems, particularly where these are organized as transit systems (e.g. the TransMilenio system in Bogota), although typically there is less emphasis on infrastructure.

Active travel modes (e.g. walking and cycling) are also increasingly important to characterize. Increasing, or in developing countries maintaining, the number of journeys made on foot and by bicycle helps achieve many local and national policy outcomes. These include improved health and well-being; creating better places for people; reducing carbon emissions; improving local air quality; and avoiding or reducing congestion (Sustrans 2014). Considerations for the development of cycling and walking facilities include:

- Pavement/sidewalk width;
- Network standards;
- Destinations;
- Incorporation of existing routes;
- Journey lengths;
- User groups.

If the project is a traffic-generating development like an office or retail park, information on any existing person/vehicle trips from the existing use of the site is also collected, so a net increase or decrease in trips within agreed assessment periods (typically weekday AM/PM and weekend peak-hour periods) can be identified. In countries where slums, hawkers, eating areas, etc. encroach on roads and footpaths, this would also need to be noted.

This data is collected for the existing situation with normal traffic flow and usage conditions (often termed the *base case*). This data should include recent traffic-counts, broken down by transport mode, including pedestrians and cyclists on given road lengths; turning-counts at junctions; queue-length surveys at signal junctions; and journey-time surveys. Counts should be taken in non-school-holiday periods and during typical weather conditions. However, implications for any regular peak traffic and busy periods (such as rush hours) should be considered, especially in urban areas. Seasonal variations should also be considered.

The future baseline without the development ("do nothing") is then determined, using local traffic-forecasting models. Further scenarios of planned interventions (such as new infrastructure or different development scenarios) can then be compared with do-nothing scenario to provide an analysis of the planned intervention. Broadly calculating the do-nothing scenario involves changing (usually increasing) the baseline levels to reflect likely future activity on the transport network. These projections are usually based on forecasts of planned increases in population and/or changes in the distribution of land uses. This change can be significant, say a 1–2 per cent increase per year. However, where the availability of resources and/or expertise dictates, this may be done manually, applying professional judgements.

Appropriate consideration is also needed of cumulative impacts arising from other committed development (i.e. development that is consented or allocated where there is a reasonable degree of certainty that it will proceed within the next, say, three years). This may require the developer to carry out an assessment of the impact of those adopted local development plan allocations and proposals that have the potential to impact on the same sections of transport network as well as other relevant local sites benefitting from as-yet unimplemented planning approval. Table 10.1 shows a simple example of how some of this data may be presented.

10.4.2 Environmental baseline

Changes to the transport system may have follow-on impacts on air quality and greenhouse gases; community and economic activity; the cultural heritage; ecology and biodiversity; health and safety; indigenous people; land acquisition and resettlement; land use, landscape and townscape; noise and vibration; soils and geology; and water resources. These will be direct impacts in the case of new transport projects, and indirect impacts where a non-transport development leads to changes to the transport system. The baselines for these impacts will need to be described. Other chapters in this book explain how this can be done.

Table 10.1 Example of baseline traffic flows on a road network (Hughes 1994)

	Ratio of flow/capacity (RFC)	Maximum queue (vehicles)	Maximum delay (min./vehicle)
Junction A	0.81	5	0.02
Junction B	0.97	20	0.50

	Peak hour 2-way flow (vehicles)	Theoretical link capacity (vehicles/hour)	Reserve capacity
Link A	2000	1700	−17.6%
Link B	2800	3200	+12.5%
Results	Junction A: RFC not close to 1.0. OK.		
	Junction B: RFC close to 1.0. Problems.		
	Link A: Over theoretical capacity. Problems.		
	Link B: Within capacity. OK.		

10.5 Impact prediction and evaluation

10.5.1 Transport impacts

Assessing transport impacts usually requires an iterative approach. This helps ensure that improvements to sustainable modes of transport are considered first. Measures to increase road or junction capacity should only be considered after this, to mitigate residual traffic impacts of proposed development. For most larger transport proposals, or to evaluate the effects of other large developments at a strategic level, one of the main internationally available software packages (e.g. SATURN, CUBE, EMME, OmniTRANS, TransCAD or VISUM) will be used. However, these are typically quite coarse-grained in terms of outputs, and therefore often do not provide good estimates of the performance of individual junctions or of delays to individual journeys due to congestion. At this level, modelling of individual junctions will often be carried out, again using internationally recognized software (such as NCAP or PICADY). Where available resources or expertise do not permit the use of computer models, manual predictions and assignment using professional judgement may be required. Figure 10.1 summarizes the key steps in transport impact assessment, which are now discussed further.

The analysis period – the future years for which impact predictions are carried out – should reflect the trip generation characteristics of the proposed development and the conditions on the transport system. It should include the construction and operation stages (normally the year of opening and at least five years after the development is built), and where appropriate the decommissioning stage of the development.

The key input to transport models is the initial impact of development in "person trips". These are then broken down by transport mode. For transport infrastructure projects, this will be done on the basis of the known transport demand predicted from planned development. For other development types, this is typically done by reference to empirical studies of similar developments elsewhere. For example, in the UK and Ireland this is normally based on the TRICS® database.[3] This enables the calculation of person/vehicle trips rates for different forms of development. Alternatively, survey data from other development sites can be used.

After this, the distribution of trips is predicted using gravity models or by reference to existing travel patterns or other empirical data. The additional trips are then added to the do-nothing scenario, to determine the total "with development" traffic levels predicted on road links and at junctions. Figure 10.3 shows an example.

Determining whether new transport infrastructure or new development will have material or significant impact on the transport network is complex and highly dependent on circumstances. In already congested urban areas, the percentage of extra traffic or passengers that can be considered significant or detrimental to the network may be relatively low (possibly below the average daily variation in flow). Elsewhere, quite large increases may be permissible in transport terms, although these may lead to significant air quality or noise impacts. Previously, some guidance has set out assessment thresholds in terms of simple percentage increases (e.g. 10 per cent or 5 per cent levels of development traffic relative to background traffic). However, this is now often no longer deemed an acceptable mechanism, since it creates an incentive in favour of locating development where high levels of background traffic already exist. Essentially, this means that local transport or traffic authorities need to

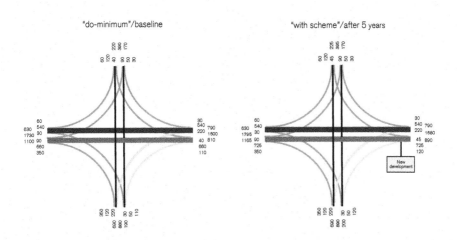

Figure 10.3

Example of baseline and with scheme junction diagrams

determine what constitutes a significant effect in their area, almost on a case-by-case basis, depending on the extent and magnitude of additional delays and/or congestion that results.

To demonstrate the potential complexity, Table 10.2 sets out the criteria adopted to assess the significance of transport effects in a recent major ESIA study undertaken in the UK.

Table 10.2 Significance criteria adopted for assessment of transport effects during the operation of the UK High Speed 2 railway (after Arup/URS 2012)

Category of effect	Criteria for significance
Public transport delay	A 10% change in a majority of journey times by any public transport mode; and A change in journey distances by bus of more than 400 m in urban areas and 1 km in rural areas
Station/interchange impacts	Increases in congestion levels arising from increased use; or changed journey patterns arising from the arrival and departure, by all available modes; and Any loss of physical linkage for the next stage of the journey. Significant journey time, interchange and accessibility changes for travellers.
Traffic flows and delays to vehicle occupants	A 10% increase in peak hour two-way traffic flows; Increases in traffic flows that cause the design capacity to become exceeded, on links that would not otherwise be congested; A 30% increase in the average off-peak hour two-way traffic flows; A permanent diversion that results in an increase in journey length of more 1 km; or Where a significant change in delay relating to junction congestion resulting is forecast.
Parking and loading	A change for more than four consecutive weeks in any 12-month period of: • A predicted increase of 10 or more vehicles, or 10%, whichever is the greater, in on-street parking demand in the vicinity of a station/interchange; • A loss of any designated on-street or off-street spaces, including spaces for disabled persons, buses, taxis, doctors, ambulances, police vehicles and car club[1] bays; • A loss of ten or more, or 10%, whichever is the greater, private off-street car parking spaces; • A loss of ten or more, or 10%, whichever is the greater, off-street station car parking spaces; • A loss of ten or more, or 10%, whichever is the greater, pedal or motorcycle parking spaces; • A loss of 10% or more designated loading bay spaces or facilities.

Table 10.2 continued

Category of effect	Criteria for significance

Vulnerable road-user delay, amenity and ambience	Significance of the impacts are based on combined assessment of:
	• the changes in journey times in proportion to the scale of the impacts being assessed, e.g.:
	  • minor (less than one minute),
	  • moderate (between one and two minutes);
	  • major (greater than three minutes).
	• And the numbers of travellers affected, e.g.:
	  • minor (less than 200 in total);
	  • moderate (between 200 and 1,000);
	  • major (greater than 1,000).

Community severance	Pedestrians at-grade crossing of a new road carrying over 16,000 vehicles per day (AADT); Journey lengths being increased by over 500 m; Total numbers of people affected across all levels of severance is greater than 1,000.
Accidents and safety	Links and junctions that have on average more than nine personal injury accidents in a three-year period, and which would be subject to an increase of 30% or more in total traffic flow for a period of more than four consecutive weeks in any 12-month period.

Note: [1] A car club is a commercial organisation offering short-term hire (i.e. 30 minutes to one day) of vehicles, typically based on-street and charged on an hourly basis.

As may be seen from this, although some effort to express significance criteria in quantified terms has been made, there is still a good deal of professional judgement required from assessment practitioners, with good knowledge of prevailing conditions in the local and strategic transport networks, to ascertain what the effects of new proposals will be.

If, on this basis, the existing networks cannot accommodate the additional transport movements, then off-site highway or public transport improvements will be considered to mitigate the travel impact of the development. Similarly pedestrian/cycle routes and public transport facilities are examined, and improvements may be required to accommodate development trips and encourage access by sustainable modes.

The operational safety of new developments is a key consideration, and a detailed appraisal of accident records on local roads is usually provided. By determining the causes, linked factors and common safety problems that may arise, this exercise may identify further mitigation measures needed to address safety as part of the development.

10.5.2 Environmental impacts

Table 10.3 summarizes the key environmental impacts of providing new transport infrastructure and increasing traffic levels. More information on assessment of these impacts is provided in individual chapters of this book, but this section discusses specific considerations relating to transport.

Table 10.3 Categorization of environmental effects from transport

Category of impact/ effect	Physical impacts/effects due to new transport infrastructure	Impacts/effects due to additional traffic
Air quality and greenhouse gases	Dust from construction. Air pollution and carbon emissions from plant operation during construction.	Air mass contaminants and carbon emissions. Movement of pollutants.
Communities and economic activity	Community severance. Loss of roadside community business and social activity. Reduced convenience of traditional and sustainable modes of transport (walking, cycling, paratransit).	New economic activity. By-passing of communities. Effects of tourism. "Culture shock" effect. Gentrification effect.
Cultural heritage	Loss of or damage to heritage resources.	Impacts/effects to the setting of heritage resources or the wider historical landscape. Loss of amenity/enjoyment.
Ecology and biodiversity	Loss of habitats and species. Damage to habitats.	Fragmentation of habitats. Hydrological/hydrogeological effects on habitats and species. Disturbance by noise and or light. Effects of air and water pollution. Roadkill.
Health and safety	Death and injuries due to accidents.	Increased deaths and health. Effects due to noise, air pollution and stress factors.
Indigenous peoples	Displacement of indigenous populations. Violation of rights to participate in development.	Loss of traditional sense of identity. Loss of livelihoods and violation of traditionally exercised land rights. Health and social problems.

Table 10.3 continued

Category of impact/effect	Physical impacts/effects due to new transport infrastructure	Impacts/effects due to additional traffic
Land acquisition and resettlement	Loss of homes, businesses and community facilities. Challenges to re-establish communities in new locations.	Lost community links and networks. Stresses in the "host" community.
Landscape and townscapes	Loss of or damage to landscape character. Loss of key visual features. Introduction of new visual features. Changes in visual quality.	Visual intrusion from vehicles. Light pollution. Maturation of landscape planting associated with project.
Noise and vibration	Noise/vibration from construction activities and operation of plant.	Noise from road or rail interface. Noise/vibration re-radiated via structures. Engine/aerodynamic noise.
Soils	Loss of productive soil. Increased erosion.	Contamination of soil. Possible landslides, slippage.
Water resources	Surface water flow modification. Groundwater flow modification. Water contamination.	Water quality degradation (surface and groundwater). Drainage modifications. Water table modification. Sensitive habitat intrusion.

Air quality and greenhouse gases

Assessment of air quality impacts is discussed in Chapter 4. **Local air quality** issues relating to transport are usually assessed by focusing on nitrogen dioxide (NO_2) and particulate matter (PM_{10}) emissions from vehicle exhausts. However, depending on the dominant fuel types in the fleet-vehicle mix, assessment of the effects associated with other pollutants such as sulphur dioxide (SO_2), carbon monoxide or polyaromatic hydrocarbons (PAHs) may also be appropriate. The difference in roadside PM_{10} and NO_2 levels is usually considered using predicted traffic flows for each option assessed. Assessments may need to take account of the presence or absence of potentially sensitive receptors close to the proposed schemes, in terms of people, agricultural land and ecological habitats. The vulnerability of groups especially prone to respiratory illnesses, such as children and elderly people, will be a particular concern.

Transport schemes can also have air pollution impacts beyond the local situation, often due to photochemical reactions of vehicle pollutants in the atmosphere. These most notably include acidification, excess nitrogen deposition, and the generation of tropospheric ozone. The potential to affect air quality on a wider regional basis will be limited but if it is likely to be an issue, a regional air quality assessment should be carried out to determine the change in emissions of nitrogen oxides.

Specific computer models are available to predict the emission and dispersion of air pollutant from road vehicles. Examples include Caline 4 and Mobile 5, available from the U.S. Environmental Protection Agency. Caline 4 is based on a Gaussian line source air-dispersal model. The models require inputs like traffic volume, vehicle speed, vehicle composition, roadway conditions, meteorological conditions, emission rate and receptor location (Banerjee and Ghose 2016).

Greenhouse gas emissions, as discussed in Chapter 5, are assessed by predicting carbon dioxide (CO_2) levels from the expected additional number of vehicle-km induced by each option. Some of the more widely used transport models, such as SATURN, include modules that will estimate CO_2 emissions from vehicles in each scenario, modelled so that alternatives may be compared on a consistent basis. Monetary values are often then calculated per tonne of carbon released into the atmosphere, using an estimated "Social Cost of Carbon" (Ackerman and Stanton 2010).

Communities and economic activity

Social interaction and economic activity is key to the vitality and sustained existence of communities, and the assessment of these is discussed in more detail in Chapters 13 and 14. Transport links are central to these continuing interactions, but introduction of a new road or railway (or the widening of existing routes) can also cause disruption to them. Transport projects can thus lead to changes in the community interactions in surrounding areas, and can influence various aspects of lifestyles and travel patterns. The nature, extent and magnitude of these changes identified in an ESIA are important considerations for decision-makers.

Sometimes, the social and economic costs may outweigh the benefits the transport investment might bring, such as lower transport costs, better access to markets, goods, jobs, or health and education services. For instance, a new road that runs straight through the centre of a farm holding, cutting this in two, will have an obvious effect on its economic viability, and the consequent loss is easy to record. Community severance – unless mitigated by re-provision of original access routes as part of the new proposal – can affect businesses, pedestrians, and users of non-motorized transport, and typically has a disproportionately adverse effect on poorer people.

The assessment of community severance should consider:

- The existing pattern of community and economic activities, the scale of movement between these, and the routes used. This will require local surveys to establish.
- The extent to which the proposed scheme (or traffic from new development) will help or hinder these movements, perhaps by introducing physical barriers, or by the increased difficulty of crossing existing routes due to higher traffic flows. This should take account of any elements of the proposals intended to avoid such effects, such as new pedestrian or cycle crossings: in some instances these may improve access to facilities for some.
- The extent to which the changes in time or distance required to make these movements when the proposals are implemented will impact on community facilities and businesses. For example, some communities could be denied access to a school within a reasonable journey time as a consequence, or the viability of a community facility may be severely affected because people in a large part of its catchment area can no longer reach it easily, and it may close as a consequence. However, again it is important to acknowledge beneficial effects such as better access to markets for businesses and reduced transport costs.

In reporting the outcomes of such analyses, it is very important to clearly identify who the winners and losers are in each case, and which economic and social groups in the community they represent.

Cultural heritage

Transport infrastructure or increased vehicle/train flows can affect the historic, scientific, social and amenity value of cultural heritage resources through:

- Physical damage from construction and related works (e.g. quarrying) or unregulated access to historic and cultural sites;
- Indirect damage to sensitive historic structures due to ground settlement or subsidence;
- Impacts on the aesthetic quality and setting of cultural and archaeological sites;
- Improving cultural heritage through better access to and interpretation of cultural heritage resources.

The assessment of historical and cultural heritage is discussed in detail in Chapter 12.

Ecology and biodiversity

Transport infrastructure can impact directly on ecology and biodiversity through the loss of, or damage to, habitats. Habitats can also be fragmented, reducing their value or viability. Infrastructure may also introduce a barrier to the movement of birds, bats and other animals when moving in relation to feeding or breeding, and they can be killed or injured when attempting to cross the road. There is also the potential for transport infrastructure to physically affect aquatic habitats due to changes to hydrology and hydrogeology.

In undeveloped areas, penetration by roads will facilitate access to ecologically sensitive areas. This may facilitate poaching and bio-piracy of endangered species, illegal logging, and increased people-wildlife conflict. On the other hand, the roads may also improve law enforcement and rapid response by the authority.

As well as such direct effects, pollution caused by the operation of transport can contaminate air, soil and water and lead to damage to species and habitats. Noise and night-time lighting from transport can also disturb species and affect breeding. In addition, new transport routes may lead to the spread of new diseases affecting both flora and fauna, and a consequent increase in human activity can increase the risk of fires.

More information on the assessment of effects of development on terrestrial ecology is set out in Chapter 6. However, the key steps in such an assessment typically would be to:

- Collect relevant baseline data on habitats and species likely to be affected;
- Minimize damage to habitats and species by choice of alignment;
- Select a preferred design that interferes least with animal movements, drainage and groundwater;
- Provide a full assessment of the likely significant effects of the proposals on habitats and species;
- Prepare a mitigation plan to avoid the significant effects if possible, or reduce these to an acceptable level, if necessary by translocation of species and compensatory planting of habitats.

Health and safety

Development of new transport infrastructure and increased accessibility can potentially lead to the decline in health of a local population in several ways, e.g. it can:

- Directly impact people through increased deaths and injuries due to accidents;
- Facilitate the transmission of diseases, including sexually transmitted diseases;
- Contaminate the local water supply (see previous section and Chapter 2);
- Pollute the air (see previous section and Chapter 4);
- Become a source of noise pollution (see also below and Chapter 9).

Air pollution can both cause and exacerbate respiratory and other diseases, particularly in vulnerable groups such as the elderly and children. Emissions of nitrogen oxides (NO_x) can aggravate and induce asthma, emphysema, tuberculosis and bronchitis, while PM_{10} emissions may cause eye and respiratory irritation, aggravation of asthma and some are suspected carcinogens. Effects associated with other pollutants such as CO, HC and SO_2 are similarly dangerous to health in various ways (Clapham 1981; Lee 1985). Noise may also cause or exacerbate certain medical conditions, such as heart disease, through increasing stress or sleep disturbance in local populations.

Impacts on indigenous peoples

In general, the effects of development on indigenous peoples are discussed in Chapters 12–15. Transport infrastructure increases in accessibility and movement are a key source of change in the lifestyles of indigenous peoples, often making it difficult for them to maintain customs and traditions, and exposing them to increased external influences and pressures.

Land acquisition and resettlement

Transport projects by their nature typically require the compulsory purchase of private land by public institutions. In exercising this right, public bodies need to ensure that this is undertaken equitably, and with proper compensation provided to all affected parties. This is particularly important where the loss of all or part of a landholding will impact on its functional or economic viability, resulting in the necessary resettlement of the previous owners or tenants. If widespread, the social and psychological impacts can be large, with whole communities disrupted or destroyed. More usually though, the effect is to cut across communities and often run through different governmental jurisdictions due to the linear nature of most transport projects (Tsunokawa and Hoban 1997).

Landscape and townscapes

Structures associated with new transport infrastructure can often be very large. This may require the removal of large areas of landscape or townscape features that can have significant effects on visual quality. These new structures can themselves be the cause of significant changes, often negative, but sometimes positive. Transport operations can also introduce new visual elements. Vehicle flows, particularly including high volumes of large freight vehicles, or longer intercity and freight trains, may be highly visible and damaging in rural or urban settings.

The assessment of impacts on the visual and amenity value of landscape and townscapes necessarily includes a great deal of subjective judgement. This not only needs to take account of the ubiquity, or rarity, of the affected areas and the different values that individuals place on the aesthetic quality of rural and urban areas, but also requires consideration of cultural and economic factors, which shape the way we all see the world. It also needs to incorporate the presence or absence and the sensitivity of people who will experience the changes caused by a project. More information on this is provided in Chapter 11.

Noise and vibration

Noise impacts depend on the time of day, flow, and type of traffic. Typically, assessment is based on the concept of noise annoyance and involves calculating the difference in the estimated population who would be annoyed by noise – e.g. experiencing changes of greater than 3 dB(A). The population annoyed under the "with scheme" and "without scheme" scenarios are compared. This will involve taking account of the societal aspects of noise impacts, as these vary significantly depending on the population affected.

A number of internationally available computer models may be used to estimate noise from road traffic, such as FHWA, STAMINA 2.0 or OPTIMA (USA), Microbruit (France), "Calculation of Road Traffic Noise" (UK CORTN), SonRoad (Switzerland), RMW2002 (Netherlands), RL90 (Germany) and ASJ RTN-Model 2003 (Japan). For instance, the FHWA TNM 2.5 model computes noise level based on a series of adjustments to a reference sound level considering traffic flow, distance and shielding effects (FHWA 1998). A similar algorithm used by CORTN performs several adjustments to reference sound level based on criteria like traffic flow, composition, road gradient, road surface, distance and barriers (Banerjee and Ghose 2016). Other specific models are available to provide estimates of railway and aviation noise.

Soils and geology

As Chapter 3 recognizes, soils and geology are an important component of the natural environment, with considerable commercial and economic importance. The protection and maintenance of the quality of soils, or adequate compensation where these are lost or denuded, should be a key element of an ESIA for transport infrastructure. This is important not just in its own right, but also in relation to effects on ecology and biodiversity (see above and Chapter 6) and water resources (see below and Chapter 2).

The key issues for transport infrastructure proposals are:

- Loss of productive soils due to the presence of new structures, which has socio-economic effects as well as effects on ecology and biodiversity;
- Erosion, which may be due to the destabilization of slopes, tipping of excavated materials, and diversion of natural surface water flows;
- Soil contamination from drainage run-off containing metals, rubber and hydrocarbons, and from local deposition of contaminants contained in vehicle exhausts;

- The cumulative effects of one or more of these.

A particular concern in hilly or mountainous areas is the risk of mass movements. Flows, slides and falls can lead to the frequent need for road reconstruction, and pose a considerable safety risk for road users. Cuttings excavated from the base of slopes can lead to collapse, with ramifications not only for structures under construction, but far upslope as well. The building of transport structures and tipping of excavated materials at the tops of slopes can result in excessive slope-loading, causing failure downslope. Alteration of the drainage regime of slopes, brought about by excavation and/or the introduction of new structures, can cause instability through erosion or increased pore pressure.

Surveys of proposed alignments should identify areas prone to mass movements, which should be avoided where possible. Care is required not to undercut or overload steep slopes, and particular attention should be paid to the implementation of adequate drainage measures. GIS-aided landslide susceptibility mapping can help with this, using remotely sensed imagery, historical landslide-inventory database, and auxiliary data (such as geological, hydrological, meteorological and geomorphological data and land-use and land-cover data) (Banerjee and Ghose 2016).

Water resources

Change to drainage regimes is an almost inescapable effect of providing new transport infrastructure. The operation of transport may also be a serious source of water pollution. Chapter 2 examines potential effects on water and flooding in detail, but in relation to transport, the key issues to consider in any ESIA will include:

- Modification of surface water flows where alignments intersect drainage basins and potentially concentrate flows so that flow speeds increase, with consequent increases in flood risk; soil erosion; channel modification; increased silting up of channels where increased flow speeds exacerbate erosion upstream; and possible changes to river and estuarine ecology.
- Groundwater flow modifications, through the lowering of the water table or restrictions below ground. This can lead to the deterioration of vegetation, increased risk of soil erosion, reduction in the volume of aquifers and a consequent loss of water supply for drinking, agricultural or industrial uses, and possible changes to river and estuarine ecology.
- Reduction in the quality of both surface and groundwater due to contamination by silt, hydrocarbons, rubber, metals, salts, deiceants, chemical and other toxic spills.

As with some of the other categories of environmental effects, these impacts are particularly important to consider in relation to proposals that introduce new infrastructure or increase the flow of transport in hilly or mountainous rural areas, which will be particularly prone to such impacts and where the consequent effects may affect extensive geographical areas and large numbers of people.

Reporting

A transport assessment (or transport chapter in ESIA) could conclude by compiling all of this information into one summary table. As an example, in the UK, an Appraisal Summary Table (AST) is typically used which contains three columns for evaluating the significance of the predicted impacts (see Table 10.4) (UK DfT 2014). The first column is qualitative and allows a textual description of the impact. The next column is quantitative, and uses numbers to measure the scale of the impacts. The final column is the summary assessment and uses a monetary scale, quantitative indicators, or a seven-point scale of the impacts (large, moderate, small negative/adverse, neutral, small, moderate, or large positive/beneficial). The AST is usually accompanied by an appraisal of the achievements of local and regional objectives; the effectiveness of problem solving; and supporting analyses that cover distribution and equity; affordability and financial sustainability; and practicality and public acceptability. Again, much of this information may be usefully included in an ESIA report.

Development should only be prevented or refused on transport grounds where the residual cumulative impacts of development are severe.

10.6 Mitigation

10.6.1 Transport measures

Traditionally, the transport impacts of development were considered to be primarily mobility-related. Mitigation measures thus involved off-site highway works to reduce driver delay (e.g. junction improvements), pedestrian and cyclist delay (e.g. improved crossing facilities), or accidents (e.g. traffic calming). In the light of changes in attitudes and public policies worldwide, mitigation measures in more recent times have become more focused on reducing the need to travel, improving accessibility, providing transport alternatives and addressing environmental issues. Mitigation measures for the transport impacts of non-transport developments thus include public transport improvements, reductions in car parking, travel plans, improved pedestrian and cycling facilities, and traffic management measures, together with contributions towards more strategic transport measures, such as park-and-ride. These measures are summarized in Table 10.5.

In most developed countries, it has now become standard practice to prepare a travel plan, normally at the request of the local planning body. This plan identifies measures for encouraging the modal shift of employees or visitors from the private car to public transport, cycling and walking. These measures

Table 10.4 Example of a Transport Appraisal Summary Table (AST)

Option	Description	Problems	Present value of costs (PVCs) to public accounts £m	
Objective	Sub-objective	Qualitative impacts	Quantitative assessment	Assessment
Environment	**Noise**			Net population win/lose, Net present value £m
	Local air quality			Concentrations weighted for exposure tonnes of CO_2
	Greenhouse gases			
	Landscape			Score
	Townscape			Score
	Heritage of historic resources			Score
	Biodiversity			Score
	Water environment			Score
	Physical fitness			Score
	Journey ambience			Score
Safety	Accidents			PVB £m
	Security			Score
Economy	Public accounts		Central and local government PVCs	PVC £m
	Transport economic efficiency: Business users and transport		Users Present Value of Benefits (PVB), Transport Providers PVB, Others PVB	PVB £m
	Transport economic efficiency: Consumers		User PVB	PVB £m
	Reliability			Score
	Wider economic impacts			Score
Accessibility	Option values			PVB £m
	Severance			Score
	Access to the transport system			Score
Integration	Transport interchange			Score
	Land-use policy			Score
	Other government policies			Score

Table 10.5 Mitigation measures: transport impacts of non-transport developments (adapted from BRF 1999, Taylor and Newson 2008)

Main factor	Examples of project-level mitigation measures	Examples of more strategic measures involving or affecting more than one development project
Car parking	• Reduce car parking. • Travel plan.	• One parking area used for multiple purposes (e.g. to service office during the week, community use or recreation facility on the weekend). • Tight maximum parking standards. • Parking controls/charging.
Highway capacity	• Travel plan. • Reduce the number and length of trips. • Promote pass-by, linked, diverted or transferred rather than new trips. • Promote car sharing/pooling. • Managed access from the development onto the highway network. • Support for public transport. • Improved road junctions, road widening, new roads etc.	• Integrated land use/transport planning. • Improved public transport (see below). • Demand management, e.g. congestion charging. • Other traffic control measures over a wide area, e.g. Intelligent Transport Systems. • High-occupancy vehicle lanes. • Advanced signal systems. • Provision of new or expanded roads.
Pedestrians	• Travel plan. • Direct and desirable pedestrian routes, including crossing facilities. • Traffic calming, e.g. humps, reduced carriageway size. • Environmental and public realm improvements, e.g. wider pavements. • Good lighting. • CCTV.	• Well-located facilities vis-à-vis homes and employment sites. • Reduced traffic speeds, e.g. Home Zones.
Cycling	• Direct and desirable cycling routes, including crossing facilities. • Provision of cycle paths and/or lanes. • Restrictions on car parking. • Traffic calming. • Secure cycle parking. • Changing facilities.	• Strategic cycling network.

Table 10.5 continued

Main factor	Examples of project-level mitigation measures	Examples of more strategic measures involving or affecting more than one development project
Public transport	• Developer contributions towards strategic measures.	• Bus lanes and bus priority improvements (e.g. bus-operated traffic lights). • Real time bus information. • Upgrading of facilities. • Introduction of new bus routes. • Park-and-ride. • Provision of more frequent/improved/ extended public transport.
Wider strategic		• Alternative fuels (e.g. LPG, flex-fuels). • Generation sources for electricity supply • Use of smaller vehicles for last-mile connectivity. • Tightening fuel efficiency requirements. • Ban on some types of vehicles at certain times (e.g. large diesel cars). • Restricted entry or higher tax for older/more polluting vehicles. • Multi-occupancy vehicle lanes. • Even-odd policy where only even licence plates are allowed on even days and vice-versa (Mathieson 2014).

can range from the provision of showers and changing facilities to bike loans and bus season tickets. This, together with restrictions on car parking provision, has been successful at a number of development sites in achieving a modal shift. The production of a travel plan may be a planning requirement for a new development that is likely to generate significant traffic, and/or a mitigation measure proposed in the ESIA. Where travel plans already exist, they may suggest other mitigation measures that can be included in the ESIA. Box 10.1 present some case studies of impacts and mitigation measures in transport ESIAs.

Box 10.1 Case study: Impacts of highway projects in mountainous areas of India and Nepal

Highway projects like the Puspalal Midhill highway project, the Kathmandu–Terai track road project in Nepal, and the Mehatpur–Una–Nerchowk roads project in Himachal Pradesh, India show that improvement in the transport sector can substantially contribute to the economic progress of the country. Moreover, such road projects can reduce automobile fuel consumption, man hours and vehicle maintenance cost. According to their EIA reports (HPRIDC 2007, MoPPW 2011, MoSTE 2015):

- Socio-economic factors like literacy, gender composition, occupation, income and possessions are major determinants in assessing project impacts. For instance, it is possible to communicate the benefits and impacts of projects in the form of posters or charts where there is a high proportion of literate people in the project-affected population near suburban areas. Moreover, good public participation can be expected in these areas. On the other hand, large-scale illiteracy in remote areas makes the project-affected people vulnerable to lack of information and misinformation.
- Training and capacity building in non-agricultural occupations is one way of reha-bilitating landless farmers and agricultural labourers. However, beyond occupation, agriculture is a way of life for many of these farmers. As such, transition to new occupations is very difficult and often unsuccessful. Land acquisition due to construction and consequent compensation need special attention.
- Village communities in hilly areas heavily depend on the forest for fuel wood, fodder and other forms of non-timber forest products. Displacement of such communities can cause much difficulty in finding the substitutes for these products.
- In hilly areas, a substantial number of households are headed by women. This is mainly due to out-migration of men to find work in other areas, mainly to cities. Involving the women in project and community development programmes should be gender sensitive.
- Loss of valuable soil from the forest and grazing lands and decline in carbon sequestration are other related impacts of highway projects in hilly areas which will lead to a decline in carrying capacity of these areas.
- Mitigation of construction and traffic induced noise at night time is essential as it affects study timing, sleep cycle and effective work hours in hilly areas, especially since hill people are early sleepers by their habit and culture.
- During the construction phase:
  1 Temporary traffic management is crucial to prevent traffic congestion, especially since steep hills may lead to fatal accidents.
  2 Awareness programmes and prevention of poaching of wild animals for meat by construction workers is required. Proper protein-rich food should be provided in the canteens.
  3 Proper sanitation, treatment of human waste and supply of potable water is essential to prevent contamination of water bodies in the hilly areas. Construction work during dry seasons can also reduce water contamination. Moreover, monitoring of water quality is pivotal.

4 Dust pollution due to backfilling, excavation, dumping of earth materials, construction spoil and vehicle movement along unpaved roads should be mitigated by water sprinkling and quality control on construction materials; and asphalt and hot mix plants should be at least 500 m from the road.

5 Safety measures such as replacement of culverts, construction of new culverts, bypasses, realignment of road, parking areas and truck bays are useful for proper traffic management.

10.6.2 Environmental measures

Mitigation measures for the environmental impacts of transport infrastructure relate primarily to reducing noise, air, water, light pollution and visual intrusion; protecting and enhancing wildlife and ecology; amenity and recreation; and promoting the sustainable use of natural resources. These measures are summarized in Table 10.6.

Table 10.6 Mitigation measures: environmental impacts of transport infrastructure (adapted from BRF 1999; Tsunokawa and Hoban 1997)

Category of impact/effect	Examples of project-level mitigation measures

Air quality and greenhouse gases	• Alternative fuels, e.g. LPG, and flex-fuels like ethanol, bio-CNG (compressed natural gas) and solar cell based electric vehicles.
	• Awareness campaigns re. vehicle maintenance and traffic codes.
	• Vehicle emission standards/maintenance.
	• Avoiding sensitive groups/resources.
	• Encouraging use of public transport/non-motorized modes.
	• Measures to reduce/improve traffic flow.
	• Pedestrian priority.
	• Speed restrictions.
	• Sealing high-use dirt roads; use of stronger road surfaces.
	• Providing vegetation to filter pollutants.
	• Financial compensation:
	  • cleaning buildings;
	  • health care facilities for respiratory diseases.

Communities and economic activity	• Taking account of local movements in design (signals, intersections, bridges, underpasses etc.). • Minimizing loss of road/track-side businesses. • Local access improvements. • Financial or offsite compensation.

Table 10.6 continued

Category of impact/effect	Examples of project-level mitigation measures
Cultural heritage	• Alignment choice to avoid cultural resources. • Physical and intrusive surveys pre-construction. • Site management plan for conservation actions, including salvage and relocation. • Recording and archiving of resources unavoidably lost or damaged. • Financial and offsite compensation.
Ecology and biodiversity	• Alignment choice to avoid ecological resources. • Buffer zones between works and watercourses. • Reducing corridor widths through engineering. • Translocation of species and seed banking. • Compensation planting. • Fencing. • Animal crossings (green bridges, underpasses). • Careful design of water crossings. • Traffic control measures.
Health and safety	• Health-and-safety awareness campaigns. • Safety audits of design alternatives. • Improvement of pedestrian and cyclist facilities. • Speed-limiting measures. • Improvement of visibility. • Limiting of non-motorized vehicle movements by fences or guardrails. • Improvement of crossing sites. • Regulations, education, and safety awareness training. • Avoiding standing water which may act as a breeding ground for snails and mosquitoes. • Improvements to community health services.
Impacts on indigenous peoples	• Alignment to avoid effects on indigenous groups. • Control of outside access (e.g. indigenous reserves). • Consultation and participation at appropriate levels. • Relocation. • Maximizing positive benefits (e.g. employment). • Financial compensation.
Land acquisition and resettlement	• Alignment to minimize effects on land holdings. • Land swaps. • Relocation. • Resettlement programmes. • Financial compensation.
Landscape and townscapes	• Careful alignment to integrate with landscape/townscape. • Integration of transport infrastructure with development. • Promotion/restriction of views. • Promotion of gateways. • Improvement of lighting while avoiding light pollution.

Table 10.6 continued

Category of impact/effect	Examples of project-level mitigation measures
Noise and vibration	<ul><li>Vehicle controls:<ul><li>tyres; construction; exhausts.</li></ul></li><li>Noise barriers:<ul><li>reflective barriers;</li><li>absorbent barriers;</li><li>vegetative barriers.</li></ul></li><li>Road surfacing:<ul><li>porous or open-faced asphalt;</li><li>"whisper" concrete;</li><li>thin surfacing.</li></ul></li><li>Traffic management, e.g. humps, chicanes, speed limits.</li><li>Engineering solutions:<ul><li>alignment choice/avoiding sharp bends, steep inclines etc.</li><li>cuttings;</li><li>cut and cover;</li><li>optimum junction design.</li></ul></li><li>Noise insulation to buildings with sensitive receptors.</li><li>Financial compensation.</li></ul>
Soils	<ul><li>Minimizing ground clearance.</li><li>Avoiding steep hillsides.</li><li>Structural stabilization, control of groundwater levels and site surveillance.</li><li>Balancing cut and fill requirements.</li><li>Avoiding contaminated sites.</li><li>Replanting cleared areas immediately.</li><li>Drainage improvements.</li><li>Compensatory measures (e.g. turning quarries, borrow pits and dump sites into amenity areas).</li><li>Soil remediation.</li></ul>
Water resources	<ul><li>Flow speed control</li><li>Improved runoff</li><li>Paving to reduce contamination</li><li>Settling ponds/basins</li><li>Pollution interceptors</li><li>Using wetland as treatment facilities.</li><li>Drainage and treatment systems (grasses, riprap,[1] engineered solutions).</li><li>Careful choice of car park surfaces</li><li>Financial or offsite compensation measures.</li></ul>

Note: [1] Riprap is loose stone used to form a breakwater or other structure.

10.7 Monitoring

Monitoring the transport impacts of specific developments is useful but often neglected. Where a developer has prepared a travel plan, monitoring its success is an important requirement, but where no such plans have been prepared, development-specific transport monitoring is rare. It is also very uncommon for traffic problems in an area to be attributed to a specific recent development, as typically transport issues will have been considered during the consenting process, and, tacitly at least, any problems that may arise will have been seen politically as part of the price that is to be paid for the benefits brought by that proposal. Hence, such problems occurring post hoc are more typically dealt with by the public authorities at a strategic level. In Europe, monitoring is a mandatory part of Strategic Environmental Assessment of plans and programmes and may therefore lead to the monitoring of the environmental performance of transport projects.

As with all ESIAs, transport projects (or the transport element of other types of development) need to be monitored to ensure environmental mitigation is implemented, and to determine whether they are effective. This is typically set out in an Environmental and Social Management Plan (ESMP), which is probably the most important output from the ESIA process generally – see Chapter 20. In relation to transport, the ESMP takes on additional importance because of the linear nature of transport infrastructure projects. The environmental effects of transport infrastructure need to be managed carefully, as they occur across extensive and diverse areas and often concurrently. It is therefore important that all proposed mitigation and monitoring actions are set out clearly in the ESMP, together with the programme for these, with specific responsibilities assigned. The specific types of effects that are required to be monitored are discussed elsewhere in this book in the chapters dealing with specific environmental topics. Some monitoring, particularly in relation to noise and air pollution during construction, may be requirements of specific consenting and licensing procedures. This also needs to be identified and accounted for in the ESMP where appropriate.

In practical terms, extensive monitoring may be problematic, particularly in developing countries where resources may be constrained, and new transport infrastructure is often provided in remote areas. An example of one way of addressing this problem may be found in Ethiopia, where aerial photographs and satellite images were used to monitor and analyse changes in the environment of a highway between 1980 and 1993 (Tsunokawa and Hoban 1997). The project is in an isolated and undeveloped area, and it was anticipated that population and land use would begin to change immediately after construction. It was thought that traffic flows, however, would increase only gradually. A GIS was used to analyse historic aerial photographs and more recent Landsat and Spot data. This indicated that the number and size of smaller villages increased, and major villages doubled in number. It was identified that a large bamboo forest had been cleared and the remaining forest was threatened. On this basis, it was possible to identify additional measures necessary to prevent further erosion, loss of soil fertility, and continued reduction in forest areas.

10.8 Conclusions

Transport planning exists in, and reflects, a rapidly changing policy arena. The sustainability agenda encourages land-use planning and transport planning to coordinate so as to reduce the need to travel. Developments must seek to achieve a modal shift. Methods to assess transport impacts reflect this changing agenda and seek to influence developments in a sustainable manner. These methods increasingly aim to provide a level playing field for different modes and an integrated assessment process for different types of impacts. The effectiveness of transport assessment depends on the way it is used, the proponents' intentions, and the stage in the development process at which they are used. The most widely adopted assessment approaches are typically designed to be flexible: they can be used at the strategic or feasibility stages, before detailed design, using subjective judgement. More quantitative information can be completed at the later detailed design stage. This two-tier approach enables environmental considerations to be incorporated into the decision-making process at an earlier stage and thus have a greater influence on the outcome.

Notes

1   "Home Zones" are residential streets in which the road space is shared between drivers and other road users with the wider needs of residents (including people who walk and cycle, and children) in mind.
2   Paratransit is public or group transport using smaller vehicles, such as cars, vans, minibuses, motor rickshaws, and in some cases non-motorized modes such as cycle rickshaws.
3   TRICS® is a system of trip generation analysis, founded and owned by six County Councils in the south of England.

References

Ackerman F and EA Stanton 2010. The social cost of carbon. *Real-World Economics Review* 53, 129–143.
Administrative Staff College of India (ASCI) 2010. *Environmental Impact Assessment Guidance Manual for Highways*. New Delhi: MoEF.
Arup/URS 2012. HS2 London to West Midlands: EIA scope and methodology report, Chapter 15: *Traffic and Transport*. London: High Speed 2 Ltd.
Banerjee P and MK Ghose 2016. Spatial analysis of environmental impacts of highway projects with special emphasis on mountainous area: an overview. *Impact Assessment and Project Appraisal* 34(4), 279–293.
British Roads Federation (BRF) 1999. *Old Roads to Green Roads*. London: BRF.

Buehler R and J Pucher 2009. Sustainable transport that works: Lessons from Germany. *World Transport Policy & Practice* 15(1), 13–46.

Clapham WB 1981. *Human Ecosystems*. New York: Macmillan Publishing Co. Inc.

European Investment Bank (EIB) 2013. *Environmental and Social Handbook, Version 9.0*. Luxembourg: Environment, Climate and Social Office, Projects Directorate, EIB.

European Union (EU) 2013. *Guidance on the Application of the Environmental Impact Assessment Procedure for Large-Scale Transboundary Projects*. Brussels: EU.

Federal Highway Administration (FHWA) 1998. *FHWA Traffic Noise Model-Technical Manual*, FHWA-10-96-010. Washington, DC: FHWA.

Grant J 1977. *The Politics of Urban Transport Planning*. London: Earth Resources Research Ltd.

Haq G and M Bolhuis 1998. Dutch transport policy: From rhetoric to reality. *World Transport Policy & Practice* 4(1), 4–8.

Highways England 2015. *The Design Manual for Roads and Bridges, Volume 11, Environmental Assessment*. London: The Stationery Office. www.standardsforhighways.co.uk/ha/standards/dmrb/vol11.

Himachal Pradesh Road and Other Infrastructure Development Corporation Limited (HPRIDC) 2007. *EIA for Mehatpur-Una-Amb and Una-Nerchowk roads, E-1540, Vols 1 and 2*. Government of Himachal Pradesh.

Hughes A 1994. Traffic. In P Morris and R Therivel (eds), *Methods of Environmental Impact Assessment*, 64–77. London: UCL Press.

Lee JA 1985. *The Environment, Public Health, and Human Ecology: Considerations for Economic Development*. Baltimore, MA: The Johns Hopkins University Press (for the World Bank).

Mathieson K 2014. Why licence plate bans don't cut smog. London: *The Guardian*, 20 March.

Ministry of Physical Planning and Works (MoPPW) 2011. Preparation of detailed project report of Puspalal (Mid Hill) highway project (Final Report). Planning and Designing Branch, MoPPW, Govt. of Nepal, prepared by TechStudio of Engineering, Kathmandu, Nepal.

Ministry of Science, Technology and Environment (MoSTE) 2015. *Kathmandu-Terai/Madhesh Fast Track Road Project: Environmental Impact Assessment, Final Report*. Kathmandu: Ministry of Physical Infrastructure and Transport.

Owens S 1995. From 'predict and provide' to 'predict and prevent'? Pricing and planning in Transport Policy. *Transport Policy* 2(1), 43–49.

Republic of Botswana 2001. *Planning and Environmental Impact Assessment of Road Infrastructure*. Gaborone: Ministry of Transport and Communications.

SACTRA (Standing Advisory Committee on Trunk Road Assessment) 1994. *Trunk Roads and the Generation of Traffic*. London: HMSO.

Sustrans 2014. *Active Travel Strategy Guidance*. Edinburgh: Transport Scotland.

Taylor I and C Newson 2008. *The Essential Guide to Travel Planning*. UK Department for Transport. London: HMSO.

Tsunokawa K and C Hoban (eds) 1997. *Roads and the Environment: A Handbook*. Technical Paper No. WTP 376. Washington, DC: The World Bank.

UK Department of Communities and Local Government/Department for Transport (DCLG/DfT) 2007. *Guidance on Transport Assessment*. London: The Stationery Office, (now withdrawn).

UK Department for Transport (DfT) 2014. *Transport Analysis Guidance (TAG)*. London: DfT TAG website (WebTAG). www.webtag.org.uk.

Willumsen L 2014. *Better Traffic and Revenue Forecasting*. London: Maida Vale Press.

11 Landscape and visual

....

Rebecca Knight and Riki Therivel

11.1 Introduction

> The landscape ... is an important part of the quality of life for people everywhere: in urban areas and in the countryside, in degraded areas as well as in areas of high quality, in areas recognised as being of outstanding beauty as well as everyday areas.
>
> (Preamble to the European Landscape Convention, Florence, 20 October 2000)

The landscape provides a setting for our day-to-day lives. It provides a resource to support livelihoods, a sense of place, an environment for flora and fauna, a cultural record of our past, a source of memories and associations, and opportunities for recreation and aesthetic enjoyment. It has benefits for health and well-being as well as providing economic benefits.

Projects can result in effects on landscape character or quality, and on views experienced and valued by the local population. **Landscape effects** describe changes in the landscape, its character and quality, and are assessed as an effect on an environmental resource. In this context, 'landscape' includes rural areas as well as urban 'townscapes' and marine 'seascapes'. **Visual effects** describe the appearance of these changes and the resulting effect on visual amenity, and are assessed as one of the interrelated effects on population.

This chapter begins by defining concepts and considering the legislative background. It then describes how landscape and visual data are collected and how landscape and visual effects are predicted; it highlights potential mitigation measures that can be employed; and points out the limitations of methods for landscape and visual impact assessment. It also points to sources of further information where relevant.

11.2 Definitions and concepts

For some Environmental and Social Impact Assessment (ESIA) topics, potential effects can be assessed against measurable, technical international or national guidelines or legislative standards. The assessment of potential effects on the landscape and visual resources is more complex, involving a combination of objective and subjective judgements.

11.2.1 Definition of landscape

The European Landscape Convention defines landscape as 'a zone or area as perceived by local people or visitors, whose visual features and character are the result of the action of natural and/or cultural (that is, human) factors'. This definition reflects the idea that landscapes evolve through time, as a result of being acted upon by natural forces and human beings. It also underlines that a landscape forms a whole, and its natural and cultural elements must be taken together, not separately (CoE 2000). Box 11.1 shows the factors that contribute to the landscape. Further information about these factors can be found in *An Approach to Landscape Character Assessment* (Natural England (NE) 2014).

Box 11.1 Factors that contribute to the landscape (based on NE 2014)

Natural
- Geology
- Landform
- Hydrology
- Air and climate
- Soils
- Land cover/flora and fauna

Cultural/social
- Land use
- Settlement
- Enclosure
- Land ownership
- Time depth

Perceptual and aesthetic
- Sight:
  - Colour
  - Texture
  - Pattern
  - Form
- Sounds
- Smells
- Touch/feel
- Preferences
- Associations
- Memories

11.2.2 From landscape quality to landscape character

Until the 1980s, the consideration of landscape in land-use planning generally focussed on landscape quality, or what makes one area 'better' than another. This process was described as landscape evaluation. Designated areas were often protected at the expense of the rest, so that non-designated areas could be targeted by developers. Another problem with this approach was that the concept of landscape beauty is not timeless and is dependent on fashion and taste. An emphasis on designations is now considered insufficient in ESIA.

In the 1980s there was a shift away from this approach of 'preserve the best and leave the rest', and landscape assessment evolved as a means of identifying what makes one area 'different' or 'distinct' from another. This approach acknowledges the *character* of individual landscapes, the diversity of all landscapes, and the benefits and services that landscapes provide. This changed the emphasis from landscape as 'scenery' to landscape as 'environment'.

Landscape character can be defined as 'a distinct, recognizable and consistent pattern of elements in the landscape that makes one landscape different from another, rather than better or worse' (NE 2014). Landscape character comprises particular combinations of *landscape elements* and *characteristics*, such as landform, geology, soils, waterbodies and catchments, trees and woodlands, boundary features and patterns, agriculture, habitats and species, settlement and development patterns, historic sites and features, recreation and access, and experiential qualities. This can also include description of social structures.

New Zealander Brabyn (2009) recognises the need to classify landscape character:

New Zealand's landscape is a multi-billion dollar tourism resource that has considerable quality of life value for ordinary New Zealanders. To manage this resource effectively, it is essential that it be classified so that planners have an inventory, and a frame of reference for communication and research. A national inventory enables local landscapes to be assessed within a national context ... Classification is important for communication because it provides a consistent frame of reference.

In England, the land has been classified into 159 Landscape Character Areas with names such as 'Yorkshire Dales' and 'Thames Valley' (NE 2014). A similar process is being used to characterise England's historic landscapes (Historic England 2015, Fairclough and Herring 2016) leading to character types such as 'Ancient enclosed land' and 'Settlement C20'. The concept of landscape character areas is widely used in ESIA, with character areas comprising areas of broadly similar landscape characteristics.

A Scandinavian team of Tweit et al. (2006) and Ode et al. (2008) has attempted to describe visual character using criteria such as complexity, coherence, disturbance, stewardship, 'historicity', and scale – see Box 11.2.

A key aspect of landscape character assessment is the distinction between the relatively value-free process of *characterisation* described above and the subsequent *evaluation*, or making of judgements that are based on knowledge of landscape character.

11.2.3 Spatial, static, temporal and dynamic approaches to understanding landscape

Stephenson (2010) proposes five types of landscape model, involving combinations of static-dynamic and spatial-temporal portrayals of the landscape (Box 11.3):

Box 11.2 Possible criteria for capturing visual character (Tweit et al. 2006, Ode et al. 2008)

Complexity	• Distribution of landscape attributes • Spatial organisation of landscape attributes • Variation and contrast between landscape elements
Coherence	• Spatial arrangement of water • Spatial arrangement of vegetation
Disturbance	• Presence of disturbing elements • Visual impact of disturbance
Stewardship	• Level of management for vegetation • Status and condition of man-made structures in the landscape
Imageability	• Spectacular, unique and iconic elements • Viewpoints
Visual scale	• Open area • Obstruction of the view
Naturalness	• Naturalness of vegetation • Pattern in the landscape • Water in the landscape
Historicity	• Vegetation with continuity • Organisation of landscape attributes • Landscape elements
Ephemera	• Season-bound activities • Landscape attributes with seasonal change • Landscape attributes with weather characteristics

Box 11.3 A multi-dimensional approach to the landscape (Stephenson 2010)

	Spatial portrayals of landscape qualities	Temporal portrayals of landscape qualities
Static portrayals of landscape qualities	Type A (static-spatial): Emphasis on the physical landscape	Type C (static-temporal): Emphasis on historic associations of the landscape
Dynamic portrayals of landscape qualities	Type B (dynamic-spatial): Emphasis on interactions between forms, relationships and practices, at a point in time	Type D (dynamic-temporal): Emphasis on interactions between forms, relationships and practices, over time
Type E (dynamic-spatial-temporal): Emphasis on interactions between forms, relationships and practices, over space and time		

Stephenson suggests that

> [L]andscape qualities are still overwhelmingly determined in formal landscape assessment methodologies as the qualities of 'space' rather than 'place'. They are mainly described with apparent rational and aesthetic authority by outsiders rather than as a phenomenological lived-in experience by insiders. Furthermore, they are overwhelmingly expressed in geographic or chronological terms, rather than in other ways of conceiving of and expressing temporality.
>
> (Stephenson 2010)

Although landscape assessment historically tended to focus primarily on static and spatial aspects as described by Stephenson (i.e. Type A with emphasis on the physical landscape and landscape as 'scenery'), progress is being made to recognise the more dynamic and temporal elements and the less tangible aspects of landscape. For example, cultural mapping and ecological-systems mapping may be classified as a dynamic-spatial approach (Type B), historic accounts and mapping of 'historic landscape character' may be classified as a temporal-static approach (Type C) and landscape histories and other accounts that trace people–place interactions over time may be classified as a dynamic-temporal approach (Type D) (Stephenson 2010). The UK Natural England's guidance on landscape character assessment (NE 2014) encourages recording of less tangible aspects such as time depth (history), cultural associations and memories. It also encourages consultation with local communities to understand values and sense of place. The ecosystem-services approach (see Chapter 8) also goes beyond the static and spatial by recognising the services provided by landscape, for example climate regulation, food provision and spiritual enrichment.

This chapter presents the current norm in landscape and visual impact assessment, which focuses on the static and spatial (i.e. the physical landscape), but also incorporates some dynamic elements such as human relationships with the landscape and temporal elements such as time depth.

11.2.4 Landscape value, landscape quality and visual amenity

Landscape value is concerned with the relative value that is attached to different landscapes by society (LI/IEMA 2013). This is often recognised through landscape designation, for example as National Parks. However, a landscape without any formal designation may also be valued by society, for example for its scenic quality, perceptual aspects (such as tranquillity or wildness), cultural associations, its functional role, or other conservation interests. The ecosystem-services approach (see Chapter 8) provides a framework for looking at whole ecosystems in decision-making, and for valuing the ecosystem services they provide, to ensure that society can maintain a healthy and resilient natural environment now and for future generations.

Landscape quality (or *condition*) can be defined as 'A measure of the physical state of the landscape. It may include the extent to which typical character is represented in individual areas, the intactness of the landscape and the condition of individual elements' (LI/IEMA 2013). Landscape quality can also contribute to landscape value.

The need to consider views and visual amenity as part of ESIA arises from the need to consider the effects of a project on population. The consideration of views represents the interrelationship between people and the landscape. *Visual amenity* can be defined as 'the overall pleasantness of the views people enjoy of their surroundings, which provides an attractive visual setting or backdrop for the enjoyment of activities of the people living, working, recreating, visiting or travelling through an area' (LI/IEMA 2013).

11.2.5 Links to other assessment topics

As shown above, landscape is inextricably linked to a range of other ESIA topics: geology, terrestrial and aquatic ecology, heritage, and more social/community concepts like residential amenity and social structures. The landscape chapter is thus likely to include cross-references to other ESIA chapters.

That said, it is critical that there is a distinction between which aspects of these topics are covered in which chapter. For example, it might be appropriate that the landscape and visual chapter addresses potential impacts of the proposed development on views from historically important sites (focussing on the impact on change in views experienced by visitors to site), while the cultural heritage chapter addresses the potential impact on the heritage significance of the site.

11.3 Key legislation and guidance

11.3.1 Legislation

Landscape legislation is typically limited to regulations that protect landscape designations. In the US, for instance, the so-called National Park Service Organic Act of 1916 set up National Parks with the purpose to

> conserve the scenery and the natural and historic objectives and the wildlife therein and to provide for the enjoyment of the same in such a manner and by such means as will leave them unimpaired for the enjoyment of future generations.

The UK National Parks and Access to the Countryside Act 1949 relates to National Parks and Areas of Outstanding Natural Beauty in England and Wales, and requires the conservation and enhancement of the natural beauty, wildlife and cultural heritage of their landscape. New Zealand's National Parks are designated by the National Parks Act 1980 as 'areas of New Zealand that contain scenery of such distinctive quality, ecological systems, or natural features so beautiful, unique, or scientifically important that their preservation is in the national interest'. That said, not all countries' National Parks have landscape at their heart: for instance, India's National Parks are primarily designated for wildlife.

The European Landscape Convention, also known as the Florence Convention (CoE 2000), promotes the protection, management and planning of European landscapes and organises European cooperation on landscape issues. The convention was adopted on 20 October 2000 in Florence (Italy) and came into force on 1 March 2004 (Council of Europe Treaty Series no. 176). It is the first international treaty to be exclusively concerned with all dimensions of European landscape.

11.3.2 Guidance

In the United States, Visual Impact Assessment and Visual Resource Management were developed in the 1970s by the Forest Service and the Department of Transportation, and subsequently the Federal Highway Administration. The Forest Design Service set the standard for integration of aesthetic considerations in large-scale resource management decisions. Where most other countries tend to separate the assessment of landscape effects from visual effects, in the US the focus tends to be on view management and impact on scenery.

A current key source of guidance in the UK is the Landscape Institute and Institute of Environmental Management and Assessment's *Guidelines for Landscape and Visual Impact Assessment*, Third Edition (LI/IEMA 2013). This provides guidelines to achieve consistency, credibility and effectiveness in landscape and visual impact assessment. The UK Landscape Institute (2011) has also published an Advice Note on *Photography and Photomontage in Landscape and Visual Impact Assessment* that is currently being updated. The Hong Kong Environmental Protection Department (2002) has prepared a guidance note on landscape and visual impact assessment. The New Zealand Institute of Landscape Architects has prepared similar guidance, as well as guidance on visual simulations (NZILA 2010a, b).

The IFC (2012) standards treat landscape under the heading of PS6 Biodiversity Conservation and Natural Resources. PS6 treats the landscape essentially as the area in which biodiversity conservation should take place, rather than as a topic of its own, for instance:

> For some projects, biodiversity values and ecosystem services associated with a site might be numerous … The relative importance with respect to conserving the feature as part of project operations could therefore be determined by its status in terms of these two axes: its irreplaceability in the landscape/seascape and its vulnerability in being able to remain there.

Other guidance documents on landscape and visual impact assessment relate to specific types of development. They include:

Roads:

- US Federal Highway Administration's (2015, updated from 1981) *Guidelines for the Visual Assessment of Highway Projects.*
- US National Cooperative Highway Research Program's (2012) *Evaluation of Methodologies for Visual Impact Assessment.*
- UK (former) Department of Transport's *Design Manual for Roads and Bridges* (DoT 1993).
- UK Department for Transport's *Transport Analysis Guidance: WebTAG* (DfT 2013).
- New South Wales (Australia) Roads and Maritime Service's (2013) *Guideline for Landscape Character and Visual Impact Assessment.*

Renewable energy:

- Scottish Natural Heritage's *Guidelines on the Environmental Impacts of Wind Farms and Small Scale Hydroelectric Schemes* (SNH 2001) and *Visual Representation of Wind Farms* (SNH 2017).
- BRE's (2013) planning guidance for the development of large-scale ground mounted solar PV systems, which contains a section on landscape and visual impact assessment.
- Northern Ireland Environment Agency's (2010) guidance on wind energy developments, which includes a section on landscape and visual impact assessment.

Other:

- US Department of Agriculture Forest Service's (1995) *Handbook for Scenery Management*, which focuses on preserving the value and importance of scenery in national forests.

11.4 Scoping and baseline studies

The scoping process is a crucial stage in ESIA, as it determines the extent of the landscape and visual assessment. The scoping process should identify the area in which landscape and visual effects should be considered, and the receptors or receptor groups that need to be assessed. The scope of assessment proposed should be proportional to the scale and nature of the proposed development and its likely effects on the receiving environment, and should focus on likely significant effects. For example, it is not necessary to undertake baseline studies for areas of landscape unlikely to be affected by a proposal.

The baseline studies should set out the study area and provide a digestible account of the area and project concerned, against which effects will be assessed. It is also necessary to investigate and record the landscape's likely evolution without the development.

11.4.1 Establishing the study area – the Zone of Theoretical Visiblity (ZTV)

The project area to be assessed should contain all of the likely significant effects of the proposal on any component of the landscape and visual resource. The term *Zone of Theoretical Visibility (ZTV)* is used to describe the area over which a development can theoretically be seen, and is usually computer-generated using a Digital Terrain Model (DTM) or Digital Surface Model (DSM). A DTM is a 3D computer model of the earth's surface without any surface objects, while a DSM is a computer-generated model that represents the surface height, including trees and buildings. A ZTV is also sometimes known as a Zone of Visual Influence, or a Visual Envelope Map. Figure 11.1 shows an example of a ZTV for a wind farm in Wales.

The extent of the ZTV should be agreed with the decision-making authority, particularly in the case of tall structures such as wind turbines. In undulating landscapes it is likely that landform will define the limit of the ZTV. However, in flat open landscapes or at sea the ZTV could theoretically extend as far as the curvature of the earth allows. In reality, of course, the extent of visibility will depend on atmospheric conditions and how far the human eye can 'see'. The extent of visibility will also depend on the size and prominence of the development being considered. As a general guide, for a wind farm in a remote area, it may be necessary to run the ZTV to 30 km or beyond; 10–20 km is usually sufficient for an industrial chimney stack, an industrial building or bridge. In urban environments it may not be necessary to consider such distances, since buildings are likely to limit the extent of visibility. Visibility studies of existing developments in the same type of landscape may help to determine the extent of visibility of such features.

More information on ZTV-generation may be found in the Landscape Institute's *Guidelines on Landscape and Visual Impact Assessment* (LI/IEMA 2013); and Scottish Natural Heritage's *Visual Representation of Wind Farms: Good Practice Guidance* (SNH 2017) for ZTVs specifically in relation to wind farms. Visual significance limits have been investigated further by CCW et al. (2001), including how the discernible level of detail on a landscape diminishes with distance.

Viewsheds are similar to ZTVs but map the geographical area that is seen from a specific viewpoint, rather than the area from which a specific feature can be seen. An example of a viewshed is included at Figure 11.5.

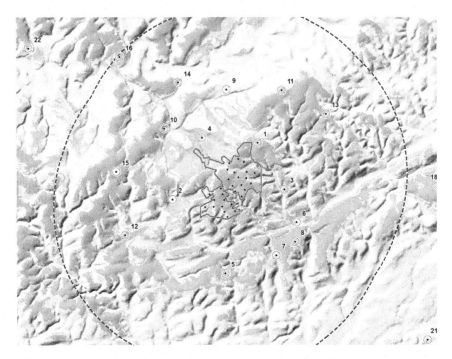

Figure 11.1

ZTV generated using a DTM for a wind farm in Wales, produced by LUC for RWE npower renewables (now RWE Innogy UK). The shades of grey (colours in the original) represent the number of turbines that could be visible from a given point. Numbered points represent assessment viewpoints.
Source: Reproduced with the permission of RWE Innogy UK

11.4.2 Landscape baseline

The key aim of the landscape baseline is to identify and evaluate the existing landscape of the site and surrounding area, including the individual elements that make up the landscape, landscape characteristics and character. Combinations of elements ('characteristics') give an area its particularly distinctive sense of place or character. The baseline should cover the underlying physical influences, such as geology and soils, land cover, the influence of human activity, and aesthetic and perceptual aspects, including how these come together to create character. Reference to historic landscape character assessments should be made where relevant.

The analysis of the baseline will start with an analysis of existing documents: maps (current and historical) of the area, maps of landscape and other designations, aerial photos, and other photos where these are available. The aim is to start identifying landscape character areas and the sensitivity of these areas. A field survey then provides an opportunity to check desk-based information in the field and to gather information on potential landscape receptors (for example, landscape elements and character types or areas) and analyse the value or importance of these.

The result is a breakdown of the study area into landscape character areas or receivers, with a description of each of these. For instance, Figure 11.2 shows a breakdown into seven landscape-character areas of a landscape that will be affected by a new road development in Australia. This example will be continued later in the chapter, focussing on Area 4, an industrial area, the character area to the north of the junction.

Photographs such as those at the bottom of Figure 11.2 and in Figure 11.3 are also useful to illustrate the baseline landscape.

As well as addressing landscape character, the description of the landscape resource should address *landscape quality (condition)* and *landscape value*. Landscape quality is based on judgements about the physical state of the landscape, and about its intactness and the state of repair of individual features and elements. Landscape designations (international > national > local > none) can be an indicator of the recognised value of a landscape. The location of landscape designations should be presented clearly, outlining the reasons for their designation. Non-designated landscapes may also be valued by different stakeholders for different reasons, and these values should be recognised as part of the baseline (see §11.2.4). Communities and individuals associated with a given landscape may conceive of and express landscape qualities in very different ways to discipline-based experts. Participative/consultative techniques involving asking people about how they use and perceive the landscape can help identify the qualities of landscapes that are important to those communities for whom landscape is a lived-in experience – particularly indigenous communities (Stephenson 2010).

Evaluation of *landscape sensitivity* should consider:

- The susceptibility of the landscape to change, i.e. the ability of the landscape receptor to accommodate the proposed development without undue consequences for the maintenance of the baseline situation and/or to achieve landscape planning policies and strategies (LI/IEMA 2013).
- The value of the landscape receptor, i.e. as set out in the baseline.

For instance, the landscape shown at Figure 11.2 is less susceptible to a road proposal than that of Figure 11.3.

The results of such an evaluation should be presented in a clear and structured way. Table 11.1 shows an example baseline description from a rail project in Hong Kong.

11.4.3 Visual baseline

Desk studies, typically including a ZTV and base maps, provide the starting point for the visual assessment. The key aim of the visual baseline is to identify and evaluate the existing visual relationships between the site and its surrounding area, and to identify representative viewpoints and key sensitive viewers. These typically include:

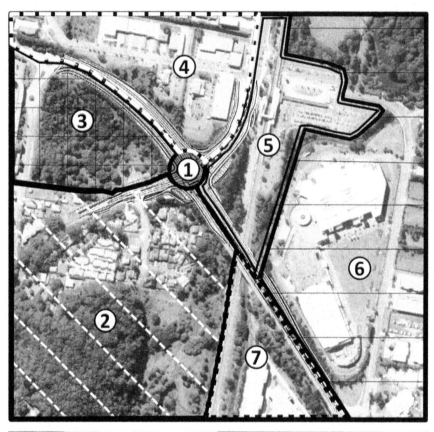

Figure 11.2

Use of landscape character areas as landscape receptors for a road junction proposal (Peter Andrews + Associates 2012)
Map data: ©Google Earth

Figure 11.3

Baseline landscape: this photo illustrates the distinctive chalk landscape of the South
Downs National Park in the UK, including the steep, wooded chalk scarp that marks the
northern edge of the downs (LUC)

- Representative viewpoints that represent the views of the visual receptors
  identified as likely to experience effects, for example a point along a
  footpath or in a local community.
- Specific viewpoints where people go to experience a particular view, such
  as promoted viewpoints or designated viewpoints where benches and
  interpretation facilities may be provided.
- Illustrative viewpoints to demonstrate a particular effect (which can
  include restricted visibility).

Viewers typically include local communities, employees at their places of work,
people taking part in recreational activities, and people passing through the
area on roads and other public rights of way. The field survey provides an
opportunity to test the ZTV, identify/confirm key sensitive receptors and
viewpoints, and record the existing visual amenity of the affected population.

Depending on the country, only publicly accessible viewpoints may need to
be considered; for instance, occupiers of residential properties do not have a
right to a view in UK law. That said, private views from individual residences
may be of concern where the change in view may be so great as to adversely
affect living conditions (for example due to the proposed development being
overbearing or dominating). Some residential visual-amenity assessments have
been undertaken in relation to UK wind farms where large wind turbines may

Table 11.1 Landscape baseline description for a rail project in Hong Kong (excerpt from MTR 2011)

Landscape character areas	Description	Sensitivity
Hong Kong Island Residential Urban Fringe LCA (Local Character Area)	This is a large area lying at the edges of Hong Kong Island on peripheral hillsides. There are medium to low-density residential development with winding roads in vegetated or wooded settings. The result is a fairly coherent residential landscape with a high coverage of vegetation, which possesses a relatively informal and tranquil character. The landscape is mature and high quality and has little ability to accommodate change.	High
Admiralty Institutional LCA	This LCA is characterised predominantly by institutional land uses including High Court, Government Office, The British Council and British Consulate General. Vegetation is limited to a few street trees and a shrub-planting area in front of the perimeter fence. This type of landscape is common with high ability to accommodate changes.	Low
Wan Chai Civic Urban Waterfront LCA	This is an urban waterfront in Wanchai. Part of the area is the future reclaimed land for preliminary open-space development. This landscape consists of a mixture of public open space and civic buildings, with some commercial land uses at close proximity to Victoria Harbour. It is a flat topography with a formal promenade or linear public space.	Medium

affect the living conditions of nearby residents. Other impacts can also affect the living conditions of residents, including noise, dust, changes to daylight/sunlight and, in the case of wind turbines, shadow flicker. These impacts can be brought together in a separate chapter of the ESIA report dealing with impacts on residential amenity.

Baseline mapping should show the location of representative viewpoints and photographs should be used to illustrate the existing visual amenity in and around the site, as well as the views from the representative viewpoints. Figure 11.4 shows an example of a viewpoint associated with a wind farm project in Scotland (note that a viewpoint location map would be produced alongside the photograph).

Evaluation of viewer *sensitivity* should consider the susceptibility of people to the change in views and the value they attach to particular views (LI/IEMA 2013). Susceptibility relates to the occupation of the receptor group and the extent to which attention is likely to be focussed on their visual environment.

Figure 11.4

An example of baseline photography for an assessment viewpoint accompanied by technical photography details, produced by LUC for RWE npower renewables (now RWE Innogy UK)

Source: Reproduced with permission of RWE Innogy UK

For example, people using footpaths or visiting tourist attractions are likely to have a higher susceptibility than people at their place of work or engaged in sport which is not dependent upon appreciation of views of the landscape. View-value may be informed by planning designations, appearance in guidebooks, or provision of facilities for their enjoyment (LI/IEMA 2013).

Table 11.2 gives examples of the assessments of sensitivity of visual receptors for a rail project in Hong Kong: note how just the change of use between the International Finance Centre (commercial) and Fenwick Pier (recreational) leads to a different judgement about sensitivity.

11.5 Impact prediction and evaluation

Successful landscape/visual impact assessment involves effectively communicated predictions of the nature, likelihood and significance of changes that may occur as a result of the proposed development, and the incorporation of good project design from the beginning.

Table 11.2 Example of baseline visual description and sensitivity judgements for a rail project in Hong Kong (excerpt from MTR 2011)

Visually sensitive receiver (VSR)	Type of VSR	Number of individuals (many/ medium/ few)	Quality of existing view (good/ fair/ poor)	Availability of alternative views (yes/no)	Degree of visibility (full/ partial/ glimpse)	Duration of view (long/ medium/ short)	Frequency of view (frequent/ occasional/ rare)	Sensitivity (low/ medium/ high)
International Finance Centre	Commercial	Medium	Good	Yes	Full	Medium	Occasional	Medium
160–169 Gloucester Road	Commercial/ residential	Medium	Good	Yes	Full	Long	Frequent	High
Fenwick Pier Street Public Open Space	Recreational	Medium	Good	Yes	Full	Medium	Occasional	High
Harcourt Road	Travelling	Many	Fair	Yes	Glimpse	Short	Occasional	Low

11.5.1 Good project design

The project's location, dimensions (especially vertical), materials, colour, reflectivity, visible emissions, access routes, traffic volumes and construction programme will all need to be described in the ESIA. Good project design and landscape/visual mitigation should be planned in at the start of the project, including: use of landscape issues as a criterion in the selection of the project site or process (e.g. landfill v. incineration); careful siting of major structures, access routes and parking, materials storage etc. in relation to visual receptors, ridgelines/valleys, etc.; sensitive choice of site levels; attention to the density, mix, height and massing of buildings; retention of special landscape features and provision of visual/ecological buffer zones; consideration of microclimates and the solar aspect of buildings; attention to materials used and details such as openings and balconies; careful design of open spaces, including plantings and fencing; and enhancement through new wildlife habitats, the restoration of derelict land, and the provision of public open space and/or beautiful new landscapes (Barton et al. 1995, Hankinson 1999).

Viewsheds can also be modelled from viewpoints of interest to inform project design and to ensure components are positioned to avoid key areas of sensitivity. Figure 11.5 is an example of a viewshed showing areas that are visible from the viewpoint at ground level. This technique can be used to inform the design and placing of objects in the landscape to minimise their visibility from important viewpoints.

11.5.2 Assessing landscape effects

The landscape impact assessment should describe the likely nature and scale of changes to individual landscape elements and characteristics, and the consequential effect on the landscape character resulting from the proposed development. This requires consideration of the characteristics of the project, the sensitivity of the area likely to be affected by the project (from the baseline analysis) and the nature of change (magnitude).

The *magnitude of change* relates to the project size or scale, geographical extent, duration and reversibility. Size/scale of change is dependent on the degree to which the character of the landscape is changed through removal of existing landscape components or addition of new ones, and is often described as large, medium, small or imperceptible/no change as illustrated in Table 11.3.

Geographical extent describes the area over which the change is experienced. The duration is short, medium or long term and the reversibility considers whether an effect could, or would, be reversed.

The changes are likely to vary between different project stages (e.g. construction and operation, or phases of mineral extraction), between seasons, and even between different times of day and different light/atmospheric conditions. Where the project involves night-time lighting, this will also need to be considered. The predictions should discuss the duration and timing of impacts, and of impacts with and without mitigation.

Figure 11.5

Viewshed from a viewpoint in Scotland produced by LUC for ScottishPower Renewables to test visibility of an access track to a wind farm from this sensitive viewpoint
Source: Reproduced with the permission of ScottishPower Renewables

The sensitivity and magnitude judgements are brought together to judge the overall level of effect and its *significance*. This process can be informed by matrices, such as the matrix in Figure 1.3, but also uses informed professional judgement. Table 11.4 shows an example of a landscape assessment for a road

Table 11.3 Possible definition of scale of landscape change

Scale of change	Definition
Large	An obvious change in landscape elements, character and quality of the landscape.
Medium	A discernible change in landscape elements, character and quality of the landscape.
Small	A minor change in landscape elements, character and quality of the landscape.
Imperceptible/ no change	No perceptible change in landscape elements, character and quality of the landscape.

Table 11.4 Example of landscape assessment – for Area 4 from Figure 11.2 (Peter Andrews + Associates 2012)

Summary	• Industrial and commercial area incorporating one- and two-storey buildings. The rear of the Petbarn, and McDonald's and Hungry Jack's restaurants adjoin Wyong Road; however vegetation screens this view.
Topographic features	• Landform slopes away from roadway.
Drainage/hydrology	• Engineered drainage associated with road corridor. • Flood prone low elevation land north-west of road corridor.
Geology/soils	• Vegetation adjoining the roadway edge generally growing on fill batters. • Adjoining low lying soils associated with broad floodplains with restricted drainage.
Vegetation type/cover	• Vegetation with Casuarinas (8–10 m) on western side of roadway. • Open character associated with development along Anzac Road includes turf grass with new planting of *Cupanopsis anacardodes* along section opposite railway station. • *Metaleuca biconvexa* located at the rear of the McDonald's restaurant site. • Mix of individual tree planting of semi-mature *Findersia sp.* and *Gazania sp.* ground cover in the median.
How development fits into setting	• Single-storey brightly painted commercial development is visually prominent on the corner of Anzac Parade and extending for a section along the northwest edge of the roadway.
Architectural form/history/mix and quality	• One- and two-storey commercial and industrial development comprising Tuggerah Straight industrial area. • Buildings front onto Anzac Road and the rear of the Petbarn, McDonald's and Hungry Jack's restaurants and associated car-parking adjoins Wyong Road.
Spatial quality of area – open or closed	• The area is generally open.
Infrastructure – scale/pattern	• Bus lane • Street lighting • Turning lane to Anzac Road • Signalised intersection at Anzac Road with pedestrian crossing and bus stop.

Sensitivity	Low	Comments
Magnitude	High to moderate	The proposal would remove most of the existing roadside vegetation and expose
Impact	Moderate	the road and proposed retaining walls.

junction proposal in Australia. This shows how the baseline landscape description and judgements on sensitivity and magnitude combine to provide an overall impact.

11.5.3 Assessing visual effects

The visual impact assessment should describe the likely nature and scale of changes in views resulting from the development and the changes in the visual amenity of the visual receptors. This requires consideration of the nature of the visual receptors (sensitivity), and the nature of the change (magnitude) to come to an overall judgement on the significance of the effect.

Magnitude of change to views and visual amenity should consider issues such as the size/scale of change, geographical extent, duration and reversibility. The scale of change relates to the extent of change in composition of the view due to loss or addition of elements, the degree of contrast or integration of new elements, distance from the proposal, and whether the view is glimpsed (perhaps due to screening) or uninterrupted. Scale of change is often described as large, medium, small or imperceptible/no change, as illustrated in Table 11.5.

Geographical extent relates to the extent of the area over which the change would be visible (for example is the change glimpsed from a small section of footpath or seen clearly from the whole length of path?) and the numbers of people affected. The duration is short, medium or long term and the reversibility considers whether an effect could, or would, be reversed.

The sensitivity and magnitude judgements are brought together to judge the overall level of effect and its *significance* in the same way as for landscape, using informed professional judgement. Table 11.6 shows an example of how magnitude of change and sensitivity of viewers combine to result in an overall impact for different visual receptor groups, for the Australian road intersection example at Figure 11.2.

Depending on the project, magnitude of change may vary between different project stages (e.g. construction and operation, or phases of mineral extraction), and between seasons. Where the project involves night-time

Table 11.5 Possible definition of scale of visual change

Scale of change	Definition
Large	The development has a defining influence on the view and becomes a key focus.
Medium	The development is clearly visible and forms an important element of the view.
Small	The development is visible, but forms a minor element of the view.
Imperceptible/no change	No perceptible change in view.

Table 11.6 Example of visual assessment for different receptor groups – for Area 4 from Figure 11.2 (Peter Andrews + Associates 2012)

	Motorist	Ped/cyclist	Resident	Comments
Sensitivity				Mature vegetation would be removed along Wyong Road as part of the proposal. The land uses in this area include a Petbarn and McDonald's and Hungry Jack's restaurants. These buildings face onto Anzac Road. The rear of these buildings and car parking adjoin Wyong Road. Some vegetation would remain between these buildings and Wyong Road reducing the impact.
• Duration	Low	Low		
• No. of viewers	Moderate	Moderate		
• Viewer sensitivity	Low	Moderate		
View sensitivity	Low	Low		
Sensitivity impact	Low	Moderate to low		
Magnitude				
View distance/ proximity	High	High		
Visibility in relation of the field of view	Low	Low		
Magnitude impact	Moderate	Moderate		
IMPACT	**Low**	**Moderate**	**N/A**	

lighting, this will also need to be considered. The predictions should discuss the duration and timing of impacts, and impacts with and without mitigation.

11.5.4 Illustrating visual effects

The overall aim of photography, photomontage, virtual reality and other forms of visualisation is to represent the landscape and visual context under consideration – and the proposed development –as accurately as is practical (LI 2011, An and Powe 2015). Photomontages are illustrations that aim to represent an observer's view of a proposed development. However, as carefully as visual representations are presented – be it through hand-drawn sketches, photographs, photomontages or virtual reality – they can never precisely match what is experienced in reality (An and Powe 2015, Corry 2011, SNH 2014). For example, photographs cannot accurately represent what is seen by the human eye, which may experience a brightness ratio of 1000:1 between the brightest and darkest shades, whereas a good-quality computer monitor is only likely to achieve a ratio of about 100:1, and a printed image is only likely to manage 10:1 (SNH 2014).

The type of visualisation produced may also depend on the amount of information available about the development, and whether it is an outline or detailed planning application. Figures 11.6 and 11.7 illustrate two types of

Figure 11.6
_____

An example of a block model to show size and location of development, taken from the outline planning application for major development at the University of Sussex, produced by LUC

Source: Reproduced with the University's permission

Figure 11.7
_____

An example of a fully rendered image to show size and location of development, as well as architectural detailing and finishing materials (image shows proposed new visitor facilities at Maeshowe in Orkney)

Source: Produced by LUC, reproduced with Historic Environment Scotland's permission

approach. The first is a simple block model, which is sufficient to illustrate the visibility and scale of a potential development, while the second shows a more detailed image including architectural details and materials.

When preparing visualisations, it is important to select the most appropriate combination of camera, lens and final presentation of image to represent the proposal in its landscape context. A camera with a fairly high resolution sensor should be selected (a 12-megapixel sensor is usually sufficient when using a 50 mm focal length standard lens producing a 40-degree field of view). A telephoto lens will provide greater detail and a narrower field of view while a wider angle lens will provide less detail and a wider field of view. Use of a fixed-focal-length lens is recommended to make sure the image parameters of every photograph are the same.

The proposal under consideration and its landscape context will determine the horizontal field of view required for photography and photomontage from any given viewpoint. This will in turn determine whether a single-frame image will suffice or whether a panorama will be required. If separate images are to be merged into a panorama it is recommended that a tripod with a panoramic head is used.

The final presentation of images should aim to produce printed images of a size and resolution that match the perspective and detail seen in the field when viewed at a certain distance. The distance (of the eye from the visualisations) at which the perspective in the photograph correctly reconstructs the perspective seen from the real-life view is called the 'viewing distance'. Guidance from the UK Landscape Institute (2011) suggests that all photographs and photomontages used in a document should have the same viewing distance, and that this should be between 300 mm and 500 mm (with larger images and longer viewing distances preferable). The viewing distance and the horizontal field of view together determine the overall printed image size.

Technology is moving forward all the time and new tools and technique are being introduced, such as video montage, animation and virtual reality. Video montages are expensive to produce and are rarely used, but animated photomontages are becoming more popular in schemes involving moving features, such as wind turbines.

Further information on methods and guidance for the production of photomontages is contained in the UK's Landscape Institute's *Advice Note 01/11 Photography and Photomontage in Landscape and Visual Impact Assessment* (LI 2011); Scottish Natural Heritage's *Visual Representation of Wind Farms* (SNH 2017); and New Zealand Institute of Landscape Architects' *Best Practice Guide on Visual Simulations 10.2* (NZILA 2010b).

11.5.5 Assessing cumulative and daylight/sunlight effects

Cumulative effects can be defined as 'the additional changes caused by a proposed development in conjunction with other similar developments or as the combined effect of a set of developments, taken together' (SNH 2012). For instance, one viewer may already be able to see one or more wind farms, in addition to the one proposed; or a proposed residential development may require new road access, both of which will affect the landscape. Today's landscape is often the result of the cumulative impact of multiple actions and developments over time.

Assessing cumulative effects is potentially a very complex area of work, and methods are still evolving. Some of the current issues that are being debated include:

- *Status of schemes that should be included in an assessment* – for instance whether other projects that have not yet received planning consent should be included (those that have received planning consent but have not yet been built should typically be included as a matter of course).
- *Type of schemes that should be included in an assessment* – cumulative assessments, particularly those for wind energy developments, often focus on the impact of a proposal in the context of other schemes of the same development type, but in some instances it may be necessary to consider the impact of a proposal in the context of *all* types of change.
- *Incremental or cumulative effects* – 'incremental' effects are those arising from the proposed project, assuming other proposed changes are already part of the baseline, while 'cumulative' effects are those arising from all the proposed changes together with the project being assessed. There are debates about terminology, when each assessment is expected, and who is responsible for assessing each – see Chapter 19.

In general, a cumulative assessment should be reasonable and in proportion to the nature of the project by focussing on likely significant effects, and should:

- Consider the incremental effects of the proposed action as well as the cumulative effect of the proposed action and other actions.
- Analyse effects using quantitative techniques, supported by qualitative discussion based on best professional judgement.
- Consider mitigation, which may require partnership working.
- Clearly state the significance of residual effects.
- Include visualisations to demonstrate effects where they are likely to be significant.

The assessment of cumulative landscape and visual effects, and the acceptability of these effects, is closely linked to concepts of landscape sensitivity, landscape capacity and the limits of acceptable change.

Assessment of a new development's impacts on the daylight/sunlight of existing properties, or of how much daylight/sunlight the residents of a new development will receive, is a specialist discipline normally carried out by a separate consultancy. It is discussed in Box 11.4.

Box 11.4 Daylight and sunlight assessment

Alistair Redler, Delva Patman Redler LLP

When a development proposal is tall or bulky, and near properties that are sensitive to light levels (notably residential properties), or if it is for apartments, or for residential properties near other tall or bulky buildings, the ESIA may need to include a daylight and sunlight assessment. Daylight and sunlight assessments calculate the change in the area of sky that can be seen from a window or within a room as a result of a built obstruction. The assessments do not calculate actual illuminance, i.e. the level of natural light in a room derived from direct sky visibility. In England and Wales, daylight and sunlight assessment is done in accordance with the recommendations set out by the BRE (2011). Standards for natural lighting are also set in places such as Hong Kong (HKG 2015) and Sydney (New South Wales 2015). Typically, daylight and sunlight assessments are carried out using specialist computer programmes.

Daylight
For assessing daylight to windows, three main methods of calculation can be used:

- The vertical-sky component (VSC) – the percentage of the total sky that can provide direct light to the centre of the face of a window when neighbouring obstructions are taken into account.
- The no-sky contour assessment (NSC) – the area of a room on a working plane (850 mm above floor level) that can receive direct light from the sky through the window.
- The average daylight factor (ADF) – the level of illuminance within a room taking account of the vertical sky component but with regard to the window size, transmittance of glazing, reflectance of room surfaces and room surface area.

The daylight standard in the BRE (2011) report states that the diffuse daylighting of a neighbouring building may be adversely affected if either the vertical sky component measured at the centre of an existing main window is less than 27 per cent and less than 0.8 times its former value, or if the area of the working plane in a room that can receive direct skylight is reduced to less than 0.8 times its former value. In simple terms, in an urban situation, it is permitted to reduce the sky visibility by 20 per cent from existing before it becomes adversely noticeable to a neighbouring occupier.

The ADF can be used to show that a room will have suitable illuminance even if the VSC or NSC are reduced below an acceptable level. This may be the case where a room has large or multiple windows where the internal illuminance will be better than the external sky visibility may necessarily indicate. The ADF is also used to assess the illuminance that will be available to rooms within a new development, and can be used to help establish an appropriate ratio of room size to window area and the placing of external obstructions such as overhead balconies.

Sunlight
The sunlight assessment calculates the annual probable sunlight hours available to the centre of a window. The annual probable sunlight hours are the number of hours in a year that a window can be expected to receive direct sunlight, taking account

of external obstructions and the likelihood of cloud cover throughout the year. For example, in London a south facing window unaffected by external obstructions can be expected to receive direct sunlight on its face for 1,486 hours in a year. That equates to an average of 4 hours a day. Three principal assessments are usually carried out:

- annual probable sunlight hours to a window;
- shadow analysis to amenity space to determine area in permanent shadow;
- sun-path analysis to a window.

The annual probable sunlight assessment is usually carried out on windows that face within 90° of due south in Northern latitudes (and vice versa) and therefore have a reasonable expectation of direct sunlight. The recommended BRE (2011) standard is that a window should be able to receive at least 25 per cent of annual probable sunlight hours including at least 5 per cent of sunlight hours during the winter months (in the UK, between 21 September and 21 March). If the available sunlight is reduced to below this minimum standard and the reduction is more than 20 per cent from the existing, then a reduction in sunlight may be unacceptable. Different standards apply in other jurisdictions. In dense urban environments it can be difficult to provide new accommodation where all new dwellings achieve these standards. The New South Wales (2015) standards, for instance, allow up to 15 per cent of apartments in a new building to receive no direct sunlight between 9 a.m. and 3 p.m. at midwinter.

The shadow analysis assesses the area of a garden or amenity area that is in permanent shadow on 21 March. The BRE (2011) recommends that 50 per cent of any garden or amenity area should be able to receive at least two hours of sunlight on 21 March in order to be adequately sunlit. The shadow is usually assessed by using three-dimensional computer modelling and running software to show the shadow cast at hourly intervals on a specified day. The total impact of the shadow can be assessed and plotted to show the area of permanent shadow in graphical form.

The sun-path assessment calculates the actual path of sunlight on any particular day or throughout a year. This can be valuable where the sunlight standard is not met but it is necessary to identify the likely periods over which sun will be received on the face of the window and whether that is at a time that will be materially useful for the occupation of the property.

Practical application of daylight and sunlight standards

Where a daylight and sunlight study is required, the developer should incorporate this assessment as part of the design process. It should not be necessary to develop a design purely to comply with daylight and sunlight standards as many other factors also need to be considered for an optimum design. However, care should be taken to avoid an unacceptable and avoidable adverse impact on neighbouring properties.

The ADF and sunlight assessments should be used to ensure that new units being designed within the development will have adequate daylight and sunlight. These can help guide appropriate size, shape and design of rooms and the placement of key habitable rooms within a building. A key consideration is the size and placing of balconies, with an increasing desire for external amenity space to be provided to apartments but with balconies inevitably blocking sky visibility, and therefore daylight and sunlight, to rooms beneath.

References
BRE (Building Research Establishment) 2011. *Site layout planning for daylight and sunlight: a guide to good practice 2011*.
HKG (Hong Kong Government) 2015. *Lighting and ventilation requirements – performance-based approach*. APP-130. Hong Kong: Buildings Department.
New South Wales 2015. *Practice note: solar access requirements in SEAPP 65*. Sydney: NSW Government.

11.6 Mitigation and enhancement

The purpose of mitigation is to prevent/avoid, reduce, and where possible offset or remedy (or compensate for) any significant adverse landscape and visual effects (LI/IEMA 2013). The preference is to prevent or avoid adverse effects in the first place. If adverse effects cannot be prevented, then they will need to be reduced as far as possible. Unavoidable adverse effects will need to be offset, remedied or compensated for.

As discussed earlier, good project design is the most effective and often cheapest way to minimise negative and optimise positive landscape/visual impacts. The landscape assessor should therefore be part of a landscape design team to ensure that the mitigation measures are designed as part of an iterative process of project planning and design.

Mitigation may fall into three categories: primary measures that intrinsically comprise part of the development design through an iterative process; standard construction and operational management practices for avoiding and reducing effects; and secondary measures designed to specifically address the remaining residual adverse effects of the final development proposals.

Common mitigation measures include:

- Sensitive location and siting – such as avoiding disturbing areas of highly valued historic or naturalistic landscapes, and siting developments to minimise visibility.
- Site layout to ensure the location of elements within the site minimise impact on valued elements, characteristics and views.
- Adjustment of site levels and use of structured planting to minimise visibility of the development – see Figure 11.8 for an example.
- Use of appropriate design, materials and finishes to ensure built features integrate with, or enhance, their surroundings (assuming it is not possible or desirable to screen or hide the development).
- Lighting design to minimise light spill and light pollution.
- Use of camouflage or disguise to minimise impacts where it is not possible to screen developments, for example the use of green roofs or living walls.

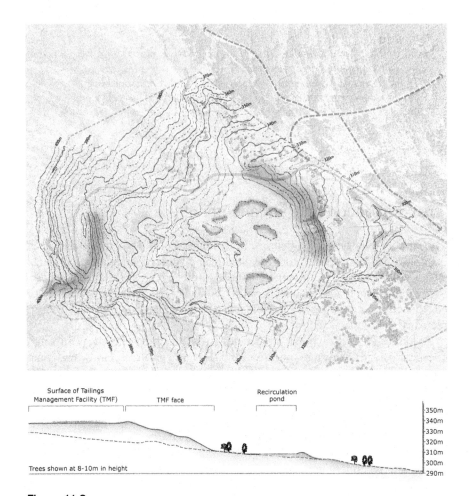

Figure 11.8

Example of restoration plan for a gold mine in Scotland, showing landscape and visual
mitigation proposals including ground modelling and tree planting to form a naturalistic
landscape, produced by LUC for Scotgold Resources Ltd
Source: Reproduced with the permission of Scotgold Resources Ltd

It is good practice to consider *landscape enhancement* in addition to
landscape mitigation. Landscape enhancement explores opportunities for the
development to contribute positively to the landscape and enhance visual
amenity through improving landscape management, restoring historic
landscapes, restoring habitats or features, or reclaiming derelict land.

Landscape/visual mitigation needs to be considered in conjunction with the
other types of impacts identified in the ESIA, and mitigation for other impacts
may have landscape implications. For instance, a noise barrier or bund will

have landscape and visual impacts. As well as how they affect the ecology, landscape plantings need to be considered in terms of whether they would thrive in the underlying soil or affect underground infrastructure such as pipes. Some of these issues are brought out by the example mitigation measures described in Box 11.5, again relating to the road junction example discussed earlier.

Box 11.5 Example mitigation measures (Peter Andrews + Associates 2012)

Landscape treatment generally:

- Enhance the road corridor with new planting where possible.
- The planting palette should be consistent with the existing planting treatment along Wyong Road on the western approach and recent work on the Pacific Highway northern approach.
- Provide an attractive, vegetated median with low-maintenance frangible species.
- Where possible provide planting to screen views of the rear of existing buildings in the north-west and north-east quadrants of the proposed works.
- Adopt appropriate plant species where new landscape works adjoin threatened ecological communities.
- Location of large trees needs to have regard to services and maintenance access requirements, and potential overshadowing of residential properties.
- Existing vegetation should be maintained and protected wherever possible noting that the proposal would result in the unavoidable loss of a considerable amount of existing roadside vegetation.

Structural elements:

- Retaining walls are a major element affecting the visual character, and the design should adopt a consistent approach to the treatment of retaining walls, including surface treatments, colours and detailing.
- Minimise the opportunity for graffiti in the selection of surface texture for the retaining walls.
- Screen plantings should be used to reduce the visual impact of the retaining walls.
- Develop a consistent approach to the detailing of barriers, street lighting, pedestrian fences and safety/privacy screens and, to the extent possible, integrate with the detailing of existing roadside element.
- Improve the quality of the existing underpass and its proposed extension ensuring it is well lit and incorporating appropriate all surface treatments, e.g. working in conjunction with local community groups and street artists as has been done with the Kariong underpass.
- If acoustic protection is required above the proposed 1.8 m privacy fence, transparent panels should be used. Architectural treatment of individual dwellings would be a preferred option if acoustic mitigation is required.

Some mitigation measures are not immediately effective. For example, planting will take time to mature and early planting in advance of the development may be necessary. Growth rates will depend on the soil type, climate, species, size at planting and seasonal weather, as well as how well they are planted and maintained. Planting semi-mature vegetation will give a more instant effect, but will be costly, will require extra care in planting and watering, and the plants are likely to take longer to establish, so that overall growth rates may be slower than younger plants. Any assumptions on planting sizes and expected growth rates should be clearly explained and documented as part of the mitigation proposals.

11.7 Monitoring

Monitoring is gradually being recognised as an essential element in environmental management and ESIA, particularly where initial predictions of landscape effects or the effectiveness of proposed mitigation measures are uncertain. Monitoring can test whether the predicted effects occur and whether any unforeseen effects arise; and it can ensure that mitigation measures are implemented and that they are effective in avoiding or reducing the predicted adverse effects. Figure 11.9 shows a photomontage prepared for a wind farm ESIA and a photo of the constructed project. In this case, the actual and predicted visual impacts are very similar.

Monitoring also provides a learning experience, which may feed directly into other projects or provide feedback on the success of assessment techniques. Both the developer and the regulatory authority should benefit from the experience – the developer by ensuring a successful project, and the authority by ensuring landscape objectives are achieved.

11.8 Conclusions

This chapter has attempted to define concepts relevant to landscape and visual impact assessment: to describe how landscape and visual data are collected; how landscape and visual effects are predicted; and to highlight potential mitigation measures. Landscape and visual impact assessment is a continually evolving process and new techniques are being developed all the time. Limitations associated with the methods have been pointed out where relevant.

In conclusion, the following are some of the key concerns of the landscape specialist in undertaking landscape and visual impact assessment:

- Given that landscape and visual impacts are inevitably qualitative, rather than quantitative, there will always be a *subjective element* in the analysis of landscape and visual effects. This makes it different to other studies carried out as part of an ESIA.

Figure 11.9

Monitoring landscape and visual impacts. Top: photomontage of the Cotton Farm Wind Farm in Huntingdonshire, UK, produced for RWE npower renewables (now RWE Innogy UK) as part of the Environmental Statement in 2007. Bottom: photograph of the Cotton Farm Wind Farm as built in 2014 (LUC).

- The greater public awareness of the effects of development on the landscape means that *consultation* is an increasingly important part of landscape and visual impact assessment (LVIA).
- Since landscape professionals are often involved in the design of the landscape, as well as undertaking the landscape and visual impact assessment, the assessment must proceed as an *integral part of the scheme design*.
- Proportionality is required to ensure the assessment focuses on the likely significant effects. The study area, scope, assessment detail and choice of visualisation techniques should be in proportion to the scale and potential effects of the project, with larger projects affecting more sensitive areas providing greater detail than smaller projects affecting less sensitive areas.
- Methods and techniques for undertaking *cumulative landscape and visual impact assessment* are still evolving – there is likely to be further guidance emerging in this area.
- Continued development is to be expected in the area of *visualisation*. A thorough understanding of the limitations of photomontages and other techniques is crucial to ensure realistic interpretation of such illustrations.

References

An K and NA Powe 2015. Enhancing 'boundary work' through the use of virtual reality: Exploring the potential within landscape and visual impact assessment, *Journal of Environmental Policy & Planning* 17(5), 673–690.

Barton H, G Davis and R Guise 1995. *Sustainable settlements: A guide for planners, designers and developers*. Luton: University of the West of England and Local Government Management Board.

Brabyn L 2009. Classifying landscape character. *Landscape research* 34(3), 299–321.

BRE (Building Research Establishment) 2011. *Site layout planning for daylight and sunlight: a guide to good practice 2011*.

BRE 2013 Planning guidance for the development of large-scale ground mounted solar PV systems. www.bre.co.uk/filelibrary/nsc/Documents%20Library/NSC%20Publications/NSC-publication-planning-guidance.pdf.

CCW (Countryside Council for Wales), Brady Shipman Martin and University College of Dublin 2001. *Guide to best practice in seascape assessment*. Maritime Ireland/Wales INTERREG 1994–1999.

COE (Council of Europe) 2000. *European landscape convention*, Florence, 20 October 2000.

Corry RC 2011. A case study on visual impact assessment for wind energy development, *Impact Assessment and Project Appraisal* 29(4), 303–315.

Department of Agriculture Forest Service 1995. Agriculture Handbook Number 701, *Landscape Aesthetics: A Handbook for Scenery Management*.

DfT (UK Department for Transport) 2013. *Transport analysis guidance: WebTAG*. Available online at: www.gov.uk/guidance/transport-analysis-guidance-webtag.

DoT (UK Department of Transport) 1993. *Design manual for roads and bridges, Vol.11: Environmental assessment*. London: HMSO.

Fairclough G and P Herring 2016. Lens, mirror, window: interactions between Historic Landscape Characterisation and Landscape Character Assessment, *Landscape Research* 41(2), 186–198.

FHWA (US Department of Transportation Federal Highway Administration) 2015. *Guidelines for the visual impact assessment of highway projects*. www.environment.fhwa.dot.gov/guidebook/documents/VIA_Guidelines_for_Highway_Projects.asp#chap62.

Hankinson M 1999. Landscape and visual impact assessment. Chapter 16 in *Handbook of environmental impact assessment*, Vol.1. J Petts (ed.). Oxford: Blackwell Science.

Historic England 2015. *Historic landscape characterisation*. www.historicengland.org.uk/research/approaches/research-methods/characterisation-2.

HKG (Hong Kong Government) 2015. *Lighting and ventilation requirements – performance-based approach*. APP-130. Hong Kong: Buildings Department.

Hong Kong Environmental Protection Department 2002. *Preparation of landscape and visual impact assessment under the Environmental Impact Assessment Ordinance*, ESIAO Guidance Note No 8/2010. www.pland.gov.hk/pland_en/tech_doc/ESIAO/GN8_2010.pdf.

IFC (International Financial Corporation, World Bank Group) 2012. Performance standards on environmental and social sustainability. www.ifc.org/wps/wcm/connect/115482804a0255db96fbffd1a5d13d27/PS_English_2012_Full-Document.pdf?MOD=AJPERES.

LI (Landscape Institute) 2011. Advice Note 01/11 *Use of photography and photomontage in landscape and visual impact assessment*. Available online at: www.landscapeinstitute.org/

PDF/Contribute/LIPhotographyAdviceNote01-11.pdf.

LI/IEMA (Landscape Institute and Institute of Environmental Management & Assessment) 2013. *Guidelines for landscape and visual impact assessment, Third Edition.* Abingdon: Routledge.

MTR 2011. *Environmental impact assessment (ESIA) study for Shatin to Central Link – Hung Hom to Admiralty Section.* www.epd.gov.hk/ESIA/english/register/open/all.html.

NCHRP (National Cooperative Highway Research Program Transportation Research Board) 2012. *Evaluation of methodologies for visual impact assessment,* NCHRP Project 741.

NE (Natural England) 2014. *An approach to landscape character assessment.* Available online at: www.gov.uk/government/uploads/system/uploads/attachment_data/file/ 396192/landscape-character-assessment.pdf.

NIEA (Northern Ireland Environment Agency) 2010. *Wind energy development in Northern Ireland's landscapes,* Supplementary Planning Guidance to accompany Planning Policy Statement 18 'Renewable Energy', Research and Development Series 10/01. Belfast: NIEA. www.planningni.gov.uk/index/policy/supplementary_guidance/ spg_other/wind_energy_development_in_northern_irelands_landscapes_spg_for_pps1 8-2.pdf.

NSW (New South Wales Government) 2013. *Environmental impact assessment practice note: guideline for landscape character and visual impact assessment.* Sydney: NSW Transport Roads and Maritime Services.

NZILA 2010a *Best practice guide – landscape assessment and sustainable management,* NZILA Education Foundation. www.nzila.co.nz/media/53268/nzila_ldas_v3.pdf.

NZILA 2010b *Best practice guide – visual simulations,* NZILA Education Foundation. www.nzila.co.nz/media/53263/vissim_bpg102_lowfinal.pdf.

Ode Å, MS Tveit and G Fry 2008. Capturing landscape visual change using indicators: touching base with landscape aesthetic theory, *Landscape Research* 33(1), 89–117.

Peter Andrews + Associates Pty Ltd 2012. Pacific Highway HW10 and Wyong Road MR335 intersection and approaches upgrade Tuggerah: landscape character and visual impact assessment. Millers Point, Australia: PAA.

SNH (Scottish Natural Heritage) 2001. *Guidelines on the environmental impacts of wind farms and small scale hydroelectric schemes.* SNH Natural Heritage Management Series. Edinburgh: SNH.

SNH 2012. *Assessing the cumulative impact of onshore wind energy development.* Edinburgh: SNH.

SNH 2017. *Visual representation of wind farms, Version 2.2.* Edinburgh: SNH.

Stephenson J 2010. The dimensional landscape model: Exploring differences in expressing and locating landscape qualities, *Landscape Research* 35(3), 299–318.

Tweit M, Å Ode and G Fry 2006. Key concepts in a framework for analysing visual landscape character, *Landscape Research* 31(3), 229–255.

New South Wales 2015. *Practice note: solar access requirements in SEAPP 65.* Sydney: NSW Government.

12 Cultural heritage

· ·

Amanda Chisholm[1] and Júlio Jesus

12.1 Introduction

For many Environmental and Social Impact Assessment (ESIA) practitioners, cultural heritage has been, and in many cases still is, interpreted as comprising the "physical expression" of culture. Environmental assessments and reports therefore tend to focus on the "historic environment" and its physical assets. Certainly this is the most straightforward aspect of cultural heritage to assess: will a proposed project result in the loss of or damage to a historic asset and, if so, how significant are these effects and what can be done to mitigate them? It is also easier to demonstrate the objectivity required of ESIA practitioners when identifying and assessing effects on physical assets, rather than the more intangible aspects of cultural heritage.

However, as ESIA systems have matured and practitioners have gained experience and expertise, this view of the "historic environment" has begun to open up to include the wider aspects of cultural heritage – the intangible, the spiritual, and the traditional, amongst others. There are strong links between cultural heritage and other environmental components, including but not limited to biodiversity, landscape, and human health and well-being. These overlaps have been made more explicit by some of the appraisals that sit alongside ESIA, such as health impact assessment, socio-economic assessment, sustainability appraisal and appraisals of effects on ecosystem services.

This chapter takes the requirements of the International Finance Corporation (IFC) Performance Standards 7 and 8 (indigenous peoples and cultural heritage, respectively) as its starting point, but also draws on approaches used by others at the international level – including the United Nations Educational, Scientific and Cultural Organization (UNESCO), the International Council on Monuments and Sites (ICOMOS), and the World Bank – and guidance produced by national governments and industry.

IFC Performance Standard 8 identifies its objectives as follows:

- to protect cultural heritage from the adverse impacts of project activities and support its preservation;
- to promote the equitable sharing of benefits from the use of cultural heritage (IFC 2012b).

World Bank Environmental and Social Standard (ESS) 8 adds the following to these objectives:

- To address cultural heritage as an integral aspect of sustainable development;
- To promote meaningful consultant with stakeholders regarding cultural heritage (World Bank 2017).

In addition, a key aim of ESIA in this context is to ensure that information about the impacts of a development on cultural heritage is made available to the decision-maker, along with the potential for mitigation and the costs of this mitigation.

12.2 Definitions and concepts

12.2.1 Cultural heritage: definitions

UNESCO (1982) defines culture as:

> the whole complex of distinctive spiritual, material, intellectual and emotional features that characterise a community, society or social group. It includes not only arts and literature, but also modes of life, the fundamental rights of the human being, value systems, traditions and beliefs.

Cultural heritage therefore comprises these expressions of culture, described by the World Bank (1994) as "the present manifestation of the human past". It tends to be divided into two categories:

- tangible heritage, i.e. heritage that is physical, touchable, and perceptible;
- intangible, i.e. heritage that is not physical or represented by a physical object, and is of a value not precisely measurable.

Some international organisations with responsibility for cultural heritage focus on the tangible aspects of cultural heritage, while others include both tangible and intangible in their legislation and guidance (Box 12.1).

Tangible cultural heritage includes, often in combination:

- **Archaeological remains**, both above and below ground: these include buried remains as well as standing buildings, and archaeological sites as well as individual objects and artefacts. Examples range from earthworks, burial mounds, and hill forts, to the remains of Bronze Age settlements and medieval cities. They also include underwater remains and remains found preserved in peat bogs. In marine waters, underwater remains mainly comprise shipwrecks, drowned settlements or underwater landscapes. Underwater remains resulting from warfare often contain human remains; many are considered to be war graves.

Box 12.1 Descriptions of cultural heritage

The cultural heritage of a people includes the works of its artists, architects, musicians, writers and scientists and also the work of anonymous artists, expressions of the people's spirituality, and the body of values which give meaning to life. It includes both tangible and intangible works through which the creativity of that people finds expression: languages, rites, beliefs, historic places and monuments, literature, works of art, archives and libraries – Mexico City Declaration on Cultural Policies (UNESCO 1982).

Cultural heritage is an expression of the ways of living developed by a community and passed on from generation to generation, including customs, practices, places, objects, artistic expressions and values. Cultural heritage is often expressed as either intangible or tangible cultural heritage (ICOMOS 2002).

Cultural heritage is a group of resources inherited from the past which people identify, independently of ownership, as a reflection and expression of their constantly evolving values, beliefs, knowledge and traditions. It includes all aspects of the environment resulting from the interaction between people and places through time (Council of Europe 2005).

- **Monuments**: these include, for example, works of monumental sculpture and painting, inscriptions, and cave dwellings.
- **Historic buildings and sites**: individual or groups of separate or connected buildings, both in terms of their architecture and their setting or place in the landscape. These may be urban or rural in nature and can include, for example, the core of a historic city or town.
- **Structures** of historic and/or architectural merit: these include, but are not limited to, bridges, canals, aqueducts and irrigation systems, ancient roads, and engineered agricultural terraces.
- **Historic/cultural landscapes**: works of human beings or the combined works of nature and humans. Such areas may be of historic, aesthetic, ethnological or anthropological interest (see §12.2.4 for more detail and examples).
- **Movable objects**, such as historic or rare books and manuscripts, paintings, sculptures, religious items, historic costumes, jewellery and textiles, fragments of monuments, archaeological materials.

Architectural and historic merit is also a feature of landscape and townscape issues (see §12.6 on interactions). The *setting* of a monument, historic building, archaeological site or historic landscape (or groups of these) should also be included in any ESIA (setting is discussed further at §12.2.3).

In contrast to tangible heritage, which concentrates on the physical expression of cultural heritage (cultural heritage "assets"), intangible cultural heritage encompasses spiritual and social practices and traditional activities. Box 12.2 presents UNESCO's definition of intangible cultural heritage, which considers these to be manifested as:

Box 12.2 Definition of intangible cultural heritage

The "intangible cultural heritage" means the practices, representations, expressions, knowledge, skills – as well as the instruments, objects, artefacts and cultural spaces associated therewith – that communities, groups and, in some cases, individuals recognize as part of their cultural heritage. This intangible cultural heritage, transmitted from generation to generation, is constantly recreated by communities and groups in response to their environment, their interaction with nature and their history, and provides them with a sense of identity and continuity, thus promoting respect for cultural diversity and human creativity. For the purposes of this Convention, consideration will be given solely to such intangible cultural heritage as is compatible with existing international human rights instruments, as well as with the requirements of mutual respect among communities, groups and individuals, and of sustainable development – Article 2(1), Convention for the Safeguarding of the Intangible Cultural Heritage (UNESCO 2003).

- oral traditions and expressions, including language as a vehicle of the intangible cultural heritage;
- performing arts;
- social practices, rituals and festive events;
- knowledge and practices concerning nature and the universe (including traditional medicines);
- traditional craftsmanship.

The Convention on Biological Diversity 1992 notes that some cultural knowledge and practices may depend on biodiversity and its components, and therefore specifically recognises the cultural value of biodiversity alongside its ecological, genetic, social, economic, scientific, educational, recreational and aesthetic values. These cultural services should be taken into account in ESIA, particularly when considering linkages between cultural heritage and biodiversity, and between environmental and social impact assessment (including well-being). This is particularly so for some aboriginal, indigenous and other communities, whose traditional and spiritual activities depend on the maintenance of particular biodiversity features. Article 8(j) of the Convention makes provision to protect the knowledge and practices of indigenous and local communities who rely on the conservation and sustainable use of biological diversity (Box 12.3). Article 10(c) protects and encourages customary use of biological resources in accordance with traditional cultural practices. The Convention also requires that benefits from the use, for commercial purposes, of traditional knowledge and/or practices should be shared fairly and equitably with the owners of such knowledge (Article 15(7)). Both IFC Performance Standard 6 and World Bank ESS 6 recognize the cultural value of biodiversity (IFC 2012c, World Bank 2017).

Taking these definitions together, ESIA should therefore consider both the tangible and intangible aspects of cultural heritage. This is reflected in the definition of cultural heritage set out in the IFC's Performance Standard 8:

Box 12.3 Role of biological resources in cultural heritage

Each Contracting Party shall, as far as possible and as appropriate:

Subject to its national legislation, respect, preserve and maintain knowledge, innovations and practices of indigenous and local communities embodying traditional lifestyles relevant for the conservation and sustainable use of biological diversity and promote their wider application with the approval and involvement of the holders of such knowledge, innovations and practices and encourage the equitable sharing of the benefits arising from the utilization of such knowledge, innovations and practices – Article 8(j), Convention on Biological Diversity 1992.

i   tangible forms of cultural heritage, such as tangible moveable or immovable objects, property, sites, structures or groups of structures, having archaeological (prehistoric), palaeontological, historical, cultural, artistic and religious values;
ii unique natural features or tangible objects that embody cultural values, such as sacred groves, rocks, lakes and waterfalls;
iii certain instances of intangible forms of culture that are proposed to be used for commercial purposes, such as cultural knowledge, innovations, and practices of communities embodying traditional lifestyles.

12.2.2 Scope of cultural heritage

In terms of temporal scope, cultural heritage includes both the past (as expressed by, for example, archaeological remains) and the present (as expressed in traditional and/or spiritual activities and knowledge, for example). There is often a continuum between the past and the present, for example, places of pilgrimage, use of historic buildings, etc. The World Bank (1994) considers that cultural heritage, in its legacy from past to future generations, is part of intergenerational equity.

In terms of geographic scope, cultural heritage – tangible and intangible – is found in the terrestrial (including freshwater), coastal and marine environments. The Convention on the Protection of Underwater Cultural Heritage (UNESCO 2001) focuses mainly on the tangible. It defines underwater cultural heritage as meaning "all traces of human existence having a cultural, historical or archaeological character which have been partially or totally under water, periodically or continuously, for at least 100 years". This includes:

• sites, structures, buildings, artefacts and human remains, together with their archaeological and natural context;
• vessels, aircraft, other vehicles or any part thereof, their cargo or other contents, together with their archaeological and natural context;
• objects of prehistoric character.

In terms of the scope of affected communities, an increasing number of texts advocate the integration of ESIA and aboriginal and indigenous peoples' interests (e.g. the UN Declaration on the Rights of Indigenous Peoples; Agenda 21; Convention on Biological Diversity; and the Akwé: Kon Voluntary Guidelines). The assessment of impacts on cultural heritage in ESIA must include effects on aboriginal and indigenous culture where projects are being undertaken in countries where these cultures are present. These effects include effects on both tangible and intangible cultural heritage.

Aboriginal and indigenous culture often brings the past into the present through, for example:

- aboriginal traditional knowledge;
- sacred sites, important both in themselves and through the traditional rituals and practices that they support;
- lands and waters traditionally occupied or used by indigenous and local communities.

Projects that affect these communities will require the Free Prior and Informed Consent of their members before commencement. Involvement of the community in project planning and development, and in the ESIA, is therefore essential and may require different public participation methods. This requires careful planning; some aboriginal communities are experiencing consultation fatigue. Some countries provide funding to aboriginal peoples to facilitate their participation. A key issue is for communities to retain ownership, protection and control of traditional knowledge.

12.2.3 Setting

The setting of a cultural heritage asset is, broadly speaking, the surroundings in which a place is experienced, while embracing an understanding of perceptible evidence of the past in the present landscape (HA et al. 2007). ICOMOS (2005) defines the setting of a heritage structure, site or area as "the immediate and extended environment that is part of, or contributes to, its significance and distinctive character". ICOMOS notes that setting goes beyond the physical and/or visual, to include interaction with the natural environment, customs, and traditional knowledge, amongst other aspects. Setting is also listed as an attribute of *authenticity* in the World Heritage Site Operational Guidelines (UNESCO-WHC 2013).

A useful working approach to the inclusion of setting in ESIA is set out in Volume 11 of the United Kingdom's *Design Manual for Roads and Bridges* (HA et al. 2007). The manual also sets out some useful working principles:

- a cultural heritage asset's setting is its relevant surroundings;
- settings have physical factors, which can be changed by a project, but it is the effect of these changes on the character and value of the cultural heritage asset that is assessed;

- context is an aspect of setting where a relevant aspect of knowledge, belief or relationships may not be immediately visible or audible;
- professional judgement is required.

With reference to the last point, as well as analysis of views and contexts, the consideration of setting requires: the use of a multidisciplinary approach; seeking the views of both experts and the local community; and review of existing information, where available, including: formal records; oral history and traditional knowledge; and artistic and literary descriptions.

Examples of project effects on setting include:

- The setting of a listed mansion may include its extensive parkland, some of which may be out of sight of the building itself but integral to its original purpose and to its appreciation today. Changes in the fabric of the park that affect that relationship would reasonably be considered to be an impact on the appreciation of the historic building, even if the changes were not visible or audible from the house.
- Knowledge about invisible buried remains related to an upstanding monument adds to its significance, and where a project would disrupt that relationship, even if not affecting the remains themselves, this could be considered to degrade the significance of the asset.

12.2.4 Historic and cultural landscapes

The concept of cultural landscapes was internationally recognised in 1992, when the UNESCO World Heritage Committee adopted guidelines concerning their inclusion in the World Heritage List. The World Heritage Convention became the first international legal instrument to recognise and protect cultural landscapes.

The Council of Europe Landscape Convention (CoE 2000, 2016) is a landmark in the recognition of the importance of the landscape and its cultural dimension (see Box 12.4 for the definition of landscape adopted by the Convention). Its first aim is to encourage states to introduce a national landscape policy that is not restricted to the protection of exceptional landscapes but also takes everyday landscapes into consideration.

Historic landscapes are defined and protected at a national level in many countries. In the UK, the guidance for the listing of buildings (DCMS 2010) identifies the importance of considering the landscape as part of the historic environment:

> our understanding and appreciation of the historic environment now stretches beyond buildings to the spaces and semi-natural features that people have also moulded, and which are often inseparable from the buildings themselves. For example, the pattern of roads and open spaces and the views they create within historic townscapes may be as valuable as

Box 12.4 Definitions of landscape

The Council of Europe Landscape Convention defines landscape as "an area, as perceived by people, whose character is the result of the action and interaction of natural and/or human factors".

In 2011, UNESCO's General Conference adopted a Recommendation on the Historic Urban Landscape. This voluntary instrument is a tool to integrate policies and practices of conservation of the built environment into the wider goals of urban development in respect of the inherited values and traditions of different cultural contexts. It defines the Historic Urban Landscape as follows:

8   The historic urban landscape is the urban area understood as the result of a historic layering of cultural and natural values and attributes, extending beyond the notion of "historic centre" or "ensemble" to include the broader urban context and its geographical setting.

9   This wider context includes notably the site's topography, geomorphology, hydrology and natural features, its built environment, both historic and contemporary, its infrastructures above and below ground, its open spaces and gardens, its land use patterns and spatial organization, perceptions and visual relationships, as well as all other elements of the urban structure. It also includes social and cultural practices and values, economic processes and the intangible dimensions of heritage as related to diversity and identity.

10 This definition provides the basis for a comprehensive and integrated approach for the identification, assessment, conservation and management of historic urban landscapes within an overall sustainable development framework.

– Recommendation on the Historic Urban Landscape (UNESCO 2011)

the buildings. In the countryside, the detailed patterns of fields and farms, of hedgerows and walls, and of hamlets and villages, are among the most highly valued aspects of our environment.

Historic seascapes may be equally important, having been used as sites for fishing, transport and trade over centuries, as well as sometimes being the sites of wrecks.

The United States National Park Service considers historic landscapes to be one of the categories of cultural landscapes:

historic landscapes include residential gardens and community parks, scenic highways, rural communities, institutional grounds, cemeteries, battlefields and zoological gardens. They are composed of a number of character-defining features which, individually or collectively, contribute to the landscape's physical appearance as they have evolved over time. In addition to vegetation and topography, cultural landscapes may include water features, such as ponds, streams, and fountains; circulation features, such as roads, paths, steps, and walls; buildings; and furnishings, including fences, benches, lights and sculptural objects.

(Birnbaum 1994)

In Italy, a research project produced a national catalogue of historic rural landscapes (Agnoletti 2013). It is the first survey of the traditional rural landscapes of a country. This research work – a preliminary study conducted as part of the compilation of a national register of historical rural landscapes and traditional practices – presents more than a hundred case studies where the historical relationships between man and nature have generated – not deterioration – but cultural, environmental, social and economic values.

12.2.5 Authenticity and integrity

The concept of authenticity related to the restoration of the cultural heritage was addressed for the first time in an international declaration in a conference in 1964, where the Venice Charter was approved: "The process of restoration is a highly specialized operation. Its aim is to preserve and reveal the aesthetic and historic value of the monument and is based on respect for original material and authentic documents" (Article 9, ICOMOS 1964).

Thirty years after the Venice Charter, a Conference on Authenticity held at Nara, Japan, produced the Nara Document on Authenticity (Box 12.5). Until the Nara Document, authenticity was restricted to tangible aspects (design, material, workmanship, setting). The UNESCO-WHC (2013) Operational Guidelines adopt the Nara Document and summarise the conditions for authenticity:

Box 12.5 The Nara Document on Authenticity (ICOMOS 1994)

3   The Nara Document on Authenticity is conceived in the spirit of the Charter of Venice, 1964, and builds on it and extends it in response to the expanding scope of cultural heritage concerns and interests in our contemporary world.

9   Conservation of cultural heritage in all its forms and historical periods is rooted in the values attributed to the heritage. Our ability to understand these values depends, in part, on the degree to which information sources about these values may be understood as credible or truthful. Knowledge and understanding of these sources of information, in relation to original and subsequent characteristics of the cultural heritage, and their meaning, is a requisite basis for assessing all aspects of authenticity.

10 Authenticity, considered in this way and affirmed in the Charter of Venice, appears as the essential qualifying factor concerning values. The understanding of authenticity plays a fundamental role in all scientific studies of the cultural heritage, in conservation and restoration planning, as well as within the inscription procedures used for the World Heritage Convention and other cultural heritage inventories.

Depending on the type of cultural heritage, and its cultural context, properties may be understood to meet the conditions of authenticity if their cultural values (as recognized in the nomination criteria proposed) are truthfully and credibly expressed through a variety of attributes including:

- form and design;
- materials and substance;
- use and function;
- traditions, techniques and management systems;
- location and setting;
- language, and other forms of intangible heritage;
- spirit and feeling;
- other internal and external factors.

Integrity is a measure of the wholeness and intactness of cultural heritage and its attributes. The World Heritage Operational Guidelines (UNESCO-WHC 2013) establish, as conditions for assessing the integrity of World Heritage properties:

the extent to which the property:

a) includes all elements necessary to express its Outstanding Universal Value [OUV];
b) is of adequate size to ensure the complete representation of the features and processes which convey the property's significance;
c) suffers from adverse effects of development and/or neglect.

The integrity of (cultural) World Heritage properties requires that

the physical fabric of the property and/or its significant features should be in good condition, and the impact of deterioration processes controlled. A significant proportion of the elements necessary to convey the totality of the value conveyed by the property should be included. Relationships and dynamic functions present in cultural landscapes, historic towns or other living properties essential to their distinctive character should also be maintained.

(UNESCO-WHC 2013)

There is sometimes confusion between the concepts of authenticity and integrity. For example, integrity would be where the building materials and shapes of the palaces, temples, burial chambers and funerary chapels have not been altered or modified. Authenticity is where the relief, writings and painted scenes have equally preserved their original design, texture and colour (Borchi 2012).

12.2.6 Early engagement/importance of stakeholder engagement

The engagement of stakeholders at an early stage is good practice in ESIA and crucial in cultural heritage impact assessment. IFC Performance Standard 8 (IFC 2012b) requires:

> Where a project may affect cultural heritage, the client will consult with Affected Communities within the host country who use, or have used within living memory, the cultural heritage for long-standing cultural purposes. The client will consult with the Affected Communities to identify cultural heritage of importance, and to incorporate into the client's decision-making process the views of the Affected Communities on such cultural heritage. Consultation will also involve the relevant national or local regulatory agencies that are entrusted with the protection of cultural heritage.

ICOMOS (2011) also considers that "early consultation with relevant parties, including any affected community, is important".

Table 12.1 lists main stakeholders who should be involved in cultural heritage impact assessment. The techniques for the engagement of regulators, affected communities and other stakeholders will depend on the political and cultural context. International best practice principles on public participation (André et al. 2006), published by the International Association for Impact Assessment, provide useful guidance on the basic and operational principles that allow the effective engagement of stakeholders. This document points out that:

> To be effective, communication between Impact Assessment actors (e.g. public, proponent, decision-maker, regulator) should give attention to active listening and to the different actors' frame of reference and connotation of terms, their attitudes towards others, their roles and rela-tionships between roles, and the general situation in which the communication takes place and its goal, as well as their state of preparation. Effective two-way communication needs the respect of others, and of their culture, tradition and personalities.

Important issues to be taken into account are the mapping of stakeholders, the type of language, the languages used, the communication media (written, oral), and the time allocated for stakeholders, particularly indigenous peoples, to respond to consultations and to request information.

The principle of "Free, Prior and Informed Consent" (FPIC) is outlined in the United Nations (2007) Declaration on the Rights of Indigenous Peoples and in the International Labour Organization Convention 169 (ILO 1989). FPIC is becoming a general principle applicable to all projects that affect

Table 12.1 Cultural heritage stakeholders

Type of stakeholders	Stakeholders
Public authorities (national/federal, state/region, local)	• Cultural heritage authorities • ESIA/Environmental authorities • Planning authorities • Tourism boards
Indigenous Peoples and local communities	• Indigenous Peoples' representative bodies and organisations (e.g. councils of elders or village councils), as well as members of the Affected Communities of Indigenous Peoples • NGOs representing indigenous peoples or local communities
Scientific institutions and individuals	• Museums • Universities and research centres involved in cultural heritage • Academics/researchers
Religious institutions	
NGOs (national, regional, local)	• NGOs involved in issues related to cultural heritage
International organisations	• UNESCO National Office • ICOMOS National Committee • IUCN National Office

Indigenous Peoples. The IFC Performance Standard 7 (IFC 2012a) includes provisions about the circumstances requiring FPIC, including what the Standard considers "Critical Cultural Heritage" (see §12.4.2 and Box 12.6).

Box 12.6 Circumstances requiring FPIC (IFC PS7 – Indigenous Peoples)

16 Where a project may significantly impact on critical cultural heritage that is essential to the identity and/or cultural, ceremonial, or spiritual aspects of Indigenous Peoples' lives, priority will be given to the avoidance of such impacts. Where significant project impacts on critical cultural heritage are unavoidable, the client will obtain the FPIC of the Affected Communities of Indigenous Peoples.

17 Where a project proposes to use the cultural heritage including knowledge, innovations, or practices of Indigenous Peoples for commercial purposes, the client will inform the Affected Communities of Indigenous Peoples of (i) their rights under national law; (ii) the scope and nature of the proposed commercial development; (iii) the potential consequences of such development; and (iv) obtain their FPIC. The client will also ensure fair and equitable sharing of benefits from commercialization of such knowledge, innovation, or practice, consistent with the customs and traditions of the Indigenous Peoples.

As Hanna and Vanclay (2013) argue:

> companies that adopt the FPIC philosophy and fully implement it in practice, in addition to respecting the right of communities to participate in decisions that affect their lives, will probably benefit from reduced conflict, reduced likelihood of reputational damage, as well as reduced risks and costs.

In some specific situations the disclosure of information about cultural heritage may compromise the safety or integrity of the cultural heritage or may endanger sources of information. A decision on disclosure shall be made in consultation with the affected parties, including communities.

12.2.7 Need for experts

In ESIA, archaeological sites were, and still are, a relevant issue, mainly because of the uncertainties and the costs and time delays that could arise. It is therefore essential to include archaeologists in the ESIA team, where archaeological sites are present or likely. The importance of other components, tangible and intangible, of cultural heritage has created the need for the growing involvement of other experts, namely anthropologists, architects, landscape architects and historians. The professionals involved will depend on the ESIA phase, and on the geographical, historical and cultural context of the project.

Archaeology is one of the more regulated professions in a great number of countries, e.g. through membership of the Chartered Institute for Archaeologists in the UK.

Members of local communities can provide expert local knowledge, and indigenous and aboriginal peoples may also provide traditional knowledge. Such local and traditional expertise should be an integral part of cultural heritage impact assessment, at all phases of the process. Depending on the context, involvement of national scientific experts may be needed to facilitate the full consideration of traditional knowledge.

12.3 Key legislation, guidance and standards

The focus of this section is on cultural heritage legislation at the international level. This comprises treaties, mostly in the form of conventions, that set the legislative and/or policy framework to be implemented by national governments. Ratification of these conventions and charters by national governments commits them to taking forward legislative and policy measures that "balance the need for development against the requirement to protect and enhance ... national cultural heritage resource as far as is practicable" (HA et al. 2007).

The World Bank (1994) notes that cultural heritage is also protected by legislation in nearly all countries. Some elements of the tangible cultural heritage may be protected by legislation, others may not be protected by statute. In some countries land-use plans provide protection for cultural heritage assets through land-use planning policy. Protection is generally afforded to cultural heritage assets (physical resources), at the national, regional and/or local levels. The ESIA practitioner should therefore identify the regime particular to the country in which the ESIA is taking place.

The World Bank (1994) identifies the key types of legislation at the national level:

- legislation that protects types of places, or particular places, and sets out procedures for their protection;
- land management, zoning, or planning acts that provide general protection;
- acts that require the recording of data on cultural sites, e.g. listings;
- acts that protect natural areas (where cultural features may be located).

Key international and European conventions, treaties and charters are summarised in Tables 12.2 and 12.3 respectively.

The Akwé: Kon Guidelines (SCBD 2004) deal with the conduct of Cultural, Environmental and Social Impact Assessment of developments proposed to take place on, or which are likely to impact on, sacred sites and on lands and waters traditionally occupied or used by indigenous and local communities. In these cases the interactions between cultural heritage, social and biodiversity impacts are such that an integrated approach is required.

More recently, a UNESCO–CBD (2014) joint programme on biological and cultural diversity has approved a declaration recognising "the importance of the links between cultural and biological diversity, and in this context noting the concept of Biocultural Diversity and the relevance of cultural services provided by ecosystems" and one of the conclusions reads as follows:

> To better understand the dynamic interplay between biological and cultural diversity at the landscape level and its implications for livelihoods and well-being, there is need for enhanced interdisciplinary and transdisciplinary research of the links between biological and cultural diversity at the national and sub-national levels, including their historical background.

Garibaldi and Turner (2004) proposed the concept of "cultural keystone species", defining them as "culturally salient species that shape in a major way the cultural identity of a people. Their importance is reflected in the fundamental roles these species play in diet, materials, medicine, and/or spiritual practices". These authors propose a set of indicators of cultural influence for those species:

Table 12.2 International heritage conventions, treaties and charters[1]

Date	Title	Key points
1954	Convention for the Protection of Cultural Property in the Event of Armed Conflict (Hague Convention)	• provides a system of protection of cultural property in situations of international and domestic armed conflict • ratified by 128 State Parties • www.unesco.org/new/en/culture/themes/armed-conflict-and-heritage/the-hague-convention
1964	International Charter for the Conservation and Restoration of Monuments and Sites (Venice Charter)	• sets out principles for the conservation and restoration of monuments and sites • includes the principle of a monument's setting • www.icomos.org/venicecharter2004
1970	Convention on the Means of Prohibiting the Illicit Import, Export and Transfer of Ownership of Cultural Property (Paris Convention)	• prevents the trade in illegally obtained cultural objects. • ratified by 132 State Parties • www.unesco.org/new/en/culture/themes/illicit-trafficking-of-cultural-property/1970-convention
1972	Convention Concerning the Protection of the World Cultural and Natural Heritage (World Heritage Convention)	• establishes a system of collective protection of natural and cultural heritage which is of outstanding universal value • ratified by 193 State Parties • http://whc.unesco.org/en/conventiontext
1992	Convention on Biological Diversity	• recognises the cultural value of biological diversity and its components; makes provision to protect the knowledge and practices of indigenous and local communities who rely on the conservation and sustainable use of biological diversity (Article 8j); and protects and encourages customary use of biological resources in accordance with traditional cultural practices (Article 10c) • ratified by 196 State Parties • www.cbd.int
1994	ICOMOS Nara document on authenticity	• builds on and extends the principles set out in the Venice Charter • http://whc.unesco.org/archive/nara94.htm
1999	ICOMOS International Cultural Tourism Charter	• aims to make cultural tourism sustainable • provides key definitions of culture and cultural heritage • acknowledges the potential adverse effects of tourism on heritage assets/communities

Table 12.2 continued

Date	Title	Key points
2001	Convention on the Protection of the Underwater Cultural Heritage	• sets out basic principles for the protection of underwater cultural heritage; provides a detailed State cooperation system; and provides widely recognized practical rules for the treatment and research of underwater cultural heritage • ratified by 56 State Parties • www.unesco.org/new/en/culture/themes/underwater-cultural-heritage/2001-convention
2003	Convention for the Safeguarding of the Intangible Cultural Heritage	• aims at safeguarding the uses, representations, expressions, knowledge and techniques that communities, groups and, in some cases, individuals, recognise as an integral part of their cultural heritage. This intangible heritage is found in forms such as oral traditions, performing arts, social practices, rituals, festive events, knowledge and practices concerning nature and the universe, and traditional craftsmanship knowledge and techniques • ratified by 174 State Parties • www.unesco.org/new/en/santiago/culture/intangible-heritage/convention-intangible-cultural-heritage
2005	Convention on the Protection and Promotion of the Diversity of Cultural Expressions	• aims to protect and promote the diversity of cultural expressions • sets out principles around cultural diversity and the diversity of cultural expressions, including traditional knowledge • ratified by 144 State Parties + European Union • http://en.unesco.org/creativity/convention
2005	Xi'an Declaration on the Conservation of the Setting of Heritage Structures, Sites and Areas	• acknowledges the contribution of setting to the significance of heritage monuments, sites and areas and provides a definition of setting • sets out principles for the way that setting should be conserved and managed • www.icomos.org/xian2005/home_eng.htm
2008	The Quebec Declaration on the Preservation of the Spirit of Place	• aims to preserve the spirit of place through the safeguarding of tangible and intangible heritage (the spirit of places is defined as their living, social and spiritual nature) • declares that intangible cultural heritage gives a richer and more complete meaning to heritage as a whole and that it must be taken into account in legislation, conservation and restoration • www.icomos.org/quebec2008

Note: [1] Partial source: https://historicengland.org.uk/advice/hpg/coventionstreatiesandcharters/#(10)

Table 12.3 European heritage conventions, treaties and charters[1]

Date	Title	Key points
1954	European Cultural Convention (Paris Convention)	• aims to develop mutual understanding of cultural diversity, and promote national contribution to Europe's common cultural heritage. It also safeguards objects of European cultural value placed under government control and ensures reasonable access to them • 50 State Parties • www.coe.int/en/web/conventions/full-list/-/conventions/treaty/018
1985	Convention for the Protection of the Architectural Heritage of Europe (Granada Convention), formerly European Charter of the Architectural Heritage	• aims to reinforce and promote policies for the conservation and enhancement of Europe's cultural heritage • defines architectural heritage; includes requirements to develop and maintain an inventory; to take statutory measures to protect cultural heritage; and to integrate conservation policies into planning systems and other spheres of government influence • 42 State Parties • www.coe.int/en/web/conventions/full-list/-/conventions/treaty/121
1992	Convention for the Protection of the Archaeological Heritage of Europe (Revised) (Valletta Convention)	• defines archaeological heritage; includes requirements to make and maintain an inventory of archaeological heritage and to legislate for its protection. Signatories also promise to allow the input of expert archaeologists into the making of planning policies and planning decisions • 46 State Parties • www.coe.int/en/web/conventions/search-on-treaties/-/conventions/treaty/143
2000	Council of Europe Landscape Convention (Florence Convention), formerly European Landscape Convention	• recognises "landscapes" in law as "an expression of the diversity of their shared cultural and natural heritage, and a foundation of their identity". These recognised landscapes are then to be subject to policies for their management, amongst other obligations • 38 State Parties • www.coe.int/en/web/landscape/home
2005	Council of Europe Framework Convention on the Value of Cultural Heritage for Society (Faro Convention)	• links the concept of the "common heritage of Europe" to human rights and the fundamental freedoms for which the Council of Europe remains one of the historic guardians • 17 State Parties • www.coe.int/t/dg4/cultureheritage/heritage/Identities/default_en.asp

Note: [1] Partial source: https://historicengland.org.uk/advice/hpg/coventionstreatiesandcharters/#(10)

i   intensity, type, and multiplicity of use;

ii   naming and terminology in a language, including use as a seasonal or phenological indicator;

iii   role in narratives, ceremonies, or symbolism;

iv   persistence and memory of use in relationship to cultural change; and

v   extent to which its role can be replaced or substituted.

12.4 Baseline studies

The aim of a baseline study is to identify, describe the nature and location, and assess the value (or the importance) of the cultural heritage assets likely to be affected, directly or indirectly, by the development project. The baseline studies should comply with the results of the scoping phase, namely the terms of reference for the ESIA. The scope of these studies should be regularly reviewed to take into account any updates or changes in the project, regulatory, planning and management contexts, or the information on cultural heritage available (e.g. new archaeological evidence could become available at any time). Baseline studies on cultural heritage should be conducted in coordination with the other ESIA team members, particularly those working on social assessment, involuntary resettlement, aboriginal and/or indigenous people, landscape and biodiversity.

12.4.1 Data collection methods

Data collection and analysis methods include (Campbell 2009, Therivel 2009):

- rapid appraisal;
- desk-based studies;
- interviews of concerned and affected communities;
- interviews of local religious leaders;
- interviews of specialists with local knowledge;
- ethnographic studies;
- gathering of oral traditions;
- field surveys;
- building surveys;
- specialised surveys for underwater areas.

Rapid appraisal of the archaeological resources involves the collation and review of existing and easily accessible data. It includes the review of the available national, provincial or local registries and contact with local cultural heritage authorities; it may also include a site visit. Rapid appraisal is useful to enable a preliminary view of the likely nature and scale of archaeological sites or designated buildings in the study area, but is generally not adequate for other types of cultural heritage.

In general, the data in archaeological databases arise from a number of different sources, ranging from detailed surveys to chance finds. As a result there is considerable variance in the reliability of the data and the interpretation that can be placed upon it. It is often not very intelligible to non-archaeologists (and may not be a public document) and it may need professional interpretation to assess the significance or potential of archaeological sites (Therivel 2009).

Desk-based studies are more in-depth studies than a rapid appraisal and involve contact with different sources of information (see Table 12.4) and the analysis of available documentation. Types of available documentation that could be used in desk-based studies include (after Therivel 2009):

- international, national, provincial and local registries or databases;
- old maps;
- aerial photographs and satellite imagery, old and recent;
- photographs;
- books, newspapers, pamphlets, scientific articles, scientific reports, etc.;
- parish and estate records, old plans.

Aerial photographs and **satellite imagery** are an important source of data. Earthworks are often more easily recognised and interpreted from the air than from the ground. Buried archaeological remains can also be traced from the air in certain circumstances, as they can affect a growing crop. For instance, a buried wall or road surface may retard crop growth, or in a dry year create a parch mark. A buried pit or ditch may promote crop growth. The patterns that result can be interpreted as archaeological features or sites. Different soil

Table 12.4 Examples of sources of information (after Campbell 2009)

Level	Sources
National/ Provincial level	• Cultural heritage authorities • Cultural heritage registers • Universities and colleges • Public and private cultural heritage related institutions • Religious bodies • Local NGOs active in cultural heritage and socio-cultural affairs
Community and individual level	• Community leaders and individuals • Schools • Religious leaders • Private scholars • Cultural heritage specialists • Historians • Archaeologists • Cultural anthropologists

colours may also reveal archaeological sites. Aerial photographs may be found in national, local authority, and possibly private collections (Therivel 2009). A recent discovery of a possible Viking settlement at Point Rosee, Newfoundland, Canada, was based on satellite imagery revealing ground features that could be evidence of past human activity:

> The remnants of structures buried at Point Rosee alter the surrounding soil, changing the amount of moisture it retains. This, in turn, affects the vegetation growing directly over it. Using remote sensing, variations in plant growth form a spectral outline of what was there centuries earlier. The Point Rosee images were taken during the fall, when the grasses in the area were particularly high, making it easier to see which plants were healthier, drinking more water from the soil.
>
> (Strauss 2016)

Contacts and interviews with communities, including non-governmental organizations (NGOs) and individuals, provide local and traditional knowledge about cultural heritage and how it is valued by communities. In the case of indigenous peoples these contacts are essential to the identification of sites of religious, spiritual, ceremonial and sacred significance (such as sacred groves and totemic sites). As stated in the Akwé: Kon Guidelines:

> traditional knowledge, innovations and practices should be considered an important and integral component of baseline studies, particularly the traditional knowledge, innovations and practices of those who have a long association with the particular area for which the development is proposed. Traditional knowledge, innovations and practices can be cross-referenced by old photographs, newspaper articles, known historical events, archaeological records, anthropological reports, and other records contained in archival collections.
>
> (SCBD 2004)

A wide range of **field survey** techniques is available, including geophysical techniques, fieldwalking, augering, test-pitting, machine trench digging and earthwork surveys. These are described in Box 12.7. Not all of these techniques will be applicable in all circumstances. Some can act as useful preliminaries to other techniques. A phased approach to field survey is often the most sensible and cost-effective, so it is common to use a suite of techniques as the proposal develops. Many of these archaeological activities may require specific permits from authorities.

Light Detection and Ranging (LiDAR) is a technology that measures the height of the ground surface and other features in large areas of landscape, with a very high resolution and great accuracy. Such information was previously unavailable, except through labour-intensive field survey or photogrammetry. It provides highly detailed and accurate models of the land surface at metre and

Box 12.7 Techniques of field survey (after Therivel 2009)

Geophysical techniques can be used to investigate some characteristics and properties of the ground that may be altered by previous land uses. The principal techniques used are resistivity and magnetometer surveys, although others are also available. Resistivity surveys measure the ground's resistance to the progress of an electrical current. Measuring increases and decreases in the resistance can indicate the nature and location of buried features. Magnetometer surveys measure the magnetic properties of the soil and can be used to identify locations of past human activity, particularly those that involved burning or heating.

Geophysical techniques can only be applied in suitable site conditions, and an experienced geophysical operator should visit the site to assess their feasibility. Where they are appropriate, geophysical techniques have an advantage over many other field techniques in that they do not damage the archaeological resource.

Although the results of geophysical techniques can sometimes be ambiguous, these techniques often successfully identify the location and extent of archaeological sites and can give some idea of their nature. The results can therefore help to focus subsequent stages of field survey to maximise data recovery. However geophysical techniques are unlikely to provide sufficient information on their own, are not universally applicable, and are often expensive.

Fieldwalking, also known as surface artefact collection, is confined to ploughed fields. In ploughed fields there is a tendency for buried material to be brought to the surface when the plough breaks and turns over the surface soil. Where the plough intrudes into a buried archaeological site this will include archaeological artefacts. Rigorous collection and plotting of this material will enable the location, date, and extent of certain types of archaeological site to be described. The archaeological material collected can be anything that reflects human activity, like pottery sherds, worked stone, coins, building material and even stone that is not local to the area and may have been imported.

Where a site has already been located, intensive fieldwalking, called total collection, can be used to determine spatial distributions across the site.

Fieldwalking is a relatively rapid and inexpensive technique that can be applied over large areas. However, the results can be ambiguous or misleading. Where a site is located by fieldwalking it is by definition being damaged. It is hard to judge from fieldwalking results alone how intact the site is, or whether it solely survives as artefacts trapped in the plough soil. A site surviving intact below the ploughed soil will not be represented on the surface. Certain periods do not produce artefacts that would survive the ploughing action. The results of fieldwalking therefore need to be qualified by some understanding of the relationship between the depth of ploughing and the depth of the archaeology.

Augering is most frequently used in river valleys where alluvial, colluvial or peat deposits have masked the original land surface and where slightly higher ground in a wet environment may have acted as a focus for human activity. By recording the soil sequence from auger holes located over a wide area, the underlying and hidden subsurface topography can be mapped and the archaeological potential of the area can be inferred. Augering alone is unlikely to confirm the presence or absence of archaeological deposits, but can clarify the archaeological potential and so focus subsequent stages of survey. It can also be used to clarify the nature of features located by geophysical techniques, and in certain areas to assess the potential for the preservation of *palaeoenvironmental data*.

Test pitting involves the hand excavation of an array of small pits of a predetermined size. It provides a clear picture of the nature of the soil structure and the upper layers of the underlying geology. As with fieldwalking, the spacing and array of test pits usually reflect assumptions about the expected archaeological resource. Test pits can be varied in size and array in order to meet the requirements of the survey. They are usually 1m × 1m, or 1m × 0.5m for ease of excavation. The soil from test pits is often sieved through a wire mesh of a set size to ensure consistent artefact recovery, enabling a rigorous statement to be made regarding the number, type and depth of artefacts. Analysis of the different artefact recovery rates over an area gives an indication of the date, location and extent of archaeological sites. Test pitting is often used instead of fieldwalking where the land is pasture rather than arable, and in woodland where machine trenching may not be possible.

Machine trenching employs trenches, usually cut with a toothless ditching bucket, laid out in a pattern across the site. The trench pattern will attempt to maximise information retrieval, possibly on the basis of existing data such as aerial photographs, fieldwalking or geophysical results. The extent of trenching required is usually an agreed sample of the land, commonly around 2 percent. When archaeological deposits are encountered excavation continues by hand. The excavation is controlled by a supervising archaeologist at all times. Machine trenching quickly locates features cut into the subsoil but, where large amounts of earth are rapidly removed, there is limited opportunity to collect artefacts and the rate of artefact retrieval is low. Higher rates of retrieval can be achieved by hand-digging parts of the trench, equivalent to a test pit, and the use of metal detectors.

Trenching is very disruptive and intervenes directly into the archaeological levels. This has the advantage of producing unambiguous information but is potentially damaging to archaeological remains one might otherwise wish to protect. It is also not always possible to get a machine onto a site.

Earthwork surveys can be used for archaeological sites that are visible as earthworks such as banks, ditches, burial mounds, and sites of deserted or shrunken settlements. Sites that survive as earthworks are generally more intact than other sites. Ploughing can degrade earthworks, and the success of earthwork surveys is limited in fields which have been arable for a long time; generally, such land is more productively scanned from aerial photographs. Pasture can have visible earthworks surviving. When they are obviously visible they may have been recorded by the national mapping systems. They can also be identified through aerial photographs. Woodland, particularly ancient woodland, holds the greatest potential for producing previously unrecorded earthworks. The sites will often be obscured from the air by trees and on the ground by undergrowth, so it is best to undertake the survey during the winter or early spring.

The nature of the earthwork survey will depend on the aims of the evaluation. The survey can vary from sketch plotting the earthworks onto a map, through two-dimensional surveys such as plane table surveys, to a three-dimensional survey producing an accurate contour or ***hachure*** plan.

Finds are recovered artefacts. Some of these may be subject to the laws of treasure trove, where these apply. In the UK, for instance, all discoveries of gold or silver should be reported to the coroner, who will consider whether the items were

sub-metre resolution. This provides archaeologists with the capability to recognise and record otherwise hard-to-detect features (Historic England 2016a). LiDAR can be airborne, static (terrestrial) or mobile. One recent success resulting from the use of LiDAR has been the discovery of old cities in the Cambodian jungle by the Cambodian Archaeological Lidar Initiative (CALI 2016). LiDAR can also be used in shallow waters.

Drones are increasingly used for heritage surveys. They "provide a useful low-level aerial platform for recording historic buildings, monuments, archaeological sites and landscapes. They can carry a wide variety of sensors including cameras, multi/hyperspectral imaging units, and even laser scanners" (Historic England 2016b).

Box 12.8 summarises some of the problems that can be encountered with field surveys.

Box 12.8 Some problems with field surveys (Therivel 2009)

Access to the site will not be a problem where the developer already owns the land, although there may be problems where the project has off-site implications, e.g. as a result of dewatering. For projects such as road schemes a field survey may not be possible until the route is finally selected and the land acquired (in some jurisdictions this could be legally overcome). This is undesirably late because it does not allow a route to be chosen which would preserve important remains *in situ*.

The project timetable may constrain the fieldwork options. Fieldwalking is not possible in a standing crop, and can only be done after the fields are ploughed. Similarly, crop patterns show best in a well-grown crop and should be photographed just before the harvest.

The cost of archaeological surveys depends upon the extent and nature of the survey and the techniques employed. Surveys are frequently labour intensive and some elements can be expensive. Where the developer is liable to pay compensation to the landowner for damage arising from the evaluation, the scale of compensation will depend upon the techniques used. However, the costs should be seen against the background of the cost resulting from unexpected delay to the progress of the consenting process or development construction – indeed, the progress of the development – if significant archaeological deposits are located at a late stage in the process.

Underwater surveys – either in rivers, lakes or the marine environment – could require specialised techniques, such as multi-beam sonar, magnetometry or sub-bottom profiling with sonar. The results of these techniques are usually followed by observations by divers (in shallow waters) or by remotely operated vehicles (ROVs). There may be health and safety issues associated with diving in turbid or sediment-laden waters.

Procedures for the management of the data collected are important, as assessment processes can be lengthy and data sources may require periodic updates.

12.4.2 Determining the importance of the existing cultural heritage

A crucial step is to identify the importance (or non-monetary value) of the cultural heritage resource that is likely to be affected. HA et al. (2007) recommend identifying its "value to the quality and understanding of the country's cultural heritage resource, as set out in national, regional and local cultural heritage legislation, priorities and frameworks". Some of these values may be formally recognised through designation, others will require the use of professional judgement. A simple hierarchy might look like that in Table 12.5. HA et al. (2007) recommend that values be qualitative rather than numbers or ranking, as this may introduce an element of "spurious accuracy" and be misleading.

Table 12.5 Assigning value to cultural heritage resources (after HA et al. 2007)

	Factors for assessing the value of cultural heritage assets
Very high	• World Heritage Sites (including nominated sites) • assets of acknowledged international importance • assets that can contribute significantly to acknowledged international research objectives
High	• nationally protected/designated cultural heritage assets • undesignated assets of national importance or historical associations • assets that can contribute significantly to acknowledged national research objectives
Medium	• designated or undesignated assets of regional importance • designated or undesignated assets that can contribute to regional research objectives
Low	• designated or undesignated assets of local importance • assets of limited value, but with potential to contribute to local research objectives
Negligible	• assets with very little or no surviving cultural heritage interest
Unknown	• the importance of the resource has not been ascertained

Care should be taken when using the hierarchy of Table 12.5. For example, the table suggests that cultural heritage resources at the local level only have a "low" value. However, these resources may be important to the local community, particularly where they are being used by indigenous and/or aboriginal peoples, or by those using them for spiritual or religious purposes. The practitioner must exercise professional judgement in such instances, or use an alternative means of considering value in this regard.

As well as considering which aspects of the cultural heritage resource are affected, it is important to consider who may be affected by these impacts. These include:

- the owners or occupiers of historic properties and monuments;
- visitors to sites and buildings, including tourists;
- local communities and the general public, should their enjoyment of and/or their access to these resources be diminished;
- local communities who use these resources for cultural, spiritual, religious and/or traditional activities, particularly aboriginal and indigenous peoples;
- the general public who, although not directly using the resource, value its existence.

ICOMOS (2011) advises a methodology for assessing the value of the assets of cultural heritage:

> the value of heritage attributes is assessed in relation to statutory designations, international or national, and priorities or recommendations set out in national research agendas, and ascribed values. Professional judgement is then used to determine the importance of the resource. Whilst this method should be used as objectively as possible, qualitative assessment using professional judgement is inevitably involved.

The methodology includes the assessment of archaeological sites, built heritage or historic urban landscapes, historic landscapes and intangible cultural heritage or associations (with particular innovations, technical or scientific developments or movements and with particular individuals).

IFC Performance Standard 8 (IFC 2012b) identifies as critical cultural heritage:

i The internationally recognized heritage of communities who use, or have used within living memory, the cultural heritage for long-standing cultural purposes.
ii Legally protected cultural heritage areas, including those proposed by host governments for such designation.

When the cultural heritage likely to be affected by the development includes World Heritage properties, the ICOMOS (2011) guidance should be followed in preparing the baseline studies, and should clearly focus on outstanding universal value (OUV) and attributes that convey its OUV. The baseline studies should collect and collate information on the other assets of cultural heritage, to allow the full understanding of the historical development of the property, its context and setting.

> The data collection must enable the heritage attributes to be quantified and characterised, and allow their vulnerability to proposed changes to be established. It is also necessary to look at the interrelationship/s between discrete heritage resources, in order to understand the whole. There is often a relationship between a material aspect and an intangible aspect which must be brought to the fore. ... When describing [World Heritage] properties, it is essential to start by describing the attributes of OUV. This is the "baseline data" against which impacts must be measured, and includes both tangible and intangible aspects. A statement of condition may be useful for each key attribute of OUV.
>
> (ICOMOS 2011)

The assessment of the "non-economic" value and of the vulnerability to change of the cultural heritage assets is also an important component of the baseline studies. The existing pressures on the cultural heritage assets likely to be affected should also be an integral part of the baseline studies.

12.5 Impact prediction and evaluation

An impact is a change to cultural heritage, arising from a project or development. They may be new impacts or changes that exacerbate existing pressures. They can occur before the project begins (e.g. through site clearance); during project construction, as a result of construction activities; and during project operation. Their significance depends on the importance of the cultural heritage resource affected, and the magnitude of the impact.

ICOMOS (2011) identifies potential sources of impacts, including:

- large-scale development: roads, bridges, tall buildings, "box" buildings (e.g. malls), inappropriate, acontextual or insensitive developments, renewals, demolitions and new infrastructure typologies like wind farms;
- land-use policy changes;
- large-scale urban frameworks;
- excessive or inappropriate tourism.

The World Bank (1994) identifies certain sectors that it considers "particularly prone" to affecting cultural heritage:

- energy (construction of gas pipelines, utility lines etc.);
- communications (laying of cables, fibre optics);
- transport (highways, road construction or extensions, bridge construction or replacement, canal construction, railway construction);
- water (dams, irrigation and drainage schemes);
- sewerage and sanitation (waste-water treatment works, laying of sewer mains, water treatment works);
- urban development (infrastructure provision);
- industry and mining;
- agriculture (intensification and extensification) and forestry.

The impacts on tangible cultural heritage depend on the nature of the project under development and the characteristics of the cultural heritage resource. Impacts vary with the type of resource. For archaeological sites, in general, the key potential impacts comprise loss of and/or damage to archaeological remains and/or artefacts, and effects on the setting of the site. Buildings and upstanding monuments may be lost through demolition or damage during construction or operation. Historic landscapes can be affected by the introduction of new elements into the landscape, or damage to, or loss of, existing assets. Tables 12.6–12.8 summarise the project activities that may give rise to potential negative effects on tangible cultural heritage.

Negative impacts on movable cultural heritage may occur due to access resulting from a project, thus increasing the vulnerability of cultural objects to theft, trafficking or abuse (World Bank 2017).

Many of these are direct impacts but some are indirect. The latter may be caused by changes in topography, water-table levels and land-use practices, and by induced development. Although the construction phase is temporary, impacts resulting from construction can be permanent in nature, particularly where they result in loss of the cultural heritage asset. For example, topsoil stripping may cause permanent damage to archaeological remains whereas it would only cause temporary damage to a historic landscape (if proper restoration is undertaken).

The potential for positive impacts on tangible cultural heritage is summarised in Table 12.9.

Table 12.6 Potential negative impacts on tangible cultural heritage: archaeological remains (after HA et al. 2007)

Negative effects to archaeological remains	Stage/activity
Removal of archaeological deposits (loss, damage)	• Ground investigations: trial pits/boreholes • Site clearance: removal of trees and vegetation; fencing; establishment of construction compounds; chemical decontamination; construction of access and haul roads; building/structure demolition

Table 12.6 continued

Negative effects to archaeological remains	Stage/activity
	• Construction: topsoil stripping; excavations (building foundations, road/railway foundations, borrow pits, etc); removal of peat; piling; construction of buildings/infrastructure/structures/car parks/access roads; installation of drainage systems, including stormwater storage (land/road drainage); landscaping (earth mounds/bunds); installation of lighting systems/noise barriers; landscape planting (damage by roots); • Operational phase: induced development; maintenance
Impact on setting	• Site clearance: removal of trees and vegetation • Construction: construction traffic movement (noise intrusion); construction of buildings/infrastructure/ structures/car parks; landscaping (earth mounds/bunds); spoil storage/disposal; installation of lighting systems/noise barriers; landscape planting; • Operation: operational traffic movement (noise intrusion); induced development
Disturbance of underwater archaeological sites, including human remains	• Construction activities in the marine environment
Compaction of archaeological deposits/ damage through rutting of superficial deposits	• Site clearance and construction traffic movement; establishment of construction compounds • Construction: landscaping (earth mounds/bunds); spoil storage/disposal
Damage to archaeological deposits resulting from changes to hydrology and soil/water chemistry; desiccation of waterlogged archaeological deposits	• Piling; installation of drainage systems (land/road drainage) • Dewatering of excavations (building foundations, borrow pits, etc); removal of peat; installation of drainage systems (land/road drainage)
Severance of cultural heritage sites from one another and/or the community (loss of and/or changes to access)	• Site clearance: fencing; construction of access and haul roads • Construction: construction of buildings/infrastructure/ structures/car parks/access roads
Damage to monuments through trampling/erosion	• Increases in visitor numbers at the operational stage
Changes to amenity through noise and/or disturbance	• Construction: construction noise (temporary) • Operation: increased traffic movements, visitor numbers, noise from residential developments etc.

Table 12.7 Potential negative impacts on tangible cultural heritage: historic buildings/monuments (after HA et al. 2007)

Negative effects to historic buildings/monuments	Stage/activity
Loss of historic buildings or upstanding remains	• Site clearance: demolition (partial or complete)
Modifications to buildings	• Construction: changes to building interiors, individual building elements; construction of extensions; repairs using inappropriate materials
Impact on setting	• Site clearance: removal of trees and vegetation; fencing • Construction: construction traffic movement (noise intrusion); construction of buildings/infrastructure/structures/car parks; landscaping (earth mounds/bunds); spoil storage/disposal; installation of lighting systems; landscape planting; installation of noise barriers • Operation: operational traffic movement (noise intrusion); induced development
Structural damage from vibration, excavation	• Construction: piling; excavation of building foundations
Damage to building fabric from air pollution	• Atmospheric emissions from operational traffic movements/power stations, industry etc.
Dust damage to fabric of monuments	• Site clearance and construction traffic movement
Damage resulting from vehicular collision	• Movement of site clearance and/or construction traffic
Severance of one part of a property from another (loss of access; dereliction/neglect)	• Site clearance: fencing; construction of access and haul roads • Construction: construction of buildings/infrastructure/structures/car parks/access roads
Changes to amenity through noise and/or disturbance	• Construction: construction noise (temporary) • Operation: increased traffic movements, visitor numbers, noise from residential developments etc.

For intangible cultural heritage (and in particular for indigenous or aboriginal people making use of the cultural heritage resource), impacts may include (ICOMOS 2011, SCBD 2004):

- effects on continued customary use of biological resources/access to traditional sites;
- effects on the respect, preservation, protection and maintenance of traditional knowledge, innovations and practices (inappropriate behaviour in certain locations for example);

Table 12.8 Potential negative impacts on tangible cultural heritage: cultural/historic landscapes (after HA et al. 2007)

Negative effects on historic landscapes	Stage/activity
Change to landscape integrity	• Site clearance: building demolition (partial or complete); loss of open spaces
Introduction of new (intrusive) elements	• Construction: construction traffic movement (noise intrusion); construction of buildings/infrastructure/structures/car parks; landscaping (earth mounds/bunds); spoil storage/disposal; installation of lighting systems; landscape planting; installation of noise barriers • Operation: operational traffic movement (noise intrusion); induced development
Severance of landscape elements from one another (loss of continuity; dereliction/neglect)	• Site clearance: fencing; construction of access and haul roads • Construction: construction of buildings/infrastructure/structures/car parks/access roads
Changes to amenity through noise and/or disturbance	• Construction: construction noise (temporary) • Operation: increased traffic movements, visitor numbers, noise from residential developments etc.

Table 12.9 Potential positive impacts on tangible cultural heritage (after HA et al. 2007)

Positive effects	Stage/activity
Improvement of setting	• Screening provided by landscaping (earth mounds/bunds) and/or landscape planting; • Improvement of lighting ambience; • Removal of traffic from sensitive areas such as historic town centres; • Re-establishment of historic setting.
Slowing/reduction of building/monument deterioration	• Removal of traffic from sensitive areas such as historic town centres.

• effects on sacred sites and associated ritual or ceremonial activities;
• effects on the exercise of customary laws.

A key activity at this stage of the ESIA, then, is the identification of the activities associated with the project, to ascertain whether they would affect the cultural heritage resource. This should include the following project information, at a minimum:

- land take;
- amount of excavation;
- construction methods and programming;
- details of temporary and permanent works, e.g. site compound;
- predicted traffic type (e.g. heavy goods vehicles) and volumes, during both construction and operation;
- building layout and height (where appropriate).

The potential effects of these activities on the characteristics of the cultural heritage assets should be identified during the establishment of baseline environmental characteristics. A matrix can be helpful in organising this information.

Once the value of the cultural heritage resource has been ascertained (see § 12.4.2), the next stage is to identify the magnitude of the impact:

- positive/negative;
- direct/indirect;
- permanent/temporary;
- irreversible/reversible;
- short, medium or long term;
- cumulative.

These factors are usually taken together to assign a value to the magnitude of impact. Guidance is provided in Table 12.10. The magnitude of the impact relates only to the characteristics of the effect, and not to the value of the resource. So, for example, if a site of local importance is predicted to be lost to construction, the magnitude of effect is the same as it would be for the loss of a World Heritage Site: major.

With this information, the significance of the impact can then be ascertained using a significance matrix like that of Figure 1.3, which brings

Table 12.10 Assigning value to impact magnitude (after HA et al. 2007)

Major	• Changes to most or all key features, such that the resource is totally altered or lost. • Comprehensive changes to setting.
Moderate	• Changes to many features, such that the resource is clearly modified. • Considerable changes to setting that affect the character of the asset.
Minor	• Changes to features, such that the asset is slightly altered. • Slight changes to setting.
Negligible	• Very minor changes to features or setting.
No change	• No change.

together the magnitude of the impact and the value of the resource. So, for example, a major change to a very high-value resource is considered to be a very large significant effect.

As with the assignation of value, the significance matrix should be used as a guide rather than being rigidly applied. Indeed, HA et al. (2007) caution that the matrix is not intended to "mechanise judgement ... but to act as a check to ensure that judgements regarding value, magnitude of impact and significance of effect are balanced". The matrix is thus very useful in ensuring that the assessment of effects on the different aspects of the cultural heritage resource remains consistent across the board.

ICOMOS (2011) has expressed concerns about the way that ESIA is applied to World Heritage Sites, considering that it tends to be too mechanistic in its methodology – the effects on each of the heritage receptors are assessed but this is often not accompanied by an assessment of the effects on the whole site's "overall ensemble of attributes". ICOMOS also considers that ESIA tends to neglect cumulative and incremental adverse effects on World Heritage Sites. ICOMOS is therefore of the view that the effect on World Heritage Sites should be assessed in terms of effects on the attributes of OUV that underpin the site in question. In general, the factors affected would include: appearance, skyline, key views, and other different attributes that contribute to OUV. Practitioners should keep this in mind when assessing effects on World Heritage Sites (see Box 12.9).

Box 12.9 Assessment of World Heritage Sites

ESIA of World Heritage Sites should consider the following questions:

- What is the baseline?
- What is the heritage at risk and why is it important – how does it contribute to outstanding universal value (OUV)?
- How will change or a development proposal impact on OUV?
- How can these effects be avoided, reduced, rehabilitated or compensated?

– Paragraph 2-2-1, ICOMOS (2011)

At the same time, however, a cultural heritage asset may have attributes that contribute to its value both as an archaeological resource and as a historic building, or to its place within a historic landscape. Care should be taken that effects on such attributes are not double-counted.

Cumulative effects may occur where:

- one project has multiple effects on the same resource;
- more than one project affects the cultural heritage in an area;
- the proposed project will exacerbate existing pressures.

Existing pressures on cultural heritage resources may include (but are not limited to):

- streambank erosion;
- wind action;
- subsidence;
- animal burrowing;
- compaction and fragmentation of soils;
- agricultural activities, e.g. ploughing;
- grazing (wild and domestic animals);
- vehicular traffic (compaction; atmospheric emissions);
- vandalism.

The assessment of cumulative effects should use the principles of assessment set out in the rest of this section. Positive effects should not be offset against negative ones to obtain an overall neutral result.

12.6 Interactions

The assessment of cultural heritage impacts should consider the potential for interaction with other components of the ESIA. Table 12.11 provides examples of how impacts on other factors may have indirect impacts on both tangible and intangible cultural heritage. These interactions are more complex than the table suggests, as many of the impacts have complex pathways and involve several components.

From a different perspective, positive or negative impacts on cultural heritage could have positive or negative effects on other factors:

- The discovery or the enhancement of cultural heritage elements could positively affect social and economic development, including the self-esteem of local communities or indigenous people, the rise of new economic activities, and improved access to infrastructure, but it could also have negative effects associated with tourism, lifestyles and the culture of local communities.
- Negative impacts on important cultural heritage elements could have negative effects on tourism, including the economic benefits associated with tourism.
- Positive or negative impacts on cultural landscapes, including historic landscapes, have influence on the overall impacts on landscape.

12.7 Mitigation and enhancement

Once impacts have been predicted and evaluated, measures for their mitigation and/or enhancement should be identified, using the mitigation hierarchy

Table 12.11 Interaction of cultural heritage (CH) impact assessment with other components of ESIA

Components	Interactions (tangible CH)	Interactions (intangible CH)
Social and material assets	Many social changes could impact on CH, including changes in human settlements, reinstatement of communities, changes in agricultural, grazing or fishing activities, increasing of tourism, increase of vehicular traffic, presence of labour force, and vandalism. The impacts could be direct (e.g. degradation caused by vandalism) or indirect (e.g. ploughing could destroy artefacts in the top layer of the soil).	The effects of a project on the social fabric of indigenous peoples or local communities can disrupt their culture and their cultural heritage. The extreme cases could be indigenous peoples who have had no or little previous contact with modern civilization (e.g. some Indigenous peoples from Amazonia).
Health and well-being	Adverse effects on cultural heritage can affect human health and well-being, both of individuals and communities. This is particularly so for aboriginal and Indigenous peoples. Where the use of traditional sites for spiritual purposes, for example, is lost, this has been demonstrated to have consequences for health and well-being, as such uses are integral to the culture of such communities.	
Noise and vibration	Noise could affect the setting of CH assets. Vibration could damage buildings and other structures. Noise barriers may affect the setting of CH assets.	Noise and vibration could affect the setting and the use of sacred and other symbolic places.
Landscape	Landscape changes could affect the setting of CH assets or landscapes with historical and cultural value.	Landscape changes could affect the setting of and the use of sacred and other symbolic places.
Air quality	Some air pollutants could damage open-air CH assets. Smoke, dust and odours could affect the setting of CH assets.	Smoke, dust and odours could affect the setting of and the use of sacred and other symbolic places.
Climate change	Climate change effects (sea water levels, desertification, soil erosion, glacial melting, change of land-use patterns, etc.) could physically affect CH assets (archaeological sites, buildings, cultural landscapes) or the setting of those assets.	Climate change could affect local communities, including indigenous peoples, and their access to the land, natural and CH resources.

Table 12.11 continued

Components	Interactions (tangible CH)	Interactions (intangible CH)
Geology and geomorphology	Change of sediment patterns in river or along the coastline could lead to bottom, stream-bank or coastal erosion, thus affecting CH assets, onshore or offshore. Subsidence could affect CH assets.	
Water	Flooding of areas could affect or change the context of CH assets. Change in underground water levels could damage archaeological remains. Water pollution could affect the setting of CH assets.	Water pollution, flooding of areas (e.g. by dam construction) or changes in water availability (e.g. by diversion or overuse) could affect local communities, including indigenous peoples, and their use of natural and CH resources.
Biodiversity	Changes in biodiversity (e.g. trees, forests) could affect the setting or the context of CH assets. Some CH assets depend on the balance of biodiversity components (e.g. the conservation of old libraries that depend on a population to control insects, as in the Coimbra University Library, part of a World Heritage property).	Some cultural knowledge and practices, including spiritual activities, depend on biodiversity and its components. Impacts on biodiversity may affect continued customary use of biological resources and/or in the preservation, protection and maintenance of traditional knowledge, innovations and practices.
Risks of accidents and natural disasters	Projects could increase the risk of accidents or the vulnerability to natural disasters therefore affecting CH assets or their setting.	Projects could increase the risk of accidents or the vulnerability to natural disasters and therefore affect local communities, including indigenous peoples, and their use of land or of natural and cultural heritage resources.

(following the principles set out by IFC Performance Standard 1, paragraph 14). This is a key objective of ESIA.

12.7.1 Types of mitigation and enhancement

Some aspects of cultural heritage cannot be replaced, e.g. archaeological remains, historic buildings. Accordingly, at the top of the hierarchy (and the preferred method of mitigation) is the principle of avoiding or preventing adverse impacts. This includes the following actions:

- do not proceed;
- employ alternative means of achieving project goals and objectives;
- avoid physical features through project siting/design measures, e.g. relocating the development, amending its boundaries, etc.;
- use alternative construction methods, e.g. bored rather than driven piling to reduce vibration and prevent structural damage.

This will allow the preservation of cultural heritage resources *in situ*, since "cultural heritage assets are non-renewable resources and ... their physical preservation *in situ* when possible should be the primary goal of cultural resource management" (HA et al. 2007). The IFC's Performance Standards 7 and 8 also emphasise the need to avoid impacts on cultural heritage that is critical to the identity and/or cultural, ceremonial, or spiritual aspects of Indigenous peoples' lives.

The remainder of the mitigation hierarchy comprises:

- reduce adverse impacts, e.g. through reducing land-take;
- remedy adverse impacts (repair/rectify/restore);
- enhance positive effects;
- where adverse impacts cannot be avoided or residual impacts remain, compensate/offset wherever technically and financially feasible.

These measures may need to be implemented during and/or after construction, or during operation of the project. The principles underpinning this mitigation hierarchy are set out in Table 12.12.

Remedial mitigation measures can include:

- screening of a cultural heritage asset, e.g. from a road/residential development;
- building repair;
- relocation of buildings/features, e.g. milestones;
- rebuilding all or part of a building as a museum exhibit;
- improving the setting of archaeological remains/buildings;
- information panels;
- improving access and amenity to historic sites (ensuring that they are protected from damage by e.g. trampling, erosion, etc.);
- erosion control;
- rerouting of vehicular and pedestrian traffic;
- control of flora and fauna.

Where loss of a cultural heritage resource cannot be avoided, mitigation involves recording of the resource and gaining an understanding of it before its loss. This may involve an archaeological dig prior to the commencement of construction or the recording of buildings/sites before site clearance and demolition. Although the site would be destroyed, there would be an understanding of what is lost.

Table 12.12 Key terms underpinning mitigation (World Bank 1994)

Conservation	Encompasses all aspects of protecting a site or remains, so as to retain its cultural significance. It includes maintenance and may, depending on the importance of the cultural artefact and related circumstances, involve preservation, restoration, reconstruction, or adaptation, or any combination of these.
Preservation	Maintaining the fabric of a place in its existing state and retarding deterioration. It is appropriate where the existing fabric itself constitutes evidence of specific cultural significance, or where insufficient evidence is available to allow other conservation processes to be carried out. Preservation is limited to the protection, maintenance, and, where necessary, stabilization of the existing fabric.
Restoration	Returning the existing fabric of a place to a known earlier state by removing accretions or reassembling existing components without introducing new materials. It is appropriate only if (a) there is sufficient evidence of the earlier state of the fabric, and (b) returning the fabric to that state reveals the significance of the place and does not destroy other parts of the fabric.
Reconstruction	Returning a place to a known earlier state, as nearly as possible. It is distinguished by the introduction of materials (new or old) into the fabric. Reconstruction is appropriate only where a place is incomplete through damage or alteration and could not otherwise survive. Reconstruction is limited to the completion of a depleted entity and should not constitute the majority of the fabric.
Adaptation	Modifying a place for compatible use. It is acceptable where the adaptation does not substantially detract from its cultural significance and may be essential if a site is to be economically viable.
Maintenance	The continuous protective care of the fabric, contents, and setting of a place. Maintenance is to be distinguished from repair, which involves restoration or reconstruction.

For intangible cultural heritage resources, mitigation/enhancement may involve:

- ceremonies, reconsecration;
- environmental/cultural heritage management plan;
- community/benefit agreements between project proponent and affected community (including rights, duties and responsibilities of all parties);
- enhancement, including institutional strengthening;
- financial compensation.

Where there is uncertainty about whether archaeological remains would be encountered during construction, requirements for an archaeological watching brief should be included in the construction contract, as well as a requirement to cease construction in the case of unexpected archaeological finds/remains during construction. The contract may also need to include procedures regarding the notification of cultural authorities in such circumstances, how long work must cease for, what should be done with found objects, etc. (World Bank 2017).

Measures for the mitigation of effects on other environmental parameters may affect cultural heritage assets. Such measures can include landscape planting, installation of noise barriers, installation of lighting systems, etc. In addition, an archaeological survey can affect biodiversity as the land usually has to be cleared and stripped of topsoil as part of the survey work.

12.7.2 Managing mitigation and enhancement

It is important to ensure that mitigation measures proposed through the ESIA are deliverable. Accordingly, their feasibility must be explored with project planners and designers. Mitigation costs, including the costs of rescue archaeological excavation, must be provided in the overall project costs, included in the project cost–benefit analysis, and considered when reviewing and assessing project alternatives. The cost of such mitigation should be a factor included in decision-making when assessing project alternatives and choosing the preferred option. Significant excavation costs may suggest that reconsideration of the options is necessary before final decisions are made.

Practitioners should ensure that sufficient time is allocated in the project programme to undertake rescue-archaeological excavation in advance of construction commencing (it has not been unknown for road construction to commence when archaeologists are still on site).

12.8 Monitoring

The prediction and evaluation of impacts on cultural heritage and the assessment of the effectiveness of mitigation programmes (including enhancement) are strongly affected by uncertainty. Monitoring, together with adaptive management, should cope with this uncertainty and be able to detect unanticipated impacts.

One of the international best-practice principles on ESIA follow-up (Morrison-Saunders et al. 2007) is the involvement of the community: "direct community participation in follow-up program design and implementation is desirable. Benefits may flow from active community involvement in ESIA follow-up including sharing of special local knowledge, focused program design, building trust and partnerships". This is particularly important in the monitoring of cultural heritage. Bradshaw et al. (2011) present two interesting case studies: one, about the incorporation of traditional knowledge in scientific monitoring in a mining project in the Northwest Territories, Canada; the other, of a large-scale mining project in Mongolia, in which the communities defined the standards of acceptable change that were used in the monitoring of the impacts on the region's people and tangible and intangible cultural heritage.

Box 12.10 presents several types of monitoring activities, with examples, focused on the impacts on physical (or tangible) cultural heritage.

Box 12.10 Types of monitoring activities, with examples

---

1 Monitoring of potential damage to known cultural assets caused directly by the construction of the project, including ancillary activities such as new and/or improved access, construction compounds, etc.

Example: structural damage in a building caused by piling.

2 Monitoring of potential damage to known cultural assets caused directly by the operation of the project.

Examples:

- Corrosion damage caused by air pollution (e.g. sulphur dioxide).
- Damage to (historic) bridge and its piers through prevention of stream sediment recharge by an upstream dam.

3 Monitoring of the changes to the setting (visual, acoustic, odours) of a cultural asset caused by the construction or operation of a project.

Example: Increase of noise levels in the surroundings of a cultural asset that requires a silent environment to be enjoyed.

4 Monitoring of potential damage to known cultural assets caused indirectly by the construction or operation of the project.

Examples:

- The construction of new accesses could increase the number of visitors that could physically damage the cultural asset.
- The abandonment of an area due to the decrease of an activity could lead to the abandonment of a cultural heritage site and to its deterioration.
- The extraction of groundwater for a project could lead to changes in groundwater levels and thus changes in the environmental conditions of archaeological remains already known and not excavated.

5 Monitoring of mitigation, including enhancement, of cultural assets:
5a Monitoring of the state of conservation of cultural assets remaining in its original location.

Examples:

- Consolidation works on buildings or structures (e.g. bridges) to resist a long period of submersion in a reservoir. Monitoring (and maintenance) of protective fencing.

5b   Monitoring of compensation or enhancement measures

  Example: Number of visitors to a restored site or a new museum created as a result of the project.

6 Monitoring of archaeological remains revealed during the project operation, due to unexpected changes in environmental conditions.

  Example: The erosion of the margins of a reservoir due to the oscillation of the water levels can lead to the discovery of archaeological remains not detected in previous surveys.

7 Monitoring of the effects on the cultural identity of a community caused by the findings of the cultural heritage surveys and studies.

  Example: Survey to assess how the cultural identity of a specific community could be strengthened by the findings of the cultural heritage surveys and studies.

8 Monitoring of the scientific knowledge obtained with the cultural heritage surveys and studies.

  Example: The number of publications with referee or the citation indexes related to the findings of the cultural heritage surveys and studies.

12.9 Conclusions

The consideration of cultural heritage in ESIA has recently expanded to include not only the assessment of the effects of development on the tangible expressions of cultural heritage (sites, monuments and landscapes), but also on its intangible aspects. Information on these effects must be provided to the project decision-makers at the key points in the project development process, to ensure that decisions are made using the relevant knowledge, including financial information such as the cost of mitigation measures.

Note

1 The contents of this chapter reflect the professional views of the author and are not a statement of Scottish Government policy.

References

For links to conventions, treaties and charters please see Tables 12.2 and 12.3.

Agnoletti M (ed.) 2013. *Italian Historical Rural Landscapes. Cultural Values for the Environment and Rural Development*. New York: Springer.

André P, B Enserink, D Connor, and P Croal 2006. *Public Participation International Best Practice Principles*. Special Publication Series No. 4. Fargo, ND: International Association for Impact Assessment. www.iaia.org/uploads/pdf/SP4.pdf.

Birnbaum CA 1994. *Protecting Cultural Landscapes: Planning, Treatment and Management of Historic Landscapes*. Preservation Briefs 36. Washington, DC: National Park Service. www.nps.gov/tps/how-to-preserve/preservedocs/preservation-briefs/36Preserve-Brief-Landscapes.pdf.

Borchi A 2012. Status and process of approval of retrospective Statements of Outstanding Universal Value. Presentation to the Meeting of National Focal Points of Nordic, Baltic, Western and Mediterranean Europe and German Site Managers on the Implementation of the Second Cycle of the Periodic Reporting Exercise. Berlin, Germany, 24–26 September 2012.

Bradshaw E, K Bryant and T Cohen with the Rio Tinto Internal Working Group and the Centre for Social Responsibility in Mining, University of Queensland 2011. *Why Cultural Heritage Matters. A Resource Guide for Integrating Cultural Heritage Management into Communities Work at Rio Tinto*. www.riotinto.com/documents/ReportsPublications/Rio_Tinto_Cultural_Heritage_Guide.pdf.

CALI (Cambodian Archaeological Lidar Initiative) 2016. http://angkorlidar.org.

Campbell I 2009. *Physical Cultural Resources Safeguard Policy: Guidebook*. Washington, DC: The World Bank. http://documents.worldbank.org/curated/en/2009/03/16528469/physical-cultural-resources-safeguard-policy-guidebook.

CoE 2000. *European Landscape Convention*. Florence, Italy: Council of Europe.

CoE 2016. Details of Treaty No. 219. *Protocol amending the European Landscape Convention*. www.coe.int/en/web/conventions/full-list/-/conventions/treaty/219.

Council of Europe 2005. *Council of Europe Framework Convention on the Value of Cultural Heritage for Society*. Faro: Council of Europe.

DCMS (Department for Culture, Media and Sport) 2010. *Principles of Selection for Listing Buildings*. www.gov.uk/government/uploads/system/uploads/attachment_data/file/137695/Principles_Selection_Listing_1_.pdf.

Garibaldi A and N Turner 2004. Cultural keystone species: implications for ecological conservation and restoration. *Ecology and Society* 9(3): 1. www.ecologyandsociety.org/vol9/iss3/art1/

HA (Highways Agency), Transport Scotland, Welsh Assembly Government and The Department for Regional Development Northern Ireland 2007. *Design Manual for Roads and Bridges, Volume 11*, Section 3, Part 2 Cultural Heritage. HA 208/07.

Hanna P and F Vanclay 2013. Human rights, Indigenous peoples and the concept of Free, Prior and Informed Consent. *Impact Assessment and Project Appraisal* 31:2, 146–157. http://dx.doi.org/10.1080/14615517.2013.780373.

Historic England 2016a. *Lidar (Light Detection and Ranging)*. https://historicengland.org.uk/research/approaches/research-methods/airborne-remote-sensing/lidar.

Historic England 2016b. *Drones for Heritage Uses*. https://historicengland.org.uk/research/approaches/research-methods/airborne-remote-sensing/drones-for-heritage-uses.

ICOMOS (International Council on Monuments and Sites) 1964. *International Charter for the Conservation and Restoration of Monuments and Sites (The Venice Charter 1964)*. Venice: ICOMOS.

ICOMOS 1994. *The Nara Document on Authenticity (1994)*. Nara, Japan: ICOMOS.

ICOMOS 2002. *International cultural tourism charter: Principles and guidelines for managing tourism at places of cultural and heritage significance*. ICOMOS International Cultural

Tourism Committee.

ICOMOS 2005. *Xi'an Declaration on the Conservation of the Setting of Heritage Structures, Sites and Areas*. Xi'an, China: ICOMOS.

ICOMOS 2011. *Guidance on Heritage Impact Assessments for Cultural World Heritage Properties*. www.icomos.org/world_heritage/HIA_20110201.pdf.

IFC (International Finance Corporation) 2012a. Performance Standard 7 – Indigenous People. In IFC Performance Standards on Environmental and Social Sustainability, Washington, DC: IFC. www.ifc.org/wps/wcm/connect/c8f524004a73daeca09afdf 998895a12/IFC_Performance_Standards.pdf?MOD=AJPERES.

IFC 2012b. Performance Standard 8 – Cultural Heritage. in IFC Performance Standards on Environmental and Social Sustainability, Washington, DC: IFC. www.ifc.org/wps/ wcm/connect/c8f524004a73daeca09afdf998895a12/IFC_Performance_Standards.pdf? MOD=AJPERES.

IFC 2012c. Performance Standard 1 – Assessment and Management of Environmental and Social Risks and Impacts. In IFC Performance Standards on Environmental and Social Sustainability, Washington, DC: IFC. www.ifc.org/wps/wcm/connect/ c8f524004a73daeca09afdf998895a12/IFC_Performance_Standards.pdf?MOD= AJPERES

International Labour Organization Convention 1989. Indigenous and Tribal Peoples Convention (No. 169). www.ilo.org/global/topics/equality-and-discrimination/ indigenous-and-tribal-peoples/lang—en/index.htm.

Morrison-Saunders A, R Marshall and J Arts 2007. *EIA Follow-Up International Best Practice Principles*. Special Publication Series No. 6. Fargo, ND: International Association for Impact Assessment. www.iaia.org/uploads/pdf/SP6_1.pdf.

SCBD (Secretariat of the Convention on Biological Diversity) 2004. *Akwé: Kon Voluntary Guidelines for the Conduct of Cultural, Environmental and Social Impact Assessment regarding Developments Proposed to Take Place on, or which are Likely to Impact on, Sacred Sites and on Lands and Waters Traditionally Occupied or Used by Indigenous and Local Communities*. Montreal (CBD Guideline Series). www.cbd.int/doc/publica-tions/akwe-brochure-en.pdf

Strauss M 2016. Discovery could rewrite history of Vikings in New World, *National Geographic*. http://news.nationalgeographic.com/2016/03/160331-viking-discovery-north-america-canada-archaeology.

Therivel R 2009. Archaeological and other material and cultural assets. Chapter 7 in Morris P and Therivel R (eds), *Methods of Environmental Impact Assessment*, 3rd Edn. London: Routledge.

United Nations 2007. *Declaration on the Rights of Indigenous Peoples*. www.ohchr.org/EN/Issues/IPeoples/Pages/Declaration.aspx.

UNESCO (United Nations Educational, Scientific and Cultural Organization) 1982. *The Mexico City Declaration on Cultural Policies*. http://portal.unesco.org/culture/en/ev.php-URL_ID=12762&URL_DO=DO_ TOPIC&URL_SECTION=201.html.

UNESCO 2003. *Convention for the Safeguarding of the Intangible Cultural Heritage*. Paris: UNESCO.

UNESCO 2011. *Recommendation on the Historic Urban Landscape, including a glossary of definitions*. http://portal.unesco.org/en/ev.php-URL_ID=48857&URL_DO=DO_ TOPIC&URL_SECTION=201.html.

UNESCO–CBD Joint Program between biological and cultural diversity 2014. *Florence Declaration on the Links between Biological and Cultural Diversity*. http://landscapeunifi.it/images/pdf/UNESCO-CBD_JP_Florence_Declaration.pdf.

UNESCO–WHC (World Heritage Committee) 2013. *Operational Guidelines for the Implementation of the World Heritage Convention.* Paris: UNESCO World Heritage Centre http://whc.unesco.org/archive/opguide13-en.pdf.

World Bank 1994. *Environmental Assessment Sourcebook Update (Number 8): Cultural Heritage in Environmental Assessment.* Washington DC: The World Bank.

World Bank 2017. The World Bank Environmental and Social Framework. Washington, DC: The World Bank. http://documents.worldbank.org/curated/en/ 383011492423734099/pdf/114278-REVISED-Environmental-and-Social-Framework-Web.pdf

13 Socio-economic impacts 1: Overview and economic impacts

John Glasson

13.1 Introduction

Major projects can have a wide range of impacts on a locality – including biophysical and socio-economic – and the trade-off between such impacts is often crucial in decision-making. Major projects may offer a tempting solution to an area's economic problems; these however may have to be offset against more negative impacts such as pressure on local services and social upheaval, in addition to possible damage to the physical environment. Socio-economic impacts can be very significant for particular projects and analysts ignore them at their peril. Nevertheless they have often had a low profile in impact assessment, although there is a growing awareness of their importance in decision-making.

This chapter begins with an initial overview of the socio-economic impacts of projects/developments, which explains the nature of such impacts and provides important context for both Chapters 13 and 14. It explores the evolving story, covering issues of definition and semantics surrounding terms such as 'socio-economic impact assessment' and 'social impact assessment'. Economic impacts, including the direct employment impacts and the wider, indirect impacts on a local and regional economy, are then discussed in more detail. These topics are addressed before the consideration of social impacts in Chapter 14, because many of those social impacts arise from the direct and indirect economic impacts of major projects. Chapter 14 therefore focuses on related impacts such as changes in population levels and associated effects on the social infrastructure, including accommodation and services, and issues of distribution/equity and social cohesion. Several of the methods discussed straddle the two chapters and will be cross-referenced to minimise duplication. Chapters 13 and 14 draw in part on the work of the Impacts Assessment Unit (IAU) in the School of Planning (now School of Built Environment) at Oxford Brookes University, which has undertaken many research and consultancy studies on the socio-economic impacts of major projects.

13.2 Definitions and concepts

Socio-economic impacts cover a wide range of social and economic impacts and the boundaries are fuzzy between both social and economic impacts, and indeed between socio-economic and other impact areas. Thus economic impacts may range from the macro-impacts of a new project on a nation's Gross National Product to micro-impacts on wages for electricians in a town adjacent to the project. Social impacts may include impacts on local demographics, livelihoods, cultural/heritage issues and upon physical and mental health. Our focus of socio-economic impacts for Chapters 13 and 14 is as set out in Table 13.1. Cultural/heritage impacts and health impacts are covered respectively in Chapters 12 and 16.

Socio-economic impact assessment (SIA) developed in the 1970s and 1980s, mainly in relation to the assessment of the impacts of major resource development projects, such as nuclear power stations in the US, hydroelectric schemes in Canada, and the UK's North Sea oil- and gas-related developments. The growing interest in socio-economic impacts, partly stimulated by the introduction of the US National Environmental Policy Act of 1969 and subsequent amendments of 1977, generated some important studies and publications, including the works of Wolf (1974), Lang and Armour (1981), Finsterbusch (1985), and Carley and Bustelo (1984). It also led to considerable debate on the nature and role of SIA.

Some authors refer to social impact assessment; others refer to socio-economic impact assessment. Some see SIA as an integral part of Environmental Impact Assessment (EIA), providing the essential 'human elements' complement to the often narrow biophysical focus of many EISs: 'from the perspective of the social impact agenda, this meant: valuing people "as much as fish"...' (Bronfman 1991). Others see SIA as a separate field of study, a separate process (e.g. Esteves et al. 2012). Social impact assessment is also sometimes seen to be more associated with a developing-world context than socio-economic impact assessment. Chapters 13 and 14 of this text focus on the wider definition of socio-economic impacts, *within* a holistic impact assessment process (be it called IA, EA, EIA, ESIA, ESHIA, etc.) that is of relevance to all stages of development.

Wolf (1974), one of the pioneers of SIA, adopted the wide-ranging definition of SIA as 'the estimating and appraising of the conditions of a society organised and changed by the large-scale application of high technology'. Bowles (1981) has a similarly broad definition: 'the systematic advanced appraisal of the impacts on the day to day quality of life of people and communities when the environment is affected by development or policy change'. A more light-hearted but often relevant approach to definition can be typified as the 'grab bag' (Carley and Bustelo 1984) or 'Heineken' approach – with SIA including all those vitally important, but often intangible impacts which other methods cannot reach.

A subsequent study by the Interorganisational Committee on Guidelines and Principles for Social Impact Assessment (1994) defined social impacts as 'the consequences to human populations of any public or private actions that alter the ways in which people live, work, play, relate to one another, organise to meet their needs, and generally cope as members of society'. These Guidelines and Principles have had a bumpy ride since their inception, and in 2003 there was a 'parting of the ways' between a US-oriented version and a more international-oriented version. The US version differs little from the 1994 original, relating closely to regulatory requirements and with a focus on assessment in advance of development actions. In contrast, the international version argues that SIA should not necessarily be tied to a regulatory context, should not just be 'in advance' (but should be more participative and ongoing), and should 'consider how to ensure the achievement of the intended positive consequences or goals of development as well as preventing unintended negative outcomes' (Vanclay 2003). This latter focus was reinforced in the latest international version, *Social Impact Assessment: Guidance for assessing and managing the social impacts of projects* (IAIA 2015), which offers the following definition of social impact assessment as:

> the process of analysing, monitoring and managing the intended and unintended social consequences, both positive and negative, of planned interventions (policies, programs, plans, projects) and any social change processes invoked by those interventions. Its primary purpose is to bring about a more sustainable and equitable biophysical and human environment.

To conclude, and hopefully not to overly simplify, social and socio-economic impacts are the 'people impacts' of development actions. Both social impact assessments and socio-economic impact assessments focus on the human dimension of environments, seeking to identify the impacts on people, including who benefits and who loses. SIA can help to ensure that the needs and voices of diverse communities are taken into account during project planning and decision-making.

13.3 Key legislation, guidance and standards

13.3.1 Socio-economic impacts in early practice: the poor relation?

The early recognition, by some analysts, of the importance of socio-economic impacts in the EIA process was partly reflected in legislation. The definition of the environment, as included in the 1979 US CEQ regulations, addresses biophysical components and socio-economic factors and characteristics. The EU Directive 85/337/EEC (EC 1985) required a description of possible impacts on human beings and this served to highlight socio-economic issues as individual Member States sought to implement the Directive through guidance and regulations. For example, the UK government produced guidance which suggested that 'certain aspects of a project including numbers employed and where they will come from should be considered within an environmental statement' (DoE 1989). The 1999 Town and Country Planning (EIA/England and Wales) Regulations required:

> a description of the environment likely to be significantly affected by the development, including, in particular, *population* (author emphasis), fauna, flora, soil, water, air, climatic factors, material assets, including the architectural heritage, landscape and the interrelationship between the above factors.
>
> (Glasson et al. 2012)

Yet despite some legislative impetus, for many years the consideration of social and economic impacts continued to be the poor relation in EIA and in EISs (Glasson and Heaney 1993; Burdge 2002, 2003; Chadwick 2002).

There may be several reasons for this, which can be summed up by the general perceptions that:

- socio-economic impacts seldom occur;
- when they do they are covered elsewhere in the planning and development process;
- their inclusion can be used to downplay biophysical impacts;
- they are invariably negative;
- they cannot easily be quantified.

However, socio-economic effects do occur in relation to most developments; they are often positive; and their inclusion in a single document facilitates a more balanced view of the range of impacts (and of trade-offs) and provides greater transparency of process. The view that certain types of socio-economic impacts are difficult to quantify is not necessarily a reason for their complete exclusion from EIA (Newton 1995). Socio-economic impacts are important because the economic fortunes and lifestyles and values of people are important.

Socio-economic impacts merit a higher profile. A UN study of EIA practice in a range of countries advocated a number of changes in the EIA process and in the EIS documentation (UNECE 1991). These included giving greater emphasis to socio-economic impacts in EIA. In Box 13.1, two international reports from the 1990s highlight the important links between social and biophysical impacts with particular reference to developing countries. This now relates to the contemporary and important interest in ecosystem services and the impact of projects on 'free' natural resources. Economic appraisal methods, such as replacement costs and contingent valuation, can be used to assess such impacts (HM Treasury 2011). Thus, for example, the development of a mining project which strips out many hectares of old growth forest may have reduced net economic benefit as a result of the additional social cost of carbon dioxide released into the atmosphere (see Chapter 8 for further discussion of ecosystem services).

Box 13.1 Importance of social impacts in EIA

To quote UNEP (1996):

There is often a direct link between social and subsequent biophysical impacts. For example, a project in a rural area can result in the in-migration of a large labour force, often with families, into an area with low population density. This increase in population can result in adverse biophysical impacts, unless the required supporting social and physical infrastructure is provided at the correct time and place.

Additionally, direct environmental impacts can cause social changes, which, in turn, can result in significant environmental impacts. For example, clearing of vegetation from a riverbank in Kenya, to assist construction and operation of a dam, eliminated local tsetse fly habitats. This meant that local people and their livestock could move into the area and settle in new villages. The people exploited the newly available resources in an unsustainable way, by significantly reducing wildlife populations and the numbers of trees and other wood species which were used as fuel wood. A purely "environmental" EIA might have missed this consequence because the social impacts of actions associated with dam construction would not have been investigated.

The close relationship between social and environmental systems, make it imperative that social impacts are identified, predicted and evaluated in conjunction with biophysical impacts. It is best if social scientists with experience of assessing social impacts are employed as team members under the overall direction of a team or study leader who has an understanding of the links between social and biophysical impacts.

And the World Bank (1991):

Social analysis in EA is not expected to be a complete sociological study nor a cost-benefit analysis of the project. Of the many social impacts that might occur, EA is concerned primarily with those relating to environmental resources and the informed participation of affected groups.

Social assessment for EA purposes focus on how various groups of people affected by a project allocate, regulate and defend access to the environmental resources upon which they depend for their livelihood. In projects involving indigenous people or people dependent on fragile ecosystems, social assessment is particularly important because of the close relationship between the way of life of a group of people and the resources they exploit. Projects with involuntary resettlement, new land settlement and induced development also introduce changes in the relationships between local people and their use of environmental resources.

In a different context, in a survey of academics on the effectiveness of the US National Environmental Policy Act, Canter and Clark (1997) drew out five priorities for the future, one of which was the need for better integration of biophysical and socio-economic factors and characteristics. For the UK, Chadwick (2002) argues for explicit recognition by all EIA stakeholders (developers, consultants, competent authorities) for inclusion of socio-economic impact as an impact category; for further quantification; and for improved guidance on the assessment of the range of such impacts.

13.3.2 Evolving international guidance and standards (IFC, IAIA, EC)

As noted in Chapter 1, this new edition seeks to consider ESIA methods in the context of best-practice international legislation, standards and guidance, drawing in particular on the International Finance Corporation's (IFC) *Performance Standards on Environmental and Social Sustainability* (IFC 2012). Performance Standard 1 establishes the importance of integrated assessment to identify the environmental and social impacts, risks, and opportunities of projects. It also notes the importance of effective community engagement, and the management of project impacts throughout the lifecycle of the project. All these are crucial for effective socio-economic impact assessment.

Other IFC Performance Standards of particular relevance to socio-economic impact assessment include PS2 'Labour and Working Conditions' and PS7 'Indigenous Peoples'.

IFC Performance Standard 2: labour and working conditions

The provision of employment and income associated with major projects can be one of the greatest benefits from such developments for local and wider communities. As noted in Chapter 14 and in Chapter 16, on health impacts, access to good employment can be a positive catalyst for a whole range of improvements in social and health indicators. However, the conditions of that employment are of central importance, and this is the focus of Performance Standard 2. It 'recognizes that the pursuit of economic growth through employment creation and income generation should be accompanied by protection of the fundamental rights of workers'. Hence, the objectives of this standard include, inter alia:

- fair treatment and non-discrimination;
- good worker–management relationships;
- compliance with relevant employment and labour laws;
- worker protection, especially vulnerable groups such as children;
- safe and healthy working conditions;
- no use of forced labour.

The scope of the standard relates to workers directly engaged by the client, contracted workers and supply-chain workers.

IFC Performance Standard 7: Indigenous Peoples

Major projects are often located in remote and rural areas. These and other areas may be populated by Indigenous Peoples, referred to in different countries by such terms as indigenous ethnic minorities, aboriginals, minority nationalities and first nations. IFC Performance Standard 7:

> recognizes that Indigenous Peoples, as social groups with identities that are distinct from mainstream groups in national societies, are often among the most marginalized and vulnerable segments of the population. In many cases, their economic, social, and legal status limits their capacity to defend their rights to, and interests in, lands and natural and cultural resources, and may restrict their ability to participate in and benefit from development.
>
> (IFC 2012)

The standard also recognises that major projects can create important employment opportunities for Indigenous Peoples to participate in, and to fulfil their social and economic aspirations. Objectives of this standard include, inter alia, a development process that:

- respects human rights, dignity, aspirations, culture, and natural-resource based livelihoods of Indigenous Peoples;
- avoids, or at least minimises, the impacts of projects on communities;
- promotes sustainable development benefits and opportunities in a culturally appropriate manner;
- has an ongoing relationship with communities based on Informed Consultation and Participation (ICP);
- ensures the Free, Prior, and Informed Consent (FPIC) of affected communities as appropriate;
- respects and preserves the culture, knowledge, and practices of Indigenous Peoples.

Issues of mitigation and enhancement of impacts and the determination, delivery and distribution of compensation and other benefit-sharing measures to affected communities of Indigenous Peoples may be particularly important considerations in the assessment process.

As noted in §13.2, the International Association for Impact Assessment (IAIA) recently updated their guidance: *Social Impact Assessment: Guidance on assessing and managing social impacts of projects* (IAIA 2015). Key concepts in the IAIA guidance include, inter alia, the importance of: Free, Prior and Informed Consent as a procedural mechanism developed to assist in ensuring the right of Indigenous Peoples to self-determination; the use of Impacts and Benefits Agreements (or Community Development Agreements) as negotiated agreements between project developers and affected peoples; recognising different perspectives on risks associated with projects, including non-technical and social risks; human rights; shared value (i.e. with both developer and society benefits); and the Equator Principles (see Chapter 20).

The IAIA guidance promotes an increased focus in the assessment process upon enhancing the benefits of projects to impacted communities. Although the need to ensure that the negative impacts are identified and effectively mitigated remains, it recognises the value in working with the project development team to deliver greater benefits to communities. This is necessary for the project to earn its 'social licence to operate'; and also because attempting to minimise harm (the traditional approach) does not ensure that the project will be considered acceptable by local stakeholders, or that a project does not actually cause significant harm. The guidance states that enhancing benefits covers a range of issues, including: modifying project infrastructure to ensure it can also service local community needs; providing social investment funding to support local social sustainable development and community visioning processes; a genuine commitment to maximising opportunities for local content (i.e. jobs for local people and local procurement) by removing barriers to entry to make it possible for local enterprises to supply goods and services; and by providing training and support to local people.

The IAIA guidance also stresses that SIA is a process of management, not a product. Figure 13.1 illustrates the key considerations in each phase in the SIA guidance. Each of the steps is detailed in the guidance.

Also at the international level, the various and evolving EIA Directives of the European Commission (EC 1985, 1997, 2014) have been influential worldwide in shaping ESIA practice. Unfortunately however, they have been much more limited on socio-economic issues with, as noted in §13.3.1, the requirement limited to a description of possible impacts on human beings. While this could be interpreted, and used, positively, there has been a tendency, at least until recently, not to do so under this legislation and the associated national regulations in EU Member States. However, the position is changing, with growing recognition of the importance of the assessment and management of socio-economic impacts, especially in relation to major infra-structure projects, e.g. recent energy projects and their ESIA reports in the UK. It was also hoped that the latest incarnation of the EIA Directive (EC 2014), to be implemented by Member States by 2017, would grasp the initiative. Yet the revised Directive still maintains a very strong biophysical focus, with the socio-economic content limited to population, human health and cultural heritage. In Europe at least, socio-economic good practice is likely to continue to outstrip legislative good practice.

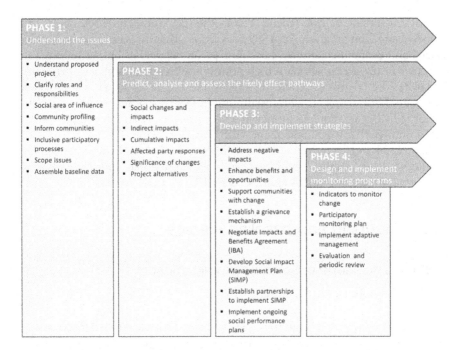

Figure 13.1

The phases of SIA as set out in IAIA guidance (IAIA 2015)

13.3.3 Evolving national legislation and guidance (examples from the UK and Australia)

As noted in §13.3.1 there is some limited reference to effects on population, employment and human beings in UK EIA regulations and guidance. Additional relevant guidance is available in *The Green Book: Appraisal and Evaluation in Central Government* (HM Treasury 2011), the official UK government guidance for public sector bodies on how to appraise proposals before committing funds to a policy, programme or project. This guidance is designed to promote efficient policy development and resource allocation across government. The objectives are to inform decision-making and improve the alignment of departmental and agency policies, programmes and projects with government priorities and the expectations of the public. While there is an economic focus to the guidance (e.g. the consideration of financial costs and benefits, discounting, and a whole range of analytical techniques), it also emphasises the need to take account of the wider social costs and benefits of proposals, equity and distributional issues, and the need to ensure the proper use of public resources.

In Australia, at the federal level, the Environment Protection and Biodiversity Conservation Act 1999 does require that decision makers consider social and economic matters in deciding project approvals but, as reflected in the title of the act, it 'affords a high priority to environmental considerations' (Hawke 2009). However, there are some interesting innovations at the level of the state. For example, in Queensland since 2008, Social Impact Management Plans (SIMPs) are required for new resource projects, as an integral part of the Environmental Impact Statement (EIS) process (Holm et al. 2013) (see Chapters 14 and 20 for discussion of SIMPs). Similarly in Western Australia, while there is weaker legislative underpinning, some innovative socio-economic impact initiatives exist (see Box 13.3).

13.4 Scoping and baseline studies

13.4.1 The scope of socio-economic impacts

The IAIA guidance (2015) considers that:

> social impacts include all the issues associated with a planned intervention (i.e. a project) that affect or concern people, whether directly or indirectly. Specifically, a social impact is considered to be something that is experienced or felt in either a perceptual (cognitive) or a corporeal (bodily, physical) sense, at any level, for example at the level of an individual person, an economic unit (family/household), a social group (circle of friends), a workplace (a company or government agency), or by community/society generally. These different levels are affected in different ways by an impact or impact causing action.

More specifically, a consideration of socio-economic impacts in ESIA needs to clarify the type, duration, spatial extent and distribution of impacts; that is, the analyst needs to ask the questions: what to include; over what period of time; over what area; and impacting whom?

An overview of **what to include** is outlined in Table 13.1. There is usually a functional relationship between impacts. Direct economic impacts have wider indirect economic impacts. Thus the direct employment of a project will generate expenditure on local services and facilities (e.g. for accommodation, fuel, food and drink). The ratio of local to non-local labour employed on a project is often a key determinant of many subsequent impacts. A project with a high proportion of in-migrant labour will have greater implications for the demography of the locality. There will be an increase in population, which may also include an influx of dependents of the additional employees. The demographic changes will work through into the housing market and will impact on other local conditions and associated services and infrastructure (for example, on health and education), with implications for both the public and

Table 13.1 What to include – types of socio-economic impacts

1. **Direct economic:**
   - local – non-local employment;
   - characteristics of employment (e.g. skill group);
   - labour supply and training;
   - wage levels.

2. **Indirect/wider economic/expenditure:**
   - employees' retail expenditure;
   - linked supply chain to main development;
   - labour market pressures;
   - wider multiplier effects;
   - effects on development potential of area.

3. **Demographic:**
   - changes in population size; temporary and permanent;
   - changes in other population characteristics (e.g. family size, income levels, socio-economic groups);
   - settlement patterns.

4. **Housing:**
   - various housing tenure types;
   - public and private;
   - house prices and rent/accommodation costs; homelessness and other housing problems; personal and property rights, displacement and resettlement

5. **Other local services:**
   - public and private sector;
   - educational services;
   - health services; social support;
   - others (e.g. police, fire, recreation, transport);
   - local authority finances.

6. **Socio-cultural:**
   - lifestyles/quality of life;
   - gender issues; family structure;
   - social problems (e.g. crime, ill-health, deprivation);
   - community stress and conflict; integration, cohesion and alienation;
   - community character or image.

7. **Distributional effects:**
   - effects on specific groups in society (e.g. by virtue of gender, age, religion, language, ethnicity and location).

private sector (see Figure 13.2). The area of health impacts has become increasingly important within the wider socio-economic field, to the extent that it has generated its own Health Impact Assessment (HIA) process often running in parallel to ESIA (see Chapter 16).

In some cases, population changes themselves may be initiators of the causal chain of impacts; new settlements would fit into this category. Development

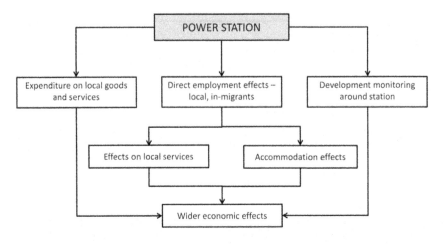

Figure 13.2

Example of linkages between socio-economic impacts for a power station project

actions may also have socio-cultural impacts. A substantial new settlement of, say, 20,000 people may have significant implications for the existing lifestyles in a host environment which is predominantly rural with small villages. The introduction of a major project, with a construction stage involving the employment of several thousand people over several years, may be viewed as a serious threat to the quality of life of a locality. Social problems may be associated with such development, which may generate considerable community stress and conflict. In practice, such socio-cultural impacts are often poorly covered in ESIA Reports, being regarded as more intangible and difficult to assess.

The question of **what period of time** to consider in a socio-economic assessment raises, in particular, the often substantial differences between impacts in the construction and operational stages of a project. Major utilities (such as power stations and reservoirs) and other infrastructure projects, such as roads, may have high levels of construction employment but much lower levels of operational employment. In contrast, manufacturing and service industry projects often have shorter construction periods with lower levels of employment, but with considerable operational employment levels over periods which may extend for several decades. The closure of a project may also have significant socio-economic impacts; unfortunately these are rarely covered in the initial assessment.

Interestingly, nuclear reactor decommissioning did become a project requiring mandatory environmental assessment under Directive 97/11/EC (EC 1997). Subsequent EISs of decommissioning projects have increasingly included a socio-economic dimension, and in the UK context the Nuclear

Decommissioning Authority Strategy gives coverage to socio-economic issues and to stakeholders (NDA 2006).

In best practice, socio-economic impacts should be considered for all stages of the life of a development. Even within stages, it may be necessary to identify sub-stages, for example peak construction employment, to highlight the extremes of impacts which may flow from a project. Only through monitoring can predictions be updated over the life of the project under consideration.

What area to cover in SIA raises the often contentious issue of where to draw the boundaries around impacts. Boundaries may be determined by several factors. They may be influenced by estimates of the impact zone. Thus, for the construction stage of a major project, a subregional or regional boundary may be taken, reflecting the fact that construction workers are willing to travel long distances daily for short-term, well-paid employment. On the other hand, permanent employees of an operational development are likely to locate much nearer to their work. Other determinants of the geographical area of study may include the availability of data (e.g. for counties or districts), and policy issues (e.g. providing spatial-impact data related to the areas of responsibility of the key decision makers involved in a project). Different socio-economic impacts will often necessitate the use of different geographical areas, reflecting some of the determinants already discussed. A focus on local area impacts may provide a very partial picture; economic impacts often have wide regional and occasionally national and international implications.

The question of **who will be affected** is of crucial importance in ESIA. The distributional effects of development impacts do not fall evenly on communities; there are usually winners and losers. For example, a new tourism development in a historic city may benefit visitors to the city and tourism entrepreneurs but may generate considerable pressures on a variety of services used by the local population. Distributional effects can be analysed by reference to geographical areas and/or to groups involved (for example local and non-local; age groups; socio-economic groups; employment groups). From the wider perspective of the World Bank (1991), Box 13.2 provides some examples of the key social differences which may be environmentally significant. Distributional issues, and the concept of environmental justice, are discussed further in Chapter 14.

There are of course many other dimensions to impacts besides the areas discussed here, including adverse and beneficial; reversible and irreversible; quantitative and qualitative; and actual and perceived impacts (see Glasson et al. 2012). All are relevant in SIA. The distinction between actual and perceived impacts, as noted in the IAIA quote above, raises the distinction between more 'objective' and more 'subjective' assessments of impacts. The impacts of a development perceived by residents of a locality may be significant in determining local responses to a project. They can constitute an important source of information to be considered alongside more 'objective' predictions of impacts.

Box 13.2 Examples of social differences which may be environmentally significant (World Bank 1991)

Communities are composed of diverse groups of people, including but not restricted to the intended beneficiaries of a development project. Organised social groups hold territory, divide labour and distribute resources. Social assessment in EA disaggregates the affected population into social groups which may be affected in different ways, to different degrees and in different locations. Important social differences which may be environmentally significant include ethnic or tribal affiliation, occupation, socio-economic status, age and gender.

Ethnic/tribal groups. A project area may include a range of different ethnic or tribal groups whose competition for environmental resources can become a source of conflict. Ethnicity can have important environmental implications. For example, a resettlement authority may inadvertently create competition for scarce resources if it grants land to new settlers while ignoring customary rights to that land by Indigenous tribal groups.

Occupational groups. A project area may also include people with a wide array of occupations who may have diverse and perhaps competing interests in using environmental resources. Farmers require fertile land and water, herders require grazing lands, and artisans may require forest products such as wood to produce goods. A project may provide benefits to one group while negatively affecting another. For example while construction of dams and reservoirs for irrigation and power clearly benefits farmers with irrigation, they may adversely affect rural populations engaged in other activities living downstream of the dam.

Socio-economic stratification. The population in the project area will also vary according to the land and capital that they control. Some will be landless poor, others will be wealthy landowners, tenant farmers or middlemen entrepreneurs. Disaggregating the population by economic status is important because access to capital and land can result in different responses to project benefits. For example, tree crop development may benefit wealthy farmers, but displace the livestock of poor farmers to more marginal areas.

Age and gender. A social assessment should include identification of project impacts on different individuals within households. Old people may be more adversely affected by resettlement than young people. Men, women and children play different economic roles, have different access to resources, and projects may have different impacts on them as a result. For example, a project that changes access to resources in fragile ecosystems may have unanticipated impacts on local women who use those resources for income or domestic purposes.

13.4.2 Baseline studies: direct and indirect economic impacts

Understanding the project/development action

Socio-economic impacts are the outcome of the interaction between the characteristics of the project/development action and the characteristics of the 'host' environment. As a starting point, the analyst must assemble baseline information on both sets of characteristics.

The assembling of relevant information on the characteristics of the project would appear to be one of the more straightforward steps in the process. However, projects have many characteristics and, for some, relevant data may be limited. In socio-economic terms, what is important is the capital investment of the project and its associated human resources for the key stages of the project lifecycle. The essential components of the project can be assembled as a flow diagram (see Chapter 9 of Rodriguez-Bachiller and Glasson 2004). The drafting of a **direct employment labour curve** is a vital initial source of information (see Figure 13.3). This shows the anticipated employment requirements of the project. To be of maximum use it should include a number of dimensions, including in particular the duration and categories of employment. In best practice, the labour curve should indicate the anticipated labour requirements for each stage in the project lifecycle.

For the purposes of prediction and further analysis, there may be a focus on certain key points or phases in the lifecycle. For example, an SIA of peak construction employment could reveal the maximum impact on a community; an analysis of impacts at full operational employment would provide a guide for many continuing and long-term impacts. The labour curve should also indicate requirements by employment type or skill category. These may be subdivided in various ways according to the nature of employment in the project concerned, but often involve a distinction between managerial and technical staff, clerical and administrative staff and project operatives. For a construction project, there may be a further significant distinction in the operatives category between civil works, and electrical, and mechanical operatives. A finer disaggregation still would focus on the particular trades or skills involved, including levels of skill (e.g. skilled/semi-skilled/unskilled) and types of skill (e.g. steel erector, carpenter, or electrician).

Projects also have associated **employment policies**, which may influence the labour requirements in a variety of ways. For example, the use/type of shift-working and the approach to training of labour may be very significant in determining the scope for local employment. An indication of likely wage levels could be helpful in determining wider economic impacts in the local retail economy. An indication of the main developer's attitude/policy to subcontracting can also be helpful in determining the wider economic impacts for the local and regional industry supply chain.

Ideally, the initial brief from the developer will provide a good starting point on labour requirements and associated policies. But this is not always the case,

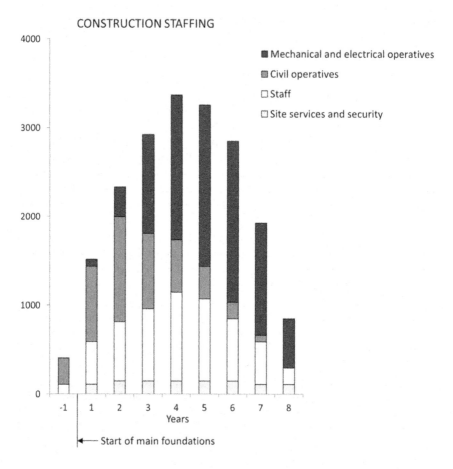

CONSTRUCTION STAFFING

- Mechanical and electrical operatives
- Civil operatives
- Staff
- Site services and security

Years

Start of main foundations

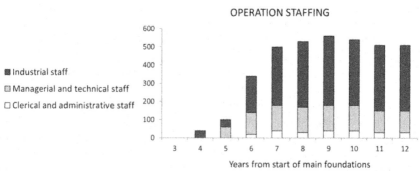

OPERATION STAFFING

- Industrial staff
- Managerial and technical staff
- Clerical and administrative staff

Years from start of main foundations

Figure 13.3

Labour requirements for a project disaggregated in time and by employment category

particularly where the project is a 'one-off' and the developer cannot draw on comparative experience from within the firm involved. In such cases the analyst may be able to draw on EIA/ESIAs of comparative studies.

However, many major projects are at the forefront of technology and there may be few national, or even international, comparators available. For instance, the EIA for the London Gateway Bridge – which would link two deprived areas of London across the River Thames – emphasised the employment benefits to local residents of being able to cross the river more easily (Transport for London 2004). Opponents of the bridge instead argued that increased access to jobs alone would not improve local residents' employment prospects without associated training and support. There were no obviously relevant UK comparators to support either side's arguments.

There may be genuine uncertainty on the relative merits of different designs for a project, and this may necessitate the assessment of the socio-economic impacts of various possibilities. Work on assessment for a possible new generation of UK nuclear power stations (2007 to date) initially had to contend with the implications of four alternative reactor types. Work on major offshore wind farms has had to contend with a rapid escalation in the size of the wind turbine generators from, for example, 3 megawatt (MW) each to as large as 15 MW, often during the period of assessment of the impacts of the project. In such cases, the focus is usually on the 'worst case' scenario.

Projects may change their characteristics through the planning and development process and these may have significant socio-economic implications. For example, the discovery during the early stages of project construction of unstable geology or contaminated land may necessitate a much greater input of civil-works operatives. Major projects also tend to have a substantial number of contractors and it may be difficult to forecast accurately without knowledge of the main contractor(s) and, especially, of the subcontractors. Such uncertainties reinforce the necessity of regular monitoring of project characteristics throughout project planning and development.

Establishing the economic environment baseline

Defining the **'host' economic environment area** depends to some extent on the nature of the project. Some projects may have significant national or even international employment implications. The construction of the Channel Tunnel had wide-ranging interregional economic impacts in the UK, bringing considerable benefits to areas well beyond Kent and the South East region of England, for example to the West Midlands (Vickerman 1987; Thomas and O'Donoghue 2012). The Three Gorges project in China is of national significance in terms of its economic impacts (Jackson and Sleigh 2001). Many projects have regional or subregional economic impacts, and almost all have local economic impacts. As noted in §13.4.1, it can be useful to make a distinction between the anticipated construction – and operational – daily commuting zones for a project. The former is invariably much larger in geographical area than the latter, possibly extending up to 90 minutes' one-way daily commuting time from the project. For these areas, and for the wider region and nation as appropriate, it is necessary to assemble data on current and anticipated labour-market characteristics, including size of labour force, employment structure, unemployment and vacancies, skills and training provision.

The **size of the labour force** provides a first guide to the ability of a locality to service a development. Information is needed on the economically active workforce (i.e. those people in the 16 to retirement age bands, or as appropriate to the country involved). This then needs disaggregation into industrial and/or occupational groups to provide a guide to the economic activities and employment types in the study area(s). An industrial disaggregation would identify, for example, those in agriculture, types of manufacturing, and services. The UN's International Standard Industrial Classification of all Economic Activities (ISIC)-Revision 4 (UN 2008) provides a template of categories (Table 13.2).

An occupational disaggregation indicates particular skill groups (Table 13.3). Data on unemployment and vacancies provides indicators of the pressure in the labour market and the availability of various labour groups. It should be disaggregated by length of unemployment, as well as by skill category and location. Data should also be collected on the provision of training facilities in an area. Such facilities may be employed to enhance the quality of labour supply. Information on earnings levels in, and gross value added (GVA) from, the host local/regional economy can provide useful indicators of prevailing economic conditions.

Such topics are conventionally covered in national programmes of labour statistics, as established by the UN International Labour Organisation Convention Number 160 (ILO 1985) on Labour Statistics. Some key ILO sources are set out in Table 13.4. Each UN member that ratifies this Convention will regularly collect, compile and publish labour statistics. For example, in the UK the National Online Manpower Information System (NOMIS) provides labour market and related population data for local areas, from a variety of sources including the Labour Force Survey, claimant count, Annual Business Inquiry, Annual Survey of Hours and Earnings, and Censuses of Population. The data are from official government sources (mostly Office for National Statistics – ONS). NOMIS includes the latest published figures and time-series data, in some cases dating back to the 1970s. The level of disaggregation is exceptional, from region-wide down to local authority wards. Data is freely available but access to Annual Business Inquiry (ABI) data requires special permission for which there is a fee.

In some areas, the sources noted may be enhanced by various one-off studies, including for example: (i) skills audits which seek to establish the current and latent skills provision of an area, and (ii) business supply-chain surveys which seek to establish the potential local supply potential to a major project. Predictably, the various data sources do not necessarily use the same

Table 13.2 International Standard Industrial Classification of all Economic Activities (UN 2008)

Section	Description
A	Agriculture, forestry and fishing
B	Mining and quarrying
C	Manufacturing
D	Electricity, gas, steam and air-conditioning supply
E	Water supply, sewerage, waste management and remediation activities
F	Construction
G	Wholesale and retail trade; repair of motor vehicles and motorcycles
H	Accommodation and food service activities
I	Transportation and storage
J	Information and communication
K	Financial and insurance activities
L	Real estate activities
M	Professional, scientific and technical activities
N	Administrative and support service activities
O	Public administration and defence; compulsory social security
P	Education
Q	Human health and social work activities
R	Arts, entertainment and recreation
S	Other service activities
T	Activities of households as employers; undifferentiated goods- and services-producing activities of households for own use
U	Activities of extraterritorial organisations and bodies

geographical unit or area for collecting data, which can cause problems for the analyst. The latter should also be aware of the influence of 'softer' data – for example, information on other major developments proposals in a locality which may have labour-market implications for the project under consideration.

Local economic impacts may also be influenced by the policy stance(s) of the host area. For many localities, the possibility of employment and local trade-gains from a project may be the only perceived benefits. There may be a desire to maximise such gains and to limit the **leakage** of multiplier benefits (see §13.5). This may result in an authority taking a policy stance on the percentage of 'local' labour content for a project. A local position may also be taken on the provision of training facilities. However, as noted by Esteves et al. (2011), 'whilst the sourcing of local labour, goods and services has obvious benefits, it cannot necessarily be assumed that local content is always a "positive" to be maximised.' For example, there may be concern about a possible local employment boom–bust scenario associated with some major projects, which may of course bring caution into the setting of high local employment ratios.

Table 13.3 Occupational classifications

International Standard Classification of Occupations 2008 (ISCO-08) (ILO 2008)

The International Standard Classification of Occupations 2008 (ISCO-08) is a four-level, hierarchically structured classification that covers all jobs in the world. ISCO-08 classifies jobs into 436 unit groups. These unit groups are aggregated into 130 minor groups, 43 sub-major groups and ten major groups, based on their similarity in terms of the skill level and skill specialization required for the jobs. This allows the production of relatively detailed internationally comparable data as well as summary information for only ten groups at the highest level of aggregation. The ten groups are shown below. The UK version is also included for information.

1 Managers
2 Professionals
3 Technicians and associate professionals
4 Clerical support workers
5 Services and sales workers
6 Skilled agricultural, forestry and fisheries workers
7 Craft and related trades workers
8 Plant and machine operators, and assemblers
9 Elementary occupations
0 Armed forces occupations

UK Standard Occupational Classification (SOC) since 2010 (ONS 2010)

1 Managers, directors and senior officials
2 Professional occupations
3 Associate professional and technical occupations
4 Administrative and secretarial occupations
5 Skilled trades occupations
6 Caring, leisure and other service occupations
7 Sales and customer service occupations
8 Process, plant and machine operatives
9 Elementary occupations

Clarifying the issues

Consideration of **project and 'host' environment characteristics** can help to clarify key issues. Denzin (1970) and Grady et al. (1987) remind us that issue specification should be rooted in several sources, and they advocate the use of the philosophy of **'triangulation'**: for data (the use of a variety of data sources), for investigators (the use of different sets of researchers), for theory (the use of multiple perspectives to interpret a single set of data) and for methods (the use of multiple methods). Thus, the use of quantitative published and semi-published data, as outlined, should be complemented by the use of key-informant interviews, working groups (e.g. including the developer, local

Table 13.4 Major international sources of labour statistics

The International Labour Office (ILO) publishes a wide range of labour information, including:

- ILOSTAT provides recent labour data for over 100 indicators and 165 economies.
- KILM (Key Indicators of the Labour Market) is a multi-functional research tool produced by the ILO Department of Statistics. It consists of country-level data on 17 key indicators of the labour market from 1980 to the latest available year. Topics include, inter alia: labour force participation, employment, hours of work, unemployment, wages, labour productivity and income distribution. The new edition includes an analysis of the link between education and access to the labour market.
- Labour Force Surveys – compiles web sites which contain data from national statistical agencies, the ILO and other sources. Includes links to source web sites and references to print publications available in the ILO Library.

planning officers, councillors/elected representatives, and specialist interest groups) and possibly focus groups and public meetings.

The adequacy of public participation can be a serious concern in specifying the issues. The scoping of a community profile at an early stage in the assessment of a project can provide a useful step in identifying the relevant public(s). There is then a need for meaningful engagement. Esteves et al. (2012) note that such participation can range from limited opportunities to comment on information supplied by the proponent, to a deliberative and active engagement process opening up governance processes to include local communities in decision-making about projects. Thorough community involvement can of course have its limitations, especially in terms of 'consultation fatigue' in communities and local governments, when it extends over long periods and involves multiple projects.

While many direct and indirect employment impacts will be specific to the case in hand, the following key questions often tend to be raised:

- What proportions of project construction and operation jobs are likely to be filled by local workers as compared to in-migrants, and what are the likely origins of the in-migrant workers?
- What is likely to be the magnitude of the secondary (indirect and induced) employment resulting from project development? What proportions of these jobs will be filled by local workers?
- How will local businesses be affected by rapid growth resulting from a major project? For example, will development provide opportunities for expansion or will local firms experience difficulty competing to attract and retain quality workers? (Murdock et al. 1986).

Box 13.3 provides some examples of economic impacts and responses to those impacts from cases of major projects in the mineral-rich state of Western

Australia, covering some of the issues identified in §13.4–13.6, and also in the following chapter.

Box 13.3 Socio-economic impacts – major projects in Western Australia (Brunner and Glasson 2015)

Context: Western Australia (WA) is a jewel box of mineral resources. Gold around Kalgoorlie and bauxite south of Perth were early resource developments, but since the 1970s it is the more northerly regions of this vast state that have seen most growth in mineral-related developments. The Pilbara region is the home of Australian iron ore extraction, which accounts for over one third of the value of Australia's mineral outputs. With major companies such as BHP Billiton and Rio Tinto, Australian dominates world trade in iron ore with over 50 per cent of production. Oil and natural gas (LNG) on the NW Shelf (off the coast of the Pilbara and Kimberley regions) are other key growth industries in the West. There has been large expenditure on major projects, such as Gorgon, and WA will soon rival Qatar as the world's largest LNG producer. Diamonds, nickel, alumina, copper and zinc are also very significant in the remoter regions of WA, and uranium is now being developed, ending a period of government restriction on its mining in the West. There have been many major mining projects, developed within the context of WA's EIA system (WA EPA 2010). These projects have generated many socio-economic issues and responses to such issues. A few are noted below.

Fly-in/Fly-out (FIFO) employment: Storey (2010) defines FIFO as follows:

'Fly-in/fly-out' is one of several terms used to refer to a set of work arrange-ments for resource operations that are typically located at a distance from other existing communities. The work involves a roster system in which employees spend a certain number of days working on site, after which they return to their home communities for a specified rest period. Typically the employer organizes and pays for transportation to and from the worksite and for worker accommo-dations and other services at or near the worksite. While most remote operations fly their workforces to and from their worksites, other modes of transport may be used.

In urbanized areas, major projects can draw their workforce from the surrounding region, perhaps supplemented by a residential site camp for more distant travelling workers. In remote regions this is more difficult; the bulk of the workforce needs to be brought to the site. Alternative models might include new towns or temporary site camps, supported by a FIFO population. In recent times, resource development projects in remote areas of countries, such as Canada and Australia, have increas-ingly shifted towards the FIFO model.

FIFO is the dominant form of travel to the remote regions of Western Australia, but there are also examples of DIDO (drive-in/drive-out) in areas such as the Bowen Basin coal mining area in Queensland. A KPMG study (2013) for the resources sector, of those commuting over 100 km to work as reported in the 2011 Census, showed that the Pilbara region had 18,703 such workers in 2011. But what are the costs and benefits of such arrangements? Drivers behind such arrangements in recent years include the spread of resource exploration into more remote and harsh environments, and the difficulty of attracting the workforce and their families to such locations. Some of the projects also have relatively short operational life spans, not justifying large capital outlays on 'new settlement' infrastructure. There have also been improvements in the quality and cost of communications, and in particular, improved efficiency, quality, flexibility and reduced cost of air travel.

However, the FIFO model is not without its problems posing a number of high profile challenges –economic; social/community; environmental and institutional – for both the host and source regions. FIFO workers usually have their main residence outside of the region in which they are working and much of their income is usually returned to the home region, partly to support the main residence and any associated family commitments (ACIL Tasman 2009). All this can be seen as a major economic leakage out of the minerals' host region, also limiting opportunities for the development of local businesses which might benefit from an increased residential presence in the host region. The high wages of FIFO workers can also bring inflationary pressures to both the host and source regions, especially with regard to the housing market, causing problems for those employed in businesses other than in the high wages minerals sector.

The social/community issues of FIFO also apply to both the host and source regions. Host regions suffer from a transient population unwilling to put down long-term roots and to commit to the local community. This may deny regional communities the access to potential local community leaders, volunteers (e.g. fire and ambulance drivers) as well as sporting coaches and participants (ACIL Tasman 2009). In addition, FIFO shift working can involve long hours, strenuous workloads in harsh and risky environments, and a potential for loneliness and homesickness (McKenzie 2010). FIFO work can put stress on family relationships in the source region, potentially leading to family break-ups.

Indigenous employment: For some major projects in remote regions, the reliance on FIFO labour has been partly offset by more employment from the local Indigenous populations. The Australian mining industry has increasingly recognised the advantages of supporting Indigenous employment. A report on such employment in the Australian minerals industry (CSRM 2006) sets out several good reasons for the industry to engage in a positive partnership with local Indigenous populations, including land access, regional workforce and industry reputation. Approximately 60 per cent of mining in Australia occurs near Indigenous land and many new mines are, or will be, subject to native title.

In particular, the High Court's Mabo decision in 1992 and the subsequent passage of the Commonwealth *Native Title Act* in 1993 have conferred on Traditional Owners a 'right to negotiate' with mining companies in relation to the granting of a mining lease. Many minerals operations in Australia are located on land where Indigenous people have had, and claim, traditional rights and interests in country. Increasingly, agreements between Indigenous groups and mineral companies require companies to engage effectively with Indigenous communities and provide assistance to help achieve long-term development objectives. Companies that are unable or unwilling to do so, or fail to follow through on undertakings, are likely to be seriously disadvantaged when it comes to negotiating future agreements with Traditional Owner groups.

Companies are also recognising the benefits to their operations of building up regional employment capacity. Supporting and promoting education, training and other local initiatives that support the skills base of Indigenous communities, can lead to the development of a skilled local workforce and a more prosperous local economy. Many communities in close proximity to mine sites have substantial Indigenous populations; Indigenous people are also less likely than other Australians to relocate to major urban areas. In addition, companies that form successful partnerships with Indigenous communities and with government are likely to have enhanced reputations, both nationally and internationally, for dealing fairly with local communities, which can support development opportunities elsewhere.

However, building up the Indigenous workforce is not without its challenges. For the Indigenous peoples, there is a lack of exposure to the mainstream workforce and industry culture, and the difficulties of balancing family and community obligations with full-time work. For the companies, there is the lack of a job ready workforce and limited appreciation of the socio-economic constraints on Indigenous recruitment (CSRM 2006). To be successful, companies need to have a long-term commitment, and tailored recruitment and retention strategies. Some companies have managed this well. For example, for their Argyle Diamond Mine in the remote Kimberley region of WA, Rio Tinto in 2014 celebrated ten years of a pioneering Partnership Agreement with the Traditional Owners of the Argyle mine lease. With work readiness programmes, employment targets and training preference schemes, the mine has managed to sustain levels of Indigenous employment, currently at around 13 per cent, despite the shift from an open mine operation to a more skill-demanding underground operation (Rio Tinto 2014).

Royalties for the Regions: Royalties for the Regions (RfR) is a WA Government initiative of 2008, under which the equivalent of 25 per cent of the state's mining and onshore petroleum royalties are returned to the state's regional areas each year. A total of around $6 bn has been budgeted for the period 2008/09 to 2014/15, with for example about $1.5 bn allocated for 2011–12. Key principles are: a priority for strategic projects; a focus on local decision-making in regional areas; and State Government administration and processes should provide for and support decision-making in regional areas. In this context, the state's nine Regional Development Commissions have a pivotal role, working with state agencies and local community and business groups, in decision-making on RfR expenditure. Key areas of expenditure include local housing, health, and education services initiatives, and some large-scale regional infrastructure projects. The latter fund has supported some major projects such as the Ord-East Kimberley Expansion project, Pilbara Cities, Mid-West Investment Plan and the Gascoyne Revitalisation programme The *Pilbara Cities* initiative was started in 2009, underpinned by a $1.2 bn RfR commitment to transform the region by 'building modern, vibrant cities and regional centres to support a skilled workforce for the major economic projects planned for the region' (WADLRD 2012).

RfR is an example of a strategic community benefits initiative which is enabling development in the WA regions through the strength of its resource base, and which has attracted strong interest from other states in Australia (Queensland now has a similar scheme, but with much less funding). It is of course not without issues. Much of the funding has been to the north of the state; there has also been an under-spending on the budget which it could be argued is prudent, as there is a need to plan for the wise use of funds. It could also be seen as indicative of the weakness of the primary delivery bodies, the multitude of very small local authorities, with limited capacity to manage the new largesse. Partly as a result of this, the WA 2012 Budget diverted some of the RfR surplus fund ($1.1 bn) into a WA Future Fund, which may be worth around $5 bn in 20 years.

13.5 Impact prediction and evaluation: direct and indirect employment impacts

13.5.1 The nature of prediction

Prediction of socio-economic impacts is an inexact exercise. Ideally, the prediction of direct employment impacts on an area would be based on information relating to the recruitment policies of the companies involved in the development, and on individuals' decisions in response to the new employment opportunities. In the absence of firm data on these and related factors, predictions need to be based on a series of assumptions related to the characteristics of the development and of the locality. These could for example include the following hypothetical:

> the labour requirement curves for construction and operation will be as provided by the client; local recruitment will be encouraged by the developer with a target of 50%; employment on the new project will be attractive to the local workforce by virtue of the comparatively high wages offered.

Predictive approaches may use **extrapolative methods**, drawing on trends in past and present data. In this respect, use can be made of comparative situations and the study of the direct employment impacts of similar projects. Unfortunately, the limited monitoring of the impacts of project outcomes reduces the potential of this source, and primary surveys may be needed to obtain such information. Predictive approaches may also use **normative methods**. Such methods work backwards from desired outcomes to assess whether the project, in its economic environment context, is adequate to achieve them. For example, the desired direct employment outcome from the construction stage of a major project may be 'X' per cent local employment.

Underpinning all prediction methods should be some clarification of the **cause–effect relationships** between the variables involved. Figure 13.4 provides a simplified flow diagram for the local socio-economic impacts of a power station development. Prediction of the local and regional (as appropriate) labour recruitment ratios is the key step in the process. Non-local workers are, by definition, not based in the study area. Their in-migration for

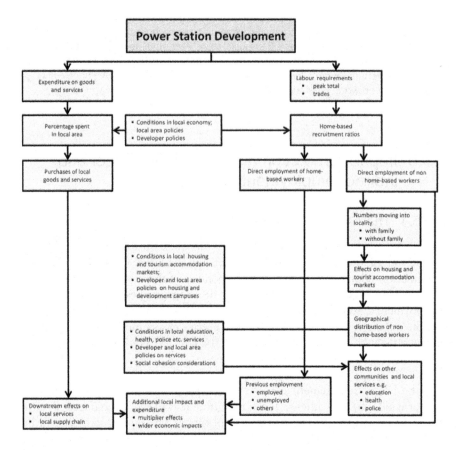

Figure 13.4

A cause–effect diagram for the local socio-economic impacts of a power station proposal (EDF 2011 based on Glasson et al. 1988)

the duration of a project will have a wider range of secondary demographic, accommodation, services and socio-cultural impacts (as discussed in Chapter 14). The wider economic impacts on, for example local retail activity, will be discussed further in this chapter. The key determinants of the local recruitment ratios are the labour requirements of the project, the conditions in the local economy, and relevant local government and developer policies on topics such as training, local recruitment and travel allowances. It is possible to quantify some of the cause–effect relationships, and various **economic impact models**, derived from the multiplier concept, can be used for predictive purposes. These are discussed further in this chapter.

Whichever prediction method is used, there will be a degree of **uncertainty** attached to the predicted impacts. Such uncertainty can be partly handled by the application of probability factors to predictions, by sensitivity analysis, and by the inclusion of ranges in the predictions (see Glasson et al. 2012, Chapters 5 and 9).

13.5.2 Predicting local (and regional) direct employment impacts

Disaggregation into project stages, geographical areas and employment categories is the key to improving the accuracy of predictions. For example, the construction stage of major projects will usually involve an amalgam of professional/managerial staff, administrative/secretarial staff, local services staff (e.g. catering, security) and a wide range of operatives in a variety of skill categories. Most projects will involve civil-works operatives (e.g. plant operators, drivers), and most will also include some mechanical and electrical activity (e.g. electricians, engineers). For each employment category there is a labour market with relevant supply and demand characteristics. Guidance on the mix of local/non-local employment for each category can be obtained from comparative studies and from the best estimates of the participants in the process (e.g. from the developer, or the local employment office). Hopefully, but in practice not often, guidance will be informed by the monitoring of direct employment impacts in practice.

As a general rule, the more specialist the staff, the longer the training needed to achieve the expertise, and the more likely that the employee will not come from the immediate locality of the project. Specialist professional staff and managerial staff are likely to be brought in from outside the study area; they may be transferred from other sites, seconded from headquarters, or recruited on the national or international market. Only a small percentage may be recruited from the local market, which may simply just not have the necessary expertise available in the right numbers. On the other hand, local services staff (e.g. security, cleaning, catering), and to a slightly lesser extent secretarial and administrative staff, can be more plentiful in local labour markets, and the local percentage employed on the project may be quite high, and in some cases very high. Other skill categories will vary in terms of local potential according to the degree of skill and training needed. There may be an abundance of general labourers, but a considerable shortage of coded welders.

Comparative analysis of the disaggregated employment categories is likely to produce broad bands for the level of local recruitment. These can then be refined with reference to the conditions applicable to the particular project and locality under consideration. For example, high levels of unemployment in particular skill categories in the locality may boost local recruitment in those categories. Normative methods may also come into play. The developer may introduce training programmes, skills audits and apprenticeships to boost the supply of local skills (see Glasson 2005). Table 13.5 provides an example of the sort of estimates that may be derived. While the predictions may still use ranges, a prediction from the disaggregated analysis is much more robust than taking employment as a homogenous category.

Table 13.5 Example of predicted employment of local and non-local labour for the construction stage of a major project

	Total labour requirements	Local labour %	range	Non-local labour %	range
Site services, security and clerical staff	300	90	250–290	10	10–50
Professional, supervisory and managerial staff	430	15	50–80	85	350–380
Civil operatives	500	55	250–300	45	200–250
Mechanical and electrical operatives	1520	40	550–670	60	850–970
Total	2750	44	1100–1340	56	1410–1650

Note: *Local labour*: Employees already in residence in the Construction Daily Commuting Zone before being recruited on-site. *Non-local labour*: All other employees.

A further level of **micro-analysis** would be to predict the employment impacts for particular localities within the study area, and for particular groups, such as the unemployed. A further level of **macro-analysis**, used in some ESIAs, would include an estimate of the total person-days of employment per year generated by the project (e.g. 10,000 employment days in 2020).

13.5.3 The range of wider economic impacts

In addition to the direct local (and/or regional) employment effects, major projects have a range of **secondary** or **indirect impacts**. The workforce, which may be very substantial (and well-paid) in some stages of a project, can generate considerable retail expenditure in a locality, on a whole range of goods and services. For example, IAU studies of the impact of power station developments suggest that retail turnover in adjacent medium and small towns may be boosted by at least 10 per cent (Glasson and Chadwick 1988–1997). In addition, the projects themselves require supplies ranging from components from local engineering firms, to provisions for the canteen. These supply-chain links can also boost the local economy.

Such demands create employment, or sustain employment, additional to that directly created by the project. As will be discussed in Chapter 14, the additional workforce may demand other services locally (e.g. health, education) and housing, which may generate additional construction. These demands will create additional employment. Training programmes associated with a project may bring other economic benefits in terms of a general upgrading of the skills in an area. Overall, the net effect may be considerably larger than the original direct injection of jobs and income into a locality, and such wider economic impacts are invariably regarded as beneficial.

However, there can be **wider economic costs**. Existing firms may fear the competition for labour that may result from a new project. They may lose skilled labour to high-wage projects. There may be inflationary pressures on the housing market and on other local services. Major projects may be a catalyst for other development in an area. A road or bridge can improve accessibility and increase the economic potential of areas. But major projects may also cast a shadow over an area in terms of alternative developments. For example, large military projects, nuclear power stations, mineral extraction projects and others may have a deterrent impact on other activities, such as tourism – although the construction stage and the operation of many projects can be tourist attractions in themselves, especially when aided by good interpretation and visitor centre facilities.

13.5.4 Measuring wider economic impacts: the multiplier approach

The analysis of the wider economic effects of introducing a major new source of income and employment into a local economy can be carried out using a number of different techniques (Brownrigg 1971, Glasson 1992, Lewis 1988, Loveridge 2004). The three methods most frequently used are (a) the economic base multiplier model, (b) the input–output model, and (c) the Keynesian multiplier, although it should be added that the percentage of ESIAs including such studies is still small.

The **economic base multiplier** is founded on a division of local (and/or regional) economies into basic and non-basic activities. Basic activities (local/regional supportive activities) are seen as the 'motors' of the economy; they are primarily oriented to markets external to the area. Non-basic activities (regional dependent activities) support the population associated with the basic activities, and are primarily locally oriented services (e.g. retail services). The ratio of basic to non-basic activities, usually measured in employment terms, is used for prediction purposes. Thus an 'X' increase in basic employment may generate a 'Y' increase in non-basic employment. The model has the advantages, and disadvantages, of simplicity (Glasson 1992).

Input–output models provide a much more sophisticated approach. An input–output table is a balancing matrix of financial transactions between industries or sectors. Adapted from national input–output tables, regional or local tables can provide a detailed and disaggregated guide to the wider economic impacts resulting from changes in one industry or sector. However, unless an up-to-date table exists for the area under study, the start-up costs are normally too great for most ESIA exercises. Batey et al. (1993) provide an interesting example of the use of input–output analysis to assess the socio-economic impacts of an airport development.

For several reasons – primarily related to the availability of appropriate data at a local level – the **Keynesian multiplier approach** has been used in several studies and is discussed in further detail here. The basic theory underlying the Keynesian multiplier is simple: 'a money injection into an economic system, whether national or regional, will cause an increase in the level of income in that system by some multiple of the original injection' (Brownrigg 1974).

Mathematically this can be represented at its most simple as:

$$Y_r = K_r J \tag{1}$$

where:
Y_r is the change in the level of income in region r
J is the initial income injection (or multiplicand)
K_r is the regional income multiplier

If the initial injection of money is passed on intact at each round, the multiplier effect would be infinite. The £X million initial injection would provide £X million extra income to workers, which in turn would generate an extra income of £X million for local suppliers, who would then spend it, and so on ad infinitum. But the multiplier is not infinite because there are a number of obvious leakages at each stage of the multiplier process. Five important leakages are:

s   the proportion of additional income saved (and therefore not spent locally)
t_d the proportion of additional income paid in direct taxation and National Insurance contributions
M   the proportion of additional income spent on imported goods and services
U   the marginal transfer benefit/income ratio (representing the relative change in transfer payments, such as unemployment benefits, which result from the rise in local income and employment)
t_i the proportion of additional consumption expenditure on local goods which goes on indirect taxation (e.g. VAT)

The multiplier can therefore be formulated as follows:

$$K_r = \frac{1}{1 - (1 - s)(1 - t_d - u)(1 - m)(1 - t_i)} \tag{2}$$

Substituting (2) into (1) then gives:

$$Y_r = \frac{1}{1 - (1 - s)(1 - t_d - u)(1 - m)(1 - t_i)} J \tag{3}$$

Thus, when applied to the multiplicand J, the multiplier K_r gives the accumulated wider economic impacts for the area under consideration, as in equation (3). The Keynesian multiplier can be calculated in income or employment terms. The various leakages normally reduce the value of local and regional multipliers in practice to between 1.1 and 1.8; in other words, for each £1 brought in directly by the project, an extra £0.10–0.80 is produced indirectly. The size of the import leakage is a major determinant, since the bigger the leakage, the smaller the multiplier. Leakages increase as the size of the study area declines, and decrease as the study area becomes more isolated. Local (county- and district-level multipliers) normally vary between 1.1 and approximately 1.4.

Table 13.6 illustrates the multiplier effects of the mining industry in British Columbia in 2010. It shows that a direct expenditure of CAD$ 5.2 bn generates a total of almost CAD$ 9 bn, with an expenditure multiplier of 1.73. The additional CAD$ 3.8 bn is made up of CAD$ 2.8 bn of indirect expenditure on supply-chain goods and services for the industry, and CAD$ 1 bn of induced expenditure from payrolls resulting from the direct and indirect activity. The employment multiplier is higher at 2.16, reflecting lower pay for the indirect and induced jobs compared with the direct mining industry jobs. This is often the case with major projects. In this case direct jobs include not only jobs in the actual mines, but also jobs related to the exploration and construction phase when mining companies are highly labour-intensive, as well as jobs involved with transporting mine output from the mine site.

In practice, ESIA studies will probably limit such analyses to gross estimates of the wider economic impacts at perhaps the peak-construction and full-operation stages. But it is possible to disaggregate also with reference to the various employee groups. A study of the predicted local socio-economic impacts of the construction and operation of the first proposal for a Hinkley Point C nuclear power station illustrated the variations, with higher multipliers associated with in-migrants with families (1.3–1.5) than with unaccompanied in-migrants (1.05–1.11) (Glasson et al. 1988). The Keynesian

Table 13.6 Summary of the economic impact of the British Columbia mining industry in 2010 (Mining Association of British Columbia/PwC 2011)

Impact	Direct	Indirect	Induced	Total
(CAD$ millions)				
Output	5,166.5	2,732.8	1,034.6	8,933.9
GDP	2,748.8	1,319.0	622.5	4,690.3
Taxes	495	253.3	190.3	938.6
Number of jobs	21,112*	16,590	8,001	45,703

Note: * Direct employment from operating mines as reported in the PwC 2010 mining industry survey was 8,195 and is included in the total of 21,112 direct jobs.

multiplier model, with modifications as appropriate, is well-suited to the assessment of the wider economic impacts of projects. But it can only be as good as the information sources on which it is based and from which both the multiplicand and the multiplier are constructed. Predictive studies of proposed developments are more problematic in this respect than studies of existing developments, although knowledge of the latter can inform prediction.

13.5.5 Assessing significance

Socio-economic impacts, including the direct employment and wider economic impacts, do not have recognised standards. There are no easily applicable 'state of local society' standards against which the predicted impacts of a development can be assessed. While a reduction in local unemployment may be regarded as positive, and an increase in local crime as negative, there are no absolute standards. Views on the significance of economic impacts, such as the proportion and types of local employment on a project, are often political and arbitrary. Nevertheless, it may be possible to identify what might be termed **threshold or step changes** in the socio-economic profile of an area: for example, impacts which threaten to swamp the local labour market, and which may produce a 'boom–bust' scenario. It may also be possible to identify likely high levels of leakage of anticipated benefits out of a locality, which may be equally unacceptable.

It is valuable if the practitioner can identify possible criteria used in the analysis for a range of levels of impacts, which at least provides the basis for informed debate. This involves, in particular, using magnitude and sensitivity criteria to attribute levels of significance from negligible through to major. Table 13.7 provides some examples of magnitude criteria from the decommissioning of a nuclear power station project. Sensitivity criteria reflect the affected receptor. A high-sensitive receptor will show evidence of severe socioeconomic challenges, underperformance and vulnerability (e.g. high unemployment), and may also be identified as a high-ranking policy priority in local and relevant policies. In contrast, a low-sensitive receptor will show evidence of good overall performance, with no particular weaknesses and with plentiful capacity to handle change; as such it is also unlikely to be identified as a policy priority as a result of economic potential or need. While the attribution of significance in social impact assessment is an imprecise exercise, it is improving in practice (for example see Rowan 2009).

In the assessment of significance, the analyst should be aware of the philosophy of triangulation noted earlier. Multiple perspectives on significance can be gleaned from many sources, including the local press, which can be very powerful as an opinion-former; other key local opinion-formers (including local elected representatives and officials); surveys of the population in the host locality; and public meetings. All can help to assess the significance, perceived and actual, of various socio-economic impacts. A very simple analysis might measure the column centimetres of local newspaper coverage of

Table 13.7 Assessing the local significance of socio-economic impacts: extracts from a nuclear power station decommissioning project

Type of impact	Negligible impact	Slight impact	Moderate impact	Major impact
Demographic				
Change in local population level	No measurable change in local population level	Change in local population of less than + or − 1%	Change in local population of + or − 1–2%	Change in local population of more than + or − 2%
Direct and indirect employment impacts				
Change in site (direct) employment levels	Change of less than + or − 10% on baseline site employment levels	Change of + or − 10–20% on baseline site employment levels	Change of + or − 20–50% on baseline site employment levels	Change of more than + or 50% on baseline site employment levels
Change in employment level in local economy	No measurable change in employment levels in the local economy	Change of less than + or − 1% of baseline employment levels in the local economy	Change of + or −1–2% on baseline employment levels in the local economy	Change of more than + or − 2% on baseline employment levels in local economy
Change in unemployment level in local economy	Change of less than + or − 2% in claimant unemployment level	Change of + or − 2–5% in claimant unemployment level	Change of + or − 5–10% in claimant unemployment level	Change of more than + or − 10% in claimant unemployment level
Local expenditure and wider economic impacts				
Change in levels of local expenditure by site employees	Change of less than + or −10% on baseline levels of local expenditure	Change of + or − 10–20% on baseline levels of local expenditure	Change of + or − 20–50% on baseline levels of local expenditure	Change of more than + or −50% on baseline levels of local expenditure

Source: BNFL (2002)

certain issues in the planning stage of the project, and a survey of local people might seek to calculate simple measures of agreement (MoA) with certain statements relating to economic impacts. MoA is defined as the number of respondents who agree with the statement, minus the number who disagree, divided by the total numbers of respondents. Thus, a MoA of 1 denotes full agreement, −1 denotes complete disagreement.

13.6 Mitigation and enhancement

Many predicted economic impacts are normally beneficial and encouraged by the local decision makers. However, some may be disputed. There may be concern about some of the issues already noted, such as the poaching of labour from local firms, the swamping of the local labour market, or the shadow effect on other potential development. In such cases, there may be attempts to build-in formal and/or informal controls, such as 'no poaching agreements'. The fear of the 'boom–bust' scenario may lead to requirements for a compensatory 'assisted area' package for other employment, with the demise of employment associated with the project in hand (Rodriguez-Bachiller and Glasson 2004). A number of studies of post-redundancy employment experiences have been undertaken in the UK. Some relate to traditional industries such as coal mining, shipbuilding and steel (Hinde 1994; Turner and Gregory 1995); others have been associated with the restructuring of the defence and aerospace sectors (Bishop and Gripaios 1993). There have also been studies of the end of construction programmes (Armstrong et al. 1998; Glasson and Chadwick 1997). An interesting study on the disputed costs and benefits of UK airport expansion, in this case the planned expansion of Stansted Airport, is provided by Ross and Young (2007). The study focuses in particular on the economic leakage out of the UK associated with an airport so closely tied to the budget airline/mass-tourism market. In this case the protagonists' proposed mitigation measure was to 'Stop Stansted Expansion'.

However, in general the focus for economic impacts is more on measures to **enhance benefits** and, in economic terms, to enhance the benefits of local content. When positive impacts are identified there should be a concern to ensure that they do happen and do not become diluted (see Chapter 14 for the potential role of Social Impact Management Plans and Chapter 20 for Environmental and Social Management Plans). The potential local employment benefits of a project can be encouraged through appropriate skills training programmes for local people. Targets for the proportion of local recruitment may be set. Various measures, such as project open days for potential local suppliers and a register of local suppliers, may help to encourage local links and to reduce the leakage of wider economic impacts outside the locality/region. For example, the UK Olympic Development Authority set up in 2008 a CompeteFor website to bring contract opportunities for the London 2012 Olympics to the attention of possible suppliers. Such initiatives may be wrapped up in the form of an Employment and Supply Chain Plan, agreed between the developer, local authorities and regulatory bodies as a requirement for the project to progress. The Hornsea Project One Offshore Wind Farm in the UK section of the North Sea, which at the time of writing is the largest such wind farm in the world, has such a plan designed to bring employment opportunities to the relatively depressed local economy of Hull and North Lincolnshire in the UK (PINS 2014). Box 13.3 provides examples of industry-community partnerships in delivering local content benefits in Australia; Esteves and Barclay (2011) provide further examples from the mining, gas and oil sectors. The related issue of Community Benefit Agreements is discussed in Chapter 14.

13.7 Monitoring

Previous stages in the ESIA process should be designed with monitoring in mind. Key indicators for monitoring direct employment impacts include: levels and types of employment – by local and non-local sources and by previous employment status; trends in local and regional unemployment rates; and the output of training programmes. All these indicators should be disaggregated to allow analysis by employment/skill category. Relevant data sources include developer/contractor returns, monthly unemployment statistics, and training programme data; these can be supplemented by direct survey information. Key indicators of the wider economic impacts include: trends in retail turnover, the fortunes of local companies, and development trends in the locality. Some guidance on these indicators may be gleaned from published data. The project developer may also provide information on the distribution of subcontracts, but surveys of, for example, workforce expenditure, and the linkages of local firms with a project, may be necessary to gain the necessary information for useful monitoring.

Monitoring has not been mandatory to date for EIA in the UK or in many EU Member States. The EC is a strong advocate, but despite good practice in some Member States (e.g. the Netherlands) others are more defensive and reactive. However, the omission was recognised in the latest review of the EU EIA Directive (EC 2014), and monitoring will become mandatory from 2017. As yet however, there are few comprehensive monitoring studies to draw upon. The work of the IAU at Oxford on monitoring the local socio-economic impacts of the construction of Sizewell B (Glasson 2005) provides one of the few documented examples of a longitudinal study of socio-economic impacts in practice. It shows the significance of direct employment and wider economic impacts for the local economy. At peak, over 2,000 local jobs were provided, but with a clear emphasis on the less skilled jobs. Local skills were upgraded through a major training programme, and while some local companies experienced recruitment difficulties as a result of Sizewell B the impact did not appear to be too significant. A group of about 30 to 40 mainly small local companies have benefited substantially from contracts with the project. Although the actual level of project employment was higher than predicted, many of the predictions made at the time of the public inquiry have stood the test of time, and the key socio-economic condition of encouraging the use of local labour was fulfilled. The study also showed the project-management advantages of monitoring, with issues being highlighted by such monitoring being managed, quickly, for the benefit of the project and the local community. The London Olympics project also had a construction monitoring programme, with monthly socio-economic returns providing detailed information on the characteristics of the workforce, including previous residence, previous employment, gender, disability and ethnicity (ODA 2011).

13.8 Conclusions

Socio-economic impacts are important in the ESIA process. They have traditionally been limited to no more than one EIS chapter, and often a small, late chapter if they have been included at all. Our focus is on the incorporation of such impacts within an ESIA process rather than as separate SIA (or even SIA and HIA) assessments (see Ahmad 2004).

This discussion has outlined the broad characteristics of such impacts and examined economic impacts in more detail, with a particular focus on approaches to establishing the information baseline and on prediction. Some predictive methods can become complex. This may be appropriate for major studies; for smaller studies, some of the simpler methods may be more suitable. The non-local/local employment ratio associated with a project has been identified as a key determinant of many subsequent socio-economic effects, and this is now taken further in Chapter 14, with the emphasis on social impacts.

References

ACIL Tasman 2009. *Fly-in/Fly-out and regional impact assessment.* Melbourne: ACIL Tasman.

Ahmad B 2004. Integrating health into impact assessment: challenges and opportunities. *Impact Assessment and Project Appraisal* **22**(1), 2–4.

Armstrong HW, M Ingham and D Riley 1988. *The effect of the Heysham 2 power station on the Lancaster and Morecambe economy.* Lancaster: Department of Economics, Lancaster University.

Batey P, M Madden and G Scholefield 1993. Socio-economic impact assessment of large-scale projects using input-output analysis: a case study of an airport. *Regional Studies* **27**(3), 179–192.

Bishop P and P Gripaios 1993. Defence in a peripheral region: The case of Devon and Cornwall. *Local Economy* **8**, 43–56.

BNFL (British Nuclear Fuels Limited) 2002. *Environmental statement in support of*

---

application to decommission Hinkley Point A nuclear power station. Magnox Electric/BNFL.

Bowles RT 1981. *Social impact assessment in small communities*. Toronto: Butterworth.

Bronfman LM 1991. Setting the social impact agenda: an organisational perspective. *Environmental Impact Assessment Review* 11, 69–79.

Brownrigg M 1971. The regional income multiplier: an attempt to complete the model. *Scottish Journal of Political Economy* 18.

Brownrigg M 1974. *A study of economic impact: the Stirling University*. Scottish Academic Press.

Brunner J and Glasson J 2015. *Contemporary issues in Australian urban and regional planning*. Abingdon: Routledge.

Burdge R 2002. Why is social impact assessment the orphan of the assessment process? *Impact Assessment and Project Appraisal* 20(1), 3–9.

Burdge R 2003. The practice of social impact assessment – background. *Impact Assessment and Project Appraisal* 21(2), 84–88.

Canter L and R Clark 1997. NEPA effectiveness – a survey of academics. *Environmental Impact Assessment Review* 71, 313–328.

Carley MJ and ES Bustelo 1984. *Social impact assessment and monitoring: a guide to the literature*. Boulder, CO: Westview Press.

Chadwick A 2002. Socio-economic impacts: are they still the poor relations in UK Environmental Statements? *Journal of Environmental Planning and Management* 45(1), 3–24.

CSRM (Centre for Social Responsibility in Mining) 2006. *Indigenous employment in the Australian mining industry*. CSRM, University of Queensland.

Denzin NK 1970. *Sociological methods: a source book*. Chicago: Aldine Publishing Company.

DoE (Department of the Environment) 1989. *Environmental assessment: a guide to the procedures*. London: HMSO.

EC (European Commission) 1985. Council Directive on the assessment of the effects of certain private and public projects on the environment (85/337/EEC). *Official Journal of the European Communities* L 175/40, 5 July 1985, Brussels: European Commission.

EC 1997. Council Directive 97/11/EC amending Directive 85/337/EEC on the assessment of the effects of certain public and private projects on the environment. *Official Journal of the European Commission* No. L073/5-21. Brussels: European Commission. www.europa.eu.int/comm/environment/eia.

EC 2014. Directive 2014/52/EU of the European Parliament and of the Council of 16 April 2014 amending Directive 2011/92/EU on the assessment of the effects of certain public and private projects on the environment. *Official Journal of the European Union*, L 124, 25 April 2014. Brussels: European Commission.

EDF 2011. *Environmental statement for Hinkley Point Two new nuclear power station*. https://infrastructure.planninginspectorate.gov.uk.

Esteves AM and MA Barclay 2011. Enhancing the benefits of local content: integrating social and economic impact assessment into procurement strategies. *Impact Assessment and Project Appraisal* 29(3), 205–217.

Esteves AM, D Franks and F Vanclay 2012. Social impact assessment: the state of the art. *Impact Assessment and Project Appraisal* 30(1), 35–44.

Finsterbusch K 1985. State of the art in social impact assessment. *Environment and Behaviour* 17(2), 193–221.

Glasson J 1992. *An introduction to regional planning*. London: UCL Press.

Glasson J 2005. Better monitoring for better impact management: the local socio-economic effects of constructing Sizewell B nuclear power station. *Impact Assessment and Project Appraisal* **23**(3), 1–12.

Glasson J and A Chadwick 1988–1997. *The local socio-economic impacts of the Sizewell 'B' PWR construction project*. Impacts Assessment Unit, Oxford Brookes University.

Glasson J and A Chadwick 1997. Life after Sizewell B: post-redundancy experiences of locally recruited construction employees. *Town Planning Review* **68**(3), 325–345.

Glasson J and D Heaney 1993. Socio-economic impacts: the poor relations in British EISs. *Journal of Environmental Planning and Management* **36**(3), 335–343.

Glasson J, D van der Wee and B Barrett 1988. A local income and employment multiplier analysis of a proposed nuclear power station development at Hinkley Point in Somerset. *Urban Studies* **25**, 248–261.

Glasson J, R Therivel and A Chadwick 2012. *Introduction to environmental impact assessment*, 4th Edn. London: Routledge.

Grady S, R Braid, J Bradbury and C Kerley 1987. Socio-economic assessment of plant closure: three case studies of large manufacturing facilities. *Environmental Impact Assessment and Review* **26**, 151–165.

Hawke A 2009. *The Australian Environment Act – report of the independent review of the EPBC Act 1999*. Canberra: Australian Government Department of the Environment, Water, Heritage and the Arts.

Hinde K 1994. Labour market experiences following plant closure: the case of Sunderland's shipyard workers. *Regional Studies* **28**, 713–724.

HM Treasury 2011. *The Green Book: Appraisal and Evaluation in Central Government*. London: The Stationary Office.

Holm D, L Ritchie, K Snyman and S Sunderland 2013. Social impact management: a review of current practice in Queensland, Australia. *Impact Assessment and Project Appraisal* **31**(3), 214–219.

IAIA (International Association for Impact Assessment) 2015. *Social impact assessment: Guidance for assessing and managing the social impacts of projects*. Fargo, USA: IAIA.

IFC (International Finance Corporation) 2012. *Performance standards on environmental and social sustainability*. Washington, DC: IFC, World Bank Group.

ILO (International Labour Organization) 1985. *Labour Statistics Convention 1985 (No. 160)*. Geneva: ILO.

ILO (International Labour Organisation) 2008. *International standard classification of occupations*. www.ilo.org/public/english/bureau/stat/isco/isco08/index.htm.

Interorganisational Committee on Guidelines and Principles for Social Impact Assessment 1994. Guidelines and principles for social impact assessment. *Impact Assessment* **12** (summer), 107–152.

Jackson S and AC Sleigh 2001. Political economy and socio-economic impact of China's Three Gorges Dam. *Asian Studies Review* **25**(1), 57–72.

KPMG 2013. *FIFO*. Melbourne: KPMG Demographics.

Lang R and A Armour 1981. *The assessment and review of social impacts*. Ottawa: Federal Environmental Assessment and Review Office.

Lewis JA 1988. Economic impact analysis: a UK literature survey and bibliography. *Progress in Planning* **30**(3), 161–209.

Loveridge S 2004. A typology and assessment of multi-sector regional economic impact models. *Regional Studies* **38**(3), 305–317.

McKenzie FH 2010. Fly-in/Fly-out: The challenge of transient populations in rural landscapes, Chapter 12 in Luck GW, D Race and R Black (eds), *Demographic change in Australia's rural landscape*, Dordrecht, The Netherlands: Springer, 353–374.

Mining Association of British Columbia, with PriceWaterhouseCoopers 2011. *Economic Impact Analysis*. MABC/PwC.

Murdock SH, FL Leistritz and RR Hamm 1986. The state of socio-economic impact analysis in the USA: limitations and opportunities for alternative futures. *Journal of Environmental Management* **23**, 99–117.

NDA (Nuclear Decommissioning Authority) 2006. *NDA strategy*. London: NDA.

Newton JA 1995. *The integration of socio-economic impacts in EIA and project appraisal*. MSc dissertation. Manchester: UMIST.

ODA (Olympics Delivery Authority) 2011. *Employment and skills update: Jan 2011*. London: ODA.

ONS (Office for National Statistics) 2010. *UK Standard Occupational Classification (SOC)*. London: ONS.

PINS (Planning Inspectorate, England and Wales) 2014. Hornsea Project One: examining authority's report of findings and conclusions, and recommendations to the Secretary of State. https://infrastructure.planninginspectorate.gov.uk.

Rio Tinto 2014. *Sustainable Development Report 2014, and Ten-year Anniversary for Participation Agreement*.

Rodriguez-Bachiller A and J Glasson 2004. *Expert systems and geographical information systems for EIA*. London: Taylor & Francis.

Ross B and M Young 2007. *Proof of evidence on behalf of Stop Stansted Expansion: economic impacts*. www.stopstanstedexpansion.com.

Rowan M 2009. Refining the attribution of significance in social impact assessment. *Impact Assessment and Project Appraisal* **27**(3), 185–191.

Storey K 2010. Fly-In/Fly-out: implications for community sustainability, *Sustainability* 2, 1161–1181.

Thomas P and D O'Donoghue 2012. *The Channel Tunnel: transport patterns and regional impacts*. Koln: IGU Urban Commission.

Transport for London 2004. Thames Gateway Bridge: Regeneration Statement. London: Transport for London.

Turner R and M Gregory 1995. Life after the pit: the post-redundancy experiences of mineworkers. *Local Economy* **10**, 149–162.

UN (United Nations) 2008. *International Standard Industrial Classification of all Economic Activities (ISIC) – Revision 4*. New York: UN.

UNECE (United Nations Economic Commission for Europe) 1991. *Policies and systems of environmental impact assessment*. Geneva: UNECE.

UNEP (United Nations Environment Programme) 1996. *Environmental impact assessment: issues, trends and practice*. Stevenage: SMI Distribution.

Vanclay F 2003. Principles for social impact assessment: A critical comparison between international and US documents. *EIA Review* **26**, 3–14.

Vanclay F and AM Esteves (eds) 2011. *New directions in social impact assessment: conceptual and methodological advances*. Cheltenham: Edward Elgar.

Vickerman R 1987. Channel Tunnel: consequences for regional growth and development. *Regional Studies* **21**(3), 187–197.

WADLRD (Western Australia Department for Local and Regional Development) 2012. *The Pilbara Cities*. Perth: WADLRD.

WA Environment Protection Agency 2010. *Environmental impact assessment administrative procedures*. Perth: EPA.

Wolf CP (ed.) 1974. *Social impact assessment*. Washington, DC: Environmental Design Research Association.

World Bank 1991. *Environmental assessment sourcebook*, Vol. 1, Chapter 3, *World Bank technical paper No. 139*. Washington, DC: World Bank.

14 Socio-economic impacts 2: Social impacts

Andrew Chadwick and John Glasson

14.1 Introduction

Economic impacts were discussed in Chapter 13, before the social impacts covered in this chapter, because many of the social impacts associated with major development projects flow straight from the direct and indirect economic impacts. The workforce involved in the construction and operation of any major project is likely to be drawn partly from within daily commuting distance of the project site and partly from further afield. Those employees recruited from beyond daily commuting distance can be expected to move into the locality, either temporarily during construction or permanently during operation. Some of these employees will bring families into the area. In-migrant employees and their families will exert a number of impacts on their host localities:

- they will result in an increase in the **population** of the area and possibly in changes to the age, gender, language, culture and other social profiles of the local population;
- they will require **accommodation** within reasonable commuting distance of the project site;
- they will place additional demands on a whole range of **local services**, including education, health and recreational facilities, police and emergency services.

They may have **other social impacts**, such as changes in the local crime rate, or in the social mix of the area's population, raising issues such as community stress, conflict, alienation and a breakdown in community cohesion.

Major projects may also have other types of population impacts, including on health and on population displacement and resettlement. The latter is of particular significance, especially in developing economies, and often for certain types of projects such as large dams and major transport infrastructure.

The Chinese Three Gorges Project provides an extreme example, with the inundation of a massive amount of farmland and the displacement and resettlement of approximately 1.25 million people over a 15–20 year period (Jackson and Sleigh 2001).

This chapter builds on the socio-economic foundations laid in the previous chapter. As such, the discussion of definitions and concepts, and the legislative and guidance base are only briefly covered to minimise duplication.

14.2 Definitions and concepts

As noted in Chapter 13, socio-economic impacts can be summarised as the "people impacts" of development actions, and socio-economic impact assessments/social impact assessments (SIA) focus on the human dimension of environments. SIA seeks to identify the impacts on people, and who benefits and who loses; it can help to ensure that the needs and voices of diverse groups and people in a community are taken into account. Social issues include population change (both growth and displacement), housing, and local services (such as education, health and emergency). They also include an important set of quality-of-life and well-being issues, reflected in social problems such as crime; ill-health and deprivation; community stress and conflict; and inequality and distributional issues. A wide definition of social impacts is included in Box 14.1. This shows the interconnectedness of social impacts with many of the other impact types covered in this book. Some of these are covered more fully elsewhere, such as cultural heritage (Chapter 12), resettlement (Chapter 15) and health (Chapter 16), and as such are only very briefly covered in this chapter.

14.3 Key legislation, guidance and concepts

14.3.1 Overview

Relevant legislation is set out in Chapter 13. Examples of guidance and guidelines on SIA include Anglo American (2012), Bureau of Minerals and Petroleum (2009), IAIA (2015), ICGPS (1995), Mackenzie Valley Environmental Impact Review Board (2007) and Ziller (2012). Aid agency guidelines for the conduct of SIA include Asian Development Bank (2008, 2009, 2012, 2014), IFC (2009, 2012b) and World Bank (2003).

Of particular significance are the International Finance Corporation's *Performance Standards on Environmental and Social Sustainability* (IFC 2012b), and the International Association for Impact Assessment's latest guidance: *Social Impact Assessment: Guidance on assessing and managing social impacts of projects* (IAIA 2015). From these two documents, supplemented by other sources, a number of key concepts of particular relevance to social impacts can

Box 14.1 A wide definition of social impacts (Vanclay 2003)

Social impacts are changes to one or more of the following:

- people's way of life – that is, how they live, work, play and interact with one another on a day-to-day basis;
- their culture – that is, their shared beliefs, customs, values and language or dialect;
- their community – its cohesion, stability, character, services and facilities;
- their political systems – the extent to which people are able to participate in decisions that affect their lives, the level of democratisation that is taking place, and the resources provided for this purpose;
- their environment – the quality of the air and water people use; the availability and quality of the food they eat; the level of hazard or risk, dust and noise they are exposed to; the adequacy of sanitation, their physical safety, and their access to and control over resources;
- their health and wellbeing – health is a state of complete physical, mental, social and spiritual wellbeing and not merely the absence of disease or infirmity;
- their personal and property rights – particularly whether people are economically affected, or experience personal disadvantage which may include a violation of their civil liberties;
- their fears and aspirations – their perceptions about their safety, their fears about the future of their community, and their aspirations for their future and the future of their children.

be identified. These include participation and Free, Prior and Informed Consent (FPIC), human rights, and environmental justice.

14.3.2 Participation and FPIC

Meaningful public participation is widely considered as critical for the basic legitimacy and fairness of impact assessment processes, including socio-economic impact assessment. A substantial literature has evolved, highlighting the potential benefits of such participation (see for example Petts 1999, O'Faircheallaigh 2010, Glasson et al. 2012). The concept of participation is a contested one, and O'Faircheallaigh (2010) takes a broad definition as "any form of interaction between government and corporate actors and the public that occurs as part of EIA processes". He further defines three possible and potentially overlapping purposes of public participation as: (i) to obtain public input into decisions taken elsewhere; (ii) to share decision-making with the public; and (iii) to alter the distribution of power and structures of decision-making.

The issue of differential power in decision-making links very much into issues surrounding the relationship between major projects, Indigenous peoples, and FPIC. Various international declarations and conventions (e.g. UN Declaration on the Rights of Indigenous Peoples 2007) recognise the fundamental rights of Indigenous peoples and support FPIC. However, the implementation of FPIC in practice can face many challenges, including for example defining: who has the right to give consent, and the extent to which there is a fundamental right to self-determination and the right to say no to a project. FPIC can involve major issues of compensation, sometimes packaged into Community Benefits Agreements (discussed in §14.6.2). However, it can also "face the risk of being treated only as a token consultation rather than being a powerful instrument to build respectful relationships among those who have a stake in the outcome" (Esteves et al. 2012).

14.3.3 Human rights

Human rights were enshrined in the Universal Declaration of Human Rights, adopted by the UN as early as 1948, but "it was not until quite recently that the UN confirmed that international human rights law holds particular relevance to business" (Kemp and Vanclay 2013). The UN Human Rights Council (UNHRC 2011) and UN Guiding Principles on Business and Human Rights (UN 2011) clarified that companies are subject to international human rights law. As a minimum this involves avoiding infringing the human rights of others: the "do no harm" principle. Human rights are wide-ranging, but those relating to labour, land and property, and freedom from discrimination are often of particular significance in impact assessment.

There have been various examples of project developments which have generated severe human rights issues. Some celebrated cases include the Marlin gold mine development in Guatemala and the Ok Tedi copper and gold mine in Papua New Guinea. The Marlin development for example raised key concerns including large-scale land disturbance, access to water, and inadequate consultation with the local Indigenous Mayan people; however, the subsequent production of a pioneering "independent human rights assessment" was arguably one valuable outcome (On Common Ground 2010). Such assessments have been limited to date, and raise the question as to whether they should be conducted as free-standing Human Rights Impact Assessments (HRIA), or more integrated assessments as part of the Environmental and Social Impact Assessment (ESIA). Another issue is the negative emphasis that characterises much of human rights assessment, with the focus largely on minimising harm, with little consideration of the potential role of business in realising rights and enhancing well-being. However, there are some positive initiatives, for example a very useful web-based tool, *Community-Based HRIA: The Getting it Right Tool*, has recently been published by Oxfam and International Federation for Human Rights (FIDH) (2016).

It should also be noted that human rights assessment is not just an issue for Indigenous peoples, often in a developing countries context. For example, in the UK, the assessment of the impacts of Nationally Significant Infrastructure Projects must have regard to the Human Rights Act 1998. This is particularly in relation to the use of compulsory acquisition powers and their potential impacts on a person's right to the peaceful enjoyment of their property, to the protection of private and family life, and to a fair and public hearing of any objections to a development (DCLG 2013).

14.3.4 Environmental justice

ESIA should also pay particular attention to vulnerable sections of the population being studied – the elderly, the poor, and minority or ethnically distinctive groups – and to areas which may have particular value to certain groups in terms of cultural or religious beliefs. In this context, a pioneering development in the USA, after long campaigning by Black and ethnic groups, was the Clinton *Executive order on federal actions to address environmental justice in minority populations and low-income populations* (White House 1994). Under this order, each federal agency must analyse the environmental effects (including human health, economic and social effects) of federal actions that fall under the National Environmental Protection Act, including the effects on minority and low-income communities. The focus is on "environmental justice", a component of the broader field of SIA; it is concerned with "fair treatment", meaning that "minority and low-income groups should not bear a disproportionate share of the negative environmental impacts of government actions" (Bass 1998). Bass provides an example of a proposal for a nuclear enrichment centre in Louisiana (US) which was refused a licence on the basis that "racial and economic discrimination played an unacceptable role in the project's planning".

Low and Gleeson (1998) define environmental justice as "fairness in the distribution of environmental well-being". In a South African context, Scott and Oelofse (2005) see environmental justice as allowing all to "define and achieve their aspirations without imposing unfair, excessive or irreparable burdens or externalities on others and their environments, now and in the future". They note that many Black communities in South African cities have inherited the impacts of poor land-use planning from the apartheid era, often living in close proximity to major projects such as landfills, airports, power stations and industries. In response they advocate a new approach to the democratization of practice in the developing world based on the principles of social and environmental justice. Their principles of procedural equity include, for example: giving power to previously "invisible stakeholders", including the poor and marginalised people; respect for diversity; equal opportunities to participate for all stakeholders; and a commitment to the integration of local and scientific knowledge.

14.4 Scoping and baseline studies

14.4.1 Scoping social issues

As for economic impacts, the scoping of social issues involves understanding the many dimensions both of the project and of the host society/community on which the project is likely to impact. Chapter 13 discusses the many dimensions of the project. With regard to the characteristics of the host society, it is important to identify an initial profile of the likely affected communities, drawing both on data and on attitudes which give social meaning to the data. This involves a thorough stakeholder analysis to identify all potential interested and affected parties, which may range from residents in the immediate impact zone; nearby communities whose livelihoods may be affected; construction workers who migrate into the area of the development; people living near where the migrant construction workers may reside; plus a whole range of agencies and developer and supply-chain links (IAIA 2015). It also involves an understanding of the socio-political context of the host community; how are local people likely to respond to the proposal and what level of trust is there in government and the project developer, possibly coloured by past experiences of other similar developments (IAIA 2015)? Figure 14.1 provides an approach to displaying social baseline information for an area, relative to national standards with performance ranging from 1 (very poor) to 3 (very good). The following sections focus on some of the key social elements from this overview, starting with population.

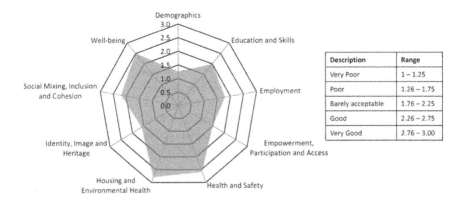

Figure 14.1

An approach to displaying the social baseline of an area (Glasson and Cozens 2011)

14.4.2 Population

The geographical extent of social impacts, i.e. **the impact area**, will depend particularly on the residential location of in-migrant workers and their families. In-migrant employees can be expected to move into accommodation within reasonable commuting distance of the project site, although the definition of what constitutes a reasonable distance will depend on the project stage (construction or operation), as well as local settlement patterns and the local transport network. Monitoring data from similar projects elsewhere should indicate the likely extent of daily commuting and thus the likely boundaries of the impact area. These boundaries can be defined in various ways, for example in terms of a fixed distance or radius from the project site or, more usually, in terms of administrative or political areas such as local authority, health authority or school catchment areas. In terms of social issues such as access to land/resources, artisanal mining and other such activities, the "footprint" of the development and associated infrastructure is also of critical importance.

The demographic impact of any development will depend on the project-related changes in population in relation to the existing population size and structure in the impact area. It is therefore necessary to establish the existing population baseline in the impact area (i.e. size and age/gender profile). The most useful source of population data in many countries will be the national census or equivalent. Other data sources include population estimates by local or regional government, proxy data such as changes in the electoral roll or doctors' registrations, and commercial market-analysis companies. In some developing countries, where there is a lack of such data, there may be a need to conduct dedicated surveys as part of the assessment process.

It is also important to project this baseline forward, ideally to the expected times of peak construction and full operational activity for the proposed development. Population projections and forecasts are often produced by national and local authorities as inputs to their land-use planning work and to estimates of future service requirements (e.g. school places). Projections for smaller areas tend to be less reliable than those for larger areas. This is because net migration is usually a more important determinant of population change for smaller areas; and migration flows are much more difficult to predict than the number of births and deaths.

14.4.3 Accommodation

As well as providing population data, censuses may provide data on the **housing stock** in an area, according to its tenure (i.e. whether it is owner-occupied, privately rented, rented with a job or business or rented from a government body). In addition, the census may identify the amount of vacant accommodation, and accommodation which is not used as a main residence, including second homes.

During the construction stage of any development, some **in-migrant employees** are likely to move into bed and breakfast establishments, hotels, caravans or other types of tourist accommodation. It is therefore necessary to establish how much of such accommodation is available in the impact area, and if possible to determine typical occupancy levels. Any unoccupied accommodation (e.g. outside the peak tourist season) could be used by in-migrant employees without affecting the availability of accommodation for other existing users. The use of spare capacity in the accommodation market, and for other services (e.g. school places), can help to offset local concerns about the impacts of projects on the host population. Tourist boards, local governments and tourist information centres often maintain databases or lists of accommodation establishments within their areas of jurisdiction, which can assist in providing a picture of the existing stock of accommodation. When combining lists prepared by different organisations for the same geographical area, care should be taken to avoid the double-counting of establishments.

To project the accommodation baseline forward, non-project-related changes in the local housing stock can be estimated most easily by using **simple trend projection methods**. These are typically based on the assumption that recent rates of growth in the number of dwellings will continue for the foreseeable future. Such methods, although easily applied, are rather crude in that they take no account of possible changes in the state of the national economy (which may affect housebuilding rates) or in local rates of population and household growth; they also fail to allow for the influence of local planning policies on the scale and location of new housebuilding.

An alternative approach is to use estimates of future population and household growth in the area to predict the likely demand for new houses. Of course, the anticipated increase in the number of households in an area may not be met by an equivalent increase in the housing stock. This is because local planning policies may not meet all of the projected increase in households.

Likely changes in the stock of tourist and other temporary accommodation are difficult to predict, although regional tourist boards and local authorities may be able to indicate the scale of any significant additional provision, either already under construction or with outstanding planning permission.

14.4.4 Local services

In-migrant employees and their families will place demands on a wide range of services including education, health, recreation, police, fire and social services. Information on these services can be obtained from the relevant department. For example, in the UK the number and type of schools within the impact area can be obtained directly from local education authorities (LEAs) or from websites such as Edubase (in England and Wales). Edubase allows users to identify all educational establishments within fixed distances of a specified location; it also provides information on the existing number of pupils on school rolls and the total available capacity for each individual school. This information can be used to determine the extent to which the available capacity in LEA schools is currently being utilised across the authority as a whole and for individual schools. Health issues are discussed in depth in Chapter 16.

14.4.5 Social cohesion

Social cohesion refers to people's sense of belonging to a community, their satisfaction with life in the community and whether or not the community is considered a social asset. In a cohesive community, a network of positive relationships is generated and maintained, and there is a level of community pride. The profiling of social cohesion is challenging; data is often limited, but it may include factors such as relative well-being and deprivation. Another key factor is a sense of personal safety and fear of crime.

The estimating of personal and area **well-being** is usually derived from questionnaire-based studies. There are international studies, such as the OECD *Guidelines on Measuring Subjective Well-being* (OECD 2013), and national studies, such as the *UK Index of National Well-being* (ONS 2015). The UK study is based on four questions, overall: how satisfied are you with your life nowadays; to what extent do you feel the things you do in your life are worthwhile; how happy did you feel yesterday; and how anxious did you feel yesterday? People are asked to give their answers on a scale of 0 to 10, where 0 is "not at all" and 10 is "completely". The data can be disaggregated down to local authority areas, and year-on-year trends can be identified. UK national scores average about 7.5 for the first three questions and 3 for the last question. A more locally based example of an innovative survey based index is the *Sociale Index* developed for 64 areas of the City of Rotterdam. It is based on four main dimensions: personal abilities (e.g. health); participation (e.g. going to work/school); bonding (e.g. mobility); and living environment (e.g. safety) (Koppelaar 2008).

Deprivation indexes provide another useful approach to social profiling. A good example from England is the Index of Multiple Deprivation (IMD) (DCLG 2016). Average levels of deprivation across the 350-plus local authority districts are indicated by rank position relative to all other English local authority districts (where a rank of 1 indicates the most deprived in England and a rank of 354 indicates the least deprived). The index includes 13 domains, for which disaggregated data is available, updated on an annual basis, and presented in both tabular and diagrammatic formats. The main domains are: income; employment; health, deprivation and disability; education, skills and training; barriers to housing and services; living environment; and crime. Figure 14.2 provides an example of a composite IMD at Lower Layer Super Output Area (LSOA) for Bradford in the UK.

An international-level example of a deprivation index is provided by the global Multidimensional Poverty Index (MPI) (Alkire and Robles 2015). The global MPI has three dimensions and ten indicators, as set out in Figure

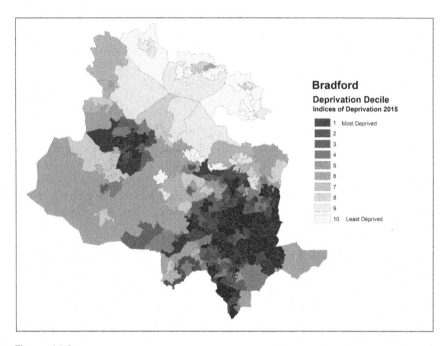

Figure 14.2

Composite Index of Multiple Deprivation for Bradford, UK (DCLG 2016)

14.3(a). Each dimension is equally weighted, as illustrated in the figure. Figure 14.3(b) provides an example of findings from the application of the index in India (OPHI 2015), clearly showing the greater problems in the rural population.

Crime and freedom from crime are high on people's agendas of most important issues in many countries worldwide. Crime and safety issues have been more thinly covered in impact assessment than many other social issues but they can be very significant. For example, a project with a large and predominantly young male in-migrant workforce is likely to represent an issue for crime and other behavioural problems in a host locality. Relevant baseline data should include data both on crime and fear of crime. Both are challenging to assemble; although local police authorities can be a valuable source of crime data. However, recorded crime may only represent a small fraction of total crime; fear of crime relies more on survey information. But the position is improving, as are mitigation responses to designing-out crime (Glasson and Cozens 2011).

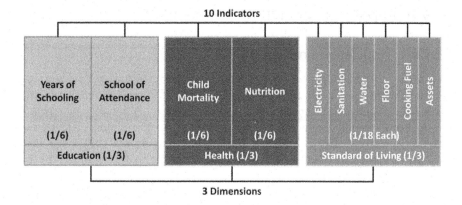

Figure 14.3(a)

Structure of the global Multi Poverty Index

Figure 14.3(b)

Percentage of the Indian population who are MPI poor and deprived
Source: (a) and (b) OPHI (2015)

14.5 Impact prediction and evaluation

14.5.1 Population changes

Changes in population caused by a major project can include both direct and indirect increases. The **direct increase** will consist of in-migrant employees and any other family members brought into the locality. A number of separate estimates are therefore required to determine the population changes directly due to the project: (a) the total number of employees moving into the impact area, during both the construction and operational stages of the development; (b) the proportion of these in-migrant employees bringing other family members; and (c) the characteristics of these families (i.e. their size and age structure).

The total number of employees moving into the impact area

Chapter 13 outlined methods for predicting the mix of local and in-migrant employees associated with the construction and operation of major projects. *During the construction stage*, the build-up in the number of in-migrant workers will reflect the build-up of the construction workforce and changes in the local labour percentage. At the end of the construction stage, most in-migrant workers will move out of the impact area and return to their original address or to construction projects elsewhere. However, a small proportion may establish local ties, especially during a lengthy construction project, and may decide to remain in the area. A construction project spanning several years may therefore result in a small permanent increase in the local population. *During operation*, the main flow of in-migrant employees will usually occur at a relatively early stage, with subsequent in-migration limited to that caused by the normal turnover of employees.

The proportion of in-migrant employees bringing their family

During the construction stage, only a minority of in-migrant employees – mainly those on long-term contracts – are likely to bring their family into the area. The precise proportion will depend on various factors:

- the length of the construction programme (for projects lasting only a few months it is likely to be negligible; for projects spanning several years the proportion may reach at least 10–20 per cent in developed economies);
- the location and accessibility of the project site, which will determine the relative merits of weekly commuting and family relocation;
- conditions in the national and local housing markets (a depressed national housing market or sharp interregional house-price differentials may discourage house and family relocation);
- availability of suitable family accommodation, schools and other amenities in the locality.

During the operational stage, the vast majority of in-migrant employees will relocate permanently to the area, although there may be some initial delay while suitable accommodation is found and, where applicable, existing properties are sold. Those employees with families can be expected to bring them into the area (with the exception of a small number of weekly commuters). The precise proportion of employees with families will depend on the age and gender profile of the in-migrant workforce. For example, a younger workforce might be expected to contain a higher proportion of single, unattached employees who will not bring families into the area.

The characteristics of in-migrant families

Once the likely number of in-migrant families has been determined, it is necessary to estimate the average size and broad age structure of these families. The usual approach to estimating the size of in-migrant families is to use detailed census data on household headship. National census data typically shows the average size of households of different types, classified according to the age, gender and marital status of the head of household. Therefore, if it was considered likely that most in-migrant families would contain a married head of household, aged 20–59 years, then the average size of this type of household – either nationally or in the impact area – could be calculated. For projects with a younger anticipated workforce, the average size of households with married heads aged, say 20–44 years, could be calculated instead. This method assumes that the household characteristics at the time of the latest census will remain largely unchanged; it also requires some knowledge (or estimates) about the age and gender profile of the in-migrant workforce.

Let us assume, for an average European case, that the method outlined above suggests that each in-migrant family will contain an average of 3.2 persons. It could then be assumed that each of these families would consist of two adults of working age (the in-migrant employee and partner) and an average of 1.2 other family members – mainly dependent children up to 18 years old, but also including a small proportion of "adult" children (over 18 years old) still living with their parents, and perhaps some elderly relatives. The precise proportion of adult children and elderly relatives should ideally be derived from monitoring data, but – in the absence of such information – a rough estimate may be required. Information on the age structure of the 0–18-year-old population is available from a number of sources, and this can be used as the basis of predictions of the ages of dependent children brought into the area. For example in the UK, the current age breakdown of 0–18-year-olds is provided by the census, the latest mid-year population estimates and local authority population estimates. The projected future age breakdown of this group can be obtained from the various population projections and forecasts outlined above. The census also provides an age breakdown of children (and others) moving into particular areas during the 12 months prior to the census date.

The precise age distribution of dependent children will of course depend on the age profile of their parents. For example, a younger workforce will tend to have a higher proportion of pre-school children than might be suggested by the data sources above, whereas an older workforce may have higher proportions of secondary school children. Some fine-tuning of the age distribution revealed by the data sources above may therefore be required, to take account of the expected age profile of the project workforce. The age breakdown of the workforce should ideally be estimated by obtaining information on the age of employees on similar projects elsewhere. Such information should be readily available to the project developer (for operation) or its contractors (for construction).

As well as the direct population increase due to the arrival of in-migrant project employees and families, the development may give rise to **indirect population impacts**. These impacts can arise in two main ways. First, some locally recruited project employees will leave local employers to take up jobs on the project. This will result in local job vacancies, some of which may be filled by in-migrants. Indirect employment may also be created in local industries supplying or servicing the project, or in the provision of project-related infrastructure. Again, some of these jobs may be taken by in-migrant employees. The scale of the resulting additional in-migration is very difficult to estimate, but its possible existence should at least be acknowledged (see Clark et al. 1981 for some possible estimation methods). A second source of indirect impacts arises from the fact that some locally recruited project employees might have migrated out of the impact area if the project had not gone ahead, especially if alternative job opportunities locally were limited. The project may therefore lead to a reduction in out-migration from the area. Again, the extent of any such reduction is difficult to predict. It is likely to be significant only in areas experiencing static or declining population, net out-migration and limited or declining employment opportunities.

14.5.2 The significance of population changes

The significance of project-related population changes will depend on three main factors: (a) the existing population size and structure in the impact area (i.e. the population baseline); (b) the geographical distribution of the in-migrant population; and (c) the timing of the population changes. Put simply, if in-migrants are few relative to the existing population; and have a similar age, gender profile and other social characteristics; are distributed over a wide area; and do not all arrive at once; then the impacts are unlikely to be significant. The first step in assessing significance is therefore to **express the estimated project-related population increase as a percentage of the baseline population** in the impact area (for example, see Table 13.7). The predicted age structure of in-migrants should be compared with the baseline age structure, and any significant differences outlined.

The next step is to estimate the likely **geographical distribution of in-**

migrants. Population changes may be quite localised, rather than being evenly distributed throughout the impact area. However, in the absence of information from monitoring studies, the precise distribution of in-migrants is difficult to predict. The simplest approach would be to assume a "gravity model" approach: the number of employees moving into a particular settlement would be a positive function of that settlement's size and a negative function of its distance from the project site. In practice, the predictions derived from this type of model would need to be modified to allow for the characteristics of the particular locality. These could include the expected location of future housebuilding in the impact area; differences in the availability and price of various types of housing; and the attractiveness of each settlement in terms of schools and other facilities and general environment. **The timing of the arrival of in-migrant employees** and the associated population changes will largely follow the expected build-up in the project workforce. However, during the construction stage, most in-migrant families are likely to arrive in the early stages, given that families will tend to be brought by those employees on long-term contracts for the duration of the project.

The nature and significance of population impacts will change as the project progresses through the various stages of its lifecycle. In-migrant employees and their families will become older. In addition, during the operational stage – which may span several decades – there may be some natural increase from the original in-migrant population. These changes can be estimated by using a simple **cohort survival** method, applying age-specific birth and death rates to the original population. Some allowance may also need to be made for the turnover of employees on the project. As older employees retire, they will tend to be replaced by younger employees, with younger families. This process will counteract, but not completely reverse, the tendency for the in-migrant population to become older.

14.5.3 Accommodation requirements

Figure 14.4 provides a simple causal flow chart of some of the key factors involved in the consideration of accommodation impacts. **The total amount of accommodation required** will be determined by the size of the **in-migrant workforce** and the extent to which accommodation is shared. Methods to estimate the total number of in-migrant employees were outlined in Chapter 13. Sharing of accommodation is likely to be minimal among the permanent **operational workforce**, since most in-migrant employees will be accompanied by their families. During the **construction stage**, sharing may be much more significant, especially among those employees using rented housing, tourism accommodation, and dedicated construction worker-site camps or hostels. Estimates of the likely extent of sharing should be incorporated into any predictions of the demand for accommodation by the construction workforce, otherwise the amount of accommodation required is likely to be overestimated.

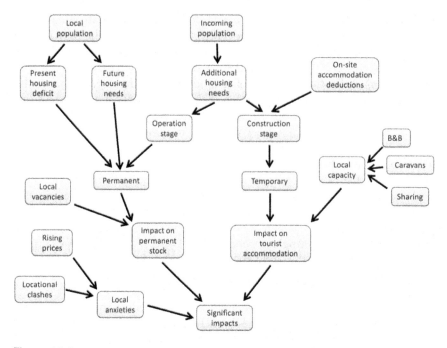

Figure 14.4

Accommodation impacts – an overview (based on: Rodriguez-Bachiller with Glasson 2004)

Published monitoring studies of recent construction projects, although limited in number, may provide an indication of the likely extent of sharing (e.g. see Glasson and Chadwick 1995).

The type and location of accommodation required will also differ in the operational and construction phases. The vast majority of in-migrant operational employees are likely to relocate permanently to the impact area. Some will purchase a property, others will use private or social rented accommodation. The likely mix between owner-occupied and rented accommodation requirements may be roughly estimated by using census data.

Predicting the likely mix of accommodation used by in-migrant construction workers is a more complicated exercise. A wider range of accommodation is likely to be suitable, including tourist accommodation. A further complication is that, for larger construction projects, the developer may decide to provide purpose-built accommodation specifically for the workforce. The extent of such provision will have important implications for the take-up of other types of accommodation. Because the local supply of different types of accommodation and the extent of developer provision will vary from one locality and project to another, the precise mix of accommodation used can vary considerably from project to project.

In the absence of developer provision, the vast majority of in-migrant construction workers are likely to use private, rented, tourism accommodation (where available) and developer-provided accommodation. The use of each type of accommodation can be roughly estimated by drawing on the available monitoring data from other construction projects, adjusted to allow for the particular supply characteristics in the impact area, i.e. the amount of each type of accommodation available, its location, cost and existing occupancy levels (see §14.4.3). For example, if the local supply of tourist accommodation is very limited, concentrated in highly priced hotels at some distance from the project site, and is usually fully occupied, the proportion of employees using such accommodation is likely to be low.

Some construction workers may wish to purchase properties in the locality. The number is likely to be minimal during construction projects lasting only a few months but may be more significant in cases where construction activity spans several years. The proportion of in-migrant employees buying properties will be closely linked to the proportion bringing families into the impact area, although some families will prefer to use rented accommodation.

In certain cases, the project developer may decide to make specific accommodation provision for the construction workforce. This may involve negotiations with the local planning authority over the provision of additional caravan sites or the expansion of existing sites. For very large construction projects in remote locations, the developer may wish to provide purpose-built hostel accommodation, located on or adjacent to the construction site. This typically consists of single bedrooms and associated catering, recreational and other facilities. To the extent that such provision is made, the proportion of in-migrant employees using other types of accommodation will be lower than would otherwise have been the case.

It may be helpful to provide estimates of the demand for different types of accommodation in various alternative scenarios, e.g. no hostel, small hostel or large hostel. Such estimates will themselves help to clarify the need for such developer provision. The precise geographical distribution of the accommodation taken up by in-migrant employees is difficult to predict: §14.5.2 outlined a possible gravity model approach.

14.5.4 The significance of accommodation requirements

The project-related demand for local accommodation is likely to result in a net change in the amount of accommodation available in the impact area. On the one hand, the availability of accommodation to local residents will be reduced by the take-up of local accommodation by project employees and their families. On the other hand, to the extent that project-related demands are met by the release of unoccupied or under-occupied accommodation and/or the bringing forward of speculative house building development, the amount of accommodation available locally will be higher than would otherwise have been the case. The balance between these types of change will represent the net change due to the project, which can be expressed as a percentage of the existing (or projected) stock of accommodation in the impact area. Similar calculations can be made for each separate type of accommodation and for particular settlements or areas within the impact area. In some cases, the net decline in the availability of accommodation due to the project may be such that the project-related and non-project demands for accommodation may outstrip the available local supply.

Good practice is to seek to minimise local accommodation impacts by not exceeding **capacity thresholds** in the various local housing markets. For example in the seasonal tourism accommodation market, project availability might be defined as the difference between peak occupancy and known bed-space capacity in various tourism accommodation sub-sectors. In the private rental market, the analyst needs to consider both (i) the percentage of properties vacant and (ii) that percentage needed to provide enough turnover (or churn) in the market for local people. In some cases there may also be a latent supply of accommodation, which the host community may be willing to put temporarily into the accommodation market, especially for the construction stage of a development. Cases in which the project still results in a shortfall in the local supply of accommodation may see some pressure on one locality relieved by the diversion of demand (both project and non-project) into adjacent localities. However it is likely that there will be a need to require the consideration of mitigation measures.

14.5.5 The demand for local services

In-migrant employees and their families will place demands on a wide range of services provided by local authorities and other public bodies, largely reflecting the age and gender distribution of the in-migrant population. For example, in the case of **health and personal social services**, the number of young children and elderly people will be a critical determinant of demand. In such cases, rough estimates of likely demand can be obtained by combining the predicted age and gender profile of the in-migrant population with age and gender-specific data on visiting rates to or by doctors, health visitors or social workers. The latter may be obtained from local and health authorities, although in developing countries this may require primary data collection.

In the case of **education services**, demand also clearly depends on the age structure of the in-migrant population, particularly the provision for children between the ages of 5 and 18. The remainder of this section provides an example of the calculations involved in estimating the number of additional state-sector primary and secondary school places likely to be required locally in response to an influx of project employees.

Predicting the demand for additional local school places requires separate estimates of: the total number of children aged 0–18 years brought into the impact area by in-migrant employees: the proportion of these children below compulsory school age (0–4 years); aged 5–16 and above school-leaving age;

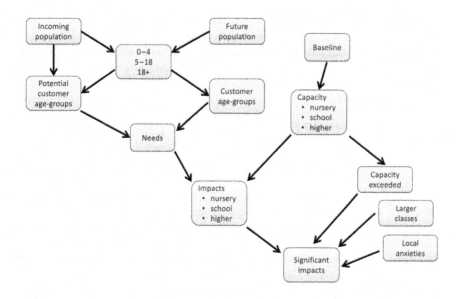

Figure 14.5

Education impacts – an overview (Rodriguez-Bachiller with Glasson 2004)

and the proportion of 5–16-year-olds attending independent (private sector) schools (Figure 14.5). The extent to which demand is geographically concentrated or dispersed in the impact area will determine the total number of schools affected and the likelihood of pressure on individual schools. The distribution of school place requirements will largely reflect the place of residence of in-migrant families. Unfortunately, the latter is difficult to predict in the absence of relevant monitoring data: §14.5.2 outlined a possible approach to prediction but it may be helpful to present a series of estimates based on different assumptions about the concentration or dispersal of in-migrant families.

14.5.6 The significance of demands on local services

An important indicator of the significance of local service impacts is again the extent to which **capacity thresholds** are exceeded as a result of the demands arising from the in-migrant population. For example, if the current accommodation capacity in a school is expected to be almost fully utilised in the absence of the project, and pupil to teacher ratios are already high, then even a small project-induced increase in pupil numbers may create a need for additional classrooms and/or extra teaching staff. By contrast, a large increase in pupil numbers in a school with a considerable amount of under-utilised capacity and low pupil/teacher ratios may be much less significant, although increases in pupil numbers in such schools may still be important, even if they do not put the available capacity under pressure. Class sizes will be larger than would otherwise have been the case, and additional staff time may need to be devoted to individual assessments of incoming pupils. Assessment of significance therefore requires information not only on the likely project-related increase in demand, but also the existing (and projected) utilisation of service capacity.

In certain circumstances, additional service demands may be seen as beneficial. For example, an influx of pupils into a small rural primary school with declining pupil numbers may help to safeguard the future of the school, either in the short term (during construction) or in the medium-to-long term (during operation). The nature and significance of local service impacts will change as the project progresses through its various stages. The in-migrant population, including children, will tend to become older, with the result that the type of services demanded will tend to change over time. For example, there will tend to be a shift away from nursery and primary school demand towards secondary school demand. This tendency will be counterbalanced to some extent by the turnover of employees (bringing new, younger families into the area) and by births in the original in-migrant families.

14.5.7 Other social impacts – including cohesion

Other social impacts can be wide-ranging and may include:

- increased crime levels locally, particularly during the construction stage, associated with an influx of (typically) young male itinerant employees into the impact area;
- changes in the occupational and socio-economic mix of the population;
- linked to the above, problems in the integration of incoming employees and families with the local community and community activities. There may be a clash of lifestyles or expectations between incomers and the existing host community;
- there may also be cultural clashes, with different cultures and languages between the host community and the in-migrant population. This can be a particular issue in relation to Indigenous peoples.

An extensive literature concerned with the assessment of such social and cultural impacts is available, much of it written from a North American perspective. Further details are provided below and a wide range of further reading is provided on the website that accompanies this book. Prediction of such impacts is difficult but is likely to require at least a comparison of the predicted age, gender and occupational profile of in-migrants with that of the existing population in the impact area. The latter can be determined largely by reference to census data, as outlined above. Monitoring studies may be helpful in indicating the likely scale of certain impacts (e.g. see Glasson and Chadwick

(1995) and Glasson and Cozens (2011) for an assessment of the impact of a major construction project on local crime levels).

14.5.8 Displacement and resettlement

The focus of our discussion has been on the social impacts on a host community of an in-migrant population associated with a major project. But there may also be social impacts from displacement of the host population, with a population forced to move out from their roots. Land acquisition and involuntary resettlement are discussed fully in Chapter 15, but are also briefly noted here.

> Physical displacement refers to the loss of housing resulting from project-related acquisition of land and/or restrictions on land-use that require the affected persons to move to another location. Economic displacement refers to situations where people's houses are not directly affected but where there is a loss of other assets or access to assets (e.g. agricultural land) that will result in a disruption of livelihoods and associated loss of income.
>
> (IAIA 2015)

Mining projects, dams, and transport projects in particular are often associated with substantial population displacement (Owen and Kemp 2015, Scudder 2005). While many countries have legislation allowing this, in the national interest, in general terms displacement should be avoided or minimised by using alternative project designs. Where there is such displacement, there must be appropriate compensation. For physical displacement, this involves a formal resettlement process; for economic displacement, compensation may be more in the form of a raft of measures to restore livelihoods, including alternative employment and support services. The payment of cash to the displaced is generally seen as inappropriate, potentially leading to issues such as localised inflation and inappropriate expenditure. Of course, such compensatory measures, especially resettlement schemes, may also have their own impacts on the communities which are host to such resettlement schemes, and many of the impacts may follow the path of impacts resulting from an in-migrant influx, as discussed in this chapter. International standards on good practice for resettlement are set out in various publications, including IFC Performance Standard 5 (Land Acquisition and Involuntary Resettlement) (IFC 2012a).

14.6 Mitigation and enhancement

14.6.1 An overview of measures

A number of approaches to the mitigation of **population impacts** are available. The most basic would be to encourage the maximum recruitment of labour from within daily commuting distance of the project site, thereby reducing the number of employees and families moving into the impact area. Possible methods to encourage the use of local labour by developers and contractors were discussed in Chapter 13. In addition, during the construction stage, developer policies on travel, accommodation and relocation allowances might be used to influence the relative attractiveness of daily and weekly commuting versus relocation. Such policies might lead to some reduction in the proportion of in-migrant employees relocating and bringing families into the area.

The mitigation of local **accommodation impacts** is likely to involve attempts either to provide additional accommodation for the workforce or to encourage the use of unoccupied or under-occupied accommodation in the impact area. The provision of accommodation specifically for the workforce, in the form of purpose-built hostels/campuses or additional caravan accommodation, has already been discussed in §14.5.3/4. The success of such provision as a mitigation measure will depend on its attractiveness in relation to the alternatives available locally, in terms of location, facilities and cost. The release of unoccupied accommodation is rather more difficult to influence. During construction, one approach might involve the placing of advertisements in the local press, requesting those willing to provide workforce accommodation to contact the developer. This may alert potential providers of accommodation to the opportunities presented by the project. In some circumstances it may be considered desirable to encourage the use of tourist accommodation (e.g. to boost occupancy levels outside a short tourist season). This could be achieved by the compilation of a directory of local accommodation establishments by the developer, and its use by contractors and individuals seeking accommodation in the area.

Impacts on **local services** can be partially mitigated by the direct provision of certain facilities by the developer. Examples might include a medical centre and fire-fighting equipment and staff located on the project site, as well as recreational facilities for the workforce, thereby internalizing some of the impacts to the project itself. Developer funding of additional local authority provision necessitated by the project is also likely to be requested. The funding of local community projects may also be offered as partial compensation for the adverse impacts of the project. The voluntary provision of community-benefit funding by developers is widespread with certain types of project (e.g. for renewable energy schemes in Scotland, see for example Highland Council 2013).

14.6.2 Community Benefits Agreements (CBAs)

CBAs are becoming an increasing element in the assessment and development of major projects, especially energy projects. Such agreements provide a range of benefits offered less to compensate for specific projected local impacts, but more in recognition of the community's participation in an activity that is perceived as being "in the national interest", contributing to the achievement of national objectives (e.g. on security of supply, CO_2 reductions/climate-change policy), rather than meeting specifically local needs. However, in addition, for some large projects there are always likely to be some indirect local disturbance effects and changes in lifestyle which are less easy to address directly.

The nature and size of CBAs is evolving. In France, the *Grands Chantiers* programme has operated for several decades in support of major infrastructure developments, such as the EDF nuclear programme. Alleged levels of CBA funding for local communities in France vary from 1 per cent to 10 per cent of development value for a wide range of local projects – especially affordable housing. For Scottish wind farm projects, the payment is currently £5,000 per MW of power produced per annum, index-linked.

CBAs do raise a host of issues, including what types of benefits should be supported, for which beneficiaries, and how they should be managed. The types of benefits include: (i) financial incentives – annual payments, lump sums, or both; and (ii) social benefit measures in kind – including additional transport improvements, affordable housing, village halls, sports facilities, improved telecoms and training schemes. Management approaches include: the use of existing agencies (e.g. local authorities) and new agencies (e.g. community trust funds); various community empowerment measures; and equity shareholding schemes (e.g. in a renewable energy project) (Glasson 2017).

CBAs are also increasingly used in contexts where there are predicted substantial development impacts on Indigenous peoples, including the acquisition of their land or rights in land. IAIA (2015) notes six ways in which a company can contribute local benefits to a host community, including: social investment funding; local content (especially employment, as discussed in the previous chapter); shared infrastructure; capacity building; supporting community initiatives; and payment of royalties to local land owners or local authorities. Social investment may go into a specially managed social investment fund, into community infrastructure such as schools and hospitals, and into micro-finance schemes to enable local people to borrow money for local initiatives.

14.6.3 Social Impact Management Plans (SIMPs)

IAIA (2015) also notes the growing significance of SIMPs, and their requirement by governments and investors in the assessment and management of social impacts. For example, in Queensland since 2008, SIMPs are required for new resource projects as an integral part of the Environmental Impact Statement process (Holm et al. 2013). The Queensland Government's (2012) *Guidelines to Preparing a SIMP* see the role of a SIMP as being to "establish the roles and responsibilities of proponents, government, stakeholders, and communities ... in mitigating and managing social impacts and opportunities ... of major resource development projects."

They also set out categories of issues which the SIMP should address: workforce management (training and employment); housing and accommodation; local content; community health and well-being (social infrastructure); and cumulative impacts. Holm et al. (2013) see benefits from the preparation of SIMPs, including holding a range of stakeholders to account for the management of social impacts, and enhancing opportunities for social capacity building. However they also note some issues including the danger of using a "one size fits all" approach, lack of differentiation between impacts and opportunities, and some confusion over stakeholder roles. But there is general support for the use of SIMPs, and there are good examples in practice; see the Bowen Gas Plant Project SIMP (URS 2012) in Australia, for example, and Chapter 20.

Box 14.2 provides some examples of social impact mitigation and enhancement measures for three major projects in the UK, Australia and Canada, partly drawing on the authors' experiences of working on two of these. It also shows that a social licence to operate does not always lead to project implementation.

Box 14.2 Social impacts mitigation and enhancement

Housing impacts mitigation: proposed Hinkley Point C nuclear station, UK
At the time of writing (June 2016) Hinkley Point C in Somerset, England, is the proposed first new nuclear power station to be built in the UK for over 20 years. It is a large project – 3.2 GW via two reactors – with a substantial EIS. Socio-economic issues, particularly for the construction stage of the project, were central in the EIS. Site labour requirements for peak construction were estimated at 5,600, with approximately two thirds of the labour forecast to be in-migrant from outside the daily commuting zone (90 minutes for this project). A key issue was where would the in-migrant construction labour live. There was a concern to not overload the local accommodation market, which included both the usual owner occupation and rental provision, but also tourism accommodation (B&B/guest houses/caravan sites) as Somerset is an important tourism location (EDF 2011).

The mitigation response was an agreement by the developer to build several construction accommodation campuses, to take approximately 1,500 workers in total at peak construction. The location of the campuses reflected a balance of developer and local requirements, with one campus very near the construction site, as preferred by the developer, and the others on the edge of local towns, to provide some economic boost for local communities and also potential local legacy-accommodation facilities after the end of construction. To be attractive to the workers, the campuses had to be of a high standard, with en suite single rooms, good dining and leisure and sports facilities, all at a reasonable price. In addition, the campuses could also be hubs for medical provision, again taking pressure off local services.

Indigenous communities' social benefits: proposed Browse LNG plant, Western Australia

The Browse Basin, 400 km off the NW Australian coast, is a major natural gas resource. A development consortium established five options for commercialising the gas fields: three onshore sites, offshore and floating. It proposed a liquefied natural gas plant at James Price Point in the remote WA Kimberley region (Beckwith 2012). The location is subject to a registered claim under the Native Title Act (1993) by the Goolarabooloo Jabirr Aboriginal people, and the surrounding area has high levels of social and economic disadvantage. In this context, the Western Australian (WA) Government saw the project as delivering potential and substantial local socio-economic benefits to fragile local communities.

The regional Aboriginal land organization was given the responsibility by the WA Government for addressing 'Indigenous impacts' as part of the overall impact assessment process. Consultations with the Aboriginal people included a range of formal and informal meetings, as preferred by local groups. Significant outcomes included a number of legally binding agreements between the Traditional Owners, the developers and the WA Government. In particular, under native title agreements reached with the claimant group in 2011, the completion of the transaction would trigger a $10 million economic development fund and a $20 million fund for Indigenous housing.

However in 2013, the onshore project was abandoned by the development consortium claiming that Australia's rising costs made the project unviable. The project is now to be developed as a floating facility with minimal construction work to be undertaken in Australia, and with potentially only very limited impact benefits reaching WA and its Aboriginal people.

Indigenous communities' socio-economic benefits: Voisey Bay nickel mine, Labrador, Canada

The Voisey Bay Nickel Company submitted a proposal in 1996 to opencast mine the large nickel deposit at Voisey Bay in northern Labrador. In 1997, the federal and provincial governments, the Labrador Inuit Association and the Innu Nation signed a Memorandum of Understanding setting out how the potential impacts of the project would be reviewed. One of the key questions for the review was 'Would the Project bring social and economic benefits to many people in northern Labrador or to only a few, and would these benefits last?' (Government of Canada et al. 1999)

Discussion on Impact Benefits Agreements between the company and the Aboriginal peoples had commenced in the mid-1990s; following the publication of the panel report a number of agreements were reached. To reduce the environmental footprint of the project and give the Innu and Inuit time to develop the capacity to participate more fully, the daily limit of production was substantially reduced down from that preferred by the developer, from 19,000 to 6,000 tonnes per day, thereby extending the project life. Provisions were made to maximise Innu and Inuit male and female employment, including pre-employment training, scholarship trusts for post-secondary education, and including supervisory and management roles (O'Faircheallaigh 2013). Part of the stakeholder success has been in local procurement and employment. Now that it is operating, and since 2005 owned by Vale, 98 per cent of the mine and concentrator workforce is from Newfoundland and Labrador – of which 50 per cent are members of either the Innu or Inuit. Similarly of the early contracts awarded, 81 per cent went to Newfoundland/Labrador companies, with the majority going to Aboriginal companies (International Mining 2006). Because of the remoteness and extreme climate it is a fly-in-fly-out worksite, where Vale employees typically work for two weeks at a time. The camp complex offers good quality accommodation to the 450 employees, including a cafeteria, a full-size gymnasium, a squash court and a cardio and weight training room.

Vale is currently completing a detailed engineering study to support the development of an underground mine. Subject to approval, construction would be completed in 2019 and an additional 400 people would be employed at the mine and concentrator when underground mining begins.

14.7 Monitoring

Monitoring of demographic and social impacts is limited, other than for large-scale energy and resource development projects. Ideally, such monitoring should consist of three key elements. The first of these is the establishment of **administrative systems to ensure a regular flow of information on key parameters**, including at the very least the total numbers directly employed on the project and the mix of local and in-migrant employees. During most construction projects, the developer is likely to request this type of information from the contractors on site as a routine part of project management, for example to monitor earnings levels, bonuses and allowances across the construction site. The provision of such information can be made a contractual requirement. Existing monitoring systems can therefore often be used with only minimal modifications. For most projects, information on the operational workforce should be directly available to the developer via its own personnel records. However, this will not be the case for certain developments, such as business parks or retail projects, where several employers occupy the floor space provided by the developer. In such cases, the developer (or perhaps the local authority) may wish to establish data collection systems covering all occupants, with the submission of information being requested on a regular basis (see Chapter 20).

The systems described above will, at best, only indicate the total number of employees moving into the impact area. Information on the number of employees who bring families; the characteristics of these families; the type and location of accommodation taken up; and the use of local services can only be obtained directly from the workforce itself. The second component of any monitoring system must therefore be a **periodic survey of the project workforce**. This is likely to involve interviewing a sample of the workforce, with care taken to ensure a representative coverage of all types of employees. Such surveys can also be used to obtain information on other issues, such as workforce expenditure and journey to work patterns. Survey work of this type might be repeated on an annual basis, at least during the initial stages of the development.

The final element in any monitoring system should be the **monitoring of various social and economic trends within the impact area**. This can range from regular monitoring of house prices or rent levels, the amount of house-building, occupancy levels in local accommodation, school rolls, doctors' list sizes or crime levels. Such trends should be compared with those in suitable control areas, including the wider region or sub-region; comparison with national trends may also be appropriate. In addition, periodic surveys of local service providers (e.g. teachers or doctors) may provide a useful source of monitoring data.

Important issues in developing monitoring systems include: who is involved in developing the system, and who manages the system and the resultant monitoring information? Local participation is important, possibly with local authorities acting as proxies for the host community. As noted in the example in Table 14.1, local authorities and other agencies, such as the police, may be well-placed to provide some relevant data.

14.8 Conclusions

The focus in this chapter has been on social issues within the wider EIA process. A number of key concepts of particular relevance to all social impacts were first identified, including participation and free, prior and informed consent, human rights, and environmental justice. Within this context, the chapter considered particular social impacts, including population change (both growth and displacement), housing, and local services (such as education, health and emergency). Also considered were an important set of quality-of-life and well-being issues, reflected in social impacts such as crime; ill-health and deprivation; community stress and conflict; and equality and distributional issues. The discussion has outlined the broad characteristics of such impacts, with a particular focus on approaches to establishing the information baseline and to prediction and evaluation. Of course, practice may not always reflect the key concepts, and may not include the range of social impacts discussed here. Nevertheless, there is some evidence of promising change over time and that social impacts are becoming more important in the assessment of the impacts of major projects in a range of developed and developing country contexts.

Table 14.1 Example of key 'vehicles/procedures' for monitoring information collection

Vehicle/ procedure	Details/timing	Issues for consideration
Induction procedures; code of conduct sign-off	Employment requirement; captive audience; very useful for collecting much relevant data on: • worker name; • age; gender/ethnicity etc; • accompanying family members (if any); • address; accommodation type; previous address (if not local); • occupation category/trade; • former economically active status; • training; apprenticeship etc; • likely duration of job; • rate of pay; • characteristics of travel to site (e.g. mode; shared; etc.).	Should be able to capture information for all workers. May be premature to capture some information at induction stage (e.g.: new address – if moving into area [may have temporary address]; mode of travel [ditto]).
Specific post–induction workforce surveys	Full or sample surveys of workforce; useful for capturing information on: • worker expenditure on food, drink, fuel, entertainment etc.; • use of local services (e.g. sports facilities; libraries); • use of local social services (e.g. medical facilities; schools attended by any accompanying children); • updating of some of induction data (e.g. on accommodation).	Need to distinguish between local and non-local workforce. Make completion of post-induction survey a requirement at induction; questionnaire could be handed out to workers then, to be returned after, say, three months (with reminder, and incentives). Consider e-version response.
Others (e.g. regular reporting by relevant agencies)	Could, for example, include: • developer weekly returns on workforce on site; • developer monthly returns on contracts awarded; • main contractor monthly data on sub-contracts awarded; • local authority traffic flows monitoring on key access routes to site; • police data on incidents/crimes; • survey of community perceptions of local impacts (annual?).	To be organized on a regular reporting basis.

References

A wide range of further reading is provided on the website that accompanies this book.

Alkire S and G Robles 2015. *Multidimensional poverty index 2015: brief methodological note and results*. Oxford Poverty and Human Development Initiative (OPHI), Oxford University Department of International Development.

Anglo American 2012. *Socio-economic assessment toolbox, Version 3*. London: Anglo American.

Asian Development Bank 2008. *Social analysis for transport projects: technical note*. Mandaluyong City, Philippines: Asian Development Bank.

Asian Development Bank 2009. *Social analysis in private sector projects*. Mandaluyong City, Philippines: Asian Development Bank.

Asian Development Bank 2012. *Handbook on poverty and social analysis: a working document*. Mandaluyong City, Philippines: Asian Development Bank.

Asian Development Bank 2014. *Guidance note: poverty and social dimensions in urban projects*. Mandaluyong City, Philippines: Asian Development Bank.

Bass R 1998. Evaluating environmental justice under the National Environmental Policy Act. *Environmental Impact Assessment Review* **18**, 83–92.

Beckwith JA 2012. A social impact perspective on the Browse LNG Precinct strategic assessment in Western Australia. *Impact Assessment and Project Appraisal* **30**(3), 189–194.

Bureau of Minerals and Petroleum 2009. *Guidelines for social impact assessments for mining projects in Greenland*. Nuuk, Greenland: Bureau of Minerals and Petroleum.

Clark BD, K Chapman, R Bissett, P Wathern and M Barrett 1981. *A manual for the assessment of major development proposals*. London: HMSO.

DCLG (Department for Communities and Local Government) 2013. *Planning Act 2008: guidance related to the procedures for the compulsory acquisition of land*. London: DCLG.

DCLG 2016. *Index of multiple deprivation 2015*. London: DCLG.

EDF 2011. *Environmental statement for Hinkley Point C new nuclear power station*. See Planning Inspectorate (PINS) National Infrastructure website.

Esteves AM, D Franks and F Vanclay 2012. Social impact assessment: The state of the art. *Impact Assessment and Project Appraisal* **30**(1), 34–42.

Glasson J 2017. Large energy projects and community benefits agreements – Some experience from the UK. *Environmental Impact Assessment Review* **65**, 12–20.

Glasson J and A Chadwick 1995. *The local socio-economic impacts of the Sizewell B PWR power station construction project, 1987–1995: summary report*. Report to Nuclear Electric plc. Oxford: School of Planning, Oxford Brookes University.

Glasson J and P Cozens 2011. Making communities safer from crime: An undervalued element in impact assessment. *Environmental Impact Assessment Review* **31**, 25–35.

Glasson J, R Therivel and A Chadwick 2012. *Introduction to environmental impact assessment*, 4th Edn. London: Routledge.

Government of Canada et al. 1999. *Environmental assessment panel: report on the proposed Voisey's Bay Mine and Mill Project*. Ottawa: Government of Canada.

Highland Council 2013. *Guidance on the application of the Highland Council community benefit policy for communities and for developers of onshore and offshore renewable energy developments*. Inverness, UK: The Highland Council.

Holm D, L Ritchie, K Synman and C Sunderland 2013. Social impact management: a review of current practice in Queensland, Australia. *Impact Assessment and Project Appraisal* **31**(3), 214–219.

IAIA (International Association for Impact Assessment) 2015. *Social impact assessment: guidance for assessing and managing the social impacts of projects*. Fargo, USA: IAIA.

ICGPS (Interorganisational Committee on Guidelines and Principles for Social Impact Assessment) 1995. Guidelines and principles for social impact assessment. *Environmental Impact Assessment Review* 15(1), 11–43.

IFC (International Finance Corporation) 2009. *Projects and people: a handbook for addressing project-induced in-migration*. Washington DC: IFC.

IFC 2012a. *Guidance note 5: land acquisition and involuntary resettlement*. Washington DC: IFC.

IFC 2012b. *Performance standards on environmental and social sustainability*. Washington DC: IFC.

International Mining 2006. Voisey's Bay – newest jewel in Inco's crown. *International Mining*, Feb 2006.

Jackson S and AC Sleigh 2001. Political economy and socio-economic impact of China's Three Gorges Dam. *Asian Studies Review* 25(1), 57–72.

Kemp D and F Vanclay 2013. Human rights and impact assessment: clarifying the connections in practice. *Impact Assessment and Project Appraisal* 31(2), 86–96.

Koppelaar P 2008. *The Sociale index*. Rotterdam, the Netherlands: Rotterdam Statistical Office.

Low N and B Gleeson 1998. *Justice, society and nature: an exploration of political ecology*. London: Routledge.

Mackenzie Valley Environmental Impact Review Board 2007. *Socio-economic impact assessment guidelines*. Yellowknife, Canada: Mackenzie Valley Environmental Impact Review Board.

OECD (Organisation for Economic Co-operation and Development) 2013. *OECD guidelines on measuring subjective well-being*. Paris: OECD Publishing.

O'Faircheallaigh C 2010. Public participation and environmental impact assessment: purposes, implications and lessons for public policy making. *Environmental Impact Assessment Review* 30, 19–27.

O'Faircheallaigh C 2013. *Community controlled impact assessment, impact and benefits agreement and World Bank policies on Indigenous peoples: briefing paper for the Bank Information Centre*. Brisbane: Griffith University.

On Common Ground Consultants Inc. 2010. *Human rights assessment of Goldcorp's Marlin Mine*. Vancouver, Canada: On Common Ground Consultants Inc.

ONS (Office for National Statistics) 2015. *UK index of national well-being*. London: ONS.

OPHI (Oxford Poverty and Human Development Initiative) 2015. *OPHI country briefing Dec 2015: India*. Oxford: Oxford University Department of International Development.

Owen J and D Kemp 2015. Mining-induced displacement and resettlement: A critical appraisal. *Journal of Cleaner Production* 87, 478–488.

Oxfam and FIDH (International Federation for Human Rights) 2016. *Community-based human rights impact assessment: the getting it right tool – training manual*. Boston, MA: Oxfam America.

Petts J 1999. Public participation and EIA. In J Petts (ed.) *Handbook of EIA: Volume 1*. Oxford: Blackwell Science.

Queensland Government 2012. *Social impact assessment: guidelines to preparing a social impact assessment management plan*. Brisbane, Australia: Queensland Department of State Development.

Rodriguez-Bachiller A with J Glasson 2004. *Expert systems and geographical information systems for EIA*. London: Taylor and Francis.

Scott D and C Oelofse 2005. Social and environmental justice in South African cities: including 'invisible stakeholders' in environmental assessment procedures. *Journal of Environmental Planning and Management* **48**(3), 445–467.

Scudder T 2005. *The future of large dams*. London: Earthscan.

UN (United Nations) 2011. *Guiding principles on business and human rights: implementing the "protect, respect and remedy" framework*. New York: UN.

UNHRC (United Nations Human Rights Council) 2011 (internet). *Special representative of the Secretary-General on human rights and transnational corporation and other business enterprises*.

URS 2012. *Bowen gas project EIS: social impact management plan*. Brisbane: URS Australia.

Vanclay F 2003. International principles for social impact assessment. *Impact Assessment and Project Appraisal* **21**(1), 5–11.

White House 1994. *Memorandum from President Clinton to all heads of all departments and agencies on an executive order on federal actions to address environmental injustice in minority populations and low-income populations*. Washington DC: White House.

World Bank 2003. *Social analysis sourcebook: incorporating social dimensions into bank-supported projects*. Washington DC: World Bank.

Ziller A 2012. *The new social impact assessment handbook: a practical guide*. Mosman, Australia: Australia Street Company.

15 Land acquisition, resettlement and livelihoods

Eddie Smyth and Frank Vanclay

15.1 Introduction

It is estimated that around 15 million people are displaced by development projects annually (Terminski 2015). Major infrastructure and natural resource projects can have significant requirements for land acquisition, and in many cases require the resettlement of any resident households and communities. Competing pressures for land have made the process of acquiring land and people's houses a controversial topic, with many reported cases of community conflict and stalled or abandoned projects (Satiroglu and Choi 2015). There is much practical evidence from case experience that the majority of people displaced by large projects have a reduced well-being after the process (Downing 2002; Cernea 2003). Large projects are often justified as being in the national interest (Hanna et al. 2014), but as Cernea (2000a) argued: 'the outcome is an unjustifiable repartition of development's costs and benefits; some people enjoy the gains of development, while others bear its pains'.

High-income OECD countries tend to have more robust environmental and social-management regulations, institutions and civil society, and generally use compulsory purchase (expropriation) in more limited circumstances. In developing and low-income countries, with weaker governance structures, governments are often under pressure to make land available for projects at the lowest cost possible, resulting in an over-reliance on low rates of cash compensation, which results in the impoverishment of impacted communities (Reddy et al. 2015).

With the increasing involvement of international watchdog NGOs, there is growing pressure on companies and governments to develop projects in a socially responsible manner (Vanclay et al. 2015). Projects are increasingly being located in more challenging and sensitive areas, including in national parks and on the lands of Indigenous peoples, leading to conflict (Cernea and Schmidt-Soltau 2006; Hanna et al. 2016a,b). There is often a lack of realism about the effort, time and cost to undertake resettlement properly, and limited capacity of communities (especially those communities hosting resettlement sites), and governments to participate meaningfully in the process. Finding replacement land for impacted farmers is especially difficult because of competition for productive land with other land uses, including industrial agriculture and forestry (Richards 2013).

The international financial institutions (IFIs), notably the World Bank, the International Finance Corporation (IFC), and the regional Multilateral Development Banks (MDBs), are increasingly recognising the shortcomings of development-induced displacement and resettlement (DIDR), and have developed resettlement policies, procedures, guidelines and handbooks to address public and private sector projects (Price 2015). Key documents include the World Bank's *Operational Policy on Involuntary Resettlement* (OP 4.12) (World Bank 2001) and accompanying handbook (2004), and the IFC Performance Standard 5 on Land Acquisition and Involuntary Resettlement (PS5) (IFC 2012a), guidance notes (IFC 2012b) and handbook (IFC 2002, with a new version being developed in 2016). The United Nations' (2007) *Basic Principles and Guidelines on Development-Based Evictions and Displacement* outlines the human rights principles that apply to resettlement. In addition, the *United Nations Guiding Principles on Business and Human Rights* (United Nations 2011) will become an increasingly important document specifying the responsibilities of companies (Vanclay et al. 2015; Götzmann et al. 2016). Increasing expectations relating to the ethical practice of practitioners, especially informed consent, will also influence future resettlement practice (Vanclay et al. 2013).

This chapter provides an overview of the key considerations in the planning, implementing and monitoring the land access and resettlement required for projects. Perhaps the primary objective of the international resettlement standards is to minimise the amount of land to be acquired and the number of people requiring resettlement. Environmental and Social Impact Assessment (ESIA) practitioners contribute to discussions about the extent of land-take in several ways. They often determine the environmental buffers that are applied to projects. Too big a buffer means more resettlement, while leaving communities too close to project infrastructure can result in poor living conditions and unacceptable exposure to hazards. ESIA practitioners therefore need to engage fully with the resettlement planning process and other stakeholders to ensure that a balance is struck. Changing project specifications can significantly reduce the amount of land required and/or the number of people affected. For example, a change in the height of a dam wall will influence the area of land inundated and the number of people displaced. Thus, the design and assessment of project alternatives must involve social, technical, and environmental considerations.

15.2 Definitions, concepts, key guidance/standards

15.2.1 Negative consequences of resettlement

The negative consequences of resettlement on people have been much documented (Downing 2002; de Wet 2005; Vandergeest et al. 2007; Oliver-Smith 2009; Penz et al. 2011; Bennett and McDowell 2012; Mathur 2013; Satiroglu and Choi 2015; Terminski 2015). Michael Cernea (1997), the World Bank's first sociologist, analysed the consequences of resettlement and produced the Impoverishment Risks and Reconstruction (IRR) Framework, which identified eight key risks arising from displacement: landlessness; joblessness; homelessness; marginalisation; increased morbidity and mortality; food insecurity; loss of access to common property; and social disarticulation. Cernea (2000a,b, 2003) argued that by focusing on these risks, projects could ensure they did not arise, which would allow shifts in project outcomes:

1   from landlessness to land-based resettlement;
2   from joblessness to re-employment;
3   from homelessness to house reconstruction;
4   from marginalization to social inclusion;
5   from increased morbidity to improved health care;
6   from food insecurity to adequate nutrition;
7   from loss of access to restoration of community assets and services;
8   from social disarticulation to networks and community rebuilding.

15.2.2 International financial institution standards and the Equator Principles

The IFIs have recognised these impoverishment risks to impacted communities and also that there are business risks to investments from poorly-planned projects (Vanclay et al. 2015). Although each bank has developed its own standards and guidance on land access and resettlement, there is increasing convergence among them (Smyth et al. 2015). The World Bank's social and environmental safeguards are most commonly used by governments, while the IFC's Performance Standards on Environmental and Social Sustainability (IFC 2012a) are used for commercial projects.

Compliance with the IFC performance standards is also a requirement of the **Equator Principles banks**. The Equator Principles is a sustainability framework for the global finance industry. Over 80 of the world's leading banks have signed up, representing over 70 percent of international project finance in emerging markets (Equator Principles 2016). Thus, even if there is not World Bank or IFC funding, the IFC performance standards may still apply to a project because of the presence of an Equator Principles bank. Internationally, the IFC performance standards are seen as the industry 'gold standard' on land acquisition and involuntary resettlement (Reddy et al.

2015), and are the basis of much of the commentary in this chapter. The IFC (2012b) has produced guidance notes for each performance standard and a range of good practice or guidance documents, including on key topics such as resettlement (IFC 2002), stakeholder engagement (IFC 2007), participatory monitoring (IFC 2010a), in-migration (2009a), grievance mechanisms (CAO 2008; IFC 2009b), cumulative impacts (IFC 2013), local procurement (IFC 2011), and strategic community investment (IFC 2010b).

The IFC and the Equator Banks do not require use of their standards in high-income OECD countries, as these are deemed to have robust environmental and social regulations and institutions. In some OECD countries, requirements similar to the IFC performance standards exist, however, in other cases practice is sadly deficient. We believe that the IFC performance standards are a largely practical and fair set of policies and procedures that are intended to ensure appropriate development outcomes, and are generally appropriate everywhere. The key objectives of IFC PS5 are given in Box 15.1. These objectives are generally similar to those of other IFI guidelines, and can be regarded as the key principles underlying best-practice resettlement (Vanclay et al. 2015). The pressure applied by the international watchdog NGOs that monitor projects around the world has the effect of ensuring compliance with these international standards, and has further contributed to positioning the IFC's performance standards as the international gold standard.

Box 15.1 The key objectives of best-practice resettlement (IFC 2012a, PS5)

- To avoid, and when avoidance is not possible, minimize displacement by exploring alternative project designs.
- To avoid forced eviction.
- To anticipate and avoid, or where avoidance is not possible, minimize adverse social and economic impacts from land acquisition or restrictions on land use by (i) providing compensation for loss of assets at replacement cost and (ii) ensuring that resettlement activities are implemented with appropriate disclosure of information, consultation, and the informed participation of those affected.
- To improve, or restore, the livelihoods and standards of living of displaced persons.
- To improve living conditions among physically displaced persons through the provision of adequate housing with security of tenure at resettlement sites.

15.2.3 Voluntary vs involuntary resettlement

Resettlement is the planned process of relocating people and communities from one place to another. **Voluntary resettlement** occurs in situations where project proponents have no recourse to government expropriation or other enforcement mechanisms and people cannot be compelled to surrender their land (IFC 2012a). This could arise, for example, where Indigenous people have constitutional rights to their land and would only agree to projects on their land when they have given their **Free, Prior and Informed Consent** (FPIC) to the project (Buxton and Wilson 2013; Hanna and Vanclay 2013). Resettlement is considered to be **involuntary** either when it occurs without the informed consent of the affected persons, or in cases where they do not have the power to refuse resettlement whether or not they approve (IFC 2012a,b).

Because of their power of eminent domain, governments can expropriate (i.e. compulsorily acquire) land when projects are deemed to be in the public interest. In effect, this means that involuntary resettlement is the norm, because almost all large projects are deemed to be in the public interest. Even purely commercial projects are often deemed to be in the public interest because they may contribute to local or regional economic development (Hoops et al. 2015). Even though the power of eminent domain potentially can be applied to any project (and certainly to large projects), governments will often leave it to the developer to negotiate an agreement directly with the impacted communities. The IFC promotes negotiation as best practice. Forced relocation is bad for business, in that it leads to protest actions by the disgruntled impacted people, frequently to legal action, and inevitably to project delays (Hanna et al. 2016a,b). Thus, even if expropriation could be applied, a negotiated agreement will be a far more effective approach to undertaking the resettlement process. This practice is termed **negotiated involuntary resettlement** and is promoted by the international standards and is viewed as best practice. Even if there is a negotiated process, under IFC PS5, the conditions pertaining to involuntary resettlement still apply and must be considered.

In any situation of resettlement, whether or not expropriation is invoked, there will be compensation for those who are displaced. In best practice, resettlement will be a carefully considered process of providing people with replacement houses and ensuring a restoration and improvement in their livelihoods and well-being. There will also generally be cash compensation for the inconvenience experienced and for the value of any assets that might have been lost. In situations of worst practice, proponents have relied on expropriation and have paid the lowest possible amount of compensation in cash to affected persons, leaving them to fend for themselves. This has typically created much harm (Cernea 2003).

PS5 requires that any compensation is paid at **full replacement cost**, which is defined as the market value of assets plus transaction costs without depreciation being taken into account (IFC 2012b). A major impact on resettled people relates to the valuation of their assets and the extent of compensation they receive. Unethical projects have undervalued the assets of relocated people. PS5 requires that people be not made worse-off, and that they are fairly compensated.

The IFC standard requires that the project not only compensate for lost assets, but goes the extra step to ensure that impacted households are able to re-establish a livelihood, and ideally that their livelihoods are improved. IFC

PS5 (2012a) defines **livelihood** as referring to 'the full range of means that individuals, families, and communities utilise to make a living, such as wage-based income, agriculture, fishing, foraging, other natural resource-based livelihoods, petty trade, and bartering'. This requirement for livelihood restoration goes beyond simply paying fair compensation for housing, land and other assets, to taking responsibility to ensure that the impacted communities are not impoverished in the process of acquiring land to develop a project.

An important concept in resettlement practice is that people can be seriously affected by a project even when they do not have to be physically relocated. When people's livelihoods are negatively affected through, for example, their loss of access to land or other productive resources, or the reduced access of their customers to their business, this is called economic displacement. The IFC performance standards specify requirements both for when people lose their housing (**physical displacement**) and/or for when people's livelihoods are affected (**economic displacement**).

The policy and procedures for land access and resettlement therefore cover not just the purchase of land, but also any action that restricts access for livelihoods through, for example, creating or extinguishing rights of way or negotiating easements. The creation of fishing exclusion zones around ports or offshore platforms are examples of a livelihood restriction on fisherfolk that would require mitigation under PS5.

15.2.4 Resettlement Framework and Resettlement Action Plan

At the beginning of a project, when the exact nature and magnitude of the land access and restrictions on land use are unknown, PS5 requires that a **Resettlement Framework** be prepared. This Resettlement Framework outlines the proponent's general approach to implementing national legislation and addressing any gap with the IFC's performance standards. Subsequently, a **Resettlement Action Plan (RAP)** is developed, providing full details of the project, resettlement process, and likely impacts and mitigation measures. Where only livelihood impacts are anticipated, and no physical displacement, a **Livelihood Restoration Framework** is prepared, which is later expanded into a **Livelihood Restoration Plan** once impacts are known and mitigation measures developed. The RAP tends to be the overarching driving document in a resettlement process.

A key concept in the resettlement process is the **cut-off date**, which is the date prescribed under national law that permits the project to freeze further development on the land to be acquired by the project. There must be a complete census of people affected by the project and an **assets inventory** of their property as per this date. In principle, people must be compensated for all improvements that occurred before the cut-off date, but do not need to be reimbursed for any improvements made after that date. Significant problems and hardships can arise when there are project delays, and/or a significant amount of time between the cut-off date and the actual relocation.

A key risk for land acquisition and resettlement projects is **speculation** and opportunistic behaviour – where local people or other speculators acquire information on where the project is planned and they buy land and/or build structures and/or plant trees or crops in order to gain compensation and/or resettlement benefits. Speculation in land is very common in projects and can increase costs, delay the project, and cause conflicts in the community as speculators buy-out local people for low amounts and then seek to sell at much higher prices to the project. The extent of opportunistic behaviour, for example planting crops, can be considerable, and careful management is needed to keep it under control.

15.2.5 A framework for thinking about the social impacts of projects

Where significant land acquisition and resettlement is required, the level of complexity involved to manage the displacement of communities from their homes requires that the process be considered as a 'project within a project' using an experienced resettlement planning and implementation team. Given the number of disciplines involved in the resettlement process, including project management, environmental, social, planning, design, engineering, livelihoods, etc., coordination can be difficult. There is a need to communicate the process internally and to communities and civil society in a language that is accessible to all. The **Social Framework for Projects** (Smyth and Vanclay 2017) is a conceptual and practical model that assists in the identification of all impacts on communities and enables the effective communication of these impacts to all stakeholders (see Figure 15.1).

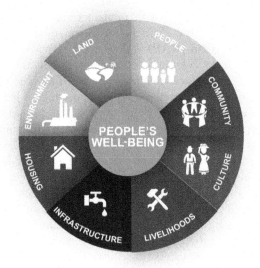

Figure 15.1

The Social Framework for Projects

At the core of the Social Framework for Projects is people's well-being. Individuals are used as the primary unit of analysis in recognition of the fact that there is considerable inequality within households and communities, and it is important to understand how some people are more vulnerable to project impacts than others. Well-being means having one's basic human needs met (e.g. adequate food and water, and being in good health) and having the ability to pursue one's goals, to thrive and feel satisfied with life (adapted from OECD 2011).

The Social Framework comprises the eight key categories that need to be considered in any project (see Box 15.2). These categories were derived from the authors' experience on several resettlement projects across the world (e.g. ICMM 2015). Our framework builds on a review of existing models, notably the Sustainable Livelihoods Approach (Scoones 1998), most of which are unnecessarily complex, unwieldy, and difficult to use in communication with other stakeholders (Smyth and Vanclay 2017). The Framework is consistent with current understandings in the field of social impact assessment (Vanclay 2002, 2003, 2012; Esteves and Vanclay 2009; Esteves et al. 2012; Vanclay et al. 2015; Mathur 2016).

Although the Social Framework is an overarching conceptual model that can be used with any project to assist in ensuring all key issues are considered, it can be used as a simple diagram, infographic or template to discuss the nature of the social issues created by the project with each stakeholder group (e.g. the affected communities, NGOs, clients). It can be used during the scoping process to enable each stakeholder group to identify what they consider is important. Each of the various social specialists (e.g. social, health, ecosystem services, human rights, RAP) can use the Social Framework to present their understanding of the positive and negative project impacts and their proposed mitigation measures. All these perspectives are combined together in a Master Social Framework which is discussed with key stakeholders enabling prioritisation based on the resources available.

Box 15.2 The eight categories of the social framework

People's capacities, abilities and freedoms to achieve their goals: The basic human rights, including family health and nutrition are the most fundamental needs to achieve a minimum level of well-being. The capacity of individuals to work inside and outside the household, their education and skills, determine how the household can exploit the livelihood resources available to it. Households with limited labour availability (e.g. that contain children, the elderly and sick) will be more vulnerable to project impacts and require special support. Women are often constrained in their freedom to fully engage in livelihood and community activities due to cultural constraints. Projects can affect people's aspirations and create fears and expectations about their future that may induce stress.

Community/social supports and political context: The household is generally part of a community, however it also exists in a social and political setting and is dependent on a combination of family, community, traditional and political networks, institutions and processes to gain access to land, housing, and livelihood resources. Communities are often divided and therefore it is very important to understand the 'politics of the project', as projects bring together a diverse group of stakeholders, each with their own agenda. If these agendas are not acknowledged, then it will be difficult to understand the true impacts of the project and the measures required to mitigate them. An understanding of the drivers of in-migration and out-migration is needed to analyse the changing context and the effects on community cohesion. The extent of safety and security in a community is a key well-being indicator and important in terms of whether people feel free to go about their daily lives. Having a free media and freedom of speech may determine whether meaningful consultation can take place. A community's past experience with projects and any legacy issues will also affect support for new developments.

Livelihood assets and activities: A household's stock of assets, including savings, food reserves and household goods is related to the household's resilience to shocks such as political instability or adverse weather. People depend on a wide range of livelihood activities to support their families, including land- and water-based, enterprise-based, and wage-based activities. They also may have supplementary livelihood supports, such as project compensation, savings, access to credit, rental income, remittances or pensions. Knowing what these are is necessary in order to restore livelihoods affected by the project. There is a vast array of informal and illegal activities, including corruption, drugs, illegal artisanal mining, sex workers, theft/crime, and smuggling, which need to be understood as these play a role in the local economic, political and security context.

Culture and religion: All societies have a shared belief system (of sorts) that frames their existence and provides psychological security. Religion and culture are important to the identity of a community and provide a basis by which households engage with and support each other. They provide insight into what the community considers important. Some societies have strong attachments to certain religious structures and shrines, which need to be considered in project planning. Tangible cultural heritage (e.g. archaeological sites) can also be affected by project developments. Intangible cultural heritage (e.g. language, oral history, music, dance and art) can be lost as a result of the social changes that accompany development. Indigenous people have a particularly strong attachment to their culture, which should be protected in accordance with their wishes.

Infrastructure and services: People's access to basic infrastructure and services such as health care, water and sanitation, energy, and social welfare is critical in determining their physical and mental well-being. Their further development and ability to exploit livelihood opportunities depends on access to education, communications, transport, agriculture and enterprise support, markets etc. The quality of both the physical infrastructure and the human resources needed to provide these services is important. Negotiating who is responsible for providing and maintaining the quality of services and infrastructure is necessary.

Housing and business structures: Having a house to live in is the most basic requirement of any household. The quality of buildings can have a big impact on the well-being of a family. Business and farm structures are necessary for conducting livelihood activities. Some households earn an income from rental properties, which also needs to be considered. Informal housing – i.e. where people build structures on land they don't own, or situations in which people are squatting – present particular challenges.

Land and natural resources: Access to land, water bodies, forests and other natural resources is necessary for conducting many livelihood activities. Such access is governed by community, traditional and political institutions, and secure tenure provides stability to enable investment and development. An in-depth understanding of land tenure arrangements that captures all the interests in land ownership and land-use is critical to the land acquisition process to minimize disputes and project delays. Projects can impact on access to land and natural resources, and the ecosystem services they provide, including crops, livestock, fish, wild food, timber, freshwater, medicinal plants, biodiversity, etc.

The living environment: Households and communities need a stable and clean environment in order to maintain their well-being. Any deterioration to the air, water or other quality-of-environment indicators can impact negatively on people's physical and mental health. Project impacts such as noise, dust, vibration, pollution, light or traffic detract from people's well-being. Aesthetic impacts are also important. Another dimension of the living environment is the way in which people rely on the weather for their livelihoods, for example, on seasonal rainfall. The impact of extreme weather events or longer term changes in the climate can have fundamental impacts on people's livelihoods. Projects need to understand the likelihood of extreme weather events and climate change and support the construction of housing and the development of livelihoods that can adapt to these changes.

15.3 Outline of the resettlement process

The act of planning and executing a resettlement is a major task. A timeframe of several years is required for each resettlement process. Depending on the number of people to be resettled and the context, the cost of the resettlement can be many millions of dollars. Errors in planning the resettlement can significantly increase project costs. Similarly, poor project planning can impact on the resettlement process. The resettlement process interacts with other aspects of planning the project, and especially with the impact assessment studies. What is done by the resettlement team and what is done by other project staff or in the EIA or Social Impact Assessment (SIA) needs to be negotiated. We note that the resettlement discourse tends to use many of the same terms that are used in SIA and EIA; however, the precise meaning of the terms may not always be the same.

In a chapter such as this, it is not possible to give full details of how to undertake a resettlement process, but we provide a brief description of the main components. More details are provided in Reddy et al. (2015). The key elements, which will be used as section headings for the rest of this chapter, are:

1   Scoping and initial planning
2   Profiling and baseline data collection
3   Development of the Resettlement Action Plan (RAP)
4   Implementation and handover
5   Livelihood restoration and enhancement
6   Monitoring and evaluation

There is no step called 'stakeholder engagement' because this is a fundamental element that is essential throughout the whole resettlement process. Typically a **Stakeholder Engagement Plan** is prepared which outlines the key stakeholders who have an interest in the project and how they will be engaged. A **Stakeholder Engagement Log** should be maintained to record all stakeholder engagement activities and any feedback received. The Stakeholder Engagement Plan will also outline the project's grievance mechanisms, showing how impacted people can lodge grievances and indicating how grievances will be resolved. Stakeholder engagement is an extensive topic that cannot be fully addressed in this chapter, however, the IFC (2007) has a useful guidance document, and many other manuals are available (Vanclay et al. 2015).

15.4 Scoping and initial planning

The extent of land acquisition and resettlement needs to be properly scoped early in the project so that the cost of compensation and mitigation measures is incorporated into the feasibility assessment of the project. It is also critical to understand the stakeholders' views about the project through effective community engagement to ensure that broad community agreement is secured to enable the project to proceed. Scoping land acquisition and resettlement should follow the general steps outlined below.

Establish a multidisciplinary team to assess the project commensurate with its scale and complexity. The social consultants need to coordinate the scoping and baseline data collection together with the environmental, resettlement and other experts. In many cases the socio-economic data collected for the ESIA process is not adequate for RAP planning due to poor coordination, leading to the need for resurveying and consequent project delays. It is important that the members of the project team have empathy with the impacted communities to ensure there is a genuine understanding of their situation and to facilitate their participation in the development of mitigation measures.

Review the project design by accessing all information available on the key environmental and social components of the project, including preliminary infrastructure designs. The project design often changes during the early stages, particularly as various options are considered. It is important that environmental and social experts provide input into the project design in order to minimise land acquisition and resettlement.

Review the existing primary and secondary socio-economic data available for the project area, including previous baseline studies and reports from other agencies, particularly any NGOs or universities that may have conducted research in the project area. Local, regional and national reports which describe the socio-economic context including development plans, the national census, and reports from the health, education, police and agricultural departments, are also important.

Get advice on the proposed **environmental buffers** from the environmental consultants on the ESIA team, and get their prescriptions about which land uses are not permitted in specific locations. For example, night-time noise buffers might determine that people cannot be resident within 100 m of a cement plant, whereas farming might still be permitted because day-time noise thresholds are higher. These buffers are important because sometimes in an effort to reduce costs or minimise resettlement, communities have been left *in situ* and exposed to excessive levels of dust, noise or other environmental nuisances, resulting in multiple complaints and demands for resettlement.

The scoping team should also **benchmark other projects** already implemented in the country to gain an understanding of how these have performed in terms of land acquisition and resettlement, and environmental and social performance. This will give the team an understanding of the issues that might arise and knowledge of whether there has been conflict around similar projects. The benchmarking will also provide information on the type of compensation paid and resettlement housing constructed, which may influence community expectations of the new project.

Develop project maps outlining the key features in separate GIS layers, e.g. project infrastructure and environmental buffer zones; project-impacted communities (including digitizing structures where possible); community infrastructure (water points, schools, health centres, etc.); key physical features such as roads, rivers and waterbodies; and key land-use categories, including arable land, pasture and forestry. Colour satellite imagery can be acquired and an initial estimate of the extent of resettlement can be based by digitizing the roofs of houses and estimating the number of structures. Imagery can also be acquired using unmanned aerial vehicles (drones), but this must be professionally processed to enable the imagery to be corrected, so that the scale is uniform to enable it to be used with GIS software. The use of drones will be evident to the community and must only be done where this has been approved by community leaders. Preliminary estimates can also be made of the areas of various crops to be compensated. It is important to acquire good satellite or aerial images for at least up to 20 km from the project area in all directions so that land-use mapping can be conducted to assist in the identification of resettlement site options.

Determine the **zones of project influence**, including the impacted communities in the primary and secondary zones of influence. The primary zone of direct impact comprises those areas where the project activities and facilities are located, including access routes, water and power supply networks, supply chains, employee accommodation, etc. The secondary zone is the larger region in which the project is located and that will encounter the main economic impacts from the project, including the influx of economic migrants, as well as employment and business opportunities.

The team should go on a **scoping visit** to the project area, being careful to tailor the level of engagement with local stakeholders to avoid creating expectations or encouraging speculative activities. The visit should follow the correct local protocols for community entry, particularly notifying the key community, traditional and political leaders. Because the final project footprint might change, the team has to be careful to not divulge too much information on the preliminary project designs and land acquisition areas, as this can cause unnecessary stress. The scoping visit gathers preliminary information on community organization and decision-making, settlement layout, housing and infrastructure, social services and livelihood activities. The scoping team should then **prepare a scoping report** which outlines how the land acquisition and resettlement process should proceed, including a staffing plan, work plan, budget and time schedule.

15.5 Profiling and baseline data collection

The Social Impact Assessment will normally have produced a social profile describing the affected communities. If done properly, the social profile should suffice for the needs of the resettlement team. However, if little thought has been given to the resettlement during the SIA, the resettlement team may need to gain more insights about the affected communities to understand what their likely concerns about the resettlement will be.

The establishment of a comprehensive baseline is critical in order to: evaluate the environmental and social impacts, risks and opportunities; determine eligibility for compensation and resettlement housing; and provide a pre-project baseline for monitoring the implementation of the RAP and any environmental and social-management plans. To ensure that they are included, the monitoring indicators for the project must be developed prior to collecting baseline data. The exact requirements of the baseline data collection are determined by the complexity of the project and the standards to which the project is adhering. Ideally, the baseline will have been compiled as part of the SIA, however the use of separate consultants, drawn-out processes, and sometimes unprofessional work frequently means that the available baseline data is inadequate and new data to suit the purposes of the resettlement will be need to be collected.

Prior to the cut-off date, a census of all persons and households in the impacted area must be undertaken. Depending on the stage of the project, it may be able to be combined with the socio-economic baseline collection. The census gathers basic demographic data (age, sex, relationship to household head, ethnicity, religion, etc.). The socio-economic baseline survey collects additional information on household access to social services, including healthcare, water, education, and skills training; and also information about livelihood activities including employment, agriculture, trading and services and natural resources. The census must record all people affected by the project regardless of their legal status – i.e. landowner, sharecropper, tenant or illegal squatter – because, according to international standards, the lack of a legal title does not disqualify people from resettlement assistance. As women are often impacted differently by resettlement, the census should disaggregate data by gender to inform the planning process so that women's interests and voices can be heard and considered.

Resettlement projects generate huge amounts of data that need to be quickly analysed and presented in ways that are meaningful and accessible to the project team. To ensure that opportunities for fraud are minimised and to respect the personal information of each affected person, it is important that the information be recorded and stored securely, with thought given as to who has access and for what purpose. The team should determine its database requirements well-prior to baseline data collection. The options range from Excel or Access databases to sophisticated SQL databases with integrated GIS that can manage the social, resettlement and environmental management processes for medium to large projects.

The project must record the assets that the impacted households and communities own or use, and will lose as a result of the project. Although adhering to international standards, the project will also need to meet the requirements specified in any relevant national legislation, which will be the basis of any future legal action. Therefore, the resettlement team must collect data and document its procedures in a manner that can be used as defence in court. Ideally, there should be recorded sign-off on all major actions by all local, government and expert authorities.

It is very important that any survey teams (e.g. building, land and crop surveys) be qualified according to any national standards and be overseen by national valuation experts, so that the process can stand up to independent scrutiny and possible legal action. The surveyors also need to undergo orientation training with the project social team prior to the survey process so that they are clear on how the assets will be recorded and the forms used. Supervision of the survey enumerators is essential to check the survey methodology in the field and ensure that the forms are checked each evening for quality control and error. The survey and valuation process should also be overseen and witnessed by government and community representatives in order to meet legal requirements and gain local support. The forms should have space to allow government and community witnesses to confirm that the survey processes have been explained to the asset owners in their own language and that their assets have been properly recorded. The asset owner, verified by appropriate ID, should also sign each form, or mark it with their thumbprint, in accordance with any national data protection guidelines.

Depending on the project context, various specialist studies may be required to provide an in-depth understanding and/or recommendations for supporting livelihood-restoration planning, e.g. agriculture, fishing, herding, hunting, artisanal mining, biodiversity/ecosystems services, etc. To avoid duplication, the specialist experts should be aware of the baseline data collection methods being used.

15.6 Development of the Resettlement Action Plan (RAP)

The RAP is the document in which the project proponent specifies the procedures to be followed and the actions taken to mitigate adverse effects, compensate losses, and provide development benefits to persons and communities affected by the project (IFC 2002). The standard components of a RAP are listed in Box 15.3. The RAP is a sub-plan of the overall Environmental and Social Management Plan (ESMP) for the project, and it is important that the ESIA and RAP teams jointly identify the impacts and mitigation measures. In actual practice, however, the ESIA is often developed earlier in the project development process in order to acquire permits, while the RAP comes later, after the project has been approved, thus limiting engagement between the two teams. It is critical that social and resettlement experts are involved at an early stage in project development so they can contribute to project design and planning. This allows for the key risks to be identified and avoided, reducing the need for costly mitigation or compensation.

The RAP is a working document that is developed over time. The ultimate version of the RAP will only be finalised during implementation of the resettlement. The initial version, however, must be developed early in the process. In effect, the Resettlement Framework required by the IFC is the initial draft of the RAP. The RAP records the outcomes of many participatory planning processes, including consideration of risk management, RAP consultations and negotiations, compensation arrangements, site selection, and the design of resettlement housing. These are discussed below.

15.6.1 Risk management

IFC PS1 requires the assessment and management of environmental and social risks, and also the adoption of the mitigation hierarchy to anticipate and avoid – or, where avoidance is not possible, to minimise and – where residual impacts remain – to compensate or offset for risks and impacts to affected communities and the environment as well as to workers. All impacts and risks related to

Box 15.3 Components of a standard Resettlement Action Plan

Project description including the specification of project buffers where no residences are permitted and/or where other livelihoods are restricted (farming, fishing, etc.).

Potential impacts including the zone of impact and alternatives considered to minimize resettlement. The magnitude of displacement should be described including numbers of persons, households, structures, community buildings, extent of land and crops affected, etc.

Legal and institutional framework: The relevant laws of the host country governing land access and resettlement, and a gap analysis against the IFC performance standards. This section should outline the legal basis for the cut-off date and any other key legal issues to be considered.

Baseline studies: Summary of studies undertaken including census, surveys, socio-economic studies, key people interviews, focus groups, etc.

Stakeholder engagement: Summary of public consultations and disclosures associated with resettlement planning, including engagement with affected households, local and/or national authorities, NGOs, and host communities. Description of grievance procedures including procedures for third-party settlement of disputes, community and traditional dispute settlement mechanisms, and judicial recourse.

Eligibility and valuation of losses and compensation entitlements: The methodology used to value losses under local laws and supplementary measures needed to achieve full replacement cost.

Livelihood restoration measures to improve and restore livelihoods of displaced people.

Housing, infrastructure, and social services: Plans to provide replacement housing, infrastructure, social services, including for host populations. The RAP should outline the special assistance to **vulnerable groups** – people who by virtue of their gender, ethnicity, age, physical or mental disability, economic disadvantage or social status may be more adversely affected by resettlement than others and who may be limited in their ability to claim or take advantage of resettlement assistance and related development benefits.

Organizational responsibilities to demonstrate that the resettlement team is made up of suitably skilled and experienced personnel.

Schedule covering all resettlement activities from preparation through implementation and how resettlement is linked to the implementation of the overall project.

Budget showing itemised cost of resettlement activities and the timetable for expenditure.

Monitoring, evaluation and reporting of resettlement implementation including performance indicators to measure inputs, outputs, and outcomes of resettlement activities. There should be provision for community involvement in monitoring, and external monitoring of the agreed resettlement indicators. There needs to be a discussion justifying the timing of the **Completion Audit**.

resettlement must be identified and integrated into the overall risk management process. For each risk, a mitigation measure or 'RAP package' must be developed, for example the design of resettlement houses, location of resettlement housing, and determination of compensation for lost assets. Using the Social Framework for Projects, the key risks for resettlement can be established (see Box 15.4).

The extent of resettlement impacts depend on the magnitude of the displacement, the characteristics of the project and its mitigation measures, as well as the characteristics of the community (Vanclay 2002). The impacts are usually assessed using a standard risk assessment process where the risks are categorised by contrasting the likelihood of the risk occurring with the consequences of that risk (Vanclay et al. 2015). For each risk, consideration is given to how to prevent the risk occurring, and what to do should the risk occur. Most projects implement a risk management system where a multidisciplinary team, including management, technical, environmental and social staff, identify the impacts and associated risks. The risks can be identified through experience with similar projects, benchmarking, specialist studies, industry best practice, expert judgement, as well as in participatory workshops with community groups.

15.6.2 RAP consultations and negotiations

Ideally, the project should obtain broad community agreement for the project. Thus, the approach to stakeholder engagement should be described in the RAP. The project team can negotiate individually with impacted households, or a group agreement can be negotiated. The IFC recommends the establishment of Resettlement Committees comprising representatives from the proponent, government, affected communities, host communities, community organizations and NGOs. It is important that these members are representative and have a legitimate mandate from the community. The project can support the process by ensuring that the impacted communities have independent advice and by providing capacity building to support them to elect suitably qualified representatives. The Resettlement Committee should be chaired by a competent chairperson with authority from all parties. The discussions should generally follow the spirit of free, prior and informed consent with the full disclosure of project plans and the establishment of clear agreements on the RAP packages, and with sign-off from the community and government representatives.

The Resettlement Committee needs to consider, and ultimately agree to, a range of issues that are addressed in the RAP. They need to consider, understand, discuss, and come to agreement on a range of matters, including: compensation arrangements, the planned resettlement site, the type of replacement housing to be provided, the services to be provided at the new site, and how the actual resettlement process will be implemented. Each of these topics is a major area of consideration and may take many hours of

Box 15.4 Typical social impacts arising from resettlement

People's capacities, abilities and freedoms to achieve their goals: The project land-take can result in a decrease in agricultural land, which together with in-migration will likely result in inflation in the cost of basic food, land and fuel, and subsequently cause food insecurity and increased morbidity, especially for vulnerable households resulting in human rights impacts. The living conditions of impacted people can deteriorate if quality resettlement housing is not provided.

Community/social supports and political context: The resettlement process can result in part or all of a community being moved into another ethnic grouping or political area. In some cases, only part of the community will be moved, resulting in a breakdown of social support and the social disarticulation of vulnerable households. The resettlement process can also result in the isolation of some households who are not moved with the main community.

Livelihood assets and activities: The loss of agricultural land and natural resources can result in landlessness, joblessness, and a loss of livelihood opportunities. The creation of jobs can benefit part of the community, but may create inequality within the community, and migrants often secure the majority of the employment or entrepreneurial opportunities.

Culture and religion: The land access can also result in the destruction and/or reduced access to cultural heritage, which can lead to a loss of identity and a lower cultural status.

Infrastructure and services: If the resettlement process is not properly planned, households can lose access to public infrastructure and services. When not properly maintained, the provision of some services, e.g. water supply and sewerage, can result in serious sanitation and/or health issues In the resettlement towns.

Housing and business structures: An over-reliance on cash compensation for housing can result in the impacted households spending their compensation and not having enough resources to rebuild quality housing. In many projects, resettlement houses are hastily constructed, resulting in substandard buildings that deteriorate rapidly.

Land and natural resources: The destruction of natural resources, often communally owned, is one of the neglected areas of resettlement planning. The loss of forests, pasture and fishing areas generally impacts vulnerable households the hardest resulting in a loss of sources of food and livelihood. A failure to understand and address individual and communal land tenure rights can result in land speculation, conflict, and/or the marginalization of vulnerable groups.

The living environment: If the resettlement is not planned properly with adequate environmental buffers, the result can be that residual residents remain, or resettled houses are built too close to project infrastructure, and thus having a poor living environment with impacts from dust, noise, blast vibrations, etc.

---

negotiation before agreement is reached. Some of these topics are discussed in more detail below.

It is important that the resettlement team allows the Resettlement Committee sufficient time to discuss these key issues and come to a decision in their own time about what they would choose, rather than forcing an artificial decision. The more 'ownership' (responsibility for decisions) the Resettlement Committee has, the greater the likely success of the resettlement process. Sometimes there will be differences of opinion between the Resettlement Committee and the project team. The resettlement staff may need to seek approval from the project management to increase what is being offered. For the negotiation process to work properly, there must be some capacity in the resettlement team to negotiate, implying some flexibility in constraints.

15.6.3 Compensation arrangements

The determination of compensation is a critical step in resettlement, and for many projects cash is still the main way of compensating for crops, land, buildings and other assets. The international standards require projects to use in-kind replacement rather than cash as much as possible, particularly for land and housing, as experience has shown that impacted communities tend not to invest cash compensation wisely and are often left landless and/or in substandard housing. The payment of large sums often distorts prices, leading to high local inflation and in rural areas encourages the transition from a subsistence to a cash economy. Projects must develop a compensation framework which outlines the rates of compensation and the eligibility rules. The compensation rates should be set by experienced valuers to determine the replacement value of each asset. Many countries have legally defined rates of compensation for crops and/or land. The valuer should assess the government-established compensation rates and adjust these as necessary to meet replacement rates. It is also recommended to support the impacted community in getting independent advice on what rates they should demand for their assets.

The IFC recommends that projects prepare an **entitlements matrix** to identify:

- Affected people (entitled persons) including property and land-right owners, tenants, squatters, sharecroppers, grazers, nomadic pastoralists and other natural resource users, vendors and other service providers, communities and vulnerable groups.
- All types of loss (impacts) associated with each category, including loss of physical assets, loss of access to physical assets, loss of wages, rent or sales earnings, loss of public infrastructure and elements of cultural significance.
- Types of compensation and assistance (entitlements) to which each category of people is entitled.

15.6.4 Site selection

Selecting a suitable resettlement site is a very important task and can have major ongoing social impacts for the resettled people. A bad site will lead to limited livelihood options and the potential impoverishment of the resettled people, and may also lead to action from human rights groups and other NGOs. This may lead to the need to resettle people again, which would be a big expense for the project. Choosing an appropriate site, however, is difficult as there are many competing demands for land, including urbanization and agricultural expansion, and finding available land can be very difficult. There will always be a host community which needs to be considered in addition to the needs of the people being relocated. Ideally, the project team would use the Social Framework to work through the following steps to identify and agree on appropriate resettlement sites.

Establish resettlement site-selection criteria. The first step in resettlement site selection is to generate a list of criteria to guide the process. The Social Framework can be used to workshop the site-selection process with project stakeholders, in language they can understand, to identify appropriate criteria. Project stakeholders must include community, traditional and political representatives, the elderly, women's groups, youth groups, ethnic minorities and disabled people. This workshop should record the initial site options proposed by these stakeholders.

Project design. The project team should establish the life-of-project land acquisition required and ensure that environmental buffers are correctly identified. It is important that buffers are determined so that resettlement site options are not located too close to project infrastructure, creating a poor living environment for the impacted communities. The project should discount resettlement site options which are in potential project expansion areas. For extractive projects, condemnation or sterilisation drilling should be undertaken to ensure that the resettlement site options do not contain any resources that would require resettlement again at some point in the future.

Land-use mapping and preliminary options assessment. The project should conduct land-use mapping of the wider project area (up to 20 km). This can be done using satellite and aerial imagery, and result in a map of the project, existing communities, community infrastructure, agriculture, forests, mountains, rivers, roads, community, political and traditional boundaries, etc. Using this map, the project can identify further site options, which can provide the extent of residential and agricultural land required based on identified criteria.

Preliminary options assessment. A preliminary assessment of water availability (quality and quantity), topography, drainage, and soil quality and stability of the preferred site(s) should be undertaken, to ensure it is suitable and feasible for the construction of resettlement housing and infrastructure, including a provision for future growth.

Present the preliminary options to stakeholders for discussion using the Social Framework to highlight the advantages and disadvantages of the different options. Stakeholders should be allowed to weigh up the often conflicting criteria and produce a shortlist. This shortlist should be subject to more detailed design and feasibility costing, and the final option(s) agreed with the project stakeholders.

15.6.5 Design of resettlement housing

The design of resettlement housing comprises several key steps. First, a baseline housing assessment in the project area should be conducted using a local architect to sketch both traditional and modern housing styles, noting how the houses are used for all house-based activities, including recreation (e.g. living rooms, verandas), sleeping, cooking, eating, bathing, toilet, and storage. Room measurements and building materials should be noted. The house plot should also be sketched, including the layout of kitchen gardens, children's play areas, fenced livestock areas, graves, fruit trees, etc.

Second, the resettlement team should engage with local people and gain an understanding of their preferred housing styles, including aspirations for modern house designs. In some countries there will be specific cultural issues associated with house designs which must be considered, such as the separation of rooms for men and women, kitchens or toilets being outside rather than inside, etc. Important considerations will be the provision of internal plumbing and electricity and sanitation. It is important to design houses that local people can maintain and expand, and which are durable and appropriate for local environmental and weather conditions. The project should also engage with local government and planning agencies on the national building standards and preference for house types. A report should be prepared with recommendations for housing sizes, styles and plot layouts.

Third, a benchmarking of other resettlement projects in the region should be undertaken, including an examination of the positives and negatives of the houses constructed at other locations.

Fourth, a draft housing policy should be developed, with proposals for house design and plot layout. Use the assets inventory to develop a basis for designing the resettlement-house categories. Ideally, there should not be any significant loss of space, and given the IFC's encouragement for improvement in well-being, some increase in total area of the house is not unreasonable. It is important to understand the implications of offering area-for-area or room-for-room (or a hybrid of both) in terms of loss of living space for the households and costs of construction for the company.

Fifth, these house design options need to be fully costed. Key cost areas that are generally underestimated are the cost of preparing the resettlement site and constructing roads and drainage. A workshop with the whole project team should be conducted to agree on the draft housing policy to propose to the communities.

Sixth, the draft housing policy should be presented to the Resettlement Committee and housing designs should be explained to the communities in feedback sessions.

Seventh, it is useful to construct life-size model houses of the design choices so that the communities can experience the houses first hand and provide input into the choice of materials, layout, etc.

Lastly, the final house designs should be prepared using feedback from the previous steps. Detailed drawings should be prepared, signed off by the company, and confirmed with the Resettlement Committee.

15.7 Implementation and handover

Implementation means putting the RAP into action. There are several steps during implementation: the construction of resettlement housing and related infrastructure; sign-off and payment of compensation; and moving and settling in. A handover process completes the physical resettlement process and support for livelihoods continues until restoration is achieved.

Construction of resettlement housing and related infrastructure This must be properly supervised to the agreed specifications and standards, in a realistic timeframe, to ensure that quality houses are provided. The housing must be ready before the community is required to move because the use of transitional or temporary housing is considered to be extremely poor practice due to the stress caused. The construction process can provide employment opportunities for impacted households and participatory community monitoring is recommended.

Sign-off and payment of compensation The impacted households should have the opportunity to review the RAP packages that have been agreed and applied to their particular situation. They should review and sign-off on house design choices, compensation for crops and other assets, and the location of their resettlement plot. The resettlement plot allocation process should include provisions for households to group together and select adjacent plots in order to maintain social cohesion. It is important that the impacted households have independent advice and that independent witnesses can explain the process in their own language and sign that the household was in agreement. The impacted households should be helped to set up bank accounts in the joint names of the household heads, in order to make payments securely and transparently.

Moving and settling in The start of resettlement moves can be marked by a ceremony appropriate to the local context, with the participation of the local community and political leaders. The impacted households should be provided with inspection visits during the construction process and sign-off on defects before the construction process is completed. The households should be provided with assistance to move their belongings to the new resettlement site, including the payment of disturbance allowances, where appropriate, to enable them to meet immediate needs while settling in. Medical support should be available for the vulnerable and elderly during the process and also to provide education/counselling to households where this need has been identified.

Handover process The project has a responsibility to assist a resettled community in becoming fully established in the new location. However, the company will want this responsibility to cease at some time (typically within a few years), and in order to manage risks and budgets it will want clarity about when that will be. In principle, a project's responsibility to a resettled community only ceases when livelihoods have been fully restored, which will be at least several years after resettlement to ensure that any issues associated with the new location have been identified. However, if the community has been resettled to a location where there are ongoing project-related impacts, such as exposure to pollution, its designation should change and it should continue to be treated as a project-impacted community for the life of the project, along with all other project-affected communities that have not been resettled. The handover process starts at the design stage for project infrastructure with the establishment of a Handover Committee involving project, government and community representation, and continues through the monitoring of construction and finally to the handover of responsibility for managing and maintaining resettlement site infrastructure. The Handover Committee comes to an agreement about when and which party will take over responsibility for the future maintenance of the resettlement site infrastructure, including roads, water and sanitation systems, schools, etc. Many issues related to handover are problematic. Governments will often require projects to pay for and sometimes provide essential services in the new community. Agreements about the handover of responsibility for these services from the project to the government or other delivery partner must be made when they are initiated. However, there are considerable examples of governments being reluctant to accept handover or of being incapable of paying for or delivering these services (van der Ploeg et al. 2017). Projects therefore need to be mindful of this possibility from the outset. Usually it is better for the community to take ownership or responsibility for ongoing services. Companies should partner with local delivery agencies as much as possible and avoid having direct responsibility for providing services.

15.8 Livelihood restoration and enhancement

The purpose of projects is generally to facilitate the economic growth of an area, which should bring improved livelihood opportunities to communities impacted by resettlement. However, experience has shown that many projects result in the impoverishment of impacted communities, usually because they have been poorly prepared to take advantage of the new economic opportunities. Livelihood restoration and enhancement is the most difficult part of resettlement because the project reduces the land available for livelihoods and communities may be resettled away from previous livelihood opportunities.

The IFC requires companies to 'improve, or restore, the livelihoods and standards of living of displaced persons' (IFC 2012a). IFC 2012b recognises that:

> Compensation alone does not guarantee the restoration or improvement of the livelihoods and social welfare of displaced households and communities. Restoration and improvement of livelihoods often may include many interconnected assets such as access to land (productive, fallow, and pasture), marine and aquatic resources (fish stocks), access to social networks, access to natural resources such as timber and non-timber forest products, medicinal plants, hunting and gathering grounds, grazing and cropping areas, fresh water, as well as employment, and capital.

Women are usually more adversely affected by, or vulnerable to, resettlement, because of the breakdown of social networks and reduced access to natural resources such as forests for the collecting of free firewood. Livelihood restoration requires the project to consider the collective resources that the community members depend on to maintain household well-being (as outlined in the Social Framework). It is important that all impacted communities (including the host community) are considered in relation to the community development support provided by the project, e.g. skills training, local employment and local procurement.

The key to successful livelihood restoration is to locate resettlement sites so that they enable access to economic opportunities for all the main groups in society, including women and men, young and old, skilled and unskilled, according to their aspirations. Training opportunities should be provided early in the process so that the affected people will have the skills needed in the project. Depending on the project context, social planners need to ensure that households have access to a mixture of land-based, wage-based and enterprise-based opportunities, although there will always be trade-offs, depending on where replacement land is available in proximity to the project and local urban centres. Specialist studies will provide recommendations for livelihood-restoration options and implementation partners should then be identified and assessed, including site visits to review their experience on the ground. Draft social-management plans should then be developed and workshopped internally so there is management buy-in to the budgets and schedules for implementation. The plans should then be discussed with project stakeholders and the most promising programmes should be piloted in the host communities before being scaled up and implemented. The livelihood programmes to be implemented will depend on the assessment of existing livelihoods, the aspirations of the impacted communities and the potential for the proposed programmes to succeed.

In rural areas, the IFC standards recommend land-for-land replacement, with security of tenure, to ensure a safety net for households, to guarantee food security, and to permit households who want to focus on farming to maximise the potential of their land for production and opportunities for processing and marketing their produce. People in rural areas in low-income countries often receive considerable income from non-farm activities, including remittances from relatives working away or entitlements from civil sector employment (Reardon 1997). Better-off households tend to diversify into non-farm business activities (trade, transport, etc.) while the poor tend to focus on casual labour, especially on other farms. The project should provide impacted households with support to acquire replacement land and also to prepare the land for production. Agriculture extension support will assist in improving productivity and storage, and the formation of producer groups to support the processing and marketing of their produce.

In urban areas, the majority of households typically depend on wage- and enterprise-based livelihoods. The informal sector can be very important to urban households and includes a wide range of services often run from home, including food processing and selling, retail, manufacturing, construction, services, and transport. Informal sector jobs are usually low-skilled, labour-intensive micro-enterprises employing small numbers of people who are often family members. The key to urban livelihoods is to maintain access to employment, markets and transport networks.

The resettlement sites should provide the opportunity for residents to access employment opportunities associated with the project or in nearby urban centres. The resettlement sites should be within the local employment zone for the project, and skills training should be provided to improve local people's access to jobs.

Where businesses are impacted, the project must provide support to re-establish the business at the resettlement site or a location of their choice. This will include reconstructing the business premises or providing accommodation to allow the business owner to construct the premises themselves. The project should also provide compensation for lost income during the resettlement process, including for employees. The local procurement plan should support the impacted businesses to provide goods and services to the project (see IFC 2011).

The IFC (2012b PS5) defines vulnerable or 'at risk' groups as:

> people who, by virtue of gender, ethnicity, age, physical or mental disability, economic disadvantage or social status may be more diversely affected by displacement than others and who may be limited in their ability to claim or take advantage of resettlement assistance and related development benefits.

IFC PS1 requires projects to identify vulnerable groups during the ESIA and provide special measures to engage with and support them through the resettlement process. The most vulnerable tend to be those with least assets, such as landless farmers or households with holdings less than 0.5 ha, no livestock, low levels of educational attainment, no savings, and limited social support

(Ellis and Allison 2004). The most vulnerable are often dependent on subsistence agriculture and labour on other farms. The concept of vulnerable households can be difficult to explain to a project-impacted community and they must understand that the project is not responsible for pre-existing poverty. However, most communities will have language to describe extremely poor households, and it is these households that need to be identified and carefully supported. Indicators developed during the baseline data collection are used to identify vulnerable households and design a **Transitional Hardship Programme** to support them through the resettlement process. For large projects, a community liaison officer or caseworker should be engaged to work closely with vulnerable households and to develop the transitional hardship programme, which should be closely linked to the overall social planning on the project.

The focus of the transitional hardship programme is to ensure that vulnerable households are provided with additional support to access the main livelihood-restoration and social-management plan benefits. Key areas of support can be: packaging their belongings for transport to their new house, including providing a team to salvage house materials where permitted, counselling so they can understand the short-term support on offer and make them aware of government services, and longer-term options. The restoration of social networks of family and friends is particularly important for vulnerable households.

An exit strategy must be developed for the transitional hardship programme with a time-limited plan for each household which outlines the support they will receive and what is expected of them in terms of using the support to improve their well-being. The support plan should be clearly explained to the household by the caseworker, in the presence of a community member, so that it is clear when the support will be phased out and what milestones are to be met along the way. A handover process to government or other agencies is also required so that vulnerable households do not develop a long-term dependence on the project.

15.9 Monitoring and evaluation

The monitoring and evaluation of RAP implementation is critical in providing feedback on problems during the physical process of resettlement, the payment of compensation, the effectiveness of consultations and the sustainability of livelihood-restoration activities. A monitoring and evaluation plan must be prepared as part of the RAP and should have three main components: performance monitoring, impact monitoring, and a completion audit. Monitoring the performance of RAP implementation depends on establishing appropriate indicators during baseline data design, which provide information on whether the programme is achieving its intended outputs, outcomes and impacts.

Key indicators include qualitative and quantitative information on the payment of compensation, resolution of grievances, community perceptions towards the project, changes in household income, agricultural productivity, and employment, etc. Ideally, a small number of key indicators should be selected rather than trying to track large numbers of minor indicators. It is important to have both internal monitoring and external expert monitoring to provide continuous feedback on RAP performance and adaptive management, so that issues can be resolved before these create community grievances and conflict. Performance monitoring will verify when measures have been implemented, and impact monitoring will assess whether the planned outcome has been achieved.

A **close-out audit** is required to assess if the RAP has been successfully implemented and that no further actions are required. The timing of the close-out audit depends on the complexity of restoring the livelihoods of the impacted communities in the project context and can take place three to five years after the land has been acquired, although experience has shown that complete livelihood restoration can take up to ten years (Smyth et al. 2015).

15.10 Conclusions

Resettlement is a major task and should be considered as a project within a project. The costs of resettlement, where this is necessary, will always be a significant budget item for a project. Done badly, the resettlement can be a major headache for projects, causing project delays, legal action, local and international protest. When resettlement is done well, with the project being committed to the use of professional experienced practitioners who have been engaged early in the project planning, the timing of the resettlement can be synchronised with other project tasks and costs properly considered in advance and included in the feasibility assessment of the project. Where the company is committed to the creation of shared value (Vanclay et al. 2015), resettlement can be a situation in which communities are given an opportunity for development (Perera 2014).

In the past, resettlement practice has often been performed badly, and companies have experienced project delays and increased costs because of poor planning and leading to the impoverishment of resettled communities. The international financial institutions and Equator Principles banks have policies and procedures to ensure better outcomes to which they require compliance from their borrowers. Although there are some differences in their requirements, there is much convergence and the International Finance Corporation's Performance Standards are positioned as the international gold standard for best practice. The IFC PS5 clearly specifies what is expected in project-induced resettlement.

Even with best practice, it is important to realise that people who are to be resettled will always experience upheaval, anxiety and stress, and will always have a degree of nostalgia about the past. Understanding and supporting people during this time of upheaval is essential.

There is a growing scholarship on resettlement and, with the various handbooks, there is a clear notion of international best practice. Training courses and summer schools are also available to improve the practice of resettlement. The growing attention being given to human rights around the world means that resettlement projects will need to be more mindful of the human rights implications of resettlement activities. Justifying compensation or RAP actions in human rights terms will likely be necessary. As land for resettled communities becomes increasingly harder to source, more attention will need to be given to host communities. While 'land-for-land' has been an important principle to ensure people's ongoing ability to maintain their traditional livelihoods, in the future much more focus will need to be given to transitioning people into new livelihoods.

References

Bennett O and C McDowell 2012. *Displaced: The Human Cost of Development and Resettlement.* New York: Palgrave Macmillan.

Buxton A and E Wilson 2013. *FPIC and the Extractive Industries: A Guide to Applying the Spirit of Free, Prior and Informed Consent in Industrial Projects.* London: International Institute for Environment and Development. http://pubs.iied.org/pdfs/16530IIED.pdf.

CAO (Office of the Compliance Advisor/Ombudsman) 2008. *A Guide to Designing and Implementing Grievance Mechanisms for Development Projects.* Washington, DC: International Finance Corporation. www.cao-ombudsman.org/howwework/advisor/documents/implemgrieveng.pdf

Cernea M 1997. The risks and reconstruction model for resettling displaced populations. *World Development* 25(10), 1569–1587.

Cernea M 2000a. Risks, safeguards and reconstruction: A model for population displacement and resettlement. In: Cernea M and C McDowell (eds) *Risks and Reconstruction: Experiences of Resettlers and Refugees.* Washington, DC: World Bank, 11–55.

Cernea M 2000b. Risks, safeguards and reconstruction: A model for population displacement and resettlement. *Economic and Political Weekly* 35(41), 3659–3678.

Cernea M 2003. For a new economics of resettlement: A sociological critique of the compensation principle. *International Social Science Journal* 55 (175), 37–45.

Cernea M and K Schmidt-Soltau 2006. Poverty risks and national parks: Policy issues in conservation and resettlement. *World Development* 34(10), 1808–1830.

De Wet C 2005. *Development-Induced Displacement: Problems, Policies and People.* New York: Berghahn Books.

Downing T 2002. *Avoiding New Poverty: Mining-induced Displacement and Resettlement.* MMSD Report 58. London: International Institute for Environment and Development. http://commdev.org/files/1376_file_Avoiding_New_Poverty.pdf.

Ellis F and E Allison 2004. *Livelihood Diversification and Natural Resource Access.* LSP Working Paper 9. Rome: Food and Agriculture Organization. ftp://ftp.fao.org/docrep/fao/006/AD689E/AD689E00.pdf.

Esteves AM and F Vanclay 2009. Social Development Needs Analysis as a tool for SIA to guide corporate-community investment: Applications in the minerals industry.

Environmental Impact Assessment Review 29(2), 137–145.

Esteves AM, D Franks and F Vanclay 2012. Social impact assessment: The state of the art. *Impact Assessment and Project Appraisal* 30(1), 35–44.

Equator Principles 2016. www.equator-principles.com.

Götzmann N, F Vanclay and F Seier 2016. Social and human rights impact assessments: What can they learn from each other? *Impact Assessment and Project Appraisal* 34(1), 217–239. http://dx.doi.org/10.1080/14615517.2015.1096036.

Hanna P and F Vanclay 2013. Human rights, Indigenous peoples and the concept of Free, Prior and Informed Consent. *Impact Assessment and Project Appraisal* 31(2), 146–157.

Hanna P, F Vanclay, EJ Langdon and J Arts 2014. Improving the effectiveness of impact assessment pertaining to Indigenous peoples in the Brazilian environmental licensing procedure. *Environmental Impact Assessment Review* 46, 58–67.

Hanna P, J Langdon and F Vanclay 2016a. Indigenous rights, performativity and protest. *Land Use Policy* 50, 490–506.

Hanna P, F Vanclay, J Langdon and J Arts 2016b. Conceptualizing social protest and the significance of protest action to large projects. *Extractive Industries and Society* 3(1), 217–239, http://dx.doi.org/10.1016/j.exis.2015.10.006

Hoops B, J Saville and H Mostert 2015. Expropriation and the endurance of public purpose: Lesson for South Africa from comparative law on the change of expropriatory purpose. *European Property Law Journal* 4(2), 115–151.

ICMM (International Council on Mining and Metals) 2015. *Land Acquisition and Resettlement: Lessons Learned.* London: ICMMs. www.icmm.com/document/9714.

IFC (International Finance Corporation) 2002. *Handbook for Preparing a Resettlement Action Plan.* Washington, DC: IFC. www.ifc.org/wps/wcm/connect/22ad720048855b 25880cda6a6515bb18/ResettlementHandbook.PDF?MOD=AJPERES.

IFC 2007. *Stakeholder Engagement: A Good Practice Handbook for Companies Doing Business in Emerging Markets.* Washington, DC: IFC. www.ifc.org/wps/wcm/connect/ 938f1a0048855805beacfe6a6515bb18/IFC_StakeholderEngagement.pdf.

IFC 2009a. *Projects and People: A Handbook for Addressing Project-induced In-migration.* Washington, DC: IFC. http://commdev.org/files/2545_file_Influx.pdf.

IFC 2009b. *Good Practice Note: Addressing Grievances from Project-affected Communities.* Washington, DC: IFC. www.ifc.org/wps/wcm/connect/cbe7b18048855348ae6cfe6a 6515bb18/IFC%2BGrievance%2BMechanisms.pdf?MOD=AJPERESandCACHEID= cbe7b18048855348ae6cfe6a6515bb18.

IFC 2010a. *International Lessons of Experience and Best Practice in Participatory Monitoring in Extractive Industry Projects.* Washington, DC: IFC. http://commdev.org/interna-tional-lessons-experience-and-best-practice-participatory-monitoring-extractive-industry.

IFC 2010b. *Strategic Community Investment: A Good Practice Handbook for Companies Doing Business in Emerging Markets.* Washington, DC: IFC. www.ifc.org/wps/wcm/ connect/f1c0538048865842b50ef76a6515bb18/12014complete-web.pdf?MOD= AJPERES.

IFC (in collaboration with Engineers Against Poverty) 2011. *A Guide to Getting Started in Local Procurement: for Companies Seeking the Benefits of Linkages with Local SMEs.* Washington, DC: IFC. www.ifc.org/wps/wcm/connect/03e40880488553ccb09cf 26a6515bb18/IFC_LPPGuide_PDF20110708.pdf?MOD=AJPERES.

IFC 2012a. *Performance Standards on Environmental and Social Sustainability.* Washington, DC: IFC. www.ifc.org/wps/wcm/connect/115482804a0255db96fbffd1a5d13d27/ PS_English_2012_Full-Document.pdf?MOD=AJPERES.

IFC 2012b. *International Finance Corporation's Guidance Notes: Performance Standards on Environmental and Social Sustainability*. Washington, DC: IFC. www.ifc.org/wps/wcm/ connect/e280ef804a0256609709ffd1a5d13d27/GN_English_2012_Full-Docu ment.pdf?MOD=AJPERES.

IFC 2013. *Good Practice Handbook on Cumulative Impact Assessment and Management: Guidance for the Private Sector in Emerging Markets*. Washington, DC: IFC. www.ifc.org/wps/wcm/connect/topics_ext_content/ifc_external_corporate_site/ifc+sus tainability/learning+and+adapting/knowledge+products/publications/publications_ha ndbook_cumulativeimpactassessment.

Mathur HM 2013. *Displacement and Resettlement in India: The Human Cost of Development*. New Delhi: Routledge.

Mathur HM (ed.) 2016. *Assessing the Social Impact of Development Projects: Experience in India and other Asian Countries*. Heidelberg: Springer.

OECD 2011. *How's Life?: Measuring Well-being*. Paris: OECD Publishing.

Oliver-Smith A (ed.) 2009. *Development and Dispossession: The Crisis of Forced Displacement and Resettlement*. Santa Fe, NM: School for Advanced Research Press.

Penz P, J Drydyk and P Bose 2011. *Displacement by Development: Ethics, Rights and Responsibilities*. Cambridge: Cambridge University Press.

Perera J (ed.) 2014. *Lose to Gain. Is Involuntary Resettlement a Development Opportunity?* Manilla: Asian Development Bank. www.adb.org/sites/default/files/publication/ 41780/lose-gain-involuntary-resettlement.pdf.

Price S 2015. Is there a global safeguard for development displacement? In: Satiroglu I and N Choi (eds), *Development-induced Displacement and Resettlement: New Perspectives on Persisting Problems*. Abingdon: Routledge, 127–141.

Reardon T 1997. Using evidence of household income diversification to inform study of the rural non-farm labour market in Africa. *World Development* 25(5), 735–748.

Reddy G, E Smyth and M Steyn 2015. *Land Access and Resettlement: A Guide to Best Practice*. Sheffield: Greenleaf.

Richards M 2013. *Social and Environmental Impacts of Agricultural Large-Scale Land Acquisitions in Africa: With a Focus on West and Central Africa*. Washington, DC: Rights and Resources Initiative. www.rightsandresources.org/wp-content/uploads/2014/01/ doc_5797.pdf.

Satiroglu I and N Choi (eds) 2015. *Development-Induced Displacement and Resettlement*. Abingdon: Routledge.

Scoones I 1998. *Sustainable Rural Livelihoods: A Framework for Analysis*, IDS Working Paper 72. http://mobile.opendocs.ids.ac.uk/opendocs/bitstream/handle/123456789/ 3390/Wp72.pdf.

Smyth E and F Vanclay 2017. The social framework for projects: A conceptual but practical model to assist in assessing, planning and managing the social impacts of big projects. *Impact Assessment and Project Appraisal* 35(1), 65–80.

Smyth E, M Steyn, AM Esteves, D Franks and K Vaz 2015. Five 'big' issues for land access, resettlement and livelihood restoration practice. *Impact Assessment and Project Appraisal* 33(3), 220–225.

Terminski B 2015. *Development-induced Displacement and Resettlement: Causes, Consequences, and Socio-Legal Context*. Stuttgart: Ibidem Press.

United Nations 2007. Basic principles and guidelines on development-based evictions and displacement: Annex 1 of the report of the Special Rapporteur on adequate housing as a component of the right to an adequate standard of living, A/HRC/4/18. www.ohchr.org/Documents/Issues/Housing/Guidelines_en.pdf.

United Nations 2011. *Guiding Principles on Business and Human Rights*, HR/PUB/11/04 www.ohchr.org/Documents/Publications/GuidingPrinciplesBusinessHR_EN.pdf.

Van der Ploeg L, F Vanclay and I Lourenço 2017. The responsibility of business enterprises to restore access to essential public services at resettlement sites. In: Hesselman M, A Hallo de Wolf and B Toebes (eds), *Socio-economic Human Rights in Essential Public Services Provision*. Abingdon: Routledge.

Vanclay F 2002. Conceptualising social impacts. *Environmental Impact Assessment Review* 22(3), 183–211.

Vanclay F 2003. International principles for social impact assessment. *Impact Assessment and Project Appraisal* 21(1), 5–11.

Vanclay F 2012. The potential application of social impact assessment in integrated coastal zone management. *Ocean and Coastal Management* 68, 149–156.

Vanclay F, J Baines and CN Taylor 2013. Principles for ethical research involving humans: Ethical professional practice in impact assessment Part I. *Impact Assessment and Project Appraisal* 31(4), 243–253.

Vanclay F, AM Esteves, I Aucamp and D Franks 2015. *Social Impact Assessment: Guidance for Assessing and Managing the Social Impacts of Projects*. Fargo ND: International Association for Impact Assessment. www.iaia.org/uploads/pdf/SIA_Guidance_Document_IAIA.pdf.

Vandergeest P, P Idahosa and P Bose (eds) 2007. *Development's Displacements: Economies, Ecologies and Cultures at Risk*. Vancouver: UBC Press.

World Bank 2001. *OP 4.12 Involuntary Resettlement*. Washington, DC: World Bank. http://go.worldbank.org/96LQB2JT50.

World Bank 2004. *Involuntary Resettlement Sourcebook: Planning and Implementation in Development Projects*. Washington, DC: World Bank. http://hdl.handle.net/10986/14914.

16 Health

•••

Marla Orenstein

16.1 Introduction

The way in which EIAs have addressed human health issues has evolved over time, from an initial focus on impacts to the biophysical environment to a more nuanced approach that includes a broad range of social, economic, institutional and environmental determinants of health.

The initial impetus for EIAs came from developments in the environmental movement that identified the potential for catastrophic effects on the health and well-being of people from anthropogenic changes in the biophysical environment. When the National Environmental Policy Act (NEPA) was passed by the United States in 1969, there was an explicit consideration of potential effects on human health. Health is mentioned six times in the Act, which states that the purpose of NEPA is to "attain the widest range of beneficial uses of the environment without degradation, risk to health or safety, or other undesirable and unintended consequences" (Human Impact Partners 2013).

Despite the legislation's original objectives, the practice of EIA became entrenched in looking at biophysical elements of the environment (e.g. air quality, water quality), rather than directly examining potential impacts to human health. While this attention to the biophysical environment ensured a certain degree of protection to people's well-being, this approach did not sufficiently address the potential for human health impacts.

In the 1970s and 80s, a number of widely publicized health disasters, such as the Love Canal and Bhopal catastrophes, turned public attention increasingly towards the potential health effects of human exposure to toxic chemicals associated with commercial and industrial developments. As a result, human health risk assessment (HHRA) began to be used as a tool within EIA/ESIA for evaluating potential health hazards resulting from the presence of toxic chemicals in the environment. HHRA is a four-step process intended to estimate the nature and probability of adverse health effects in humans who may be exposed to chemicals in contaminated environmental media (US EPA 2015). Currently, HHRA is included in many, if not most EIA/ESIAs.

However, over the past ten to fifteen years, there has been increasing acknowledgement that an HHRA is not sufficient to address the broad range of health concerns raised by stakeholders in relation to major development projects. While HHRA is an extremely useful tool for examining potential exposure to chemical contaminants, it does not address concerns around potential project effects on acute infectious diseases, injury, mental well-being or nutritional outcomes, for example. In response to this gap, the field of Health Impact Assessment (HIA) was developed. HIA focuses on a broader range of health outcomes than does HHRA. As described below, HIA uses a methodology similar to many other ESIA disciplines and is increasingly becoming the standard way in which community health concerns are addressed through the ESIA process.

16.2 Definitions and concepts

16.2.1 Defining "health"

An important first step in understanding issues relevant to community health is an attempt to define what "health" means. This can be surprisingly difficult. Different societies view health in different ways, and there is no standard definition of health that works in all circumstances. Most contemporary definitions of health acknowledge that good health comprises more than just an absence of disease; it includes physical, mental, and social well-being, and is affected by many personal, environmental, economic and societal factors.[1] Healthy people are able to cope with everyday activities and to adapt to their surroundings.

To a large extent, health is determined by where we live, the state of our environment, our income and education level, our jobs, and our relationships with friends, family and the larger community. These critical factors are often called **health determinants** (or determinants of health) because of their roles in shaping health in individuals and communities.

A widely used model describing the determinants of health was put forward by Dahlgren and Whitehead in 1991 (Figure 16.1). This model depicts the determinants of health as a series of multiple overlapping arcs. Age, sex and constitutional factors are biological or genetic characteristics of an individual and usually cannot be changed. Individual lifestyle factors refer to a person's behavioral choices, such as diet, smoking, exercise, wearing a seatbelt or practicing safe sex. Social and community networks are the supportive relationships that help buffer adverse conditions. Living and working conditions include factors such as education; jobs, income and working conditions; and housing. General socio-economic, cultural and environmental conditions describe social and physical environments that are often influenced by governments or other large social forces.

Dahlgren and Whitehead's model is particularly useful for showing that, while we can affect our health through the personal choices we make, those

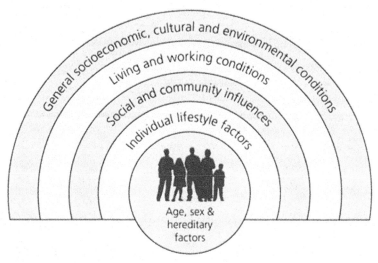

Figure 16.1

Model of health determinants (adapted from Dahlgren and Whitehead 1991)

choices are in turn influenced by the culture and environment we live in. The way the model is structured also implies that in order to achieve the greatest population health gains, action should be taken to address these upstream elements (CCSDH 2015). This upstream action includes the appropriate use of ESIA, as ESIA is one mechanism to protect or enhance these critical factors.

In a health impact assessment, "health" is generally considered to consist of these multiple dimensions that comprise and contribute to health among individuals and communities. This can be difficult to operationalize: a health impact assessment therefore needs to consider both **health outcomes**—specific types of illnesses and conditions such as infectious diseases, chronic diseases, injuries, nutrition-related conditions and mental well-being—and also the **health determinants** that influence the experience of health in affected communities.

16.2.2 Overlap and integration among ESIA disciplines

Because health outcomes are influenced so strongly by economic, institutional, social and environmental conditions, there is often a high degree of crossover and interdependence between those issues examined in an HIA and those that may be assessed by other disciplines, particularly socio-economic disciplines. This crossover is highlighted in Figure 16.2, which shows typical topic areas dealt with under the biophysical environment disciplines, socio-economic disciplines, and community health. A number of topics bridge different disciplines: for example, drug and alcohol misuse, health care infrastructure,

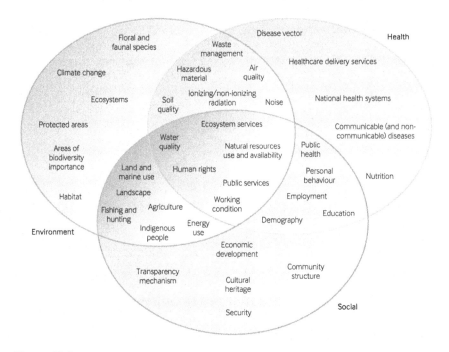

Figure 16.2

Examples of content areas for EIA, SIA and HIA (IPIECA-IOGP 2016)

housing, and subsistence activities could wind up being included as topics either under a community health assessment, or under a socio-economic assessment. Similarly, the health impacts of noise or odour, or changes to water quantity or availability may be included as topics under a specific environmental discipline, or as part of a community health assessment. In practice, these topics don't necessarily "belong" in one chapter vs. another; the choice of where they should be addressed can be flexible, and should be decided to best meet the needs of the ESIA readers.

Another important note is that a community health assessment may look at a topic that comprises a valued component for another discipline, but do so in a way that is different and more relevant to community health outcomes. For example, socio-economic disciplines may look at housing with respect to availability and the effect of increased pricing on housing affordability; community health may also look at housing, but in terms of the effect of overcrowding on disease transmission, or the effect of housing affordability on food security or homelessness. Similarly, water is usually considered as part of the biophysical environment in terms of water quality with respect to chemical contaminants. A community health perspective may also consider whether project-induced changes will result in waterborne disease transmission (e.g. resulting from the relocation of livestock or change in water sources). A traffic assessment may be conducted to estimate the volume of project-related traffic and its effects on road infrastructure; a community health assessment would also look for the potential for changes in traffic-related injuries and fatalities.

Ultimately, there is no single "best" way to identify which health issues should be examined under the umbrella of "health" or HIA, and which should be addressed in other ESIA chapters. The organization of the ESIA should respond to the specific project context and the needs of the ESIA readers. However, it is critical to ensure that health considerations are adequately and appropriately addressed in the document.

Over the years, different entities have taken different approaches to the integration of health considerations within ESIA documents. Several different approaches that have been observed by the author are listed below, along with the pros and cons of each.

a) Including health considerations only as secondary effects of primary ESIA analysis. For example, a discussion on human consumption of fish would appear in the ESIA section on fish and fish habitat; and a discussion on air quality mentions the implications to respiratory disease.

- Pros: it can be useful to read the analysis of changes in human health immediately following a discussion of changes in the health determinant that precipitate them.
- Cons: it can be very difficult to find information about human health, as it is scattered throughout a voluminous ESIA report. Second, these human health conclusions are often written by a discipline specialist who does not have sufficient public-health training. Third, important health topics are frequently missed, since they don't stem from any primary changes in the biophysical environment. And finally, health issues are addressed in isolation of one another, whereas in reality they may have a strong cross-influence.

b) Identical to the above, but providing a matrix showing where different health topics are addressed in the ESIA report.

- Pros: same as above.
- Cons: addresses the problem of not being able to find information as easily; all other cons remain the same.

c) Providing an HIA (either a stand-alone document or as part of the ESIA) that summarizes the analyses housed under separate disciplines.

- Pros: enables all the health information to be easily located and understood. Enables further discussion of health issues in relation to one another.

- Cons: may or may not address health issues that have not been considered under other ESIA disciplines.

d) Stand-alone HIA technical report.

- Pros: all health issues can be fully discussed. Not constrained by the format of the ESIA, which can sometimes be organized in a way that is difficult for "human disciplines" such as health and socio-economic considerations.
- Cons: health issues may not receive adequate attention in the ESIA' itself or in the discussion that comprises the regulatory process. After review by legal teams, the HIA results may not be made public.

e) HIA comprises one chapter of the ESIA.

- Pros: health issues have the chance to be fully explored. Health issues become an explicit part of the regulatory decision-making process.
- Cons: good narrative may be constrained by the format of the ESIA report.

Of the alternatives described above, d) and e) have the greatest potential to help project proponents, regulators and other stakeholders to really understand how the project is likely to influence health, and what actions may be taken to minimize potential harms and to enhance potential benefits.

No matter which approach is taken, it is critical to maintain strong integration between the health and other discipline teams throughout the ESIA process, and not only at the end when the report is being compiled (Birley 2003). Appropriate integration among discipline teams will assist with: appropriate scoping of topics to include in the ESIA; less burden on stakeholders if joint approaches to information gathering (e.g. focus groups) are used; cross-informing analyses of different topic areas; and—perhaps most importantly—"a set of integrated recommendations for safeguarding and enhancing the health of stake-holders, environmental quality, and social well-being" (Birley 2003).

16.2.3 "Inside the fence" vs. "Outside the fence"

As should be clear by this point, health impact assessment—like all other disciplines within environmental assessment—looks at the potential effects of a project "outside the fence": that is, on populations who live, work or play in communities near to and potentially affected by the project. Health impact assessment is not intended to look at health and safety among the project workforce itself. Occupational health and safety is a separate discipline: equally important, but not dealt with under the auspices of an ESIA.

Nonetheless, the line between the two can be blurry, as there are some issues, such as the fatality of a locally based worker, where health issues can spill over and affect community members who are not themselves workers. There are also some occupational health issues, such as malaria, where the best prevention for the workforce involves improving the status of health in the community at large. And finally, there are a large number of social health issues, such as sexually transmitted infections, in which the workforce acts as a "vector" for affecting disease rates in the local population. Despite these grey areas, the purpose of HIA is to identify and mitigate the impacts of a project on the health and well-being of local communities.

16.3 Key legislation, guidance and standards

For the most part, health impact assessment has not been entrenched in legislation at a national or regional level. Although health is often an important driver for the regulatory review of projects and the preparation of an ESIA, there are very few firm rules around when health must be considered, and to what extent. The exception is for HHRA: it is relatively common for regulatory guidance to require a project proponent to identify the potential for human exposure to chemical contaminants. However, in terms of requiring an ESIA to include a broader consideration of health outcomes and health determinants, regulations are seldom the driving factor.

There are exceptions. Mongolia has recently adapted its legislation to include "health" as part of the ESIA process (Byambaa et al. 2014). Similarly, Thailand, Korea, Ghana and a number of Australian states have—to varying degrees—entrenched HIA as part of the ESIA process (Kemm 2012). In these jurisdictions, HIA is used as a tool mainly for large development projects that are anticipated to result in extensive changes to the biophysical and social environments and thus potentially to result in substantial changes to health determinants and health outcomes. In all three regions, a mandated requirement for HIA was added to pre-existing Environmental Impact Assessment (EIA) legislation, rather than being developed as part of the original legislation. Within the United States, the existing NEPA legislation is, in some circumstances, being interpreted as requiring an explicit examination of human health impacts. The regulatory landscape continues to change and the change is often towards the inclusion of a broad range of community health effects as part of the ESIA process. Nonetheless, at the time of writing this is still the exception rather than the norm.

So why, if it isn't required by legislation, is health impact assessment ever conducted? There are a number of strong reasons, and there are different drivers for different circumstances and contexts.

The first of these is lending agency requirements. Lending institutions such as the International Finance Corporation, African Development Bank, Asia Development Bank and Inter-American Development Bank require that projects receiving the institution's funding include a health effects assessment as part of the environmental assessment process, to ensure sustainability and minimal impact on the environment and on local populations (e.g. AfDB 2003; IFC 2012). This consideration of potential health impacts is an explicit part of the IFC's Performance Standard 4: Community Health, Safety, and Security (IFC 2012). The objectives of Performance Standard 4 are:

- To anticipate and avoid adverse impacts on the health and safety of the Affected Community during the project life from both routine and non-routine circumstances.
- To ensure that the safeguarding of personnel and property is carried out in accordance with relevant human rights principles and in a manner that avoids or minimizes risks to the Affected Communities.

In practice, the implementation of HIA as part of ESIAs for projects funded by lending institutions is inconsistently applied. In an attempt to provide guidance, the International Finance Corporation has produced a document entitled *Introduction to Health Impact Assessment* (IFC 2009). This document, intended for government agencies, project proponents or HIA/ESIA practitioners, sets out technical guidance for both the process and the content of the HIA, with specific consideration of the types of issues that would be encountered in non-OECD countries (that is, those countries outside of Europe, North America, Japan/Korea and Australia/New Zealand).

A second driver for HIA is a strong business case. A number of multinational extractive industry corporations have internal requirements for conducting an HIA, even when there are no external regulatory or lending requirements applicable. For example, as part of its environmental stewardship efforts, Chevron incorporates environmental, social and health impact assessment in the planning and implementation of its capital projects to identify, assess and manage potentially significant impacts of the project (Chevron 2014). Other industry leaders, including Shell, ExxonMobil, BP, Barrick and Rio Tinto have also developed these internal requirements for HIA because they find that this approach minimizes their risk.

Several industry organizations promote the use of HIA to their members based on the business value it appears to provide. Both the International Council on Mining and Metals (ICMM) and the International Petroleum Industry Environmental Conservation Association (IPIECA) have developed guidance documents encouraging their members to use HIA as part of their ESIA approach. The International Council on Mining and Metals frames the business case for HIA in terms of its ability to:

- Identify and maximize the positive community health and well-being impacts and opportunities that a mining and metals project can bring.
- Identify, avoid and minimize, through changes to the project design and implementation, the unintended negative community health and well-being impacts that can arise.

- Identify existing community health problems, which could amplify the impact of a proposed project and affect its viability.
- Identify country-specific health regulations, which may affect the proposed project.
- Provide a process through which the project can work in partnership with local health, social care, and welfare services to jointly alleviate these health problems.
- Form one part of a broader involvement and engagement process that can build trust, draw out any community concerns and generate a dialogue about the best ways that the project can benefit local communities.
- Help to make explicit the potential trade-offs between community health and well-being and other economic, environmental and social objectives of the proposed project.
- Provide an equitable, transparent and evidence-based approach to planning and funding community health infrastructure and development activities, to protect and enhance sustainable local livelihoods.
- Help to jointly negotiate those aspects of community health and well-being that are the responsibility of the project and those aspects that are the responsibility of local government and local public services.
- Help to manage project sustainability and obtain a long term license to operate.

(ICMM 2010)

A third driver for the inclusion of HIA or a community health effects assessment is the desire of project proponents to get ahead of potential problems and concerns that may be raised by stakeholders, residents, local governments or intervenors in the permitting process. Increasingly, public focus is turning to potential health effects of major projects; HIA is a way to directly engage with these concerns at a time in the decision-making cycle when they can best be addressed without causing disruption or delay to the project.

There exists little in the way of absolute standards for how HIA must be conducted, in terms of either process or output. Several guidebooks and toolkits have been produced by lending agencies (e.g. IFC 2009) and industry organizations (e.g. ICMM 2010; IPIECA-IOGP 2016), as well as textbooks (e.g. Birley 2011; Ross et al. 2014) and practice standards (e.g. Bhatia et al. 2014). However, these documents provide only suggestions for best practice and do not represent an absolute or enforceable standard.

16.4 Screening and scoping for HIA

The methods used for assessing health considerations within an ESIA are parallel to those used for most other disciplines, such as social impacts, terrestrial ecology or air quality. As shown in Figure 16.3, these comprise

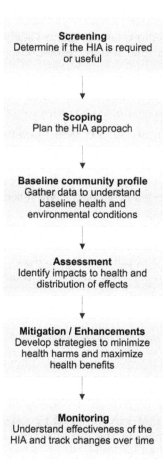

Screening
Determine if the HIA is required
or useful

Scoping
Plan the HIA approach

Baseline community profile
Gather data to understand
baseline health and
environmental conditions

Assessment
Identify impacts to health and
distribution of effects

Mitigation / Enhancements
Develop strategies to minimize
health harms and maximize
health benefits

Monitoring
Understand effectiveness of the
HIA and track changes over time

Figure 16.3

The HIA process

screening, scoping, development of a baseline community profile, assessment, development of recommendations, evaluation and monitoring, and reporting.

16.4.1 Screening

Screening refers to the process of deciding whether an HIA would be useful or required. Similarly, the decision of which elements, components or disciplines will be included in the HIA is made during the screening stage.

16.4.2 Scoping

Once it has been determined that an HIA will go forward, the **scoping** stage begins. During scoping, the project description is reviewed, and the geographic and temporal boundaries of the assessment are set. The ways in which the project could potentially interact with human health are identified, the health issues or valued components that will be carried forward for assessment are determined, and a priori definitions for characterizing and determining the significance of impacts are developed.

As described earlier, there is no single "best" way to categorize the health issues that should be examined within an HIA. Rather, the health areas of importance for a particular project should be determined by considering the elements of the project that act as drivers of change and ensuring areas of potential health impact are systematically identified.

Nonetheless, several guidance documents have attempted to set out an approach for categorizing health issues for the purposes of impact assessment. Tables 16.1 and 16.2 show different approaches taken by the International Finance Corporation and the International Council on Mining and Metals. In both cases, these frameworks work with a set of health outcomes and health determinants that are most likely to be potentially affected by extractive industry projects and other major development activities. The two examples differ in the way in which the specific health outcomes and health determinants (e.g. malaria, traffic injuries, employment/income) are combined into various categories.

16.4.3 Baseline/community profile

The third step of the assessment is to develop a **baseline community profile** in order to characterize current conditions relevant to health for the affected populations. In a baseline community profile, the following types of information are generally collected:

- Demographic information such as population size and distribution by age and sex.
- Information on health outcomes such as life expectancy, rates of specific diseases, rates of injury and mental health status.
- Information on health-related behaviors such as smoking, physical activity and diet.
- Information on social, environmental or institutional health determinants such as housing, access to health care services, income and education levels.

This information is usually gathered through a combination of secondary data sources such as local or national health datasets, reports, etc. and interviews with key informed sources both inside and outside the health sector.

Table 16.1 Health area categories used by International Finance Corporation (IFC 2009)

Health area	Examples
Vector-related diseases	Malaria, schistosomiasis, dengue, onchocerciasis, lymphatic filariasis, yellow fever, and so on.
Respiratory and housing issues	Acute respiratory infections (bacterial and viral), pneumonias, tuberculosis; respiratory effects from housing, overcrowding, housing inflation.
Veterinary medicine and zoonotic issues	Brucellosis, rabies, bovine TB, bird flu, and so on.
Sexually transmitted infections (STI)	HIV/AIDS, syphilis, gonorrhea, chlamydia, hepatitis B.
Soil- and water sanitation-related diseases	Giardiasis, worms, water access and quality, excrement management.
Food- and nutrition-related issues	Stunting, wasting, anemia, micronutrient diseases (including deficiencies of folate, vitamin A, iron, iodine); changes in agricultural and subsistence hunting, fishing, and gathering practices; gastroenteritis (bacterial and viral); food inflation.
Accidents and injuries	Road-traffic related, spills and releases, construction (home- and project-related) and drowning.
Exposure to potentially hazardous materials	Pesticides, fertilizers, road dust, air pollution (indoor and outdoor, related to vehicles, cooking, heating, or other forms of combustion or incineration), landfill refuse or incineration ash, and any other project-related solvents, paints, oils or cleaning agents, by-products, or release events.
Social determinants of health (SDH)	Including psychosocial, social production of disease, political economy of health, and ecosocial issues such as resettlement or relocation, violence, gender issues, education, income, occupation, social class, race or ethnicity, security concerns, substance misuse (drug, alcohol, smoking), depression and changes to social cohesion, and so on.
Cultural health practices	Role of traditional medical providers, indigenous medicines, and unique cultural health practices.
Health services infrastructure and capacity	Physical infrastructure, staffing levels and competencies, technical capabilities of health care facilities at district levels; program-management delivery systems; coordination and alignment of the project to existing national- and provincial-level health programs (for example TB, HIV/AIDS), and future development plans.
Non-communicable diseases (NCDs)	Hypertension, diabetes, stroke, cardiovascular disorders, cancer, and mental health.

Table 16.2 Health area categories used by the International Council on Mining and Metals (ICMM 2010)

		Examples
Health outcomes		
Infectious diseases		Malaria, HIV and influenza
Chronic diseases		Heart disease, cancer, bronchitis and asthma
Nutritional disorders		Malnutrition, vitamin deficiencies and obesity
Physical injury		Accidents, heavy metal and chemical poisoning and community violence
Mental health and well-being		Suicide, depression, stress and anxiety
Modifiable determinants of health		
Individual and family	**Physiological**	Vaccination status Nutrition
	Behavioral	Lifestyle and daily routines Physical activity Use of tobacco, alcohol and other drugs Acceptability of health services Risk-taking behavior
	Socio-economic circumstances	Income and wealth Education and learning Employment and economy
Environmental	**Physical**	Housing and shelter Transport and connectivity Exposure to chemicals Agriculture and food supply Land and spatial—air, water and soil
	Social	Community infrastructure Crime and safety Leisure and recreation Arts and culture Faith, spirituality and tradition Social capital and community cohesion
	Economic	Employment and economy Investment Access to goods and services Affordability of health services
Institutional	**Organization of health care**	Availability of health services Accessibility of health services Adequacy of health services Quality of health services

Table 16.2 continued

	Examples
Other institutions	Social care
	Police service
	Emergency services
	Judiciary
	NGOs
	Local government
Policies	Governance and public policy— industrial, health, transport, housing, etc.
	Private sector norms
	Third sector norms

In some HIAs that have been published as part of ESIAs, the approach has appeared to be "more is better"—that amassing large volumes of baseline data somehow validates the efforts of the HIA practitioner. This approach is rarely justified. Instead, the baseline community profile should be fit for purpose and should include only that information that is relevant and illuminative. Within an HIA, a baseline community profile can serve several distinct purposes:

- It helps with understanding the nature of the population in the area of the project: What is the population size and distribution? Are they rich or poor? Do they have many personal and societal resources and are they resilient to change, or are they highly vulnerable to environmental or social stressors? This information is essential for making accurate predictions and developing appropriate mitigation measures, since a project being built in suburban California is likely to have very different effects on the population than the same project being built in rural Mozambique.
- It helps with understanding the burden of disease in the population and the way in which health care exists or operates. Is the burden of disease in the population mostly chronic conditions like hypertension and cancer, or is there a large problem with mosquito-borne illnesses and nutritional deficiency? Are health care systems in place to take care of people? This is important for ensuring that the project does not exacerbate current difficulties and, where possible, leverages the opportunity to improve health.
- It helps you identify potentially vulnerable subsets of the population, such as infants and children, people with pre-existing disabilities, people who are food-insecure or those with specific medical conditions that may be challenged.
- It can help you to create a reference point for measuring or gauging future change in health status.

Some HIA practitioners (notably Francesca Viliani) have made a distinction between a **baseline** and a **community profile**. A community profile is intended to create an overall picture of community health to help understand the health context of affected populations. A baseline, in contrast, is intended to collect a limited set of replicable data indicators that will be monitored over time to identify change related to a specific project/policy. Whether a baseline or a community profile is more appropriate will depend on the intended use of the data. Currently, most HIAs appear to favor the community-profile approach, but use the term "baseline" nonetheless (Ross et al. 2014).

16.4.4 Ethical issues around health data collection

Some unique ethical issues arise in an HIA that don't appear in most of the other disciplines involved in an ESIA: these have to do with privacy and confidentiality of health-related data. Like all ESIA disciplines, HIA relies on gathering information about the local context in order to make predictions. In some cases, information on the health status of individuals may be gathered deliberately through the use of surveys or interviews of local residents. In other cases, interviews with key knowledge holders such as medical personnel may turn up information that is specific to the health of a single or small number of individuals. Unlike information about fish, vegetation or air quality, information about people's health is extremely sensitive. There is a substantial risk that this information could be used inappropriately; for example, HIV or immunization status has been used to discriminate against individuals in the hiring process; sensitive information has been accidentally released to a person's community or employer; and in some cases the health status of an individual has been unintentionally revealed because numbers are so small that people can be identified in other ways. This information release doesn't necessarily happen on purpose; nonetheless, when individuals' health information is held by a private company, the potential exists for it to be used inappropriately and/or unethically.

In many countries laws exist around the collection, storage and use of individual health information, in particular by government agencies or academic institutions. These laws are intended to protect the privacy of individuals and ensure anonymity when health data is used. Although these laws may not apply to the collection of data for the purpose of an ESIA, the same principles should be applied within an ESIA in order to achieve the same results of protecting individuals from privacy breaches. These include (but are not limited to):

- Obtaining ethics approval from an appropriate academic, governmental or non-affiliated institutional review board.
- Obtaining consent to participate from the individual from whom information will be collected. Note that the concept of "consent" here is different than that used for Free, Prior and Informed Consent (FPIC),

which refers to the consent of communities as a whole to projects that may affect their lands, rather than the consent of an individual for use of their private information.

- Not using collected data for any purpose other than that for which it was originally intended.
- Not collecting any information for which there is not a direct justification and "planned" use (i.e. not collecting something because it might eventually be useful or could be interesting).
- Only presenting health data in the aggregate and not at the individual level; and suppressing actual numbers when they are small enough that individuals could be identified (e.g. "two members of the village are HIV-positive").
- Keeping individual identifiers (name, date of birth, etc.) separate from the health data.

The ethical approach is to ensure that individual-level health data is not collected by, or directly given to, a private company such as a consulting company or a project proponent, or even others such as a lending institution. Ideally, this information should be collected in partnership with the local health authority, and results, as needed, provided in the aggregate to the ESIA writers. The health authority would maintain ownership, protection, control and access to the data.

16.5 Assessment for HIA

In the assessment, information from a variety of sources is used to help predict the **potential effects** of the project on the selected health issues. Sources of information include published literature, discussions with key informed sources, information gleaned from engagement with various stakeholder groups, and evidence gathered from the effects of past projects in the area. Importantly, the assessment will look at both the potential for positive health benefits to accrue, as well as the potential for the project to result in adverse health consequences. While potential adverse effects can be easily imagined (e.g. increases in exposure to contaminants; decreases in subsistence food sources; increases in sexually transmitted infections due to the presence of moneyed mobile workforces), it can be more difficult to picture how potential health benefits may result from major development projects. Table 16.3 shows several examples of health benefits that may arise from large infrastructure-based projects such as mining, oil and gas development, or dams. These are examples only and would not arise in all cases: as with any other issue in ESIA, details of the project and the local context are critical in identifying the nature of effects that are likely to occur.

There is no single "magic bullet" approach to accurately identifying the likely effects of the project on specific health outcomes. Similarly, there are

---

Table 16.3 Examples of health benefits from development projects

Project activity	Potential positive community health outcomes	Explanation
Building/ upgrading roads	Reduced fatalities and injuries	In some cases, building or upgrading roads can result in improved access for emergency vehicles such as ambulances, firefighting equipment or police. The timely arrival of these emergency services increases the chances that an adverse event, such as a collision, fire, domestic assault or heart attack, will not result in fatality.
Occupational health strategy	Reduced rates of infectious disease	All companies employ an occupational health strategy to minimize health harms among their workforce and maintain productivity. For some diseases the most effective approach to protecting the workforce involves improving conditions in the surrounding community. For example, malaria is best reduced by mosquito abatement and the use by all members of the community of bed nets treated with insecticide. HIV among workers is best prevented through a community-wide education and transmission prevention approach. These activities, although instigated by occupational health concerns, can result in lower infectious disease rates for the local community as a whole.
Jobs and income	Multiple physical and mental well-being outcomes	Almost all major development projects promise (and ideally result in) the creation of local job and procurement opportunities. Jobs and income are considered to be the most influential of all health determinants. Specific health outcomes that have been associated with income include birth weight and infant mortality; self-rated health; adult mortality; chronic and acute infectious diseases; mental well-being; social pathologies; and health service utilization. To the extent that a project provides employment to people who were previously underemployed or unemployed, or whose employment conditions may improve from their previous job, improvements in these health outcomes may occur. Benefits extend not only to the person employed, but also to their family members.

very few topic areas for which quantitative models have been developed that can be applied in the context of a major development project. Rather, the assessment of most health areas is approached in a qualitative manner. This qualitative assessment is, however, based on a variety of sources that include peer-reviewed research literature, key-informant technical interviews and monitoring studies. This allows the development of evidence-based conclusions that provide valuable information to decision makers—identifying avoidable risks and providing a suite of recommendations to ensure that adverse health effects are mitigated and beneficial health effects are enhanced.

Due in part to the uncertainty inherent in qualitative predictions, it is particularly important to ensure that strong definitions are developed to characterize the anticipated effects. It is not sufficient merely to provide alternatives of "low," "medium" and "high" for parameters such as likelihood, severity, magnitude, etc. Rather, descriptions must be developed that transparently allow the ESIA reader to understand why the conclusions were reached. An example of appropriately specific definitions are shown in Box 16.1.

Box 16.1 Example of severity ratings from Alaska HIA technical guidance

Low:	Effect is not perceptible.
Medium:	Effect results in annoyance, minor injuries or illnesses that do not require intervention.
High:	Effect results in moderate injury or illness that may require some intervention.
Very high:	Effect results in loss of life, severe injuries or chronic illness that requires intervention.

(State of Alaska HIA Program 2011)

In order to highlight some of the health issues and findings that can occur within an HIA, Box 16.2 presents an overview of the results of an HIA that was conducted on a project to repair and upgrade the Nacala Dam in Mozambique (Divall et al. 2010). As shown in this example, a number of important health issues emerged that would not necessarily have been examined if a health assessment had not been conducted.

16.6 Mitigation/enhancement and monitoring for HIA

In this stage, appropriate **mitigations and enhancements** are determined, with the objective of minimizing potential adverse health effects and enhancing health benefits, where possible. These mitigations/enhancements may take the

Box 16.2 HIA case example

[This case example was summarized from the original HIA report (Divall et al. 2010) by the chapter's author.]

Project description and context:
Nacala Dam is the principal water source for Nacala City, Mozambique. However, due to the collapse of the bottom outlets of the dam, the Nacala Dam is not able to meet either the current or future water needs of Nacala City. An infrastructure project to repair and upgrade the dam was proposed.

The beneficiaries of the project are the population living in Nacala City, about 35 km from the dam, who will experience improved water supply. The populations in the small communities near the dam will experience negative impacts of the dam development activities. This population is mainly poor, with high levels of malaria and infectious diseases. Few houses have access to safe drinking water or sanitation. Subsistence farming is the dominant occupation, and many people fish for food.

Health issue	Existing conditions	Anticipated effects	Potential changes in health outcomes
Communicable diseases	Housing is generally basic, poorly ventilated, and overcrowding is common. TB & HIV are major public health challenges.	The project may increase risks mainly through potential in-migration. Makeshift housing and housing inflation may increase overcrowding. The construction camp workforce accommodation could place pressure on local housing through potential for rental income.	Increased transmission of TB and other viral and bacterial disease spread.
Malaria	Malaria is the biggest public health threat. Knowledge and health-seeking behavior is adequate but not consistent, and prevention activities are limited and rare.	Project will increase and decrease likelihood of vectors in different areas, but is unlikely to play a significant role in the transmission of malaria in the short to medium term.	Possible change in malaria rates.

Health issue	Existing conditions	Anticipated effects	Potential changes in health outcomes
Access to safe drinking water	Limited access to improved water supplies has made diarrheal, eye and skin diseases common.	There is the potential for inequality in services if the intended recipients (Nacala Port) receive all the benefits of improved water supplies and the rural communities (Muhecule) do not.	Health impacts, including diarrheal, eye and skin diseases.
Sanitation and waste management	Few sanitation facilities, and indiscriminate defecation. Sharing of latrines is common. Soil-transmitted helminths (parasitic worms) and schistosomiasis, linked to sanitation and waste, are highly prevalent.	Decaying organic material, algae, increased pollution in the dam water are potential impacts. In-migration and pressure on limited services may have an impact.	Health impacts, including exacerbating of existing helminths and schistosomiasis problems.
Sexually transmitted infections and HIV/AIDS	HIV/Aids and STIs are a significant problem. Opportunistic sex work is common and accepted practice, with women and children particularly vulnerable. Muhecule serves as truck stop for drivers to Nacala Port.	More disposable income and in-migration may lead to an increase in casual sexual relationships and support existing commercial sex activities.	Increases in sex work and accompanying health impacts, including HIV/AIDS and STIs.

Health issue	Existing conditions	Anticipated effects	Potential changes in health outcomes
Malnutrition and food security	Lack of food and availability of a balanced diet, poor feeding practices, and malnutrition are a concern: local nutritional support programs aim to address this. Access to land is important to sustaining local livelihoods.	Rising dam levels and access to land may affect current wetland fields. Downstream communities rely on water supply for manual crop irrigation.	Malnutrition and food insecurity.
Accidents/ injuries	The community is located on the main road between Nacala and Nampula and is thus exposed to high loads of traffic.	Construction traffic, slow moving vehicles, and use of heavy machinery in and around the community pose risks. There is a risk of dam failure with the potential for flooding in downstream communities.	Road traffic accidents, other accidents/injuries related to machinery, health impacts of flooding.
Air pollution, noise and odors		Exposure to dust (PM10) is the major air borne pollutant risk from vehicle entrained dust. SO_2/CO from construction vehicles, and chemical risks are unlikely.	Health impacts.
Domestic violence, alcohol and drugs	Domestic violence is a challenge in the area linked to alcohol and substance abuse.	Disposable income and influx of workers may increase domestic violence.	Health impacts from domestic violence, with women and children as victims.

Health issue	Existing conditions	Anticipated effects	Potential changes in health outcomes
Social cohesion and well-being		Resettlement and ability to access different communities after increased reservoir size may lead to social disruption. Influx may play a role.	Health impacts including mental health consequences and social pathologies.
Health systems	Adequate health services at local level. Limited referral capability results in delays for transfers.	In-migration may stress the system. Workplace accidents and physical risks are most significant, and limited emergency response activity and health care services to address this.	Health impacts stemming from long waiting periods for care, possible decrease in quality of care.

Examples of proposed mitigations:

- To reduce infectious disease transmission, avoid hiring from locations with higher TB prevalence and potentially drug resistant TB strains (China, Philippines, South Africa and southern Mozambique).
- Develop community and occupational health and safety programs addressing HIV/AIDS, TB, STI, and malaria policy/programs; influx management; emergency response planning; code of conduct for engagement with local community; and a workplace medical support plan.
- The project will need to develop its own water, sanitation and waste management systems. These need to be of sufficient capacity so as not to impact on the community water supplies. Provide camps and work areas with proper and sufficient toilet facilities.
- Perform a baseline parasite prevalence survey in the communities. This should ideally occur at the end of the rainy season, and must be followed up at similar times of the year to ensure consistent comparisons. This will determine the burden of disease in the community, and also serve as an indicator to monitor the impact of the disease and interventions.

form of management plans, worker policies, recommendations for working with external agencies, or other forms. Good public-health practice principles should be employed, with the understanding that it is more effective to prevent adverse health consequences than to treat problems after they arise (Public Health Leadership Society 2002). The emphasis of mitigations should be on preventing or avoiding harm, rather than managing its consequences. The "mitigation hierarchy" states that the preferred order of addressing potential impacts is:

- To avoid the impact altogether.
- To minimize the impact (through decreasing the duration, severity, extent, etc.).
- To repair the adverse consequences of impacts.
- To compensate people for impacts that cannot be avoided or mitigated.

Importantly, the development of mitigations must ensure protection for the most vulnerable members of the community, and not merely for the community's "average experience." In this, the baseline community profile is useful, as these vulnerable groups should have been identified early in the HIA process.

As part of mitigation, a plan for **monitoring** may be developed. Because health issues are tied so closely to other social and environmental issues addressed in the ESIA, the health monitoring plan may be developed as a joint social and health monitoring plan, or as part of a broader environmental monitoring plan that includes consideration of health, social and economic issues.

In practice, monitoring is often not conducted and when it is the results are often not made public. For HIA, as well as all other ESIA disciplines, practice will improve as the results of monitoring—and comparing these results to the original predictions set out in the ESIA—are more widely disseminated and incorporated into subsequent ESIAs.

16.7 Human Health Risk Assessment (HHRA)

The methods used to conduct a human health risk assessment (HHRA) are very specialized and quite different than those used for HIA. As described earlier, the purpose of HHRA is to assess the potential for human exposure to environmentally mediated chemical contaminants, and to predict the change in health risk that could occur. HHRA brings together information on the population that would be exposed, the dose to which they would be exposed, and the toxicity of the substances, to provide an estimate of the human health impacts that could be expected from a particular development scenario. HHRA can be used to model increases in the number or rate of specific diseases (e.g. cancer, respiratory disease), or it can provide information to determine if regulatory thresholds for exposure to specific contaminants will be exceeded.

HHRA assessment is similar to the approach to risk assessment described in Chapter 18. It is based on a modeling approach that involves four steps (US EPA 2015):

- *Hazard identification* is used to identify whether a given project will produce chemical substances that have the potential to cause harm to humans.
- *Dose-response assessment* pulls together information from studies in humans and animals to identify how the amount of exposure to a given chemical substance (the "dose") is related to specific adverse health effects (the "response").
- *Exposure assessment* comprises the process of identifying the population that is likely to be exposed to the substances as a result of the project, including the timing, frequency, and duration of exposure, as well as the characteristics of the exposed population.
- Finally, *risk characterization* brings this information together to draw conclusions about the additional burden of health risk faced by the population due to the exposure.

Many jurisdictions have specific guidance and/or technical tools for the conduct of HHRA and the standards, benchmarks and thresholds against which acceptable risk is measured. While there are some differences among the guidance approaches, the same basic four-step approach is, for the most part, consistent.

16.8 What capabilities are needed to undertake HIA?

At this point, very few practitioners specialize in HIA for resource development projects. As a result, some ESIAs have either used a local health practitioner or someone in a different discipline who is already working on the ESIA. While this approach may be adequate, it has frequently resulted in a mess: either because the local health practitioner doesn't understand how an ESIA works and as a result doesn't provide a useful assessment or implementable mitigations, or the specialist from a different discipline doesn't understand health very well. The following capabilities are needed for robustly conducting an HIA. These capabilities may reside in one individual, or be housed within a team, but all of them need to be in place in order for the HIA to be successful:

- Understanding of the project.
- Understanding of the local context, particularly with respect to how health is conceptualized locally, what the burden of disease is, and what factors most strongly influence health.
- Understanding of the ESIA process overall, and the desired outcomes and outputs of the HIA process.
- Understanding of the specific health topics that are going to be important in a particular context (e.g. malaria, nutritional issues, etc.).

16.9 Conclusions

Health Impact Assessment is relatively young as a discipline compared to EIA/ESIA; however, its underlying premise—that human health concerns are important to consider when permitting major projects—is a concept that dates back to the founding of EIA. As noted in the first section, the effects of human exposure to contaminants have been under consideration in decision-making processes longer and more thoroughly than for other health issues. However, over the last fifteen years, there has been an increasing focus on identifying and mitigating the full range of effects that major projects can have on health determinants and health outcomes. Because human health is strongly influenced by economic, social, institutional and biophysical factors, there will always be a substantial overlap between those issues that are addressed in HIA and those that are considered within other ESIA disciplines. As ESIA continues to grow and evolve, the place of HIA within the ESIA process will continue to be an important topic of discussion as government agencies, the media and the general public focus continue to draw attention to a broad range of legitimate health concerns.

Note

1   See, for example, the Preamble to the Constitution of the World Health Organization as adopted by the International Health Conference, New York, 19–22 June, 1946.

References

AfDB (African Development Bank) 2003. *Integrated environmental and social impact assessment guidelines*. Available from: www.afdb.org/fileadmin/uploads/afdb/ Documents/Policy-Documents/Integrated%20Environmental%20and%20Social%20 Impact%20Assesment%20Guidelines.pdf. [18 June 2013].

Bhatia R, L Farhang, J Heller, M Lee, M Orenstein, M Richardson and A Wernham 2014. *Minimum elements and practice standards for health impact assessment, Version 3.* http://hiasociety.org/wp-content/uploads/2013/11/HIA-Practice-Standards-September-2014.pdf.

Birley M 2003. Health impact assessment, integration and critical appraisal. *Impact Assessment and Project Appraisal* 21(4), 313–321.

Birley M 2011. *Health impact assessment: principles and practice.* London: Earthscan.

Byambaa T, M Wagler and C Janes 2014. Bringing health impact assessment to the Mongolian resource sector: a story of successful diffusion. *Impact Assessment and Project Appraisal* 32(3), 241–245.

CCSDH (Canadian Council on Social Determinants of Health) 2015. *A review of frameworks on the determinants of health.* http://ccsdh.ca/images/uploads/Frameworks_ Report_English.pdf.

Chevron 2014. *Environment.* www.chevron.com/globalissues/environment.

Dahlgren G and M Whitehead 1991. *Policies and strategies to promote social equity in health.* Stockholm: Institute for Future Studies.

Divall M, A Ibiejugba and M Winkler 2010. *Health impact assessment Nacala Dam study.* Available from: www.terratest.co.za/files/downloads/Nacala%20Dam/EIA%20to %20PDF/Appendix%20I/Nacala%20Dam%20Health%20Impact%20Assessment%20 -%20%20June%202010.pdf.

Human Impact Partners 2013. *Frequently asked questions about integrating health impact assessment into environmental impact assessment.* www.humanimpact.org/new-to-hia/faq/#EIA.

ICMM (International Council on Mining and Metals) 2010. *Good practice guidance on health impact assessment.* London: ICMM.

IFC (International Finance Corporation) 2009. *Introduction to health impact assessment.* Washington DC: IFC.

IFC 2012. *IFC performance standards on environmental and social sustainability.* Washington DC: IFC.

IPIECA-IOGP 2016. *Health impact assessment: a guide for the oil and gas industry.* London: IPIECA. www.ipieca.org/sites/default/files/publications/Health_impact_assessment_ LR_2016_05.pdf. Figure 16.1 is Figure 1 from that report.

Kemm JR 2012. *Health impact assessment: past achievement, current understanding, and future progress.* Oxford: Oxford University Press.

Public Health Leadership Society 2002. *Principles of the ethical practice of public health, Version 2.2.* http://phls.org/CMSuploads/Principles-of-the-Ethical-Practice-of-PH-Version-2.2-68496.pdf.

Ross C, M Orenstein and N Botchwey 2014. *Health impact assessment in the United States.* New York: Springer.

State of Alaska HIA Program 2011. *Technical guidance for health impact assessment (HIA) in Alaska.* Anchorage: Alaska Department of Health and Social Services.

US EPA (United States Environmental Protection Agency) 2015. *Conducting a human health risk assessment.* Available from: www.epa.gov/risk/conducting-human-health-risk-assessment.

17 Resource efficiency

••

Riki Therivel

17.1 Introduction

Resource efficiency is about how carefully resources are used and wastes are managed during project construction, operation and decommissioning. Resource efficiency is not typically covered as a separate chapter in Environmental and Social Impact Assessment (ESIA). Rather, it tends to be covered in the project description and as mitigation measures for various impacts. However, this may change with the recent evolution of legislation and standards worldwide, notably IFC Performance Standard 3 on resource efficiency and pollution prevention (IFC 2012).

Air, water and soil pollution, and greenhouse gas emissions have already been covered in Chapters 2–5. Issues like what proportion of net primary production should be appropriated by humans, and whether land should be used for production of food versus biomass (e.g. BIO Intelligence Service 2012), are strategic concerns that go beyond individual project ESIAs, and are not covered here. This chapter covers the use of materials, land, energy and water; and waste management.

17.2 Definitions and concepts

Resource efficiency is a measure of the inputs needed to produce a required product or project, or the waste associated with that product or project, for instance:

- Amount of energy needed to construct a building that performs certain functions (e.g. an airport terminal).
- Litres of water needed to produce 1 million m^3 of natural gas through fracking.
- Number of homes per hectare in a new urban extension.
- Tonnes of spoil per tonne of mineral extracted.

The fewer inputs needed and the less waste produced per 'unit' of project or product, the more resource-efficient the project or product is.

17.2.1 Life cycle assessment and embodied energy

Concepts of resource efficiency can apply to the *whole life cycle* of the project and of the products it produces. In both cases, the ESIA considers the 'upstream' impacts of the project (i.e. where and how the original materials are produced, processed and manufactured; how they are packaged and transported; how they are used) and its 'downstream' impacts (i.e. how resulting wastes are managed). Figure 17.1 shows a typical product life cycle. The materials for a product are extracted, possibly processed, then packaged and transported to where they are used. The resulting waste is either reused, recycled or disposed of, typically through incineration or landfill. Different phases of the development process would use different inputs/materials and generate different outputs/wastes. The aim is, over the entire life of the development project, to minimize the use of inputs and generation of wastes.

Historically, life cycle assessment (LCA) has been carried out for individual products like building materials or vehicles. Comparisons of products using LCA have shown, for instance: that electric vehicles are often no more environment-friendly than internal combustion vehicles (Owaro 2012); that moving goods using ocean-going vessels produces significantly fewer greenhouse gases than moving them by rail or lorries (Nahlik et al. 2015); that electric hand driers are frequently more environment-friendly than paper towels (Brady 2012); and that the environmental impacts of eating frozen food are significantly worse than eating fresh food when that food is locally in season, but are roughly the same as eating fresh imported food when that food is not in season (Centre for Environmental Strategy 2008). That said, all LCAs must be viewed with caution, since many of them are carried out by the producers of products that are marketed as being 'greener' than others, using assumptions that could skew the LCA significantly to their benefit.

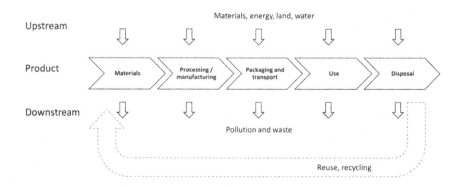

Figure 17.1

Typical product life cycle

For ESIA purposes, one project would use multiple 'upstream' inputs during construction, operation and possibly decommissioning; for instance steel, concrete, aggregates and asphalt. During operation, the inputs will vary widely, from sows, pig food and water for a pig farm; gas, electricity and water for a housing development; to ceramic uranium oxide fuel and water for a nuclear power plant. The project may itself subsequently produce products such as energy (from a power station), pigs (from a pig farm) or chemicals (from a chemicals factory). 'Downstream' wastes would include unused construction material, outflow water and greenhouse gas emissions.

LCA in ESIA considers the amount of resources used at different stages of a product or project. For instance, fewer aggregates may be needed to build a road that cannot cope with harsh winter conditions or heavy lorries, but in the long term it may be more resource-efficient to build a more robust road which requires less frequent repairs. Similarly, during project construction, it may be very energy efficient to build a steel structure once the steel is at the site, but production of the steel may require much more 'embodied energy' than an alternative building material such as bricks or concrete. LCA is a technically complex subject: further information is available from, for example, Curran (2012) and Jolliet et al. (2015).

Figure 17.2 shows the typical stages of LCA for a building. The indicators used in the table could include millions of litres of water, GJ of energy, hectares of land, and tonnes of waste.

Embodied energy is a component of LCA analysis which refers to the energy used to produce a product or building, as opposed to 'operational energy' used during the operation of the building. Embodied energy can comprise a substantial proportion – roughly between 15 per cent and 50 per cent – of a building's whole-life energy use (Pullen et al. 2014). For instance, the embodied energy of a typical house is equivalent to roughly 15 years of operational energy use (Adams et al. 2006). A LCA for a building at Western Australia's Curtin University showed that about 7 per cent of the building's 'cradle to use' greenhouse gas emissions could be cut by replacing 30 per cent of cement with fly ash, and substituting new aluminium and steel with recycled aluminium and steel (Biswas 2014). As the operational efficiency of buildings improves over time in response to environmental concerns and increasing fuel costs, and especially where building lifetime is reasonably short (e.g. warehouses), the relative importance of their embodied energy will increase (Allwood and Cullen 2012).

17.2.2 Energy and mass balance

An important component of resource efficiency assessment is *energy and mass balance*. This means that the sum of mass going into a system must equal the sum of mass going out or being stored in the system; and also that the sum of energy going into a system must equal the sum of energy going out of it plus any changes in the composition of that system. This ensures that all inputs and

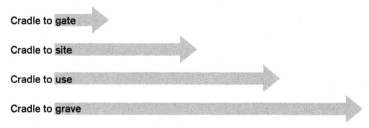

	Building life cycle (BLC)													Beyond BLC
	PRODUCT stage			CONSTRUCTION PROCESS stage		USE stage					END OF LIFE stage				POST PROJECT
	A1 Raw material supply	A2 Transport	A3 Manufacturing	A4 Transport	A5 Construction installation process	B1 Use	B2 Maintenance	B3 Repair	B4 Replacement	B5 Refurbishment	C1 Deconstruction demolition	C2 Transport	C3 Waste processing	C4 Disposal	D Reuse – Recovery – Recycling potential
Materials															
Land															
Energy															
Water															
Waste															

Cradle to gate

Cradle to site

Cradle to use

Cradle to grave

Figure 17.2

Assessment of resource performance of buildings (adapted from British Standards Institution 2012 and CPA 2012)

outputs are accounted for. Figure 17.3 shows a water balance figure for an Indian pulp and paper mill. Figure 17.4 shows a mass balance figure for a German waste-to-energy plant.

17.2.3 Closed loop systems

Closure of materials loops is another important component of resource efficiency. Where a waste product can be used as a resource, either in the same process or elsewhere, resource efficiency increases. For instance, where a previously developed site is being redeveloped, it is much more resource-efficient to break up the concrete and asphalt from the site and recycle it as aggregate in the new development, than to send it away to landfill and use virgin aggregates. Similarly, water from hydraulic fracturing operations can be

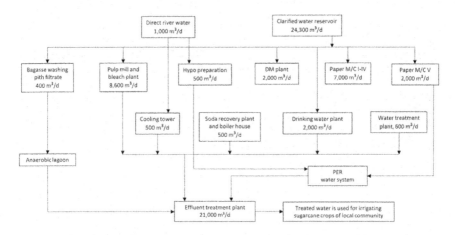

Figure 17.3

Water balance figure for a pulp and paper mill (Based on CPPRI 2008)

treated and reused for other fracking operations (EPA 2016). Figure 17.5 shows how oil, coke and gas from pyrolysis (the process of decomposing materials in the absence of oxygen) of tyres can be used to power the pyrolysis process itself.

17.2.4 Waste hierarchy, self-sufficiency and resilience

The *waste hierarchy* puts the principles above into practice and is a frequent component of ESIA mitigation. The waste hierarchy encourages first reduction of the amount of waste produced, then reuse of materials, then recycling of

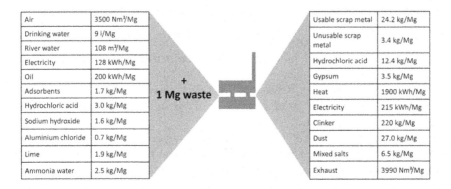

Figure 17.4

Materials balance for a waste-to-energy plant (based on MVR 2016a)

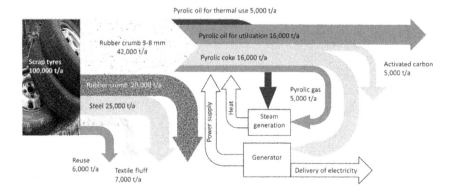

Figure 17.5

Closure of materials loops in a tyre recycling factory (based on Mobius 2016)

materials into new products, then recovery of energy from waste, and only then the landfill of any remaining waste.

Resource efficiency also links to concepts of *self-sufficiency* and *resilience*. It allows more outputs (e.g. housing, food, energy) to be achieved from a given amount of inputs or, in United Nations terminology, it decouples economic growth from resource use and environmental degradation. Examples include the 'Green Revolution' of the twentieth century, which achieved a step change in the food produced per hectare through the use of high-yield plant varieties, irrigation, and use of fertilisers and pesticides (the latter, of course, increased 'waste'); and LED lightbulbs which allow many more lumens (measure of light produced) per watt than traditional incandescent bulbs. In turn, this has made it possible to combine LED lighting with photovoltaic panels, allowing people to have lighting where they would previously not have been able to (Adkins et al. 2010).

Of course, there are inherent conflicts in implementing resource efficiency. For instance, many old buildings are energy-inefficient, so reusing these buildings (good for embodied energy and historical reasons) could go counter to operational energy use. Reused and recycled material may not achieve the quality standards needed, or may not be available in the quantities needed for some ESIA developments.

However, an emphasis on resource efficiency can achieve multiple gains. For instance, BIO Intelligence Service (2012) found that increasing settlement and infrastructure density can decrease material consumption, support the continuous recycling of construction materials, decrease energy use of infrastructure construction, and decrease the use and sealing of land. Redesigning products for longevity and recycling can increase the use of biodegradable, repair-friendly and reusable products; reduce energy use in the production and use of products; and reduce the use of toxic materials (BIO Intelligence Service 2012).

17.3 Key legislation, guidance and standards

IFC Performance Standard 3 is about 'resource efficiency and pollution prevention'. The standard states:

> The client will implement technically and financially feasible and cost effective measures for improving efficiency in its consumption of energy, water, as well as other resources and material inputs, with a focus on areas that are considered core business activities. Such measures will integrate the principles of cleaner production into product design and production processes with the objective of conserving raw materials, energy and water. Where benchmarking data are available, the client will make a comparison to establish the relative level of efficiency...
>
> The client will avoid the generation of hazardous and non-hazardous waste materials. Where waste generation cannot be avoided, the client will reduce the generation of waste, and recover and reuse waste in a manner that is safe for human health and the environment. Where waste cannot be recovered or reused, the client will treat, destroy, or dispose of it in an environmentally sound manner.
>
> (IFC 2012)

The United Nations Environment Programme (UNEP 2015) has also for the past 25 years promoted sustainable consumption and production as an overarching theme to deal with environmental and developmental challenges. Its goals for the theme include:

- By 2030 to achieve the sustainable management and efficient use of natural resources.
- By 2020 to achieve the environmentally sound management of chemicals and all wastes throughout their life cycle, in accordance with agreed international frameworks, and significantly reduce their release to air, water and soil in order to minimize their adverse impacts on human health and the environment.
- By 2030 to substantially reduce waste generation through prevention, reduction, recycling and reuse.
- To encourage companies, especially large and transnational companies, to adopt sustainable practies and to integrate sustainability information into their reporting cycle (UNEP 2016).

Chapter 3 of the Johannesburg Plan of Implementation at the World Summit on Sustainable Development was devoted to changing unsustainable patterns of consumption and production, and the 2012 UN Conference on Sustainable Development (Rio +20) adopted a ten-year framework of programmes on sustainable consumption and production patterns. UNEP's Global Partnership on Waste has published a database of waste management guidelines, which give further information on how to manage specific types of wastes (UNEP 2016).

Implementing these concepts through Environmental Impact Assessment (EIA), the impressively far-sighted United States National Environmental Policy Act 1969 requires environmental statements to include information on '(iv) the relationship between local short-term uses of man's environment and the maintenance and enhancement of long-term productivity, and (v) any irreversible and irretrievable commitments of resources which would be involved in the proposed action should it be implemented' (42 USC Sec. 4332).

The European EIA Directive 2014 requires, as part of the Annex III screening process, information on the characteristics of the project, including the use of natural materials, in particular land, soil, water and biodiversity, and the production of waste. Annex IV requires an environmental statement to include:

> 1(c). A description of the main characteristics of the operational phase of the project (in particular any production process), for instance, energy demand and energy used, nature and quantity of the materials and natural resources (including water, land, soil and biodiversity used);
> 1(d)... quantities and types of waste produced during the construction and operation phases...
> 5. A description of the likely significant effects of the project on the environment resulting from, inter alia:.... (b) the use of natural resources, in particular land, soil, water and biodiversity, considering as far as possible the sustainable availability of these resources, (c) the emission of pollutants, noise, vibration, light, heat and radiation, the creation of nuisances, and the disposal and recovery of waste.

Some countries also have legislation on sustainable production and consumption, resource efficiency or waste avoidance. For instance, in Australia the Victorian Environment Protection (Resource Efficiency) Act 2002 aims to 'promote the establishment of voluntary sustainability covenants to improve the efficiency of resource use and to reduce impacts on the environment', and the New South Wales Waste and Resource Recovery Strategy 2014-21 sets ambitious targets for recycling waste. Other resource-related legislation includes levies on disposal of waste by landfill, and increasingly demanding requirements for the energy efficiency of vehicles, appliances and lighting. Chapter 3 discusses legislation related to land and soil.

17.4 Scoping and baseline studies

The aim of the ESIA scoping and baseline study stage is to identify which issues relating to resource efficiency may be significant during project planning and decision-making. Table 17.1 shows possible scoping questions. The resource-related questions correspond to the sensitivity of the receiving environment, and the impact-related questions correspond to the magnitude of the project's effects. Much of this information will be available from the developer, manufacturers' specifications, or comparison with other similar existing projects.

These indicators might be used to not only scope issues in and out, but also to suggest, at an early stage, ideas for how the project's resource impacts can be minimized. For instance, a project will have very different land-take impacts if it is built on greenfield or previously developed land, and different resource requirements if it uses recycled or virgin materials.

17.5 Impact prediction and evaluation

The ESIA impact prediction and evaluation stage involves, for issues identified as being significant in the scoping phase, making more detailed predictions (or at least active consideration of whether mitigation measures may be needed). These could include:

Materials

- The quantity of materials/resources consumed, by type, for every project stage.
- Where the materials/resources will come from.
- Whether there will be significant impacts at any 'materials donor' sites, for instance borrow pits for aggregates.
- How far the materials will need to be transported and by what mode.

Land (see also Chapter 3)

- The quantity of undeveloped and previously developed land required for every project stage, including any 'materials donor' sites.
- Whether land (and greenfield land) is scarce in the area.
- What the alternative land use would be, e.g. agriculture, derelict previously developed land.

Energy

- The quantity of energy used by the proposed project, by type (e.g. gas, electricity, oil), for every project stage – including load-profile data if available. The load profile is the energy needed over time (day, week, year).

Table 17.1 Possible ESIA resource and waste-related scoping questions (informed by EU 2016)

Topic	Resource-related questions (sensitivity of receiving environment)	Impact-related questions (magnitude of impacts)
Materials	• Will the project use any rare or otherwise limited materials during construction or operation, and what are these? • Will the project use any materials that are harmful or require special handling?	• Will the project use large quantities of any materials during construction or operation, and what are these? • Will the project materials be reused, recycled or from virgin sources?
Land	• What proportion of the country's (or local area's) land is built-up? • What is the current land use of the proposed site?	• What is the footprint of the project? • What is the greenfield (undeveloped land) footprint of the project?
Energy	• How is electricity produced in the country (ratio of fossil fuel, nuclear, other renewable)? • Does spare capacity exist in the existing energy infrastructure? • How resilient is the existing energy infrastructure?	• Will the project use large quantities of energy during construction or operation? If yes, how much and can this be reduced? • Will the energy come from the electricity grid? If not, what is the energy source? • Can waste energy/heat be reused on or offsite, e.g. district heating schemes from waste incineration?
Water	• Is the proposed site in a national or regional area of water shortage or over-abstraction? • Does spare capacity exist in the existing water infrastructure? • How resilient is the existing water infrastructure?	• Will the project require much water during construction or operation? If yes, how much, and can this be reduced?
Waste	• What facilities are there for reuse, recycling and use of waste for energy? • Does spare capacity exist in the different waste management facilities?	• Will the project generate significant quantities of waste during construction or operation? If yes, how much and can this be reduced? • Will the waste be reused, recycled, used for energy or sent to landfill? Can the waste hierarchy be used to change this? • Will the project generate any waste that will be difficult to manage or dispose of?

- The quantity, if any, of energy produced by the proposed project, by type (e.g. anaerobic digestion, solar, coal).
- The embodied energy of the project, for every project stage.
- Where the energy will come from, and whether there is capacity for this in the existing infrastructure.

Water

- The quantity of water used by the proposed project, for every project stage and for fire service testing if appropriate – including load-profile data if available.
- Where the water will come from, whether there is capacity for this in the existing infrastructure (including cumulatively), and whether the proposed project is in an area of water scarcity.

Waste

- The quantity of waste generated by the proposed project, by type (e.g. construction/demolition, solid, hazardous), for every project stage.
- Where the waste will go, whether there is capacity for this in the existing infrastructure (including cumulatively).

Some project stages may have much greater impacts on a resource than others, and some stages may be scoped-out from further analysis at this point. Construction often has major impacts on material use and waste production: up to 40 per cent of waste in some areas is due to construction and demolition (Hoornweg and Bhada-Tata 2012).

Various approaches can be used to identify and present this information. Maps can show the location of waste management sites and other infrastructure near a proposed development site; the location of aquifers; and whether land has been previously developed or whether it is 'greenfield'. Materials, energy or water balances can be shown using Sankey diagrams, such as that of Figure 17.6 for a waste-to-energy plant. This approach helps to ensure that all materials and wastes are accounted for. It can also give an indication of where losses occur, and so trigger consideration of whether these losses could be reduced.

Table 17.2 shows a prediction of materials used during the construction of Western Sydney Airport in Australia, and where those materials might come from. The table shows that the materials are not rare and gives an indication of quantities. For a full life cycle assessment, the resources needed to produce each material (e.g. the water and energy needed to abstract, process and transport the material) would also be carried out. Table 17.3 shows a similar example for waste management, from the ESIA for the proposed South Stream natural gas pipeline between Russia and central Europe.

Embodied energy has, to date, rarely been addressed in ESIA, if at all. Table 17.4 shows typical *embodied energy* for some common materials in Australia and

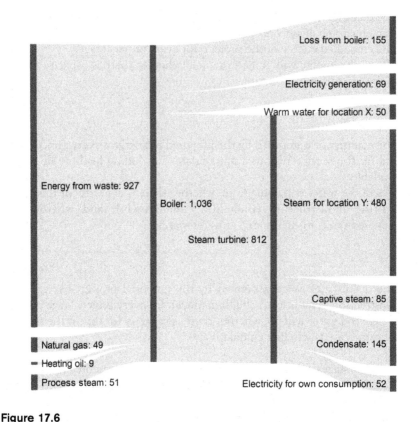

Figure 17.6

Example Sankey diagram for energy from waste plant (broadly based on MVR 2016b)

the UK, along with associated carbon emissions for the UK. The table shows, for instance, that timber, concrete and bricks have lower embodied energy than steel and aluminium. It also shows that embodied energy can vary significantly from country to country (and indeed from project to project) depending on the type of materials, the energy mix used to produce them, how far they are transported, and the assumptions made in the calculations etc.

Where the project requires a significant quantity of materials, the ESIA may need to consider impacts at the donor sites. For instance, roads require large quantities of aggregates, which often come from nearby 'borrow pits'. The ESIA may also need to consider how the products produced by the project are used 'downstream', along with the impacts of these products. For instance, most oil and gas extraction ESIAs do not, at present, consider the greenhouse gas emissions that will result when their product is used, although they are indirect impacts of the projects.

Table 17.2 Materials consumed during construction of Western Sydney Airport (extract from Australian Government 2016)

Activity	Material	Quantity (daily)	Quantity (total)	Potential sources
Earthworks	Water	1.36 ML	650 ML	Existing surface water, farm dams and sediment basin. Potable water supply pipes and temporary storage dams.
Asphalt	Aggregates (63%)	822 T	450,000 T	Gunlake Marulan Quarry Holcim Lynwood Quarry Boral Peppertree Quarry
	Sand (8%)	380 T	57,000 T	Calga Quarry Kurnell Quarry
	Lime filler (2%)	27 T	14,000 T	Various
	Crusher dust (22%)	279 T	157,000 T	Various
	Bitumen (5%)	70 T	36,000 T	Camellia
Concrete	Cement (13%)	128 T	60,000 T	Boral Cement Australia
	Sand (38%)	373 T	174,000 T	Calga Quarry Kurnell Quarry
	Aggregate (44%)	434 T	200,000 T	Gunlake Marulan Quarry Holcim Lynwood Quarry Boral Peppertree Quarry
	Fly ash (1%)	42 T	19,300 T	Various
	Admixture (0.1%)	1 T	460 T	Various
Machinery operation	Fuel/diesel	55,000 litres	—	Banksmeadow Silverwater

Note: T = tonnes

There are no commonly accepted standards for evaluating the significance or acceptability of resource or waste impacts. For resource use, significance will comprise a combination of the quantities of resources needed versus whether the resource is scarce and/or replaceable. Where resources are replaceable or grow (for instance timber or aquifers), then sustainable management involves resource withdrawals remaining, over time, less than the speed of regrowth of the resource. Where resources are irreplaceable or do not grow (for instance land, minerals, biodiversity) then sustainable management involves making the best, most careful use of that resource. For waste, significance could

Table 17.3 Wastes generated during construction of the South Stream offshore pipeline, and potential management approaches (adapted from South Stream Transport B.V. 2014)

Waste code	Waste description	Source of waste	Estimated quantity Offshore	Onshore	Potential management route	Potential facilities
02 01 07		Site vegetation clearance waste	—	10 to 100 tonnes	Utilized on site for habitat improvement or composted	Ecoinvest, Varna
12 01 01	Ferrous metal filings and turnings	Scrap from preparing pipes for welding	100 to 1,000 tonnes	1 to 10 tonnes	Recycling	Ekomax Shumen
12 01 05	Plastics shavings and turnings	Scrap from preparing pipes for welding by abrasion of polypropylene coating	10 to 100 tonnes	1 to 10 tonnes	Recycling (if outlets are available) or landfill disposal	Shumen or Varna Municipality Landfill
12 01 13	Welding wastes	Waste from pipe welding	10 to 100 tonnes	10 to 100 tonnes	Landfill disposal	Varna Municipality Landfill
13 04 13 (hazardous waste)	Bilge oils from other navigation	MARPOL Annex I waste from vessels	100 to 1,000 tonnes	—	Treat using oily water separator if en route, else retain on board for subsequent discharge to dedicated vessels or port waste reception facilities.	

Table 17.4 Embodied energy in some common materials (based on Hammond and Jones 2011; Lawson 1996)

Material	Energy MJ/kg (Australia)	Energy MJ/kg (UK)	Carbon CO₂/kg (UK)
Aggregate		0.083	0.0048
Concrete (1:1.5:3 e.g. *in situ* floor slabs, structure)	1.9	1.11	0.159
Bricks (common)	2.5	3.0	0.24
Concrete block (medium density 10 N/mm²))	1.5	0.67	0.073
Aerated block	3.6	3.50	0.30
Rammed earth (no cement content)	0.7	0.45	0.023
Cement mortar (1:3)		1.33	0.208
Steel (general – average recycled content)	38	20.10	1.37
Stainless steel		56.70	6.15
Timber (general – excludes sequestration)		10.00	0.72
Glue laminated timber	11.0	12.00	0.87
Sawn hardwood	0.5–2.0	10.40	0.86
Glass fibre insulation (glass wool)		28.00	1.35
Rockwool (slab)		16.80	1.05
Expanded polystyrene insulation		88.60	2.55
Polyurethane insulation (rigid foam)		101.50	3.48
Wool (recycled) insulation		20.90	
Straw bale		0.91	
Aluminium (general and incl. 33% recycled)	170	155	8.24
Bitumen (general)		51	0.38–0.43
Hardboard	24.2	16.00	1.05
MDF	11.3	11.00	0.72
Plywood	10.4	15.00	1.07
Plasterboard	4.4	6.75	0.38
Gypsum plaster	2.9	1.80	0.12
Glass	12.7	15.00	0.85
PVC (general)	80	77.20	28.1
Linoleum		25.00	1.21
Vinyl flooring		65.64	2.92
Terrazzo tiles		1.40	0.12
Wool carpet		106.00	5.53
Copper (average incl. 37% recycled)	100	42	2.60
Lead (incl. 61% recycled)		25.21	1.57

comprise a combination of the amount and type of waste (inert, non-hazardous, hazardous); whether it can be reused or recycled; and whether suitable facilities are available to manage the quantities of waste (South Stream Transport B.V. 2014).

17.6 Mitigation

Although resource efficiency is best considered in the early stages of ESIA – during the consideration of alternatives and in design iterations – the principles of resource efficiency discussed in §17.2 can also be used at the mitigation stage to minimize impacts on resources and reduce waste. The 'hierarchy of alternatives' discussed at §1.2.3 can help to structure the consideration of mitigation measures: first need or demand, then mode or process, then location, and then timing and detailed implementation. The higher up the hierarchy the decisions are made, the more sustainable they tend to be.

At the risk of being contentious, the *need or demand* stage allows the developer or consenting authority to consider whether a factory that makes disposable cameras, rapid-turnover fashion, sugary drinks or cheap chickens raised in poor conditions, is 'needed'. This is typically done by comparing the project against the 'no action' alternative. Arguably the world does not 'need' more disposable or short-use items, more things that are bad for people's health, or more things that are produced in a clearly non-ethical way.

The choice of *mode or process* can help to minimize the resource requirements of a project whose need has been shown. At this stage, the developer should consider: whether other materials can be used that will have fewer impacts or are less scarce; whether the project could use less land; whether more energy-efficient technologies or approaches can be used, or materials with lower embodied energy, or whether waste products from the processes can be used as energy; whether more water-efficient technologies or approaches can be used, or whether rainwater can be used, or if water can be recycled within the project or elsewhere; and whether technologies or approaches can be used that generate fewer wastes or less toxic wastes, including during the packaging and transportation of the product.

Improved technologies can maximize the efficiency of resource use and minimize pollution. For instance, retrofitting a new paper machine at a pulp and paper plant could allow the plant to expand with minimal additional impacts:

> The new paper machine will be equipped with state of the art technology for water conservation and will use water more efficiently than paper machines #4 and #12. These efficiencies will make it possible to increase paper production at the Existing Mill plus the proposed Project by 197 percent, while only increasing water consumption by 10 percent. The company intends to implement water conservation projects on the existing paper machines #4 and #12 to further increase the overall water efficiency of the Existing Mill. Currently, the Existing Mill as a whole uses 17,400 gallons of water per ton of paper. After startup of the new paper machine, the Existing Mill plus the proposed Project as a whole will use 6,500 gallons of water per ton of paper.
>
> (Minnesota Pollution Control Agency 2009)

Closure of materials loops also minimizes the need for new resources. Examples are:

> Water from dewatering of the underground mine and from removal of (incidental mine gas) will be transferred to the (Goonyella Riverside and Broadmeadow) mine water management system. This will be used for reuse in coal handling and processing and dust suppression ... Waste from (the mine) in the form of rejects and dewatered tailings will also be disposed of in waste disposal areas for the open-cut operations at the GRB mine complex (BMA 2013);

and 'Mitigation measures include: ... Investigating whether the waste such as stripped pavement material can be reused in the project or by other interested parties ... Reuse of waste materials wherever possible' (Islamic Republic of Afghanistan 2013).

Different *locations* may have different resource requirements if, for instance, they are located close or far from suppliers or markets, or on greenfield versus previously developed land.

When considering *timing and detailed implementation*, the waste hierarchy can be applied, as well as consideration of what will happen to the materials at the decommissioning stage (Pullen et al. 2014; Australian Government 2013):

Use existing buildings/infrastructure

↓

Use parts of existing building/infrastructure;
modify or refurbish instead of demolishing;

↓

Avoid building a larger development than is needed

↓

Use durable, long-life, flexible designs;
design for disassembly, ensure that materials can be easily separated;
use materials that can be reused or recycled easily at the end of their lives

↓

Use prefabricated components with standard sizes to minimize
the need for additional materials as filler;
use low embodied energy materials (including materials with a high
recycled content and/or manufactured using renewable energy);
use locally sourced materials (including materials salvaged on-site) to
minimize transport

↓

Reuse and recycle materials on-site

↓

Reuse and recycle materials off-site

↓

Dispose of materials as landfill

17.7 Conclusions

Resource efficiency is an economic and environmental win-win. It may have short-term costs for the developer in terms of more planning, more expensive equipment, and implementation of appropriate management systems (see Chapter 20). However, in the longer term the developer wins from requiring fewer inputs and needing to dispose of less waste. The environment wins from having fewer resources withdrawn and from needing to accommodate fewer outputs. To date, resource efficiency has typically been considered outside the ESIA process, if considered at all. However, application of the techniques in this chapter allows the consideration of resource efficiency and waste management to be more robust and better documented in the ESIA.

References

Adams E, J Connor and J Ochsendorf 2006. *Embodied energy and operating energy for buildings: cumulative energy over time*. Cambridge, MA: Civil and Environmental Engineering, Massachusetts Institute of Technology.

Adkins E, S Eapen, F Kaluwile, G Nair and V Modi 2010. Off-grid energy services for the poor: Introducing LED lighting in the Millennium Village Project in Malawi. *Energy Policy* 38, 1087–1097.

Allwood JM and JM Cullen 2012. *Sustainable materials with both eyes open*. www.withbotheyesopen.com.

Australian Government 2013. *Embodied energy*. yourhome.gov.au/materials/embodied-energy.

Australian Government 2016. *Western Sydney Airport draft environmental impact assessment. Chapter 25: Resources and waste*. http://westernsydneyairport.gov.au/resources/deis/files/2015/29-volume-2-chapter-25.pdf.

BIO Intelligence Service, Institute for Social Ecology and Sustainable Europe Research Institute 2012. *Assessment of resource efficiency and targets*. Final report prepared for the European Commission, DG Environment. http://ec.europa.eu/environment/enveco/resource_efficiency/pdf/report.pdf.

Biswas WK 2014. Carbon footprint and embodied energy consumption assessment of building construction works in Western Australia. *Intern J of Sustainable Built Environment* 3(2), 179 yourhome.gov.au/materials/embodied-energy 186.

BMA (BHP Billiton Mitsubishi Alliance) 2013. Red Hill Mining lease environmental impact statement, www.statedevelopment.qld.gov.au/resources/project/red-hill-mining-lease/rhml-eis.pdf.

Brady RC 2012. *Throw in the towel: High-speed, energy-efficient hand dryers win hands down*. http://continuingeducation.bnpmedia.com/article.php?upgrade=new&L=199&C=637&P=3.

British Standards Institution 2012. *BS EN 15804:2012 Sustainability of construction works*. London: BSI.

Centre for Environmental Strategy, University of Surrey 2008. *Life cycle assessment (LCA) of domestic v. imported vegetables*. www.surrey.ac.uk/ces/files/pdf/0108_CES_WP_RELU_Integ_LCA_local_vs_global_vegs.pdf.

CPA (Construction Products Association) 2012. *A guide to understanding the embodied impacts of construction products*. London: CPA.

CPPRI (Central Pulp and Paper Research Institute) 2008. *Report on water conservation in pulp and paper industry*. Report to Cess Grant Authority. Saharanpur (UP) India: CPPRI. www.dcpulppaper.org/gifs/report17.pdf.

Curran MA (ed.) 2012. *Life cycle assessment handbook: A guide for environmentally sustainable products*. London: John Wiley & Sons.

EPA (Environmental Protection Agency) 2016. *The hydraulic fracturing water cycle*. Washington DC: EPA. www.epa.gov/hfstudy/hydraulic-fracturing-water-cycle.

European Union 2016. *EU resource efficiency scorecard 2015*. Brussels: European Union.

Hammond G and C Jones 2011. Inventory of Carbon & Energy (ICE) V2.621, cited in Anderson J 2016. *Embodied carbon & EPDs*, GreenSpec. www.greenspec.co.uk/building-design/embodied-energy.

Hoornweg D and P Bhada-Tata 2012. *What a waste: A global review of solid waste management*. Washington DC: The World Bank.

IFC (International Finance Corporation) 2012. *IFC performance standards on environmental and social sustainability*. Washington DC: The World Bank.

Islamic Republic of Afghanistan Kabul Municipality 2013. *Kabul urban transport efficiency improvement (KUTEI) project: Environmental and social management plan*. www-wds.worldbank.org/external/default/WDSContentServer/WDSP/IB/2013/06/06/00044 2464_20130606141701/Rendered/PDF/E42310Afghanis0050201300Box377327B.pdf.

Jolliet O, M Saade-Sbeih, S Shaked, A Jolliet and P Crettaz 2015. *Environmental life cycle assessment*. London: CRC Press.

Lawson B 1996. *Building materials, energy and the environment: Towards ecologically sustainable development*. Red Hill: Royal Australian Institute of Architects.

Minnesota Pollution Control Agency 2009. *Environmental assessment worksheet for the proposed Sappi Cloquet New Paper Machine Project*. www.pca.state.mn.us/sites/default/files/sappi-eaw.pdf.

Mobius Enviro-Solutions Inc. 2016. *Process flow*. www.mobiusenviro.com/tire-technologies-pyrolysis-retreading/tyre-recycling-process-flow.

MVR (Müllverwertungsanlage Rugenberger Damm) 2016a. *Umweltaspekte und -Auswirkung*, www.mvr-hh.de/Energieflussdiagramm.76.0.html.

MVR 2016b. *Material- und Stoffflüsse der MVR*. www.mvr-hh.de/Mengenstroeme-bezogen-auf-1-Mg-Muell.70.0.html.

Nahlik MJ, AT Kaehr, MV Chester, A Horvath and MN Taptich 2015. Goods movement life cycle assessment for greenhouse gas reduction goals. *Journal of Industrial Ecology* 20(2), 317–328.

ODPM (Office of the Deputy Prime Minister), Scottish Executive, Welsh Assembly Government and Department of the Environment for Northern Ireland 2005. *A practical guide to the Strategic Environmental Assessment Directive*. London: ODPM.

Owaro, N 2012. When green turns toxic: Norwegians study electric vehicle life cycle. http://phys.org/news/2012-10-green-toxic-norwegians-electric-vehicle.html.

Pullen S, K Chiveralls, G Zillante, J Palmer, L Wilson and J Zuo 2014. *Minimising the impact of resource consumption in the design and construction of buildings*. www.anzasca.net/wp-content/uploads/2014/02/p32.pdf.

South Stream Transport B.V. 2014. *ESIA Bulgarian sector – Waste management – Chapter 19*. www.south-stream-transport.com/media/documents/pdf/en/2014/07/ssttbv_bg_esia_19_waste_web_en_v2_284_en_20140728.pdf.

UNEP (United Nations Environment Programme) 2015. *Goal 12. Ensure sustainable consumption and production*. https://sustainabledevelopment.un.org/sdg12.

UNEP 2016. *Waste management guidelines.* Global Partnership on Waste Management. www.unep.org/gpwm/KnowledgePlatform/WasteManagementGuidelines/tabid/104478 /Default.aspx.

18 Risk and risk assessment

Garry Middle

18.1 Introduction

This chapter covers the broad notions of risk and risk assessment, and how they are applied within the practice of Environmental and Social Impact Assessment (ESIA). It starts with a brief history of risk assessment followed by a discussion of quantitative risk assessment. The limitations and criticism of quantitative risk assessment are then discussed, leading into an examination of qualitative risk assessment. The chapter finishes with a discussion of risk in a broader context and its relevance to ESIA.

Risk is unlike most considerations in ESIA. Risk is not something that is readily experienced in the usual sensory ways. We can't smell it, taste it or feel it. Risk is more an idea or concept that is perceived rather than felt in any real way, or as Elliott (2003) puts it, risk is a 'judgment'. However, when a risk is realized, impacts occur, which in most cases are negative impacts. Clearly, the notion of risk has negative connotations: we talk about the risk of getting an injury when we play sport but we don't talk about the risk of getting healthier if we exercise.

Central to the idea of risk is uncertainty: for example, if I go surfing I know there is a risk that I will be attacked by a shark, but whether I will actually be attacked by a shark each time I go surfing is unknown. I know it is a possibility because other surfers have been attacked by sharks but I don't know the likelihood that I will be attacked. To put this into more precise terms, there is a *probability* that I will be attacked by a shark, and whether I go for a surf will depend on my personal assessment of the risk of an attack and my personal view about whether that probability of an attack is acceptable.

There is both an objective and subjective element to my decision-making about whether I go surfing or not. The probability of being attacked by a shark will depend on a range of factors, notably whether there has been a shark attack at this location previously. It should be possible that by collecting a range of relevant data – for example, marine habitat types present, fish species present, and the preferred diet of shark species – the probability of a shark attack could be quantified. The subjective element is the level of risk that is acceptable to an individual. Some surfers may decide that they will never surf in a location where there has been a previous attack, whereas other surfers will make the judgment that, while there is a risk, it is sufficiently low that they will 'take that risk' and go surfing. For some their decision will involve making a trade-off. If the quality of the surf is not that good, the risk of a shark attack may well be enough for some surfers to decide not to go surfing. Whereas, if the quality of the surf is exceptional, some surfers will accept a level of risk they wouldn't normally accept and go surfing anyway. In this case, they are deciding that the reward of catching good waves outweighs the risk of a shark attack. They have traded risk for reward.

This is a long introduction, but it has introduced some of the key concepts of risk and risk assessment which will be expanded in the rest of this chapter. Before that, it is necessary to examine the formal definition of risk.

An old and broad definition of risk is that by Cooper and Chapman (1987), which is, 'Exposure to the possibility of economic or financial loss or gains, physical damage or injury or delay as a consequence of the uncertainty associated with pursuing a course of action.'

Chapman refers to economic risk, for example trading on share markets where there is always uncertainty as to whether the price of a share will go up or down, thus the risk is of losing money. While financial or economic risk may not seem relevant to ESIA, it will be argued later in this chapter that ESIA has a role to play in the financial risk of a proposal. Chapman's reference to physical damage refers in particular to property damage because of natural events – for example storms or an earthquake – but also damage (injury or death) to human-caused disasters, for example a chemical explosion or a car accident.

Also missing from Chapman's definition is reference to the damage to the natural environment: to rephrase Chapman's definition and to put it within the context of ESIA, risk is exposure to the possibility of damage to humans and the natural environment as a consequence of the uncertainty associated with pursuing a course of action.

A final introductory point that needs to be made here is the reference to the risk to humans. As will be seen below, the early history of risk assessment had a primary focus on human health and safety. This is highly relevant to ESIA as most definitions of 'environment' make reference to humans specifically. For example, the New South Wales (Australia) Environmental Planning and Assessment Act 1979 defines environment as: 'environment' includes all aspects of the surroundings of humans, whether affecting any human as an individual or in his or her social groupings'.

The United States National Environmental Policy Act defines the 'Human environment' as: '"Human environment" shall be interpreted comprehensively to include the natural and physical environment and the relationship of people with that environment'.

The United Kingdom Environmental Protection Act 1990 defines environment to be: 'all, or any, of the following media, namely, the air, water and land; and the medium of air includes the air within buildings and the air within other natural or man-made structures above or below ground'. And environmental harm as 'harm to the health of living organisms or other interference with the ecological systems of which they form part and, in the case of man, includes offence caused to any of his senses or harm to his property'.

Clearly, when the notion of environmental risk is considered it relates to both the natural environment – best considered as ecological risk – and to its impact on humans.

18.2 Definitions and concepts

18.2.1 Definitions

Some key words and phrases are used in association with the notion of risk, as described below.

Risk is an outcome of *hazards* or *hazardous events*. A *hazard* is the inherent property of an object to create damage to humans or the environment. These objects can be inherently dangerous and exposure to them can cause injury, health problems or loss of environmental quality. Hazards and can be natural (cliffs, jellyfish, crocodiles etc.) or caused by humans or human activity (for example water spilt on the ground producing a slippery surface to walk on, or an explosion in a chemical factory).

These hazards can cause harm in one of two ways: through long-term exposure at low levels or very short-term exposure at high levels.

Hazards that cause harm over a period of time through long-term exposure typically include toxic chemicals, for example pesticides, and low-level radioactive material, for example isotopes used to create X-rays. These objects cause harm to humans by extended exposure but at relatively low levels. These can be considered *chronic* hazards.

Some chemicals can cause ecological damage, for example fertilizers that end up in wetlands and cause algal blooms. In this sense, fertilizers are a natural hazard that can either be chronic, where the export of nutrient to a wetland is gradual, or acute where a spill of fertilizer into a wetland causes rapid ecological changes.

Some hazards can cause a *hazardous event*, which is an event where humans or other species are exposed to a hazard at high levels and harm to humans or the environment occurs very quickly. In these cases, the hazard results in an accident where very high levels of a toxic chemical or radiation are produced. For example, natural gas is transported and distributed through, and in, urban areas via underground pipelines that protect people from direct exposure to the gas. In this way, residents are not exposed to any chronic hazard. However, occasionally accidents occur where a pipeline is ruptured, causing a gas leak. If the rupture is caused by a metal object (for example, by a workman digging a hole with a shovel) then the spark created by metal on metal contact can cause the gas to ignite and explode. This can be considered an *acute* hazard.

Accidents like the 1984 leak of deadly methyl isocyanate gas at a Union Carbide pesticide plant in Bhopal, India, is an extreme example of such an accident. The leak caused 2,800 people to die and tens of thousands of others to be injured. The 2010 explosion of the Deepwater Horizon oil platform in the Gulf of Mexico resulted in over 1,000 km of coast being affected directly by a mixture of oil and chemical dispersant, leading to degradation and loss of beaches, mudflats and coastal wetlands, and habitats for a range of fauna, notably waterbirds (Henkel et al. 2012).

Risk is the likelihood that harm to a human or the environment will occur as a result of exposure to a hazard. Typically, risk is seen as a combination of the two factors: frequency and consequences. *Frequency* for chronic hazards is about the length of time of exposure to a hazardous chemical or the number of times that exposure occurs. As the length of time increases, or the number of exposures increases, the risk of harm increases. For acute hazards, frequency is more generally about how many times a certain type of accident occurs, or how often things go wrong. *Consequences* refers to the seriousness of the harm. Slipping on a wet surface would likely cause a sprained ankle, whereas falling off a tall cliff would lead to serious injury and even death. Exposure to gases like sulphur dioxide can cause respiratory problems, whereas exposure to carcinogenic compounds can lead to serious illness and even death.

The level of risk, therefore, increases with the frequency of exposure and the severity of consequences. A frequently used technical definition of risk is that it is a product (multiplier) of the *probability* of an event and the potential *damage* of that event (Adams 1995). In short,

Risk = Frequency × Consequences.

Great Britain's Health and Safety Executive (2001) categorises hazards into two different broad categories: individual concerns and societal concerns. *Individual concerns* relate to how individuals perceive risk from a particular hazard and how it affects them directly, and how it affects the things they personally value. This can be quantified as *individual risk*. *Societal concerns* is a broader concept which relates to those hazards that impact on society rather than an individual, and is usually associated with an accident or an event that leads to significant and widespread impacts causing multiple fatalities or injuries. If such an event occurred, there would be a significant political backlash against the institutions responsible for ensuring practices and policies are in place to protect people. The risk of multiple fatalities is known as *societal risk*.

The process of estimating the level of risk that a hazard poses is called *risk assessment*. It is generally seen within a broader risk management strategy, where the risk assessment provides an estimate of the likelihood of an adverse outcome, which then informs the industry carrying out the risk assessment whether the level of risk needs to be reduced to a more acceptable level. The risk assessment provides useful data on which components of the operations need to be addressed to best reduce risk.

This way of conceptualizing risk suggests that it is possible to quantify risk in some way. This quantification part of the process is described in more detail later in this chapter.

Modern risk assessment began in the early part of the twentieth century and has aimed at finding ways to properly quantify risk. This is termed *quantitative risk assessment* – QRA. Below is a brief history of risk assessment, following a discussion of the broader risk management framework.

18.2.2 Risk assessment and a broader risk management framework

The broader risk management framework has two elements: risk assessment and post-assessment.

Risk assessment involves four broad steps, as shown in Figure 18.1. The post-assessment follows from the last step in risk assessment. If the risk level is found to be acceptable, then no further action may be required at this point and the proposal can proceed, especially if the risk is negligible. If the risk is deemed to be unacceptable, then either the industry will be required to modify the proposal to make the risk acceptable (further reducing the risk), or the regulator will not approve the proposal if it is not practicable to lower the risk to acceptable levels. If the proposal is approved, then the proponent will be required to monitor the activities to ensure risk levels remain acceptable through ongoing control and monitoring. Readers familiar with the ESIA process will note there are clear parallels to the broader steps in the ESIA process, and this is discussed in detail in Section 18.6.

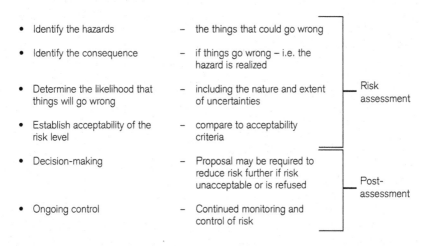

Figure 18.1

Steps in risk assessment

18.2.3 Modern quantitative risk assessment (QRA) and its history

Quantitative risk assessment (QRA) is the process of evaluating and assigning a numerical value to the risk associated with an industry, and also includes the practices of probabilistic safety assessment and probabilistic risk analysis (PRA). QRA is the most commonly used term.

The origins of the current practice of risk assessment (including QRA) can be traced back to the turn of the twentieth century where health experts and scientists began to raise concerns about the health impacts of the industrial revolution, especially the emergence of unique diseases observed in particular industrial workplaces (Paustenbach 2002). It was becoming clear that ongoing exposure to certain chemicals being produced or used in industrial production was leading to a range of adverse health outcomes. By the early 1930s the notion of risk assessment began to emerge, in that governments began to set acceptable levels of exposure to toxic chemicals in the workplaces. It was recognized that these chemicals were hazardous and that exposure to them posed a risk of poor human health outcomes.

The use of QRA in industrial-based activities that were the source of accidents causing injury or death to workers began with the US aerospace industry, following the death of three astronauts in 1967 from a fire in an Apollo flight test (Cooke 2009). This led to the first PRA, and, although it was of limited value at the time, led to the use of risk assessments for the space shuttle program and improved safety following the Challenger accident of January 1986.

PRAs were also used in the nuclear industry in the 1960s and 1970s, and following the Three Mile Island accident in March 1979, the existing methodology for doing PRAs was reassessed, leading to more rigorous assessments. The chief concerns arising from this accident were adverse health outcomes, especially the risk of cancer.

Exposure to radiation wasn't the only risk of concern to the wider community: exposure to a range of toxic substances that could cause adverse health impacts, especially from the petrochemical industry, emerged as a key concern. One of the first QRAs carried out in the UK was in 1978: of the petrochemical installations in the Canvey Island industrial area near London.

In 1983, the US Congress requested that the National Academy of Sciences establish an expert committee to examine the state of risk assessment (Committee on the Institutional Means for Assessment of Risks to Public Health 1983) and it set three objectives for the study:

- To assess the merits of separating the analytic functions of developing QRAs from the regulatory functions of making policy decisions.
- To consider the feasibility of designating a single organization to do QRAs for all regulatory agencies.
- To consider the feasibility of developing uniform QRA guidelines for use by all regulatory agencies.

The committee concluded that the key problem with existing QRAs was lack of data and too much reliance on 'inference and informed judgment to bridge gaps in knowledge'. Another important recommendation of the committee was to separate risk assessment from risk management (the policy and decision-making required, taking into account the risk assessment). This led to much more transparent and credible risk assessments. The report set the tone for the evolution of QRA over the next few decades, with an emphasis on rational and objective methodologies using robust data and arriving at numerical values as a measure of the risk.

Importantly, the committee noted that risk assessment needs to be seen within a broader risk management framework, albeit remaining a significant part of that framework. This is discussed in the next section.

18.3 Prediction and evaluation

18.3.1 Outcomes of QRA

The key outcome of a QRA is a series of risk contour lines around the hazardous facility. Each contour line represents a line of equal risk for any individual standing on that line. These contours are generated from sophisticated computer models which generally include data on:

- The nature of the chemical hazard and the levels (concentrations) at which impacts are experienced;
- The quantity of the chemical on-site;
- The nature of the technology used and the known occurrences of accidents, leaks, explosions etc.;
- The local meteorological conditions;
- The local site characteristics; and
- The nature of emergency response proposed.

These contours are *individual risk* contours, which is the likelihood (probability) that a person will be fatally injured by a hazardous event to which that person may be exposed. It is expressed numerically as the expected frequency of fatalities per year. Figure 18.2 shows what a typical QRA risk contour map looks like. The contours show the modelled probability of fatality: for example, the innermost figure is 1×10^{-5} – that is, a one in 100,000 chance of an individual fatality anywhere on that line.

One industry where QRAs are a normal part of the design process is the liquefied natural gas (LNG) liquefaction industry. For example, in a study of floating production, storage and offloading LNG plants the following process was carried out to produce a QRA (Dan et al. 2014). First, the likely consequences of an accident was determined, which involved three broad steps. The most likely point in the process where leakages are expected to occur were

Figure 18.2

Hypothetical risk contour map around a factory

identified, which, in this case, were five expansion valves. Then, three leakage scenarios were established, based on different leakage holes sizes: 5, 50 and 100 mm. The smaller hole would most likely result in an immediate ignition and what is called a jet fire. The larger holes would more likely lead to a delayed ignition, and, therefore either an explosion or a less severe flash fire, depending on the concentration of the gas in the air. Finally, modelling was carried out to establish the likely nature and extent of the consequences of an accident. The modeling assumed a 10-minute delay in ignition. The local weather was a key input into the modeling, most significantly wind speed. The faster the wind speed, the greater the dispersal of the gas, the less likelihood of an explosion, and the lower the severity should an explosion occur. Modelling was based on a worst-case scenario of 1.5 m/s, or a light wind.

The first output of the modelling are consequence maps. For example, Figure 18.3 shows the intensity of a jet fire from a 50 m hole with increasing distance from the leak.

The final step is frequency analysis, which is estimating the probability of the accident occurring. This includes consideration of known historic accidents involving this process: for example in 2004 an explosion involving an LNG plant in Algeria killed 27 people, with an additional 72 people

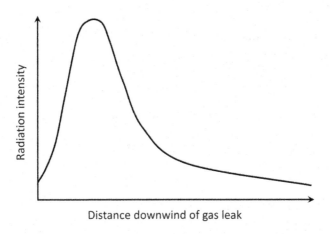

Distance downwind of gas leak

Figure 18.3

Hypothetical modelling consequences of a jet fire from a gas leak in a LNG plant –
radiation strength vs distance from leak

injured. Often, relevant data has also been collated by a research institution or
an industry body, for example the International Association of Oil & Gas
Producers, detailing the frequency of gas releases from a range of equipment
types associated with LNG plants (OGP 2010).

The consequence and frequency data is combined in another computer
model to generate the risk contours, for example those shown in Figure 18.2.

18.3.2 Increasing complexity and uncertainty in QRAs

This example of a QRA can be criticised for being narrow in the way it
identified the nature of the hazards. It primarily focused on equipment failure
and the frequency with which technical things can go wrong. It ignored the
human element, for example where a leak is caused by human error rather than
equipment failure.

Ashrafi et al. (2015) argue that there are three types of QRA, which have
evolved over time. The first type is technically focused, and focuses on
estimating the risks of structures and technical parts failing – for example the
case study discussed above. The assumption was that accidents are caused only
by the failure of equipment. The next generation of QRAs recognized that
human error also contributes to accidents, in particular, human errors in
carrying out assigned tasks or maintenance, or humans doing unplanned or
unintended activities. The final type is called safety/organizational.
Investigations into major accidents like Bhopal revealed that the factors

behind the accident went beyond technical and simple individual human error: organization-wide systematic factors also contributed to the accident.

This evolution in thinking about QRA has added more complexity to the calculations. It is relatively easy to calculate the chances of equipment failure, but much more difficult to estimate human error. Estimating safety/organizational problems is even more difficult. Even if a company's accident record is known, other factors can influence overall safety practice, for example the legislative framework of the country within which a company operates.

This added complexity results in more uncertainty in predicting risk levels, which has resulted in increasing criticism of QRA. This discussion is picked up later in this chapter.

18.3.3 Determining acceptable risk

A key question that emerges from the above discussion of QRA is 'what level of risk is acceptable?' Within the context of this discussion, policy makers distinguish between the notions of tolerable and acceptable. A level of risk may be seen as tolerable but this should not be seen as implying acceptability. Tolerable means: 'a willingness by society as a whole to live with a risk so as to secure certain benefits in the confidence that the risk is one that is worth taking and that it is being properly controlled' (Health and Safety Executive 2001).

Clearly, establishing the level of risk that is considered tolerable to society is a political process and involves trading off benefits against the risk of an adverse outcome. Risk levels lower than the tolerable level can be considered acceptable. This does not mean that once the level tolerable risk has been established every individual will accept this, especially not those individuals who would be subject to the risk.

Elliott (2003) argues that four factors affect how individuals judge the level of risk:

- If risk is being imposed involuntarily on a community or an individual. If so, then the level of risk will be perceived as higher, and the level of acceptable risk will be lower.
- If risk is seen as being allocated in an unequal way between communities. Communities already exposed to existing risks are more likely to consider additional risk as being both unfair and unacceptable, and will argue that communities not already being exposed to elevated levels of risk should carry the new risk.
- The nature of the potential consequences, especially where the consequences of an accident are perceived as being catastrophic.
- The level of uncertainty, including the level of confidence that risk levels can be determined and how well the potential consequences of an accident are known and understood.

Notwithstanding these socio-political and technical complexities in doing QRAs, many jurisdictions require QRAs to be carried out for hazardous industries and have set tolerable/unacceptable risk levels. Two approaches are adopted in considering tolerable/unacceptable risk levels. Some jurisdictions (for example the UK) set two levels of risk as aids to decision-making – so-called *de minimus* risk and *de manifestus* risk. *De minimus* risks are risks that are clearly insignificant and no further consideration of risk is needed as part of any assessment and subsequent decision-making. *De manifestus* risk are significant and are either considered unacceptable or further remediation is required to lower the risk levels. The UK (through the Health and Safety Executive) and the Netherlands have been world leaders in developing criteria for tolerable/unacceptable risk, but they use different levels of risk. This is summarized in Table 18.1.

Risk in-between the two values is often called the *ALARP* region – *as low as reasonably practical*. While it is preferred that risk levels are reduced to the *de minimus* level, risk within this range can be acceptable if reducing the risk level further is either impractical or so costly that any benefits are minimal compared to the improvement in risk levels.

As can be seen from Table 18.1, there are considerable differences between the two jurisdictions as to what is tolerable/unacceptable risk, which reflects the significant role that the socio-political context plays here. One particular QRA in the Netherlands was important in how the Netherlands arrived at its criteria for tolerable/unacceptable risk. In the late 1980s the planning authorities in the Netherlands were considering a proposal for a new housing development near a large existing chemical plant. A QRA was carried out, but there was no standard at that time for what was acceptable risk. After considerable public debate, the Dutch parliament finally debated the issue in 1990, which led to the setting of the 1×10^{-8} and 1×10^{-6} criteria (Pasmana and Reniers 2014).

Other jurisdictions adopt a single figure, with the most commonly used level of acceptable risk being 1×10^{-6} for residential areas. In these cases, industries are expected to keep risk below the tolerable/unacceptable level.

Table 18.1 Summary of tolerable/unacceptable risk levels for the UK and the Netherlands (Marszal 2001)

	de minimus risk level for residential areas (public) – insignificant risk	de manifestus risk level for residential areas (public) – considered unacceptable or that further remediation is required to lower the risk levels
UK	1×10^{-6}	1×10^{-4}
Netherlands	1×10^{-8}	1×10^{-6}

Nations that have adopted criteria for tolerable/unacceptable risk levels include Australia, Brazil, Canada, China (Hong Kong), Czech Republic, Denmark, France, Hungary, Netherlands, Norway, Singapore, Switzerland, and the UK (Travis 2013). These countries are exceptions in that most countries do not adopt government-endorsed tolerable/unacceptable risk criteria and allow the individual industries to set their own risk criteria (Marszal 2001).

Those jurisdictions that have set criteria for tolerable/unacceptable risk have set different levels depending on the type of land use, for example industrial, commercial or residential. It is generally accepted that the level of risk in residential areas should be lower than in places where people work. This is in part due to the fact that people spend less time at work than at home. The Environmental Protection Authority (EPA) in Western Australia has adopted criteria for what it considers to be acceptable individual risk (EPA 1994) for five main types of land uses:

- Sensitive development (hospitals, schools, child care facilities and aged care housing developments) – 0.5×10^{-6}
- Residential development – 1×10^{-6}
- Commercial development – 5×10^{-6}
- Non-industrial activities – 1×10^{-5}
- Industrial facilities – 50×10^{-5} at boundary and 1×10^{-4} from all sources at plant.

A typical approach is for an authority to adopt standards used elsewhere, test these in local circumstances, and modify as required. For example, in 1987 the Western Australian EPA proposed to use the 1×10^{-6} criterion for residential areas when it 'took note of how decisions are taken in other parts of the world' (EPA 1987). Five years later, in reviewing its policy on risk, the EPA concluded that 'Experience since that time has demonstrated that the criterion adopted has received broad support across the community and has worked well in practice' (EPA 1992).

This is a common approach to this issue, as it provides a certain credibility to the decision on tolerable/unacceptable risk levels which helps manage community concerns about risk. Authorities can argue that if other jurisdictions use a certain number (i.e. 1×10^{-6}) then it must be valid.

Brazil has a less conservative level of tolerable/unacceptable individual risk associated with industry: individual risks greater than 1×10^{-5} per year are considered unacceptable. Kirchhoff and Doberstein (2006) note that while the more recognized 1×10^{-6} figure was considered, the regulatory authorities adopted the less conservative figure, '(f)earing that restrictive RA criteria would make some activities unfeasible, the technical team responsible for adopting the criteria decided to be less restrictive'.

In this case, economic and political considerations played a significant part in determining tolerable/unacceptable risk levels.

Notwithstanding all of this technical work at arriving at tolerable/unacceptable risk levels, the affected communities often have a different perception of what is acceptable risk. This will be covered in more detail in a later section where the concerns and limitations of QRA are discussed.

18.3.4 Societal risk

Section 18.2.1 noted that there are two broad categories of risk: risk of injury or fatality of an individual (individual risk) and risk of multiple fatalities or injuries. This latter category is known as societal risk. To be precise, societal risk is the relationship between the frequency of an adverse event occurring and the number of people suffering a specific level of harm, often fatal, in an affected population. As can be seen above, a considerable amount of work has gone into quantifying individual risk but less work has been done on societal risk.

Part of the reason for this is that calculating societal risk is much more complex than individual risk, primarily because of the extra variable of the number of fatalities, which makes representing the risk in a spatial way, as can be done with individual risk, very difficult. Societal risk is general shown as a F–N (frequency–number) curve, as shown in Figure 18.4. The x-axis shows the number of fatalities using a logarithmic scale, and the y-axis shows the frequency of accidents. The general concept behind this curve is the combination of the two variables and the total number of deaths per year. For accidents that occur rarely, it is more tolerable to have a higher number of fatalities, whereas for accidents that occur more frequently a fewer number of deaths would be tolerable.

As can be seen in Figure 18.4, it is possible to identify three areas of risk: risk insignificant, therefore acceptable; risk significant, therefore unacceptable; and an in-between zone where further assessment of the risk is required and where the ALARP principle applies. This is similar to the *de minimus* risk and *de manifestus* approach to individual risk discussed in §18.3.3.

While it is possible to develop tolerable/unacceptable criteria using the F–N curve, fewer countries have done so compared to their assessment of individual risk criteria; in part because of the complexities, but also the likely political difficulties in deciding how may fatalities can be considered acceptable – for example is 10, 100 or 1,000 an acceptable number, and how often?

18.3.5 Concerns about QRAs

QRA gained considerable support as a tool to help in decision-making in the 1980s and 1990s, especially as part of ESIA. Proponents of QRA made claims like this from Carpenter (1995): '(environmental risk assessment) is objective advisory information that can enhance the participative political process where values and preferences are properly integrated into the final administrative program for protecting lives and ecosystems'.

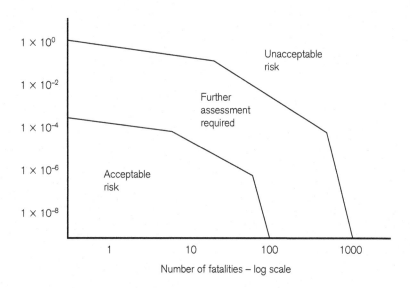

Figure 18.4

Conceptual representation of societal risk using an F–N curve

In recent years, QRA has attracted criticism from a range of experts, communities and green groups. In part, this criticism is based on the increasing complexity of environmental problems and the inability of QRA to deal with this complexity. Ashrafi et al. (2015) put it this way:

> Traditional risk assessment and modelling approaches are architected based on the Newtonian paradigm and reductionism. This paradigm is based on the belief that systems are composed of independent elements and can be easily understood by breaking down to their smallest elements and describing interaction manner of their elements. In other words, following Newtonian paradigm of analysis, traditional risk assessment methods are incapable of modelling interactions among elements of complex systems.

The increasing power of computing and the evolution of more powerful models and tools to apply to QRAs led many to believe that QRAs were foolproof, which created 'a spurious sense of infallibility and overconfidence in the results of the assessment, engendering an elitist culture among "experts" while excluding and alienating the recipients of the assessment' (Eduljee 2000).

Corburn (2002) raises two criticisms of QRA. First, he argues that where QRAs are carried out in relation to chemical pollution they only deal with the impacts of single chemicals and cannot deal with multiple chemical exposure where thresholds for impacts may well be lower. This is in part because exposure to one chemical may well make an individual more sensitive to exposure to another chemical, or more than one chemical. Second, he argues that the data used in risk assessment is primarily highly technical and does not include local and non-expert sources of knowledge. Importantly, the exclusion of local knowledge leads to an exclusion of local values in carrying out the assessment.

The implication of Corburn's last criticism is that QRA is not the fundamentally objective tool that its proponents claim it to be, but that it is inherently subjective and, as a result, should involve the values of the communities being impacted. He argues for the democratization of risk assessment: 'the assessment process may shift the risk discourse away from an expert-only model of analysis and decision-making toward a more democratic model – where local knowledge can improve both the technical assessment and procedural'.

As well as QRA's objectivity, its use in more complex environmental settings has also been questioned. To apply QRA in complex ecological systems requires that there is a substantial amount of reliable data to draw on to enable modelling to be carried out. However, the complexity of many ecological systems means that even the best QRA can't accurately predict risks to ecosystems. For example, the northern cod fishery in Canada was well-studied and researched with a significant data set available to fishery managers to carry out modelling and a risk assessment. Notwithstanding, the fishery experienced a significant collapse in number in the early 1990s (Astles 2006).

Finally, ordinary members of the community generally perceive risk in quite a different way than experts in the field, seeing risk as more descriptive and qualitative rather than quantitative (Kajenthira 2012). Also, communities can have a different perception of what is acceptable risk.

In summary, QRA is facing a range of challenges, and its use in ESIA is becoming less and less frequent. The inherent complexity of many ecological systems and the lack of good data to explain these complex processes makes modelling responses to stresses – the consequences – more and more difficult. Further, QRA's fundamental objectivity has come into question. The decision on what data to use as an input to modelling is made by the modellers, and is inevitably based on their own subjective assessment of what is good data, as is their assessment of which scenarios are to be modelled. The reliability of the outputs (e.g. contour lines) of QRA are increasing being challenged for both their accuracy and credibility. As a result, risk assessment in these complex situations is becoming qualitative rather than quantitative. *Qualitative risk assessment* fully embraces the inherent uncertainty and subjectivity of modern environmental problems where risk of an adverse outcome is central to decision-making.

18.3.6 Uncertainty: the rise of qualitative risk assessment

Typically, qualitative risk assessments do not generate numerical assessment but are descriptive and categorical, which still allows for the comparison of options (WHO/FAO 2009). Because environmental data is limited, inputs to qualitative risk assessments typically include the views of experts, and sometimes the affected stakeholders and their assessment of the risks based on the available data.

Clearly, qualitative risk assessment will lead to different perceptions of what the risk level is and whether that risk is acceptable or not. Disagreement between experts and laypersons about risk levels is, in part, based on the interpretation of the data (experts may well argue that laypersons cannot understand the data and will, therefore, arrive at a different and inaccurate assessment of risk) but also on what data to include as part of the assessment (experts may well dismiss local and indigenous knowledge as being irrelevant).

The same general approach is adopted for qualitative risk assessment – that is Risk = Frequency × Consequences – but in this case a descriptive approach is used to define the range of both elements of risk. A risk matrix is often used, with a two-dimensional graphical representation of risk, with likelihood of an adverse outcome on one axis and the possible consequences of a hazard being released on the other axis. Figure 18.5 shows a typical matrix.

This table allows for a visual and conceptual representation of risk, with the individual boxes in the table being a measure of risk – i.e. likelihood × consequence. The greater the consequence (down the table) the less acceptable that outcome is. Similarly, the more likely that an adverse outcome will occur, the less acceptable that will be (across the table). Arguably, an outcome of minor consequence may be acceptable no matter how likely it is, whereas an outcome of extreme consequences may only be acceptable if it is at

Likelihood of adverse outcome	Remote	Unlikely	Possible	Likely
Consequence level				
Minor				
Moderate				
Major				
Extreme				

Figure 18.5

Typical qualitative risk assessment matrix

least unlikely, and preferably remote. Based on this, the darker the shading in the table, the less acceptable the risk will be.

Another approach is to allocate numbers to each step on each axis, the risk being the product of the two numbers (a literal interpretation of the Risk = Frequency × Consequences equation). This is shown in Figure 18.6. This method provides a way to compare the risk of different mixes of likelihood and consequence. This matrix approach can be used to assess a single project and its risks of an adverse outcome – that is, project ESIA – or it can be used in complex ecosystem management approaches where there are a number of influences at work that could impact on the ecosystem – that is, strategic environmental assessment.

Various methods are used to identify where a particular hazard sits within this matrix. Fletcher (2015), working for a government department that manages fisheries, recommends using workshops where the key stakeholders are invited to use the matrix to identify risk levels for a range of management concerns. Participants are first presented with the existing data on the fishery, and then a facilitated discussion takes place to arrive at the appropriate risk level. Fletcher recognizes that while such an approach can lead to a higher level of acceptance of the outcomes of the risk assessment, having stakeholders with such diverse backgrounds and interests in the fisheries can lead to large discrepancies in the scores and risk levels. He argues that there are four reasons for this. First, different stakeholders may have different objectives for the fishery, for example recreational fishers and commercial fishers. Second,

Likelihood of adverse outcome		Remote	Unlikely	Possible	Likely
		1	2	3	4
Consequence level					
Minor	1	1	2	3	4
Moderate	2	2	4	6	8
Major	3	3	6	9	12
Extreme	4	4	8	12	16

Figure 18.6

A qualitative risk assessment matrix using numbers to provide some level of quantification

different stakeholders will have different views on what is an acceptable impact, for example a conservationist might see any reduction in a fish species population as unacceptable whereas a commercial fisher will naturally accept some loss as a commercial take. Third, notwithstanding that each participant is provided with the same initial information, their understanding and interpretation of the implications of that information may well vary. Finally, participants will vary in their capacity or willingness to change their minds on some key issues as learning takes place, for example to accept a different level of tolerable risk for a certain species.

This example of using a risk assessment matrix in managing a fishery is a good example of a strategic environmental assessment. Twelve key elements of the fishery were identified for assessment, possible consequence defined, and the likelihoods of each consequence determined using the workshop approach (Fletcher 2015). The twelve elements were:

- Target species;
- By-catch species;
- Protected species;
- Ecosystem structure;
- Habitat;
- Economic;
- Social structures;
- Food security;
- Social amenity;
- Reputation and image;
- Occupational health and safety;
- Operational effectiveness.

The complexity of the assessment becomes obvious, as does the value to the ongoing management of the fishery if done competently. The choice of which elements to include in the assessment highlights the inherent subjectivity of the process, for example the inclusion of items like reputation and image, and social amenity.

The Environmental Protection Authority in Western Australia has carried out qualitative risk assessment on a case-by-case basis to assess individual proposals where there is a risk of environmental damage, and where that risk cannot be easily quantified. It has developed a modified version of the matrix shown in Figures 18.5 and 18.6 and is shown in Figure 18.7.

The EPA has set the tolerable/unacceptable risk level as follows: only proposals that pose either a very low or low risk are considered acceptable. This matrix raises two interesting issues. Firstly, if any of the possible consequences of a proposal can be considered catastrophic then the risk can never be below medium and the proposal automatically becomes unacceptable. This was the case for a proposed underground coal mine about 200 km south of Perth (the capital city of Western Australia) in the Margaret River region. The coal

Consequence category						
	6	5	4	3	2	1
	Negligible	Minor	Moderate	Major	Massive	Catastrophic
1 Almost certain	Low	Medium	High	Extreme	Extreme	Extreme
2 Likely	Low	Low	Medium	High	Extreme	Extreme
3 Possible	Very low	Low	Low	Medium	High	Extreme
4 Unlikely	Very low	Very low	Low	Low	Medium	High
5 Remote	Very low	Very low	Very low	Low	Low	Medium

Figure 18.7

The Western Australian EPA's qualitative risk assessment matrix (EPA 1992)

reserves are within the Sue groundwater aquifer, which is connected to another aquifer above it, the Leederville aquifer, which in turn is known to discharge groundwater into the Margaret River. This year-round discharge ensures that there are permanent pools in the river bed, providing important habitat for a range of flora and fauna. The EPA was primarily concerned that there was a possibility that the mine could impact directly on the Sue aquifer, and indirectly on the Leederville aquifer and, consequently, the Margaret River. The EPA (2011b) determined that the loss of the permanent pools was possible, albeit remote, and that this would be a catastrophic outcome (consequence).

As a result, the EPA (2011b) concluded:

> Even though some risks may be capable of being managed so as to reduce them to acceptable levels, the environmental consequences of some low probability events may be such (i.e. serious, widespread or irreversible) that the proposal, taken as a whole, on balance presents unacceptable risks to important environmental values. In conclusion, it is the EPA's judgment that should the proposal be implemented, the serious risks to important environmental values in the Margaret River region, especially surface and groundwater and the consequential impacts on the social surroundings, render this proposal environmentally unacceptable.

While this assessment by the EPA was welcomed by many in the local community, industry was critical of the EPA's assessment, especially the aspect that where a proposal could result in a catastrophic outcome, no matter how remote the likelihood, that proposal could never be made environmentally acceptable.

The second issue is, what process did the EPA use in determining where a proposal sat within the matrix? Unlike the workshop approach used by Fletcher above, the EPA itself made the determination. This is a somewhat less democratic process than the Fletcher approach but has the advantage of fewer people having input, so that reaching a consensus will be much easier – the EPA is a board made up of only five people.

The EPA also tried to apply this matrix to a few other projects and in each case it was the EPA itself that determined where a proposal was on the matrix. However, none of its assessment reports made reference to the matrix, and recent advice from the EPA indicates that it is no longer using this approach and has adopted a broader risk management framework with a focus on uncertainty rather than risk. The EPA has found that the use of the matrix caused all those involved in the ESIA to focus on the detail of inputs into the matrix rather than outcomes of the ESIA (Taylor 2015, personal communication).[1]

To summarise this section, while QRA is still a useful tool in ESIA in some circumstances, notably industrial proposals, it has significant limitations in assessing risks to complex ecological systems. Further, the assumed inherent objectivity of QRA has come into question, including decisions around what is tolerable/unacceptable risk. In these complex ecological and social contexts, qualitative risk assessment emerged as a tool in ESIA, where the inherent subjectively and uncertainty of risk assessment is recognised.

18.4 Mitigation

Risk assessments, especially QRAs, are meant to inform the decision-making of both policy makers and the affected industry. For policy makers, QRAs inform land-use planning decisions: the location of new industries and industrial areas and their proximity to existing residential areas, and the location of new residential areas. ESIAs of these new proposals will consider the acceptability of risk to human health and safety.

Traditional QRAs led to the application of a mitigation hierarchy in this way:

- *Reduce* – if the risk is too high then the proposal will need to be modified to reduce that risk to acceptable levels. This will primarily mean introducing stronger management measures at the source of the hazard: the industry itself. If the proposal is for new residential areas near existing industries, the proposal could be modified to keep more sensitive land uses away from the industry and have less sensitive land uses near the boundary.
- *Avoid* – if the risk is too high, then an alternative location could be considered for the industry, or no sensitive land uses could be permitted within the tolerable/unacceptable risk contour.

Offsetting residual risk is not an option for QRA.

For industry, QRA is a risk management tool for the protection of its own workforce and surrounding industries. The risk assessment will highlight areas of operation that need to be better-managed to keep risks to acceptable levels.

18.5 Beyond QRA and qualitative risk assessment?

More recent approaches to qualitative risk assessment may well flag up a new approach to managing risk, away from the idea of determining tolerable/unacceptable risk toward one of managing uncertainty. The approach can be summarised as follows. First, the EPA identifies objectives that need to be met for each environmental issue (called a factor) relevant to a proposal being assessed. The EPA then considers the predicted impacts of the proposal and any additional management measures that are to apply, and can arrive at one of three possible conclusions:

- The EPA could find that the proposal meets its objectives. In this case, the proposal is acceptable.
- The EPA could find that the proposal cannot meet its objectives. In this case, the proposal is unacceptable.
- There is uncertainty as to whether its objectives can be met. In this case the EPA would recommend a precautionary approach to implementation, involving a staged approval process or adaptive management.

In some cases, the EPA objective involves meeting quantifiable criteria: for example, air-quality standards, water-quality standards, no loss of critical environmental assets (e.g. nature reserves) and even QRAs of industry. In other cases, these objectives are general, and involve the EPA carrying out a qualitative assessment of the proposal. For example, in a recent assessment of a proposed LNG processing plant and associated port facilities, the EPA noted there would be impacts on tidal benthic habitats, largely because of dredging activities for the port, but the extent of those impacts was uncertain: there was a risk that the impacts would be unacceptable (EPA 2011a).

The EPA set the following objective in assessing this proposal against impacts on this issue: 'to maintain the abundance, diversity, geographic distribution and productivity of flora and fauna at species and ecosystem levels through the avoidance or management of adverse impacts and improvement in knowledge'.

There is no obvious quantitative way to show that this objective has been met, so the EPA carried out a qualitative assessment. In summary it concluded as follows:

Implementation of the proposal would result in unavoidable impacts to subtidal Benthic Primary Producer Habitat ... Impacts are predicted based on modelling results for which there is always some degree of uncertainty

... In view of these uncertainties, the EPA has recommended detailed conditions ... with the aim of limiting loss of subtidal benthic habitat to as low as reasonably practicable and to ensure as high a standard of monitoring and management as practicable.

(EPA 2011a)

In short, there is significant uncertainty about the nature and extent of impacts on benthic habitat: there is a risk that impacts would be unacceptable. To reduce that risk, additional management measures were required, including the monitoring of impacts, and adaptive management in response to the results of the monitoring.

To summarise this section, a possible recent development in ESIA is the shift in emphasis from determining tolerable/unacceptable risk to a focus on managing and reducing uncertainty through more adaptive approaches. Such approaches put greater emphasis on the follow-up in ESIA (i.e. post-assessment) to ensure that impacts are acceptable.

18.6 QRA and ESIA

The links between QRA and ESIA should now be quite apparent. First, both QRA and ESIA are about assessing the impacts of new industrial proposals, with QRA focusing on human health and safety aspects of the proposal, whereas ESIA looks at a broader range of environmental impacts. QRA can be seen as a tool within the ESIA process. Second, on a more conceptual level, these two processes have similar theoretical underpinnings. Historically, both QRA and ESIA evolved at about the same time, emerging in the 1960s as rationalist approaches to decision-making (Jay et al. 2007). Practitioners of both processes saw both assessments as being objective tools based on good science and arriving at clear outcomes about the acceptability of the proposal being assessed. Finally, just as the assessment part of ESIA needs to be seen within a broader framework of environmental planning and management, so too QRA needs to be seen within a broader framework risk management. Figure 1.1 summarised the key steps in ESIA. The risk management framework has parallels with this.

The discussion of risk above sees risk assessment as an internal process within ESIA that informs both the overall assessment process and the subsequent decision-making and follow-up based on an ESIA. A broader view of risk and ESIA is how the ESIA process affects risk assessment outside the ESIA process. Graetz and Frank (2013) see this as integrated risk management, which 'considers the relationships between the generation of risks for social groups or individuals and the consequent transmutation of those risks into business risks'.

This is the process where the outcomes of the risk assessments within an ESIA, particularly the social risks, and the broader assessment process, impact on the overall risk to the business. If the social risks have been managed and reduced, then the proposal is likely to receive community acceptance. This can be broadened into consideration of all of the ESIA issues, including environmental. In short, the quality and credibility of an ESIA can impact on risks to the business and the proponent. There is also an element of political risk here.

Arvanitis et al. (2015), in an article examining the role of multilateral banks in funding projects in developing nations, noted that the banks added value to these projects by moderating both political and financial risks. Political risk was seen as the potential for governments to intervene negatively in the projects by either nationalizing the project or reneging on contracts. Financial risk related to the financial viability of the project. These risks are higher in countries that have weaker institutions, including legislation. Clearly, given that these projects would normally go through an ESIA, those countries with strong ESIA legislations and processes will have lower political and financial risk associated with them.

Most of these banks have adopted the Equator Principles (EP), which are a 'risk management framework, adopted by financial institutions, for determining, assessing and managing environmental and social risk in projects' (www.equator-principles.com). These principles are used to guide the ESIA processes in borrowing countries. The EP note that the assessment processes in the borrowing nation need to be complied with, but also require additional standards of assessment where the local assessment process is inadequate. While the EP aim to achieve sustainable environmental and social outcomes, there is a clear recognition that in delivering these better outcomes, the financial risks of the project are reduced.

Banhalmi-Zakar and Larsen (2014) identify another type of risk for banks financing new projects – reputational risk. Most new projects attract some level of public concern about their environmental and social impacts, and the reputations of both the proponent and the lenders come under scrutiny. Banhalmi-Zakar and Larsen argue that a robust and credible assessment process, in this case a strategic environmental assessment, will address these concerns and help reduce public concerns. In this way, the risk to the reputation of the lending institution is also reduced.

Reputational risk became a significant issue around the assessment and approval of a large coal mine in Queensland, Australia. The site for the proposed Carmichael coal mine is within the Galilee Basin in central Queensland, and would involve the export of 60 million tonnes of coal per year. One of the key environmental issues related to the construction and operation of the port, which would be near the Great Barrier Reef, is the possible loss of coral through construction dredging and the ongoing movement of ships during the operational phase. The project has been vigorously opposed by environmental groups and some community groups, including farmers. Two of Australia's biggest banks announced that they will not be providing financing for the project – the NAB and the Commonwealth Bank. The NAB's announcement noted that: '(the) NAB looked at all financing project proposals on a case-by-case basis and assessed a potential borrower's capacity to repay based on management, reputation, industry funda-mentals, legal, tax, insurance and environmental, social and government risks' (Haxton 2015).

Clearly, the NAB identified financial, reputational and political risk from being involved in the proposal, which were in addition to the environmental and social risks.

18.7 Conclusions

This chapter explores the broad notions of risk and risk assessment, and how it is applied within the practice of ESIA. Key words and concepts around risk and risk assessment include hazards, hazardous events, acute risk, and chronic risk. Central to the notion of risk is uncertainty. Risk can be summarized in the equation: Risk = Frequency × Consequences.

Two broad types of risk assessment were identified – quantitative risk assessment (QRA) and qualitative risk assessment. Both ESIA and QRA can be seen as part of a broader planning and management framework, where each assessment process provides a measure of likely impacts, which can then be compared to established criteria as part of decision-making and follow-up (post-assessment). Mitigation hierarchies can be applied to both assessment processes.

The key outputs of QRAs are risk contours, which are lines of equal risk, mostly individual risk, or the likelihood (probability) that a person will be fatally injured by a hazardous event to which that person may be exposed. It is expressed numerically as the expected frequency of fatalities per year. In quantifying risk in this way the issue of tolerable/acceptable risk is raised. Two approaches were identified:

- A single number approach – for example risk above 1×10^{-6} for residential areas is considered unacceptable in many jurisdictions.
- A tiered approach, involving two criteria, with an upper level where risk above it is considered unacceptable, and a second level where risk below it is considered insignificant. ALARP is applied for proposals that have risk between the two levels.

In more recent times QRA has come under criticism and its limitations recognised, especially in dealing with complex ecological and socio-political contexts. This led to the evolution of qualitative risk assessment, where the inherent subjectivity and uncertainty of risk assessment is recognised. Risk matrices are often used either for single project based ESIAs or, where a complex ecological system is involved, in a strategic environmental assessment. It was suggested that there may be a new trend in ESIA away from an emphasis from determining tolerable/unacceptable risk to a focus on managing and reducing uncertainty through more adaptive approaches.

Finally, this way of looking at risk can be seen as an internal process within ESIA, where the risk assessment is a key tool in ESIA and the ultimate decision-making. The chapter finished with looking at ESIA and external risk, where the ESIA can reduce the financial and political risks of a project, and can also enhance the reputations risk of both proponents and lending institutions.

Note

1   Kim Taylor was the General Manager of the Office of the Environmental Protection Authority in WA.

References

Adams J 1995. *Risk*. London: Routledge.

Arvanitis Y, M Stampini and D Vencatachellum 2015. Balancing Development Returns and Credit Risks: Project appraisal in a multilateral development bank. *Impact Assessment and Project Appraisal* 33(3), 195–206.

Ashrafi M, H Davoudpour and V Khodakarami 2015. Risk Assessment of Wind Turbines: Transition from pure mechanistic paradigm to modern complexity paradigm. *Renewable and Sustainable Energy Reviews* 51, 347–355.

Astles KL, MG Holloway, A Steffe, M Green, C Ganasin and PJ Gibbs 2006. An Ecological Method for Qualitative Risk Assessment and Its Use in the Management of Fisheries in New South Wales, Australia. *Fisheries Research* 82, 290–303.

Banhalmi-Zakar Z and S Vammen Larsen 2014. How Strategic Environmental Assessment Can Inform Lenders about Potential Environmental Risks. *Impact Assessment & Project Appraisal* 33(1), 68–72.

Carpenter RA 1995. Risk Assessment. *Impact Assessment* 13(2), 153–187.

Committee on the Institutional Means for Assessment of Risks to Public Health 1983. *Risk Assessment in the Federal Government: Managing the process*. National Research Council, Washington DC: National Academy Press.

Cooke RM 2009. A Brief History of Quantitative Risk Assessment. *Resources*, Summer: 8–9.

Cooper DF and C Chapman 1987. *Risk Analysis for Large Projects: Models, methods, and cases*. 1. Chichester: John Wiley & Sons.

Corburn J 2002. Environmental Justice, Local Knowledge, and Risk: The discourse of a community-based cumulative exposure assessment. *Environmental Management* 29(4), 451–466.

Dan S, CJ Lee, J Park, D Shin and ES Yoon 2014. Quantitative Risk Analysis of Fire and Explosion on the Top-side LNG-liquefaction Process of LNG-FPSO. *Process Safety and Environmental Protection* 92(5), 430–441.

Eduljee GH 2000. Trends in Risk Assessment and Risk Management. *The Science of the Total Environment* 249(1–3): 13–23.

Elliott M 2003. Risk perception frames in environmental decision making. *Environmental Practice* 5(3): 214–222.

EPA (Environmental Protection Authority) 1987. *Environmental Protection Authority*

Guidelines: Risks and hazards of industrial development on residential areas in Western Australia. Bulletin 278. Perth, Western Australia: EPA.

EPA 1992. Criteria for the Assessment of Risk from Industry: Environmental Protection Authority guidelines. Bulletin 611. Perth, Western Australia: EPA.

EPA 1994. Risk Criteria – On-site Risk Generation for Sensitive Developments. Perth, Western Australia: EPA.

EPA 2011a. Report and Recommendations of the Environmental Protection Authority: Wheatstone Development – Gas Processing, Export Facilities and Infrastructure. Report 1404. Perth, Western Australia.

EPA 2011b. Vasse Coal Project, Vasse Coal Management Pty Ltd: Report and Recommendations of the Environmental Protection Authority. Report 1395. Perth, Western Australia: EPA.

Fletcher WJ 2015. Review and Refinement of an Existing Qualitative Risk Assessment Method for Application within an Ecosystem-Based Management Framework. ICES Journal of Marine Science 72(3), 1043–1056.

Graetz G and DM Franks 2013. Incorporating Human Rights into the Corporate Domain: Due diligence, impact assessment and integrated risk management. Impact Assessment and Project Appraisal 31(2), 97–106.

Haxton N 2015. NAB the latest to rule out funding Adani's $16 billion Carmichael coal mine. ABC News. 3 September. www.abc.net.au/news/2015-09-03/nab-rules-out-funding-adani's-$16bn-carmichael-coal-mine/6747298.

Health and Safety Executive 2001. Reducing Risks, Protecting People: HSE's decision-making process protecting. Sudbury: HSE Books.

Henkel JR, BJ Sigel and CM Taylor 2012. Large-Scale Impacts of the Deepwater Horizon Oil Spill: Can Local Disturbance Affect Distant Ecosystems through Migratory Shorebirds? BioScience 62(7), 676–685.

Jay S, C Jones, P Slinn and C Wood 2007. Environmental Impact Assessment: Retrospect and prospect. Environmental Impact Assessment Review 27(4), 287–300.

Kajenthira A, J Holmes and R McDonnell 2012. The Role of Qualitative Risk Assessment in Environmental Management: A Kazakhstani case study. Science of the Total Environment 420, 24–32.

Kirchhoff D and B Doberstein 2006. Pipeline Risk Assessment and Risk Acceptance Criteria in the State of Sao Paulo, Brazil. Impact Assessment and Project Appraisal 24(3), 221–234.

Marszal EM 2001. Tolerable Risk Guidelines. ISA Transactions 40(4), 391–399.

OGP (International Association of Oil & Gas Producers) 2010. Risk Assessment Data Directory. OGP.

Pasmana H and G Reniers 2014. Past, Present and Future of Quantitative Risk Assessment (QRA) and the Incentive it Obtained from Land-Use Planning (LUP). Journal of Loss Prevention in the Process Industries 28(1), 2–9.

Paustenbach DJ 2002. Primer on Human and Environmental Risk Assessment. In Human and Ecological Risk Assessment: Theory and practice, DJ Paustenbach, ed., 3–85. New York: John Wiley & Sons.

Travis R 2013. Quantitative Risk Assessment in Chevron: Use in Decision-Making Involving Major Risks. Safety and Chemical Engineering Education. http://sache.org/workshop/2013Faculty/presentations.asp.

WHO/FAO (World Health Organisation and Food and Agriculture Organisation) 2009. Risk Characterization of Microbiological Hazards in Food: Guidelines. Microbiological Risk Assessment Series, Volume 17. WHO/FAO.

---

19 Cumulative effects

••

Martin Broderick, Bridget Durning and Luis E. Sánchez

19.1 Introduction

Cumulative impacts or effects 'refer to the accumulation of changes in environmental systems over time and across space in an additive or interactive manner' (Spaling 1994). Cumulative effects can result from multiple effects from the same project or the combined effect of multiple developments giving rise to multiple effects. Whilst the effects from a single development may not be significant on their own, when combined with others the resultant effect could be significant.

Cumulative Effects Assessment (CEA) is a systematic procedure for identifying and evaluating the significance of the effects from multiple activities within the framework of the Environmental and Social Impact Assessment (ESIA) process. The acronyms CEA and sometimes CIA (Cumulative Impact Assessment) are used (often interchangeably). In North America, 'monitoring and management' is also considered and then the acronym Cumulative Effects Assessment and Management (CEAM) is used. In this chapter the term CEA also encompasses CIA and CEAM.

The purpose of CEA in ESIA is to consider the incremental contribution of the project together with effects from other past, present and reasonably foreseeable future activities. Assessing cumulative effects is complex. It has been described as a 'dark art', a 'wicked problem' and by Hegmann and Yarranton (2011) as: 'like forecasting weather or climate (as) the system under examination is complex and often responds to disturbance in a non-linear fashion'.

CEA practice has been described as inadequate and unsatisfactory across all industry sectors, with few assessments at any level adequately considering cumulative effects (IEMA 2011, Canter and Ross 2010, Warnback and Hilding-Rydevik 2009). More serious is the absence of consideration of cumulative effects in ESIA-based government approvals (Neri et al. 2016). The absence of effective assessments of cumulative effects has been cited (Cooper and Sheate 2002) as being a function of lack of guidance and particularly the absence of comprehensive definitions: without a clear definition it is not possible to ensure an assessment demonstrates adequate consideration of all characteristics of the effects including spatial and temporal scale.

However, practical experience is building up internationally and CEA is increasingly supporting project decision-making, as exemplified by it being required under the International Finance Corporation (2012) Performance Standards on Environmental and Social Sustainability. Additionally, a growing body of knowledge now exists in support of practice (e.g. Canter 1999, 2015; Eccleston 2011).

19.2 Definitions and concepts

19.2.1 Cumulative additive and synergistic effects

Many definitions of cumulative effects/impacts exist (Cooper and Sheate 2002, Duinker et al. 2013). A definition from an early piece of guidance from the USA is: 'the impact on the environment which results from the incremental impact of the action when added to other past, present, and reasonably foreseeable future actions' (USCEQ 1997).

A similar definition was also adopted in early European guidance (EC 1999): 'impacts that result from the incremental changes caused by other past, present and reasonably foreseeable future actions together with the project'.

Among other definitions provided by the scholarly literature and practical guidance, the International Finance Corporation's (IFC) is comprehensive: 'cumulative impacts are those that arise from the successive, incremental, and/or combined effects of an action, project, or activity when added to other existing, planned, and/or reasonably anticipated future ones' (IFC 2013).

However, there are many definitions of cumulative effects, depending on the context of the publication in which they appear. In their review, Duinker et al. (2013) include ten 'unique' definitions extracted from a range of peer reviewed literature in order to highlight this.

Since the early guidance, the language used to define cumulative effects/impacts has caused considerable confusion. We consider that best practice is to clearly set out the concepts and definitions of CEA and to that end we propose this short definition to guide practice and to help bring some transparency to the complex 'dark art' of CEA. Cumulative effects are:

> those that result from additive effects caused by other past, present or reasonably foreseeable actions together with the plan, programme or project itself and synergistic effects (in combination) which arise from the interaction between effects of a development plan, programme or project, on different components of the environment.
>
> (Durning and Broderick 2015)

This definition is also used in:

- RUK (Renewables UK) (2013) guiding principles document for offshore wind farms;

- Natural England (2014) framework (which is based on evaluation of best practice);[1]
- British Standard (BSI 2015) guide to environmental impact assessment (EIA) for offshore renewable energy projects;
- Report on recent research into CEA practice in the offshore wind farm industry.

(Durning and Broderick 2015)

Adoption of the terms 'additive' and 'synergistic' as in the above definition will, we believe, help provide some clarity, as these are often termed differently in European regulations and in selected guidance documents (see Table 19.1).

Additive effects are possibly the most obvious to recognise and most likely to be encountered. CEA for additive effects is conceptually simple: it is the sum of the magnitudes of the same effect (e.g. noise + noise) and the combined effect is assessed as if resulting from a single project (Eccleston 2011).

Synergistic effects are less obvious to recognise and therefore more complex to assess (as they can arise from the combination of different effects, e.g. noise + visual) and as the effects can be intra (occurring within one project, programme or plan) as well as inter (occurring due to different projects, programmes or plans).

IEMA (2011) distinguishes intra and inter as:

- Intra-projects effects: these effects occur between different environmental topics within the same proposal, as a result of that development's direct effects.
- Inter-project effects: this form of cumulative effect occurs as a result of the likely impacts of the proposed development interacting with the impacts of other developments in the vicinity.

Table 19.1 Examples of differing terminology to describe cumulative effects

Term	Examples of related terms in European regulations and selected guidance
Additive effects	• European EIA Directive (2011) refers to these as 'cumulative effects' • EC (1999) guidance refers to these as 'cumulative impacts' • European Strategic Environmental Assessment Directive (2001) refers to these as 'cumulative impacts' • European Habitats Directive refers to these as 'in-combination' effects
Synergistic effects	• EIA Directive (2011) refers to these as 'interrelationships' and effect 'interactions' • EC (1999) guidance refers to these as 'impact interactions' • Strategic Environmental Assessment Directive (2001) refers to these as 'synergistic' impacts • Habitats Directive does not refer to these separately

In the wider scientific community there are also different interpretations or understandings of the term 'synergistic', which has the potential to be a challenge for some practitioners in recognising synergistic effects in an ESIA context. As Eccleston (2011) also points out, synergistic effects often cannot be expressed quantitatively in CEA (as some are non-quantifiable, e.g. visual) which is a further challenge in the assessment process. The term synergistic in the context of cumulative effects first appeared in the CEQ guidance on CEA in 1997 (USCEQ 1997). It continues to be used by many CEA practitioners (Eccleston 2011) to generally describe the position when two or more effects work in concert causing an overall effect where the whole is greater than the sum of the parts. From our review of the literature, it is clear that CEA practitioners have a consistent understanding of the term synergistic in ESIA. The schematic at Figure 19.1 gives examples of what inter-project additive effects might like and what intra-project synergistic effects might look like. Case studies 1 and 2 expand on these examples.

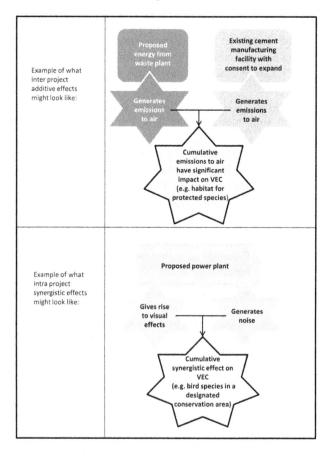

Figure 19.1

Schematic to illustrate examples of additive and synergistic effects (based on EC 1999)

Case study 1 – Example of inter-project additive effects with potential to exceed a threshold (PINS 2015a)

The proposed development was a new thermal (gas-fired) generating plant on land at the Hirwaun Industrial Estate, in Aberdare, South Wales. The Blaen Cynon Special Area of Conservation (SAC) is located 0.25km to the north east of the site. The designated feature of the site is the marsh fritillary butterfly *Euphadryas (Eurodryas, Hypodryas) aurina*. Blaen Cynon contains an extensive complex of damp pastures and heaths supporting the largest meta-population of marsh fritillary on the southern edge of the Brecon Beacons National Park. This SAC is considered to be the one of the best areas in the UK for marsh fritillary butterflies. The Blaen Cynon SAC consists of two Sites of Special Scientific Interest (SSSI), Cors Bryn-y-Gaer SSSI and the Woodland Park and Pontpren SSSI. Cors Bryn-y-Gaer SSSI is of special interest for its lowland bog and for areas of soligenous flush, marshy grassland, dry neutral grassland and lowland acid grassland. These habitats occur in a complex with wet heath, swamp and semi-improved grassland.

The lowland-raised bog feature is the most sensitive to the effects of nitrogen deposition and consequently has the lowest minimum critical load of the habitats present on site, which is 5 kgN/ha/yr. The operation of the Hirwaun Power Plant had the potential to increase levels of nitrogen and acid deposition on the Blaen Cynon SAC, which could affect the broad habitat types used by marsh fritillary.

The air quality modelling undertaken by the applicant demonstrated that, for a 30 m stack (chimney), deposition onto suitable marsh fritillary habitats alone would be less than 1 per cent of the relevant critical load. However, in the immediate vicinity of the application site there are three other developments which were of direct relevance to this application:

- Waste processing facility comprising materials recovery, anaerobic digestion and gasi-fication. The facility is located 300 m from the project site on a separate industrial park. Planning consent was given in 2010 and development of first phase had begun.
- Proposed biofuel plant within the Hirwaun Industrial Estate. Application for planning consent was submitted in 2013 but has not yet been determined.
- 20 MW liquid fuel power generation plant is situated adjacent to the project's laydown area and gas connection. This was operational and it was assumed it could remain operational during both construction and operation of the project. The plant comprises 52 × 440 kW diesel generator engines which are fired on liquid fuel. The plant only operates at times of peak demand or to balance the grid.

These three developments have the potential for cumulative additive effects with the project via influences on air quality. Mitigation and monitoring measures to address this comprised:

- Implementation of industry standard methods and procedures to ensure air quality impacts are minimised throughout all phases of the project.
- In any calendar year the operation of the gas turbine generators shall not exceed 1500 hours in total. This is the maximum number of hours assessed in the environmental statement and so this requirement ensures that the authorised development is operated in accordance with the worst-case scenario assessed.

Developer C.GEN Killinghome Ltd proposed to develop a new thermal generating power station (together with associated development) that will operate either as a Combined Cycle Gas Turbine (CCGT) plant (gas-fired) or as an Integrated Gasification Combined Cycle (IGCC) plant (coal-fired), with a total electrical output of up to 470 MWe, at North Killinghome, north-east England. The proposed development would lie partly on the south bank of the Humber Estuary, which is designated under European law as an important site for nature conservation and forms part of the Natura 2000 network of sites.[1]

A worst-case assessment of the delivery of coal from Immingham Docks by half-trains noted that a maximum of 16 half-train deliveries would be required each day over a 12-hour period to provide coal to the gasification plant. These trains would run full length within five metres of the North Killingholme Haven Pits (NKHP) SSSI.

The boundary of the Special Protection Area (SPA)[2] and the Ramsar site extends inland to take in the NKHP SSSI. The site comprises three pits of differing size and salinity, both factors which contribute to its national and local importance. The Black Tailed Godwit (*Limosa limosa*) (a species of bird) is a qualifying species for the Humber Estuary SPA and is therefore protected by EU Directives. Black-tailed godwits migrate from Iceland and the Faroe Islands (north of Scotland) to the Humber Estuary to undergo their post-breeding moult. Within the SPA, foraging Black-tailed godwit numbers peak in the autumn, coinciding with peak usage at the adjacent NKHP where foraging birds move to roost. Upon making the return migration from their breeding grounds, these two connected foraging and roosting sites are the preferred locality for Black-tailed godwits in the Humber Estuary SPA.

From the first, it was clear that the maintenance of the integrity of the Humber Estuary SPA was predicated on measures to mitigate the **cumulative intra synergistic effects of the noise and visual aspect of the trains** on the qualifying features within the NKHP and in particular the Black-tailed godwits. The Applicant, in its early submissions, argued that the Black-tailed godwit and other NKHP bird species would become habituated to the noise and visual disturbance associated with the trains.

The Examining Authority[3] and the statutory nature conservation body Natural England believed there was a reasonable scientific doubt surrounding the habituation of birds, in particular for the Black-tailed godwits, to the cumulative synergistic visual and noise disturbance from train movements. *Natural England considered that additional mitigation measures were required in order to reliably conclude that there will not be an adverse effect on roosting Black-tailed godwits in NKHP.*

The Applicant sought to resolve the habituation issue by isolating and addressing the aspects of the interaction of train movements and their cumulative synergistic effects on birds at NKHP. The two components of the synergistic effects of train movements on NKHP are noise and visual disturbance. The Applicant therefore sought to determine how those impacts could be addressed via mitigation measures. Mitigation was secured and delivered through the following Development Consent requirements:

- Requirement 14 – Construction Environment Management Plan;
- Requirement 46 – Train speed at NKHP;
- Requirement 47 – Acoustic hoarding;
- Requirement 48 – Visual attenuation of train movements;
- Requirement 49 – Control of construction noise at NKHP.

1   Natura 2000 is a network of core breeding and resting sites for rare and threatened species, and some rare natural habitat types which are protected in their own right. The aim of the network is to ensure the long-term survival of Europe's most valuable and threatened species and habitats, listed under both the Birds Directive and the Habitats Directive.

2   Special Protection Areas (SPAs) are strictly protected through the EC Birds Directive. They are classified for rare and vulnerable birds (as listed on Annex I of the Directive), and for regularly occurring migratory species

3   Martin Broderick was member of the Examining Authority who were charged with examining the application and then making a recommendation to the UK government.

19.2.2 Valued ecosystem components

Besides the concept of cumulative effect/impact, the notion of 'valued ecosystem component' (VEC) is seminal to CEA. The term was first used in a major review of the state of the art of EIA in Canada (Beanlands and Duinker 1983) and later adopted for CEA practice. Canadian guidance defines VEC as: 'any part of the environment that is considered important ... on the basis of cultural values or scientific concern' (Hegmann et al. 1999).

The acronym VEC is also adopted by IFC (2013), but renamed as 'valued environmental and social components' and defined as 'environmental and social attributes that are considered to be important in assessing risks'.

An equivalent term is simply 'valued component', used in new draft guidance in Canada (CEAA 2014) and defined as: 'environmental features that may be affected by a project and that have been identified to be of concern by the proponent, government agencies, Aboriginal peoples, or the public'.

Whatever the term, the fundamental aspect is the need to scope-in relevant issues for consideration in CEA and to scope-out less relevant issues so as to focus the assessment in a limited, but important, number of matters.

19.2.3 Cumulative effects mechanism and typology

The mechanisms which cause cumulative effects can be categorised in a number of different ways according to their spatial and temporal attributes. Table 19.2 is a typology of cumulative effects which incorporates temporal and spatial attributes in the context of offshore wind farms.

19.3 Key legislation and guidance

19.3.1 Key legislation

European Union

The first European EIA Directive included the need for a description of the likely significant effects of the proposed project on the environment, including

Table 19.2 A typology of cumulative effects – with examples relating to offshore wind farms (adapted from Spaling 1994, DEAT 2004)

Cumulative effect mechanisms	Description	Examples for offshore wind farms for temporal and spatial attributes
Time crowding	Frequent, repetitive or simultaneous effects on an environmental resource	Piling noise Transport of materials Decommissioning activities
Time lag	Long delays between cause and effect	Sediment degradation Effects exacerbated by climate change
Space crowding	High spatial density of effects on an environmental system	Several wind farm developments in a single coastal zone; high use maritime areas – navigation, fishing, military
Cross boundary movement	Effects occur some distance away from source	Impacts on migratory birds Impacts on migratory marine mammals
Fragmentation	Change in landscape or seascape pattern	Habitat fragmentation of fisheries
Triggers	Fundamental changes in system behaviour or structure	Scouring triggers loss of VECs, bird collisions cause population decline
Nibbling	Incremental or decreasing effects	Gradual loss of natural areas at project margins, permanent incremental effects of scour Reduction in species density due to reduction in feeding grounds Population resilience may suffer as a result of collisions

'direct effects and any indirect, secondary, **cumulative** ... effects of the project' (Directive 85/337/EEC). This requirement has continued to be included in subsequent amendments to the Directive. The most recent amendment in 2014 (Directive 2014/52/EU) strengthens the need for consideration of cumulative effects:

- Annex III – criteria to determine whether the projects listed in Annex II (of Directive 2011/92/EC) should be subject to an environmental impact assessment include:
  - '1. The characteristics of the project must be considered with particular regard to (amongst other factors) (b) **cumulation** with other existing and/or approved projects'
  - '3. The likely significant effects of projects on the environment must be considered ... taking into account (amongst other factors) (g) the

cumulation of the impact with the impact of other existing and/or approved projects'

- Annex IV – The environmental impact statement should include:
  - '5. A description of the likely significant effects of the project on the environment resulting from, inter alia:
    - (e) the **cumulation** of effects with other existing and/or approved projects, taking into account any existing environmental problems relating to areas of particular environmental importance likely to be affected or the use of natural resources'
  - 'The description of likely significant effects ... should cover the direct effects and any indirect, secondary, **cumulative**, transboundary ... effects of the project'.

Similarly, EC Directive 92/43/EEC (the 'Habitats Directive') requires that, where a plan/project is likely to have a significant effect on a Natura 2000 site (i.e. Special Areas of Conservation or Special Protection Areas), either individually or **in combination** with other plans/projects, then the proposed plan/project needs to have an Appropriate Assessment made of its implications for the site. The Strategic Environmental Assessment Directive (2001/42/EC) also requires information to be provided on 'the likely significant effects [these effects should include secondary, cumulative, synergistic ... effects] ... on the environment.'

USA

In the United States, cumulative effects should be addressed in all environmental impact statements prepared under the National Environmental Policy Act. In the early implementation of the Act, consideration of cumulative effects was scarce and the practice only emerged in the 1990s. Publication of guidance by the Council on Environmental Quality in 1997 (USCEQ 1997) was seminal in advancing practice.

Canada

The assessment of cumulative effects is required by the 2012 Canadian Environmental Assessment Act (and the original 1992 Act). Guidance was first published in 1999 (Hegmann et al. 1999) and was under revision at the time of writing. The Act requires EIAs to take into account 'any cumulative effects that are likely to arise from the designated project in combination with other physical activities that have been or will be carried out'.

19.3.2 Guidance and standards

There is a large body of international academic literature on CEA. Early generic guidance focused on defining and explaining the process, and the EC

(1999), US (USCEQ 1997) and Canadian (Hegmann et al. 1999) guidance is still often cited. More recent generic guidance builds on an evidence base of practice and has evolved from guidance on the assessment of particular receptors (e.g. birds) or valued attributes (e.g. protected habitats), to guidance for particular sectors – e.g. nature conservation organisations (Natural England 2014), emerging markets (IFC 2013), hydropower in Turkey (World Bank/ESMAP 2012), coal mining in Australia (Franks et al. 2010a) and impacts on Australia's Great Barrier Reef (Anthony et al. 2013). Natural England (2014) provides a useful accessible summary of best practice in CEA in relation to Marine Protection Areas. Durning and Broderick (2015) review current UK practice in relation to CEA and offshore windfarms, with inclusion of case studies.

Duinker et al. (2013) highlight a number of areas for improvement in CEA practice. They recommend that future guidance needs to focus on a number of factors:

- Defining concepts of cumulative effect (they suggest simple sentences are not sufficient and conceptual frameworks are needed);
- The use of scenarios and particularly an expanded definition of what are 'reasonably foreseeable projects';
- The evolution of analytical methods (which needs to be reflected and made transparent in records of the outcome of assessment);
- The importance of collaboration of relevant stakeholders and implementation of appropriate governance models (which needs to be acknowledged and addressed);
- The use of thresholds and balancing the precautionary approach;
- Strengthening follow-up and monitoring post consent;
- Sharing of knowledge accumulated.

The importance of such factors has been identified by industry and led to the development of the Renewables UK Guiding Principles (RUK 2013; see Box 19.1) which were developed collaboratively by industry, regulators and stakeholders.

As noted, guidance is now being developed on CEA for specific sectors (e.g. World Bank/ESMAP guidelines on hydropower projects in Turkey). Recent EIA guidance for offshore renewables (not just offshore wind farms – BSI 2015) draws on the RUK Guiding Principles document in establishing the terminology and key principles for CEA. BSI (2015) does not go so far as to give specific guidance on CEA methodology but cross-refers to other guidance (such as Natural England 2014). However, it does emphasise the importance of 'confidence assessments' to ensure that evidence used in any assessment is robust and provides an evaluative process to ensure that evidence is 'fit for purpose' when being used to inform decision-making. The Planning Inspectorate, which are part of the decision-making process for national infrastructure in England and Wales, have produced an advice note on CEA for

Box 19.1 Guiding principles for cumulative impacts assessment in offshore wind farms (RUK 2013)

1   CEA is a project-level assessment, carried out as part of a response to the requirements of the European EIA, Habitats and Wild Birds Directives, designed to identify potentially significant impacts of developments and possible mitigation and monitoring measures.

2   Developers, regulators and stakeholders will collaborate on the CEA.

3   Clear and transparent requirements for the CEA are to be provided by regulators and their advisers.

4   CEA will include early, iterative and proportionate scoping.

5   Boundaries for spatial and temporal interactions for cumulative effects assessment work should be set in consultation with regulators, advisers and other key stakeholders, in line with best available data.

6   Developers will utilize a realistic Project Design Envelope#.

7   Developers will consider projects, plans and activities that have sufficient information available in order to undertake the assessment.

8   The sharing and common analysis of compatible data will enhance the CEA process.

9   CEA should be proportionate to the environmental risk of the projects and focused on key impacts and sensitive receptors.

10 Uncertainty should be addressed and where practicable quantified.

11 Mitigation and monitoring plans should be informed by the results of the CEA.

#The Project Design Envelope (or 'Rochdale Envelope' as it is also known) is an acknowledged way of dealing with an application comprising EIA development where details of a project have not been resolved at the time when the application is submitted. Based on UK case law, it is mainly utilised for major infrastructure developments. An explanation on its use is given by PINS (2011).

proponents seeking consent for infrastructure projects (PINS 2015c). The Advice Note drew heavily on the analytical framework contained in Durning and Broderick (2015).

Durning and Broderick (2015) provide a useful step-by-step guide to ensuring that all relevant issues have been included in the CEA process and it is itself based on a range of existing guides and literature on CEA practice. Table 19.3 gives a summary of the framework. As it sees CEA as an integrated part of the ESIA process, selection of receptors/VECs is undertaken during the scoping and baseline assessment stage. The IFC (2013) guidance is directed to private projects in emerging markets but it also provides a useful approach to CEA in any country. It draws heavily on North American experience and recommends a stepwise approach. Reflecting that assessment of cumulative effects is, in some circumstances, carried out independently of the ESIA process, it includes scoping and consideration of baseline conditions in its framework. Canter (2015) is a comprehensive study which builds upon two decades of North American experience and reviews several generic and specific stepwise processes for conducting CEAM studies. Canter and Ross (2010) provide a condensed framework in six steps. These two frameworks are also provided in Table 19.3 and show how there is similarity in the three processes proposed. The IFC is notable in containing additional sections which look at wider issues of stressors and resilience.

Although European and North American legislation require that, when assessing the impacts of an individual project, consideration be given to cumulative effects, some practical applications may require the assessment of several spatially concentrated and simultaneous projects. Figure 19.2 shows one framework used to assess ten mining and related projects using the pressure-state-response approach (Neri et al. 2016). In this case, the scoping phase started by selecting projects to be assessed, but the framework can equally be applied by selecting VECs first.

19.4 Scoping and baseline

Scoping should be undertaken as early in the process as possible; early scoping helps to focus on key effects and makes the CEA process more efficient and proportionate. However, at this early stage there may not be enough information to scope issues effectively and subsequent iterative reviews may need to be undertaken where appropriate.

The working definition of cumulative effects should be clearly defined at this stage. Temporal and spatial limits should be established through consultation with regulators, advisors and other key stakeholders, and be in line with the best available data. The RUK (2013) Guiding Principles advise that spatial boundaries should take account both of the relevant spatial scales for individual receptors (such as foraging distances, migratory routes) and the spatial extent of environmental changes introduced by developments (area of influence), so that all potential impact pathways can be identified in line with the source–pathway–receptor model (see Box 19.2).

Temporal boundaries should take account of the project life cycle, the duration of environmental changes introduced by the project at different phases of the life cycle, and the life cycles and recovery times of potentially affected receptors. The temporal nature of impacts can be significant, e.g. the impact of three projects built concurrently over two years may be different to three projects built without overlap over ten years. The temporal scale of the assessment should include the lifetime of the project (construction, operation and decommissioning) and consider the cumulative effects of constructing, operating and decommissioning any reasonably foreseeable and current projects and activities within that timeframe. No single spatial or temporal scale is generally appropriate for all resources or issues (Eccleston 2011).

Other past, present and 'reasonably foreseeable future projects' (RFFP) should be identified in the form of a 'long list', i.e. sources that could affect

Table 19.3 Comparison of CEA frameworks

Canter and Ross 2010	IFC 2013	Durning and Broderick 2015 (summary)
1 Identify the incremental direct and indirect effects of the proposed project on selected VECs. Once the VECs have been selected, they should be subject to each of the following five steps.	Step 1: Scoping Phase I • Identify and agree on VECs in consultation with stakeholders • Determine the time frame for the analysis • Determine spatial boundaries (study area)	1 Clearly explain your definition of cumulative effects – identify likely significant CEA elements associated with the proposed project/plan and define approach.
2 Identify appropriate spatial and temporal boundaries for each VEC.		2 Establish temporal and spatial limits – spatial can be different for noise, air quality, landscape and visual impact assessment etc.
3 Identify other past, present and reasonably foreseeable within the space and time boundaries that have been, are, or could contribute to cumulative effects on the VECs.	Step 2: Scoping Phase II • Identify other past, existing or planned activities within the analytical boundaries • Assess the potential presence of natural and social external influences and stressors	3 Identify other past, present and Reasonably Foreseeable Future Projects (long list of sources) that could affect receptors (VECs, humans, resources).
		4 Establish sensitivity of receptors and define thresholds (regulatory or other).
4 'Connect' the proposed project and other actions in the study area to the selected VECs and their indicators.	Step 3: Baseline • Define existing conditions of VECs • Understand their potential reaction to stress and resilience • Assess trends	5 Define and describe the baseline.
For selected VECs, assemble appropriate information on their indicators and describe and assess their historical to current and even projected conditions.		6 Establish source–pathway–receptor (short list). Consult on this list giving reasons for dropping projects from stage 3 above long list.

Table 19.3 continued

Canter and Ross 2010	IFC 2013	Durning and Broderick 2015 (summary)
5 Assess the significance of the cumulative effects on each VEC over the time horizon for the study	Step 4: Assess cumulative impacts on VECs • Identify potential environmental and social impacts and risks • Assess expected impacts • Identify any potential additive, countervailing, masking and/or synergistic effects Step 5: Assess significance of predicted cumulative impacts • Define appropriate thresholds and indicators • Determine impact and risk magnitude and significance • Identify trade-offs	7 Assess significance of effects (cumulative – additive and synergistic) which established the likely future state of receptors.
6 For VECs that are expected to be subject to negative incremental impacts from the proposed project and for which the cumulative effects are significant, develop appropriate action or activity-specific mitigation measures	Step 6: Management of cumulative impacts: design and implementation • Design management strategies to address significant cumulative impacts on selected VECs • Engage other parties needed for effective collaboration or coordination • Propose mitigation and monitoring programs • Manage uncertainties with informed adaptive management	8 Modify or add alternatives or propose mitigation to avoid or reduce cumulative effects. 9 Detail the uncertainty and limitations in the assessment. 10 Monitoring and management via an Environmental Management Plan

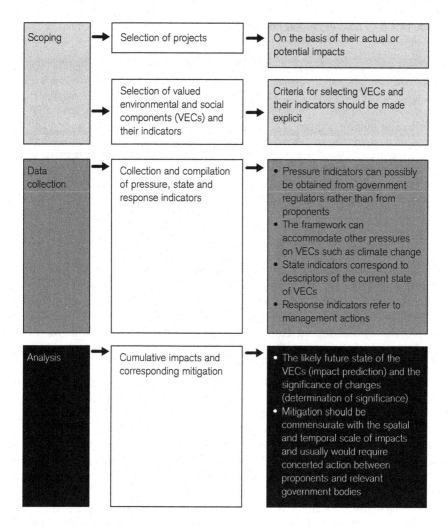

Figure 19.2

A framework for CEAM of multiple projects (modified from Neri et al. 2016)

receptors such as VECs, humans or resources. Practitioners should use broad screening criteria to identify possible projects and other activities for inclusion, and use the scoping process to focus on the most relevant projects. However, the detail of which projects and human-use activities should be included in a CEA will need to be discussed and agreed with regulators and their statutory advisers. Broadly, RFFP are projects that are currently known to the planning system or already within the consenting process. Table 19.4 lists examples of activities that might be included, based on practice in the USA (Eccleston 2011) and UK (PINS 2015a).

Box 19.2 The source–pathway–receptor model

The source–pathway–receptor model is frequently used in CEA. It identifies the links (pathways) between the source of an impact and the receptor of the impact. For example highway construction (source) generates noise which follows a pathway to a sensitive bird species (receptor) in adjacent habitat. The model is important in CEA because multiple sources (e.g. highway + housing development + factory) can affect one receptor, or one source can affect a receptor via several different pathways (e.g. sensitive bird species affected by highway noise + fragmentation of their habitats + roadkill).

The model can also be used to identify possible mitigation measures: can the source or pathway be removed? Can the receptor be made more robust?

Physical proximity of other actions is not a key reason for including RFFP, but their potential to influence selected VECs is important (Canter 2015), i.e. considering the source–pathway–receptor reasoning. Sensitivity of receptors should be established and thresholds (regulatory or other) defined in order to appropriately describe the baseline. For an assessment to be meaningful it has to be based on evidence. World Bank/ESMAP (2012) uses a scale of three levels to screen other actions (Figure 19.3).

A key challenge in CEA is to keep the assessment reasonable and in proportion to the nature and scale of the development. Consultations have an important part to play in reaching agreement about the scope of the assessment. All stakeholders have to exercise their judgement about what is appropriate and proportionate and be able to justify the approach taken. It is always important to remember that the emphasis is on identifying and assessing potentially *significant* effects rather than on comprehensive cataloguing of every conceivable effect that might occur. Carefully thinking through the potential significant cumulative effects that are likely to be generated by the development should allow a sensible decision to be reached at the scoping stage. CEA should be proportionate, focusing on key effects and sensitive receptors – i.e. a limited number of VECs – to ensure a useful assessment of the environmental risks and impacts.

19.5 Prediction and evaluation

Eccleston (2011) points out that the criteria for judging the significance of cumulative effects are no different from those for direct and indirect effects, but thresholds are frequently of greater concern in CEA. Glasson (2008) defines thresholds as: 'discrete points that must be exceeded to begin producing a given effect or result to elicit response' but notes that how they are defined can be hugely variable, from legislated thresholds to vague guidance to those based on

Table 19.4 Activities that might be included as reasonably foreseeable future projects

Categories of 'reasonably foreseeable future activities' that should be included in the CEA (modified from Eccleston 2011)	'Other development' for inclusion in CEA (modified from PINS 2015a)	
• Projects directly associated with the project under review • Projects currently undergoing regulatory review with a reasonable possibility of approval • Projects that have been formally approved	**Tier 1**	• Under construction • Permitted application(s), but not yet implemented • Submitted application(s) but not yet determined
• Projects not directly associated, but which would likely be induced as a result of the project's approval	**Tier 2**	• Projects on the Planning Inspectorate's (or other regime) Programme of Projects where a scoping report has been submitted
• Projects identified in a development plan (such as a comprehensive plan or master plan) for the area • Projects officially announced by a proponent	**Tier 3**	• Projects where a scoping report has not been submitted • Identified in the relevant Development Plan (and emerging Development Plans – with appropriate weight being given as they move closer to adoption) recognising that much information on any relevant proposals will be limited • Identified in other plans and programmes (as appropriate) which set the framework for future development consents/approvals, where such development is reasonably likely to come forward

'general societal values and preferences'. There are no commonly accepted definitions or criteria as to what constitutes an acceptable threshold. Case Study 1 (§19.2.1) provides an example of inter-project additive effects with the potential to exceed a quantitative threshold. In this case the threshold was the critical load for nitrogen deposition for the particular habitat at risk. Critical loads are a quantified estimate of exposure to a pollutant based on scientific data and an understanding of the pollutant emission and deposition process (APIS 2016). In contrast, case study 3 provides a short example of the use of theoretical thresholds and highlights the challenges to decision makers that uncertainty over project-level effects has in the CEA process.

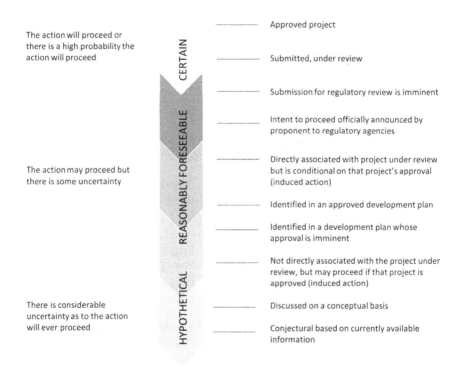

The action will proceed or there is a high probability the action will proceed

CERTAIN

The action may proceed but there is some uncertainty

REASONABLY FORESEEABLE

There is considerable uncertainty as to the action will ever proceed

HYPOTHETICAL

Approved project

Submitted, under review

Submission for regulatory review is imminent

Intent to proceed officially announced by proponent to regulatory agencies

Directly associated with project under review but is conditional on that project's approval (induced action)

Identified in an approved development plan

Identified in a development plan whose approval is imminent

Not directly associated with the project under review, but may proceed if that project is approved (induced action)

Discussed on a conceptual basis

Conjectural based on currently available information

Figure 19.3

Categories of other actions to be included in a CEA (Modified from World Bank/ ESMAP 2012)

Case study 3 – Outcome of use of theoretical thresholds (*Daily Telegraph* 2012)

In the UK in 2012, an offshore wind farm in the southern North Sea, Docking Shoal, was assessed against the theoretical thresholds for Sandwich tern *Sterna sandvicensis* harvest rates. Docking Shoal compromised the 'maximum planned capacity' for the Greater Wash Strategic Area when assessed against the cumulative effects of two other proximate offshore wind-farms (Race Bank and Dudgeon Sands). Evaluated against a precautionary theoretical threshold, Docking Shoal was refused consent, the first refusal of consent of its kind in the UK.

One criticism of CEA has been that the methodology used is often not transparent. Using a paradigm such as the source–pathway–receptor approach (see Box 19.2) to develop a short list of projects to consult on, aids in ensuring transparency. In developing the short list, it is important to explicitly give reasons for deleting projects from the 'long list'. The significance of both additive and synergistic effects should be assessed. Case study 4 provides an example of the use of this paradigm.

Case study 4 – Example use of the source–pathway–receptor approach (SmartWind 2013, DECC 2014)

An application was made for a development of up to 240 turbines in the southern North Sea, approximately 100 km off the east coast of England. The proposed development would have a gross electrical generating capacity of 1200 MW and cover 407 km². The proposals were that either two or three wind farms would be constructed within a project 'envelope' (see Box 19.1 for explanation of 'project envelope').

The EIA report included an annex on the cumulative, trans-boundary and interrelated effects of the offshore elements of the development. The onshore elements were addressed separately, although the document noted that this 'does not negate the requirement of the offshore EIA to consider onshore projects and plans, where they may have cumulative effects with the offshore elements'. The annex is an example of good practice CEA because it:

- States what has been included in the baseline assessment and explains how the 'long list' of plans, projects and activities to be considered has been devised;
- Explains how the spatial and temporal ranges have been identified;
- Uses a clear methodology for 'screening' projects in or out of the assessment process and presents the results in a transparent format (see Figures 19.4 and 19.5, which are based on the methods used in the case study).

The development was given planning consent in December 2014 with the restriction that consent for 'Work No. 3 is subject to the limitations that they cannot be built if more than 80 wind turbines are constructed as part of Works Nos 1 and 2'.

The final matrix would consider all the EIA topics that had been scoped-in to the assessment and present the screening in relation to the full RFFP long list.

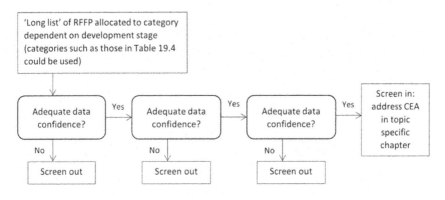

Figure 19.4

Proposed method for screening projects/plans in or out of the CEA process using a source–pathway–receptor model (adapted from SmartWind 2013)

I	Potential effect receptor pathway and potential for overlap between effect and receptor – screened in
O	Potential effect receptor pathway but no potential for overlap – screened out
OO	No effect receptor pathway – screened out
C	Low data confidence – address on a case-by-case basis

RFFP 'long list'	Status of RFFP	Data confidence	Distance from project	EIA topic A	EIA topic B
Facility 1	Tier 1 – under construction	High	< 1 km	I	OO
Development 2	Tier 3 – identified in development plan	Low	5 km	C	C

Figure 19.5

Example of matrix to present screening findings (adapted from SmartWind 2013)

19.5.1 The 'proximate cause' test of significance

Eccleston (2011) suggests using a three-phase procedure for performing a 'defensible' CEA:

1 Determining 'whether a proposal would 'proximately' cause a significant effect on a given resource';
2 Investigating 'how the affected environment has reached its current condition and projected trajectory as a result of the aggregate effects of past actions (i.e. the affected environment) and defining an appropriate spatial and temporal scale for the impact assessment';
3 Determining 'what happens to the condition and trend of this resource if no action is taken, including effects of other present and reasonably foreseeable future actions and what happens to this maintained or altered trend if the proposed or alternative actions are taken (i.e. the incremental and resulting cumulative effects of the action alternatives)'.

Eccleston's 'proximate cause test' for effect prediction is based on that of Magee and Nesbit (2008), who related the basic concepts of CEA to the US case law principle of 'proximate cause'. Proximate cause can be defined as: 'An act from which an injury results as a natural, direct, uninterrupted consequence and without which the injury would not have occurred' (Gale Group 2008).

Substituting the terms project for 'act', pathway for 'direct, uninterrupted consequence' and effect for 'injury' in the above definition (i.e. a project from which an effect results as a pathway, and without which the effect would not have occurred) and using it in concert with the source–pathway–receptor model, provides a robust evidential base and transparent methodology for identifying effects.

Proximate cause is the primary cause of an injury/effect. It is not necessarily the closest cause temporally or spatially, or the first event that sets in motion a sequence of events leading to an injury/effect. Proximate cause produces particular, foreseeable consequences without the intervention of any independent or unforeseeable cause. To help determine the proximate cause of an injury in legal cases, courts have devised the 'but for' or '*sine qua non*' rule, which considers whether the injury would not have occurred but for the act/project. A finding that an injury would not have occurred but for an act, establishes that the particular act is the proximate cause of the harm (Gale Group 2008). Simply stated there needs to be a *reasonably* close causal relationship; and a *pathway*, in place and time, between the source and the receptor for there to be a perceptible cumulative effect.

19.5.2 Dealing with uncertainty

Where uncertainty is considered to be significant, there is merit in looking at these issues in more detail as it is important to always explicitly detail the uncertainty and limitations in the assessment. Where there is insufficient evidence, this will necessarily preclude a meaningful quantitative assessment, as it could not be appropriate for developers to make assumptions about the detail of future projects in such circumstances. However, applicants should make some attempt to address cumulative effects (even if only qualitatively), even when information and data may be missing or sparse, or when it is difficult to analyse the impacts of future actions. When information is missing, sparse or unavailable, it is important to ensure that the situation and rationale for assessment conclusions are adequately documented. However, the focus of the assessment will therefore be on those project or activities for which sufficient relevant information exists.

When several similar projects are considered, or possible for a given region, average or 'typical' characteristics of a class of projects can be used to derive scenarios of additional pressure on VECs. Evans and Kiesecker (2014) used such an approach to study the cumulative impacts of a combination of wind farms and shale gas developments in the North-Eastern United States. They projected the future number of wells and turbines and described each kind of project with a set of variables and estimated their average direct and indirect footprint. In a study area of 12 Mha, they projected drilling about 122,000 new gas wells and building about 11,000 new wind turbines to assess their cumulative impact.

Improving CEA practice which is based on transparent methodology and robust evidence will aid decision makers, although as Hegmann and Yarranton

(2011) observe CEA 'can assemble geographical and biological information and present it in a manner useful to decision makers, but it is not itself a method of decision-making'. Consideration could be given to the need for a 'tool' to aid decision makers: whilst a wealth of 'tools' already exist in some industry areas (e.g. to support marine spatial planning), research by Stelzenmuller et al. (2013) noted that many of the current tools 'are designed to be used by scientists, programmers or strategic planners, with only a few that could be used by case officers (regulators)'.

Kelly et al. (2014) experimented in developing a GIS-based tool to aid decision-making in marine spatial planning around the Shetland Isles (Scotland). They 'scored' or 'weighted' the cumulative effects of existing marine activities around the islands using different criteria. They noted many challenges with doing this, including: the need for local evidence and the importance of local stakeholder and expertise input; that applying 'scores' to different impacts/pressures could be seen as subjective and therefore open to challenge; and that it focused on direct additive effects only, so did not take into account synergistic or indirect effects. CEA will continue to be a challenging process.

19.6 Mitigation

Mitigation should be proportionate to the nature of cumulative effects, e.g. landscape-scale biodiversity offsets. Practitioners should be prepared to modify or add alternatives, or propose mitigation to avoid or reduce cumulative effects (see case study 5 below). Examples of mitigation for cumulative effects have been provided in case studies 1, 2 and 5. Measures included:

- Reducing construction and operational aspects of the development;
- Provision of measures to manage the effects;
- Restricting the extent of the proposed development.

Case study 5 – Example of additive, synergistic (inter and intra) effects; the use of thresholds to manage cumulative effects and the role of SEA in CEA in mitigating effects at project level (Cole and Broderick 2007)

The Khalifa Port and Industrial Zone (KPIZ) development in Abu Dhabi UAE comprises a new port, to be developed on an artificial island made from seabed material dredged to create the port entrance channel and manoeuvring basins. The port is located approximately 3.5 km offshore and will be connected to land by a bridge. The entrance channel is approximately 15 km long in order to reach deep water.

The proposed landward Industrial Zone will occupy a land area of approximately 1,000 km² extending from the shoreline to Ras Sadr. It will have a heavy industry zone (including a steel mill and aluminium plant), a medium and light industry zone, an auto and truck transport zone, an office and retail zone, a housing community and all associated ancillary and support facilities.

Khalifa Port was officially inaugurated in December 2012; it now handles all Abu Dhabi's container traffic. The port is part of a wider industrial development project for which a Strategic Environmental Assessment (SEA) was prepared and submitted to the Environment Agency of the UAE. The SEA considered the environmental aspects of the project during the master planning and design of the port and the industrial zone. The KPIZ SEA was undertaken voluntarily rather than following a regulated process such as that of the European SEA Directive (Cole and Broderick 2007). The balance of the SEA process led to significant environmental values being considered, identified and protected as part of the master planning work.

An important part of the SEA was to consider the interactions between the KPIZ development and other plans or programmes proximal to the Taweelah area comprising large-scale, multi-use developments proposed, under construction or recently completed including:

- *Taweelah Power Plants and Desalination Facilities*: along the shoreline west of the proposed KPIZ site, which share a common intake/outfall channel arrangement. Taweelah 'B' was to be extended and there is a proposal for a Taweelah 'C' power plant generating 2.0 GW;
- *Palm Islands, The World, the Dubai Marina, and the Dubai Waterfront*: along the coast to the east.

The environmental impacts of these developments (particularly on the marine environment) were considered cumulatively with those impacts predicted to result from the KPIZ development. A study was carried out by consultancy firm Halcrow to assess the potential impacts of KPIZ on the existing power plants at Taweelah as well as the marine environment. It considered that the offshore developments along the Dubai coastline were unlikely to contribute to cumulative effects due to their considerable distance from the study. However, Taweelah power plant and desalination facilities were shown clearly to interact with KPIZ.

There are many examples of 'nibbling' impacts on mangrove stands, in the Gulf and the UAE through large-scale developments (industrial and recreational). In most cases mangrove deterioration has been in excess of 50 per cent in less than a decade. Although many of the projects involved in causing the deterioration have been relatively small in size (and arguably with small effects), it is their frequency which is now causing significant additive and synergistic effects. Inevitably, the end result is that virtually all the mangrove stands have been subjected to the gradual decimation of what was once a widespread ecosystem in the Gulf. The mangrove stands are part of an ecosystem succession of mangrove-corals-seagrass.

The SEA demonstrated that KPIZ would have significant cumulative additive (with Taweelah NOx) and synergistic effect (NOx from KPIZ and high-temperature liquid discharges from Taweelah B) on the sensitive and important coral reef habitat north of the proposed project area. As a result of these findings, to mitigate the effects, the port location and orientation were altered. The port design subsequently included the construc- tion of an environmental permanent breakwater to protect the reef from construction and operational effects, thereby avoiding potential effects to known VECs in the vicinity of the port.

It was important that the concept of environmental 'headroom' be established in a project of this scale. Headroom is defined as being the difference between the assessment criteria and the prevailing ambient pollutant concentration. The headroom allowed is a balance between the requirements of the industry and the protection of the environment, with an overriding consideration to ensure the cumulative effects (both current and future) do not exceed assessment criteria, e.g. effectively thresholds for air quality, noise, water quality and ground conditions. For some pollutants, such as nitrogen dioxide (NO_2), which will be emitted by a number of industries in the industrial zone, it is clear that there will be cumulative additive effects on VECs.

Hydrodynamic modelling included in the SEA led to major changes in the design and location of the industrial area effluent outfall. A stringent continuous monitoring program was set up to facilitate minimal impact of the project on the water quality in the region. These mitigation measures have significantly reduced the environmental impact of the project, thereby ensuring safe and sustainable operation of the port.

Therivel and Ross (2007) give examples of other project-level mitigation measures from Canada including:

- Wildlife management programmes;
- Compensation measures;
- Monitoring programmes and appropriate corrective action.

CEA presents particular problems for the implementation of mitigation proposals in terms of spatial scale and the need for collaboration. The regional spatial nature of any mitigation proposals is an important consideration in implementing mitigation plans. The potential role of Strategic Environmental Assessment (SEA) in managing and mitigating adverse effects is illustrated in case study 5. Cole and Broderick (2007) in case study 5 have shown that a SEA provided an opportunity to highlight environmental threshold levels, reveal headroom capacities, suggest guideline emissions levels, and recommend potential mitigation measures to the Industrial Zone Master Planners when appointed, so that they might incorporate these factors at the earliest point in the development of their options, with the aim of progressing their proposals within the environmental limits delineated within the Strategic Environmental Framework studies. Approaching environmental issues (and opportunities) in a strategic manner is vital.

Franks et al. (2010b) propose a hierarchy of mitigation of increasing complexity, effort and coordination for cumulative effects:

- The first level of mitigation is simply information exchange. Although this could not be simpler, many developers are reluctant to disclose information that may be relevant for managing cumulative impacts.
- Mitigation efforts evolve through pooling of resources from different developers in support of specific programmes, to coordination of response to cumulative effects of high concern to stakeholders.
- At the very end of the scale, CEA would facilitate 'proactive management of the timing and location of developments'.

The availability of qualified information (i.e. good quality and/or reliable information) contributes to reduce uncertainty (§19.5.2) and facilitates the assessment and management of cumulative impacts (Neri et al. 2016).

19.7 Monitoring and management

Effective CEA will involve some form of monitoring to assess the actual environmental outcomes that result from a development, and to provide a check on the quality of the predictions made within such assessments. There is a need for improved coordination and strategic thinking on how best to monitor wider CEA impacts, which often occur on a regional scale.

There are two main forms of CEA monitoring:

1 Compliance monitoring in line with consent conditions (including efficacy of mitigation measures and environmental impacts) undertaken by the developer;
2 Monitoring/research of wider environmental impacts, which is not necessarily the responsibility of the developer and may be undertaken collaboratively with a range of parties (wider research).

Clear objectives for the monitoring programme are essential to ensure that appropriate monitoring is implemented. They should be appropriate and proportionate to the magnitude of observed effects.

Monitoring and management of individual projects should be via an Environmental Management Plan (see Chapter 20 and Perdicoulis et al. 2012). Developers should monitor the impacts of their own developments; however, any wider monitoring proposals need to be considered in collaboration with regulators, stakeholders and other developers. When management is necessary at regional level (e.g. pooling of resources from different developers to implement joint programmes), other private and governmental parties should also be involved.

19.8 Conclusions

All stakeholders involved in the ESIA process are aware of the need to address cumulative effects, but also that a lack of consistency in terminology and lack of transparency and robustness in the methodology used to assess the effects are key areas that need to improve. There is evidence that CEA practice is improving and evolving. Key to driving practice forward has been the requirements made by decision makers and statutory stakeholders at the scoping stage. The IFC (2012) Environmental and Social Performance Standards and corresponding guidance provides much-needed orientation to developers, consultants and financial institutions. Practitioners are being more transparent in their methodologies, which will aid others to reflect on their own practice and innovate. The generation of examples of good practice will also aid this. Improving assessment practice which is based on transparent methodology and robust evidence will also help decision makers.

Several challenges remain. Uncertainty over project-level effects is likely to always be an issue. Kelly et al. (2014), in attempting to address cumulative effects assessment in marine spatial planning around the Shetland Islands, note that trying to resolve the relationship between human use of the marine environment and its ecosystem components is 'widely acknowledged' as difficult because 'several human activities have the same or similar effects on the marine environment and its ecosystems. Attempting to attribute or distinguish each effect to a single use in multi-use areas has not been achieved convincingly to date'. However, using the 'source–pathway–receptor' model – in concert with the principle of 'proximate cause' as part of the assessment methodology – can aid in identifying where there may be site specific issues.

It is hoped that the use of a clear definition, a robust/transparent analytical framework, the continuing development and publication of good practice case studies will all aid in continuing to improve and shine light on, the 'dark art' of cumulative effects assessment.

Note

1 Natural England reviewed all the latest CEA international best practice (USCEQ (1997), Hyder (1999), English Nature (2001, 2006), Natural England (2007), Canter (2008, 2012), King et al. (2009), SNH (2012), RUK (2013), IFC (2013) and in their view this definition of CEA is 'the most comprehensive and appropriate definition of cumulative impacts'.

References

Anthony KRN, JM Dambacher, T Walshe and R Beeden 2013. *A framework for understanding cumulative impacts, supporting environmental decisions and informing resilience-based management of the Great Barrier Reef World Heritage Area*. Canberra: Commonwealth of Australia.

APIS 2016. Air Pollution Information System. www.apis.ac.uk.

Beanlands GE and PN Duinker 1983. *An ecological framework for environmental impact assessment in Canada*. Halifax: Dalhousie University.

BSI (British Standards Institution) 2015. *Environmental impact assessment for offshore renewable energy projects – guide*. PD6900:2015. http://shop.bsigroup.com/forms/PASs/PD-6900/

Canter L 1999. Cumulative effects assessment. In J Petts (ed.) *Handbook of environmental impact assessment v.1*. Oxford: Blackwell Science, pp. 405–440.

Canter L 2008. Conceptual models, matrices, networks and adaptive management-emerging methods for CEA. Texas: Environmental Impact Training. Presented at Assessing and Managing Cumulative Environmental Effects, Special Topic Meeting,

International Association for Impact Assessment, November 6–9, 2008, Calgary, Alberta, Canada.

Canter L 2012. Guidance on cumulative effects analysis in environmental assessments and environmental impact statements. Texas: Environmental Impact Training.

Canter L 2015. *Cumulative effects assessment and management: principles, processes and practices*. Horseshoe Bay: EIA Press.

Canter L and W Ross 2010. State of practice of cumulative effects assessment and management: the good, the bad and the ugly. *Impact Assessment and Project Appraisal* 28(4), 261–268.

CEAA (Canadian Environmental Assessment Agency) 2014. *Technical guidance for assessing cumulative environmental effects under the Canadian Environmental Assessment Act, 2012*. Draft. www.ceaa-acee.gc.ca/default.asp?lang=en&n=B82352FF-1&offset =&toc=hide.

Cole P and M Broderick 2007. Environmental Impact Assessment (EIA) and Strategic Environmental Assessment (SEA): An exploration of synergies overseas. Third International Conference on Sustainable Development and Planning, April 2007.

Cooper L and W Sheate W 2002. Cumulative effects assessment: A review of UK environmental impact statements. *Environmental Impact Assessment Review* 22, 415–439.

Daily Telegraph 2012. Centrica criticises policy as seabirds block Docking Shoal wind farm. www.telegraph.co.uk/finance/newsbysector/energy/9382527/Centrica-criticises-policy-as-seabirds-block-Docking-Shoal-wind-farm.html.

DEAT (Department of Environmental Affairs and Tourism) 2004. *Cumulative effects assessment*, Integrated Environmental Management, Information Series 7. Pretoria: DEAT. www.environment.gov.za/sites/default/files/docs/series7_cumulative_effects_assessment.pdf.

DECC (Department of Energy and Climate Change) 2014. *Planning Act 2008 – application for the Hornsea One Offshore Wind Farm Order*. Secretary of State decision letter, 10 December 2014. http://infrastructure.planninginspectorate.gov.uk/wp-content/ipc/uploads/projects/EN010033/3.%20Post%20Decision%20Information/Other/Hornsea %20Offshore%20Wind%20Farm%20Notice%20of%20Secretary%20of%20State%20 Decision%20and%20Statement%20of%20Reasons.pdf.

Duinker PN, EL Burbridge, SR Boardley and LA Greig 2013. Scientific dimensions of cumulative effects assessment: toward improvements in guidance for practice. *Environmental Reviews* 21, 40–52.

Durning B and M Broderick 2015. *Mini review of current practice in the assessment of cumulative environmental effects of UK Offshore Renewable Energy Developments when carried out to aid decision making in a regulatory context*. Report to Natural Environment Research Council. http://bit.ly/1XPGvHt.

EC (European Commission) 1999. *Guidelines for the assessment of indirect and cumulative impacts as well as impact interactions*. Brussels: EC DGX1 Environment, Nuclear Safety and Civil Protection. http://ec.europa.eu/environment/archives/eia/eia-studies-and-reports/pdf/guidel.pdf.

Eccleston CH 2011. *Environmental impact assessment: a guide to best professional practices*. London: CRC Press.

English Nature 2001. *Habitats regulations guidance note (HRGN) no.4. Alone or in combination*. Peterborough, UK: English Nature.

English Nature 2006. A practical toolkit for assessing cumulative effects of spatial plans and development projects on biodiversity in England. English Nature Research Reports No. 673. Peterborough: English Nature.

Evans JS and JM Kiesecker 2014. Shale gas, wind and water: assessing the potential cumulative impacts of energy development on ecosystem services within the Marcellus Play. *PLoS One* 9(2): e89210.

Franks DM, D Brereton, CJ Moran, T Sarker and T Cohen 2010a. *Cumulative impacts – a good practice guide for the Australian coal mining industry*. Brisbane: Centre for Social Responsibility in Mining & Centre for Water in the Minerals Industry, Sustainable Minerals Institute, University of Queensland.

Franks DM, D Brereton and CJ Moran 2010b. Managing the cumulative impacts of coal mining on regional communities and environments in Australia. *Impact Assessment and Project Appraisal* 28(4), 299–312.

Gale Group 2008. *West's encyclopedia of American law*, 2nd Edn. Farmington Hills, MI: Gale Group.

Glasson J 2008. Principles and purposes of standards and thresholds in the EIA process. In Schmidt M, Glasson J, Emmelin L and Helborn H (eds) *Standards and thresholds for impact assessment Vol 3., Environmental Protection in the European Union*. Berlin, Heidelberg: Springer.

Hegmann G and GA Yarranton 2011. Alchemy to reason: Effective use of cumulative effects assessment in resource management. *Environmental Impact Assessment Review* 31, 484–490.

Hegmann G, C Cocklin, R Creasey, S Dupuis, A Kennedy, L Kingsley, W Ross, H Spaling and D Stalker 1999. *Cumulative effects assessment practitioners guide*. Ottawa: Canadian Environmental Assessment Agency.

Hyder Consulting 1999. *Guidelines for the assessment of indirect and cumulative impacts as well as impact interactions*. Brussels: EC DGX1 Environment, Nuclear Safety and Civil Protection.

IEMA (Institute for Environmental Assessment and Management) 2011. *Special report – the state of environmental impact assessment practice in the UK*. Lincoln, UK: IEMA. www.iema.net/iema-special-reports.

IFC (International Finance Corporation) 2012. *IFC Performance Standards on Environmental and Social Sustainability*. Washington DC: World Bank.

IFC 2013. *Good practice handbook – cumulative impact assessment and management: guidance for the private sector in emerging markets*. www.ifc.org/wps/wcm/connect/ 3aebf50041c11f8383ba8700caa2aa08/IFC_GoodPracticeHandbook_Cumulative ImpactAssessment.pdf?MOD=AJPERES.

Kelly C, L Gray, RJ Shucksmith and JF Tweedle 2014. Investigating options on how to address cumulative impacts in marine spatial planning. *Ocean & Coast Management* 102, 139–148.

King S, I Maclean, T Norman and A Prior 2009. *Developing guidance on ornithological cumulative impact assessment for offshore wind farm developers*. COWRIE. Crown Estate. www.thecrownestate.co.uk/media/5975/2009-06%20Developing%20Guidance%20 on%20Ornithological%20Cumulative%20Impact%20Assessment%20for%20Offshore %20Wind%20Farm%20Developers.pdf.

Magee J and R Nesbit 2008. Proximate causation and the no action alternative trajectory in cumulative effects analysis. *Environmental Practice* 10(3), 107–115.

Natural England 2007. *Natural England guidance on assessing the cumulative effects of development*. Peterborough, UK: Natural England.

Natural England 2014. *Development of a generic framework for informing cumulative impact assessments (CIA) related to Marine Protected Areas through evaluation of best practice*. NECR147. http://publications.naturalengland.org.uk/publication/6341085840277504.

Neri AC, P Dupin and LE Sánchez 2016. A pressure-state-response approach to cumulative impact assessment. *Journal of Cleaner Production* 126, 288–298.

Perdicoulis A, B Durning and L Palframan 2012. *Furthering EIA – towards a seamless connection between EIA and EMS*. Cheltenham: Edward Elgar.

PINS (The Planning Inspectorate) 2011. *Advice note 9: using the 'Rochdale Envelope'*. http://infrastructure.independent.gov.uk/wp-content/uploads/2011/02/Advice-note-9.-Rochdale-envelope-web.pdf.

PINS 2015a. Hirwaun Power Station. https://infrastructure.planninginspectorate.gov.uk/projects/wales/hirwaun-power-station.

PINS 2015b. North Killingholme Power Project. https://infrastructure.planninginspectorate.gov.uk/projects/yorkshire-and-the-humber/north-killingholme-power-project.

PINS 2015c. *Advice note 17: cumulative effects assessment*. http://infrastructure.planninginspectorate.gov.uk/legislation-and-advice/advice-notes.

RUK (Renewables UK) 2013. *Guiding principles for cumulative impact assessments in offshore wind farms*. www.nerc.ac.uk/innovation/activities/infrastructure/offshore/cumulative-impact-assessment-guidelines/.

SmartWind 2013. *Hornsea offshore wind farm project one. Environmental statement volume 4, Introductory annexes – Annex 4.5.1 Cumulative, transboundary and inter-relationships document*. http://infrastructure.planninginspectorate.gov.uk/wp-content/ipc/uploads/projects/EN010033/2.%20Post-Submission/Application%20Documents/Environmental%20Statement/7.4.5.1%20Cumulative%20Transboundary%20and%20Inter-related%20Effects%20Document.pdf.

SNH (Scottish Natural Heritage) 2012. *Assessing the cumulative impact of onshore wind energy developments*. Edinburgh: SNH. www.snh.gov.uk/docs/A675503.pdf.

Spaling H 1994. Cumulative effects assessment: concepts and principles. *Impact Assessment* 12(3), 213–251.

Stelzenmuller V, J Lee, A South, J Foden and SI Rogers 2013. Practical tools to support marine spatial planning: a review and some prototype tools. *Marine Policy* 38, 214–227.

Therivel R and B Ross 2007. Cumulative effects assessment: does scale matter? *Environmental Impact Assessment Review* 27, 365–385.

USCEQ (United States Council on Environmental Quality) 1997. *Considering cumulative effects under the National Environmental Policy Act*. Washington, DC: CEQ.

Warnback A and T Hilding-Rydevik 2009. Cumulative effects in Swedish EIA practice: difficulties and obstacles. *Environmental Impact Assessment Review* 29, 107–115.

World Bank/ESMAP 2012. *Sample guidelines: cumulative environmental impact assessment for hydropower projects in Turkey*. www.esmap.org/node/2964.

20 Environmental and social management plans

● ●

Bridget Durning and Martin Broderick

20.1 Introduction

The environmental effects of development are mitigated from two distinct perspectives – *ex ante* (before), through ESIA, and *ex post* (after), through environmental management systems (EMS). The need to ensure that the outcomes of environmental assessment are carried through into environmental management, and that management practices are adaptive, was first recognised in the 1970s (e.g. Holling 1978). The use of management plans, particularly Environmental and Social Management Plans (ESMPs), to ensure that mitigation measures identified during ESIA are implemented and monitoring is carried out, has been a steadily evolving area of ESIA practice since the 1990s. A study by the Institute for Environmental Assessment and Management (IEMA) into the state of EIA practice in the UK reported that 80 per cent of its survey respondents would welcome inclusion of the requirement to develop Environmental Management Plans (EMPs)[1] as mandatory in EIA legislation (IEMA 2011). Bennett et al. (2016) observed that 95 per cent of the practitioners interviewed for their UK based study dealt with EMPs 'either all the time or fairly regularly'. Interest in and use of ESMPs is increasing globally with impetus coming from International Financial Institutions (e.g. the World Bank, and the European Bank for Reconstruction and Development (EBRD)), which include the requirement for management plans within their policies and performance requirements when providing project finance.

ESMP practice has evolved through a 'bottom-up approach' where practitioners have shaped and developed the arena of practice, as opposed to a 'top-down' legislative process which sets boundaries to what should be addressed. Consequently, practice is diverse, terminology can vary and there are a range of 'management plans' which might be used differently in different contexts or geographical settings. Management plans developed at project level can have a broad remit applicable to a number of stages in the development life cycle, or they may have a narrow focus addressing specific

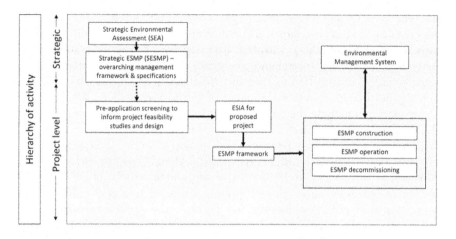

Figure 20.1

ESMPs within environmental assessment and management processes (based on Lochner 2005)

stages or issues (such as a construction management plan, ecological management plan, etc.). ESMPs can also exist at strategic or institutional level and link into overarching management systems (see Figure 20.1). Whilst recognising the range of variations that exists, this chapter focuses on the key principles associated with the development and implementation of project level ESMPs.

20.2 Definitions and concepts

An ESMP is a document that sets out the actions needed to manage environmental and social effects associated with the construction and operation of a development (additional ESMPs may be developed for demolition/restoration phases). It includes details on when each action should occur and who is responsible for its delivery. ESMPs are frequently included in the outcome of the ESIA process (i.e. the ESIA Report/Environmental Statement), summarising the mitigation, compensation, monitoring and enhancement measures required to effectively manage the predicted environmental and social effects of a development if it gains development consent (IEMA 2016).

20.2.1 Linking impact assessment and environmental management

Whilst the *ex ante* impact assessment of proposed development is now an established environmental policy instrument, there is ongoing debate about how effective it really is (e.g. Sadler 1996, Carroll and Turpin 2009). However, effectiveness studies have largely focused on procedural aspects within national jurisdictions (Wood 2003), with less attention paid to the more substantive outcomes, i.e. the operational and post-closure environmental performance of projects. ESIA is a key integrative and anticipatory environmental decision-making aid, but it is only one element of the policy toolbox (Wood 2003). Other elements include the monitoring and evaluation of project impacts and the subsequent management of environmental performance.

EIA in a formal sense first appeared in the USA in 1969 through the National Environmental Protection Act (NEPA) and has since spread globally (Glasson et al. 2012). Environmental management also originated through legislative processes, although with primary focus on aspects of public health protection and pollution prevention. Through the uptake of institutional voluntary environmental management systems (EMS) (e.g. ISO14001 (ISO 2015) and EMAS (2015)), environmental management now incorporates a wider range of issues of relevance to ESIA. Whilst there are differences and similarities between ESIA and EMS, a growing body of literature considers how to link the two so that environmental assessment and management are seen as a continuous process (Broderick 2012, Perdicoulis et al. 2012). As Holling (1978) argues:

> If assessment continues into the future, then prediction loses its status as a goal and assessment merges into environmental management. Prediction and traditional 'environmental impact assessment' suppose that there is a 'before and after' whereas environmental management is an ongoing process.

More recently, an international study into the effectiveness of environmental assessment (Sadler 1996) emphasised that without going beyond the EIA phase 'the process risks becoming a pro forma exercise rather than a meaningful exercise in environmental management'.

20.2.2 Early 'practice-led' initiatives: The World Bank and Environment Agency (England)

The 1990s witnessed the evolution of practice-led initiatives to integrate environmental assessment and management. For example, in order to manage the environmental impacts of its investments, in 1991 the World Bank introduced 'environmental mitigation plans' into its environmental assessment operating directive (World Bank 1993). The concept evolved within World Bank documentation over subsequent years, being referred to as 'environmental mitigation or management plans' and then 'Environmental Management Plans' (EMP), with guidance on what should be covered in an EMP being issued in January 1999. The broad aim of an EMP was to 'provide an essential link between the impacts predicted and mitigation measures specified within the EA report, and implementation and operational activities' (World Bank 1999).

During the early 1990s the Environment Agency (EA) (key environmental regulatory body in England) was also developing its policies and tools to integrate environmental assessment and management. Hickie and Wade (1997) document the development of the concept of 'environmental action plans' (EAP) as a way to strengthen the Environment Agency's environmental assessment process. Prompted partly by the findings of the EIA effectiveness study reported in Sadler (1996), the Environment Agency EAPs detailed 'how the protection, conservation, mitigation and enhancement measures for the project will be delivered by the Environment Agency and its contractors' (Hickie and Wade 1997). More specifically, EAPs:

- confirm details of environmental parameters and constraints for working in nature conservation areas;
- summarise environmental issues and constraints for the project design team;
- clarify implementation and monitoring of environmental constraints and mitigation measures by Environment Agency staff;
- identify how post-EIA report changes would be assessed and approved;
- indicate how objectives and targets for successful post-project appraisal would be identified.

Table 20.1 summarises key aspects of the processes in these two early examples. Although titled slightly differently and evolving in response to differing operational needs, some commonalities are apparent between the World Bank and Environment Agency processes.

Other examples from this period include approaches developed by the Hong Kong Environmental Protection Department (Sanvicens and Baldwin 1996) and the use of 'environmental management programmes' in Western Australia (Bailey 1997).

20.2.3 Contributions from research

Following the arrival of formalised (but still voluntary) environmental management systems in the 1990s, research examining EIA-EMS links began to appear in the academic literature. Sheate (1999) observes that the abundance of academic articles on the integration of EIA with other tools served to highlight the growing interest in research associated with extending EIA beyond its traditional boundaries. Of particular note is the work of Eccleston (1998), who developed a conceptual framework showing the potential for synergy between EMS and the EIA process, arguing that integrating the two could 'lead to more effective planning and enhanced environmental protection, while streamlining compliance'. New concepts also started to appear in the literature, particularly the need for EIA 'follow-up'

Table 20.1 Comparison of requirements within early guidance on EMP (Durning 2012)

World Bank (1999) – Aspects to be typically addressed within Environmental Management Plans	Hickie and Wade (1997) – Elements of an EAP in EIA* in England and Wales
	A: Management and monitoring for final design and delivery of the project in accordance with the EIA report
Summary of impacts	(1) Summary of the EIA process and the environmental constraints to be taken into account in terms of protection, conservation, mitigation and enhancement measures.
Description of mitigation measures	(2) Management of change in project design and implementation in relation to environmental impacts.
Institutional arrangements	(3) Communication programme to network in-house staff; engineering consultants and contractors; residents; landowners; public; user groups; and conservation bodies.
Description of monitoring programme	(4) Commitment to staff resourcing and procedures, normally a project EIA Officer (as an independent member of the project team) and an Environmental Clerk of Works (as part of the supervising Resident Engineering team); Environmental Protection Schedules (EPS) to be checked by the Environmental Clerk of Works on a weekly basis; and Environmental Incident Forms and an associated reporting and follow-up system.
	(5) EIA quality assurance system.
	B: Objectives and targets for each environmental constraint
	(1) Objective.
	(2) Implementation statement.
Implementation schedule and reporting procedures	(3) Targets for objective (to be reviewed at post-project appraisal stage and remedial works instigated if necessary).
	C: Summary of environmental specifications required in engineering contract
	(1) Contractors' workmanship including procedures and limitations.
Cost estimates and sources of funds	(2) Materials specifications.
	D: Drawing showing all constraints and comments

Note: * In the UK, EIA is called 'environmental assessment' and the EIA report is called 'environmental statement'. We have changed the terms to EIA and EIA report for simplicity

after project approval decision-making. According to Arts et al. (2001), environmental management is one of the four key activities of EIA follow-up (which also include monitoring, evaluation and communication). Morrison-Saunders and Arts (2004) describe management in this context as: 'making decisions and taking appropriate action in response to issues arising from monitoring and evaluation activities' and as such can be considered a key to coordinating other follow-up activities.

However, whilst there is a body of academic work on theoretical ways of 'coupling' assessment and management, research that focuses upon practical implementation is scarce. In Canada, Barnes and Lemon (1999) describe a proponent-driven EMP process for transport infrastructure that extended beyond the construction phase to encompass the full development life span. Other examples of the use of EMP in practice include Marshall (2004) and Broderick and Durning (2006). Further consideration of the theoretical and practical ways of extending EIA into management is detailed in Perdicoulis et al. (2012) which includes an example on the current use of EAPs by the Environment Agency (Fuller et al. 2012). More recently, research on stakeholder perceptions of EMPs in the UK by Bennett et al. (2016) notes that the use and effectiveness of EMPs varies in practice.

20.2.4 Incorporating the social dimension: from EMPs to ESMPs

Social impact assessment (SIA) (see Chapters 13 and 14) is, in some geographical regions, undertaken as a separate assessment to EIA. Research on the use of management plans to implement the outcomes of SIA has only started to appear relatively recently. Franks and Vanclay (2013) provide a useful comparison of social impact assessment processes (see Table 20.2), including approaches developed by project funders, project proponents, and decision makers. Most recently Vanclay et al. (2015) have updated guidance on the development of Social Impact Management Plans, which has been published by the International Association for Impact Assessment (IAIA) (see §20.3.3).

More widely, since the beginning of the millennium, the development of an explicitly social aspect within EIA has been driven by the World Bank, International Finance Corporation (IFC), and other influential funding institutions, including the EBRD and the Asian Development Bank (ADB 2003) (see §20.3.2). EIA became ESIA for many impact assessment practitioners working on projects financed by financial institutions operating under the Equator Principles (see §20.3.2). This change in practice and language was also mirrored in the move from EMPs to ESMPs.

Whilst EMP/ESMPs have now become well-established in practice, the exact nature of their application is highly variable and therefore any examples will only serve to highlight a particular aspect of practice. Box 20.1 provides a short case study of an ESMP.

The Rwimi Small Hydro Power Project is a proposal for 6.6 MW of installed generating capacity to be built and operated by Eco Power Holdings Ltd. By harnessing the hydro power potential of the Rwimi river in Uganda, the proposal will generate an annual energy output of 24.8 GWh.

An ESMP has been developed which flows from the ESIA, the Resettlement Action Plan, and other statutory permits and licenses issued by the Ugandan authorities. Guided by the IFC Performance Standards, the ESMP aims to ensure the adoption of a strategic approach to mitigate the environmental, health and safety impacts of the project according to accepted guidelines and best practice. The ESMP (a 220+ page-long document) is intended to be used during the full project life cycle to serve as a 'ready reference' for the developer and key stakeholders (namely contractors, sub-contractors, external consultants and other statutory agencies). Major responsibility for implementing the ESMP lies with the developer and the civil contractors (during the construction phase). The ESMP is structured in four parts:

Part 1: Introduction to the project, impacts, the organisational structure for the management of the ESMP, plus a detailed description of the roles and responsibilities of respective parties.

Part 2: Provides details of the Environmental and Social Impact Mitigation Action Plans.

Part 3: Explains requirements for compliance monitoring and reporting; provides guidelines and format to be used to report on mitigation actions as required by various lead agencies.

Part 4: Supplemental management plans to meet additional requirements of the IFC PSs, including:

a   Public Consultation and Disclosure Strategy/Plan (PS1)
b   Employment Policy (PS2)
c   Explosive Handling and Blasting Procedure (PS2)
d   Slope Protection and Soil Conservation and Erosion Control Plan (PS3)
e   Construction Waste/Spoils Disposal Management Plan (PS3)
f   Hazardous Materials Management Plan (PS3)
g   Occupational Health and Safety Management Plan (PS4)
h   Community (Public) Safety Management Plan (PS4)
i   Traffic Management Plan (PS4)
j   Chance Find Procedure (unearthing of unexpected graves or sites of heritage significance) (PS8).

20.3 Key legislation, policy and guidance

ESMP practice has mainly been driven forward through the use of performance standards and voluntary codes or principles, particularly those developed by major development bodies (International Financial Institutions) and

Table 20.2 Comparison of three forms of social impact management processes (Based on Franks and Vanclay 2013)

Criteria	IFC ESMP (2006–current; revised 2012)	Anglo-American 'Socio-Economic Assessment Toolbox' SEAT (2002–current; revised 2007, 2012)	Queensland SIMP (2008–2012)
Scope	Integrated (social, environmental, economic)	Socio-economic (link to environment)	Socio-economic
Driver	Required as a condition of an IFC loan (at project development)	Voluntary corporate policy (operational projects) to meet public expectations	Required for project development approval (prior to construction)
Application	All IFC-funded development projects with significant social and environmental risks	All operational mining projects managed by Anglo American require a SEAT Assessment every three years	New or expanded mining and petroleum projects within the identified resource provinces requiring an approved EIS
Public availability of plan	Encouraged	Yes (SEAT report public, SMP internal)	Yes
Public involvement during preparation	Government and public involvement encouraged	Stakeholder involvement in issue identification, prioritisation and development of management options	Stakeholder involvement and public comment period on draft plan
Issue prioritisation	Wide coverage	Prioritisation of key stakeholder identified issues	Prioritisation of key socio-economic issues and rated by consequence
Plan aligned with government or regional planning	Government and regional planning should be taken into account and responsibilities differentiated	No explicit requirements but engagement required with government stakeholders to encourage alignment	Negotiated with government – designed to align with community plans (local) and regional plans (State)
Plan linked to internal management systems	Yes – an Environmental and Social Management System is also required	Yes – direct link to Anglo American Social Way and management systems	Not explicitly – up to project proponent

Table 20.2 continued

Criteria	IFC ESMP (2006–current; revised 2012)	Anglo-American 'Socio-Economic Assessment Toolbox' SEAT (2002–current; revised 2007, 2012)	Queensland SIMP (2008–2012)
Review process and link to assessment	Plan linked to ESIA. Triggered by loan application. Unspecified assessment process during review of plans or as part of management system	SEAT Assessment is triggered every three years. This is a formal opportunity to update the SMP. Includes opportunity for stakeholder review	Plan initially developed as part of project level impact assessment. Annual reporting during construction. External review usually after two years, then every three years. Process undetermined
Coordination with other developments	Developments required to understand and respond to impacts of third parties but no specific requirement to coordinate	Not explicitly but stakeholder engagement requirements may encourage consideration	Plans include collaborative activities across multiple developments and must consider other known projects
Framework for activities within plan	Mitigation, avoidance and offset of negative impacts	Mitigation and benefit delivery through core business activities. Legacy planning for project closure	Predominantly mitigation framework for social infrastructure and a focus on benefit delivery
Partnerships for the delivery of programs	No explicit requirement	Yes – tool within SEAT to encourage identification of partnerships	No explicit requirement but usually conditioned as part of approval
Reporting and social impact monitoring	Monitoring process and grievance handling required. Public participation encouraged	SEAT public reporting and grievance handling process required. Impact monitoring encouraged. An output monitoring framework for community development sits alongside SEAT	Process for government reporting and requirement for monitoring plan and grievance handling
Articulation of all proponent community commitments	Lists plan commitments	Lists plan commitments and company policies	Lists all project related commitments conditioned as part of approval

investment banks. There is no legislative requirement to produce ESMP explicitly, although various pieces of legislation have or have had requirements for monitoring to be carried out post-EIA. The following section first considers selected examples of national legislation that links EIA/ESIA with follow-up activities, then outlines key policies and standards, followed by a brief overview of relevant guidance.

20.3.1 Selected examples of related legislation

Examples of countries that are acknowledged as having more sophisticated systems of environmental controls, particularly in relation to EIA follow-up, include the following:

- In Hong Kong, the Environmental Monitoring and Auditing component of the EIA Ordinance includes a requirement for an 'Independent Environmental Checker' to systematically verify that mitigation measures proposed in the EIA report are fully implemented (EPD 2002).
- In South Africa, EMPs are usually prepared for applications submitted within the regulations promulgated under the Environment Conservation Act (Act 73 of 1989). EIA regulations are being revised and will be replaced by regulations promulgated in terms of the National Environmental Management Act (Act No. 107 of 1998).
- In the Netherlands, the Environment Management Act 1994 and the EIA Decree were for many years regarded as some of the most effective systems in Europe (Glasson et al. 2012, Sadler 1996, Wood 2003). They included a requirement for follow-up, although the process was slow to commence, with only a 'handful' being carried out in the first few years (Arts 1998). The legislation has recently undergone 'modernisation' (Runhaar 2011) that has removed some of the mandatory review requirements.
- In Canada, Noble (2010) notes that, informally, EIA has increasingly become a routine part of the environmental management and auditing systems of municipalities and corporations. Early and sustained public involvement throughout the process is a key part of Canadian law together with monitoring and managing the actual outcomes, although the effectiveness of follow-up monitoring and enforcement has been criticised (see Gibson 2012). The Canadian Environmental Assessment Act was substantially amended in 2012 and whilst this has been seen as a retrenchment, one positive outcome is the inclusion of an enforceable decision statement that sets out mitigation and follow-up monitoring requirements (Gibson 2012).

The continuing absence of formal follow-up (monitoring) requirements in the European EIA Directive was felt to limit the effectiveness of EIA in the EU and to be a 'cause for concern' (EC 2009). However, the new EU EIA Directive (EC 2014) includes the introduction of mandatory monitoring for significant adverse effects at Article 8.

20.3.2 Policies and standards: International Financial Institutions (IFIs)

Whilst most multilateral development banks now have policy requirements and guidance on ESIA, the **World Bank** first developed its environmental guidelines in 1988. These were superseded in 1999 by the *Pollution Prevention and Abatement Handbook* (World Bank 1999) and its first operational directive on ESIA. The World Bank Group's Operational Policy (OP) 4.01 'Environmental Assessment' (World Bank 2013) and related documents support the preparation of World Bank Group projects. EMPs are listed in OP4.01 as an instrument used to comply with the Bank's environmental assessment requirements. EMPs are 'integral' to the ESIA of projects classed as likely to have significant impacts that are 'adverse, sensitive or unprecedented'. The EMP details measures to mitigate the environmental impacts and the actions needed to implement these measures. The borrower has to report on implementation of the EMP during project implementation.

The **International Finance Corporation**'s (IFC's) Policy and Performance Standards on Social and Environmental Sustainability (see Chapter 1, Box 1.1) are applied to all IFC-funded investment projects in order to minimise their impact on the environment and affected communities.

Performance Standard 1 highlights the importance of:

i   integrated assessment to identify the environmental and social impacts, risks, and opportunities of projects;
ii   effective community engagement through disclosure of project-related information and consultation with local communities on matters that directly affect them;
iii the client's management of environmental and social performance throughout the life of the project.

(IFC 2012)

Performance Standards 2 to 8 establish requirements to 'avoid, minimize, and where residual impacts remain, to compensate/offset for risks and impacts to workers, Affected Communities, and the environment' (IFC 2012). Importantly, where 'risks and impacts' are identified 'the client is required to manage them through its Environmental and Social Management System (ESMS) consistent with Performance Standard 1' (IFC 2012).

Monitoring and follow-up are integral to the performance standards, and are intimately linked to the IFC Environmental, Health and Safety Guidelines (Rankin et al. 2008), which are:

technical reference documents with general and industry-specific examples of Good International Industry Practice (GIIP) ... Reference to the EHS Guidelines by IFC clients is required under Performance

Standard 3 whilst IFC uses the EHS Guidelines as a technical source of information during project appraisal activities.

(IFC 2016)

Linked to this is the IFC's Environmental and Social Review Procedures Manual (IFC 2013), which describes the approach used for 'monitoring and recording client performance' against the performance standards. The performance standards are also used by other financial institutions on a voluntary basis.

The **European Bank for Reconstruction and Development** also has 'performance review standards' (EBRD undated) that require all companies/institutions in receipt of project funding to have a systematic approach to managing the associated environmental and social issues and impacts to ensure compliance with the Bank's Environmental and Social Policy (EBRD 2014). EBRD Performance Requirement 1 (Assessment and Management of Environmental and Social Impacts and Issues) requires the development of a programme of actions to address the environmental and social impacts which can consist of a 'combination of documented operational policies, management systems, procedures, plans, practices and capital investments, collectively known as Environmental and Social Management Plans (ESMPs)'.

The **Equator Principles** (EPs) (2013) provide a financial industry benchmark for determining, assessing and managing social and environmental risk in project financing. The EPs are a voluntary scheme, based on World Bank and IFC guidelines, that oblige participating financial institutions to finance projects only if it can be guaranteed that the social and ecological impact of projects are assessed. There were 84 signatories (from 36 countries) to the EPs in June 2016, covering 70 per cent of project finance debt in emerging markets. These EP Financial Institutions have consequently adopted the EPs in order to ensure that the projects they finance are developed in a manner that is socially responsible and reflects sound environmental management practices.

There are ten principles, but Principle 4 (Environmental and Social Management System and Equator Principles Action Plan) is the key one relevant to ESMPs. For projects screened as requiring assessment (based on the IFC environmental and social categorisation process), the client is required to develop or maintain an Environmental and Social Management System (ESMS) to manage risks on an ongoing basis. The EPs indicate that the ESMS includes 'policies, management programs and plans, procedures, requirements, performance indicators, responsibilities, training and periodic audits and inspections with respect to environmental or social issues, including Stakeholder Engagement and grievance mechanisms' (Equator Principles 2013). The ESMS provides the overarching framework through which an ESMP is implemented in order to address issues identified during the assessment process.

The ESMP:

> summarises the client's commitments to address and mitigate risks and impacts identified as part of the Assessment, through avoidance, minimisation, and compensation/offset. This may range from a brief description of routine mitigation measures to a series of more comprehensive management plans (e.g. water management plan, waste management plan, resettlement action plan, indigenous peoples plan, emergency preparedness and response plan, decommissioning plan). The level of detail and complexity of the ESMP and the priority of the identified measures and actions will be commensurate with the Project's potential risks and impacts.
>
> (Equator Principles 2013)

Under Principle 7 (Independent Review), and as part of the due diligence process, the 'assessment documentation' (including the ESIA, ESMPs, the ESMS, or Stakeholder Engagement process documentation) is reviewed by an independent environmental and social consultant. Where the documentation does not meet the required standards to the satisfaction of the Equator Principle financial institution, then an Equator Principles Action Plan is required to prioritise actions to address the gaps and secure compliance. The independent consultants also play a role in proposing the contents of the Action Plan.

20.3.3 Good-practice guidance

A number of different organisations have produced good-practice guidance which addresses ESMPs and a selection is now briefly outlined.

International Council for Mining and Metals (ICMM)

ESIA (i.e. *assessment*) is typically associated with the exploration and feasibility stages of the mining project cycle, whereas EMS (i.e. *management*) is more closely associated with operations and mine closure. The ICMM guidance emphasises that the systems, tools and processes of ESIA and EMS are applicable at any and all stages of the mining development life cycle, i.e. development, operation and closure (ICMM 2010). For example, they highlight the relevance of implementing an EMS during exploration to provide a framework for identifying and managing impacts at this early stage. Similarly, the determination of the significant aspects for an EMS may require the application of the evaluation and assessment stages of an ESIA. ICMM views ESIA as a process for managing environmental and social impacts rather than an exercise solely linked to permitting requirements. Whilst the ICMM do not provide specific guidance on how to produce EMPs, they highlight their use as good practice and as a related part of an EMS. The ICMM also identifies Anglo American's Socio-Economic Assessment Toolbox (SEAT) (see also

Table 20.2) as a good-practice method for assessing and managing social impacts (Anglo American 2012).

Design Manual for Roads and Bridges (DMRB)

The DMRB was introduced in 1992 in England and Wales and subsequently in Scotland and Northern Ireland (DTHA 2011). It provides a comprehensive system which accommodates all current standards, advice notes and other published documents relating to the design, assessment and operation of trunk roads (including motorways). Environmental management/action plans (EMPs) are a key component in the operational aspects of the environmental performance of roads and highways (DTHA 2011). DMRB specifies the requirement for an EMP to be developed prior to any work being carried out on a project and defines the key purpose of the EMP as being to guide the environmental management during the implementation of the project. In 2014 an Interim Advice Note to guide the development of EMPs was published (DTHA 2014). This introduced a new EMP stage – the Handover Environmental Management Plan – which should be drawn up once the project is completed. The aim is for essential environmental information to be conveyed to the client and 'crucially to the body responsible for the future maintenance and operation of the asset'.

The International Association for Impact Assessment (IAIA)

IAIA have issued guidance on assessing and managing the social impacts of projects (Vanclay et al. 2015) which includes a typical contents list for a Social Impact Management Plan (SIMP) (see Table 20.3). IAIA do not currently provided similar guidance on EMP or ESMP.

The Institute for Environmental Management and Assessment (IEMA)

As part of the IEMA 'Best Practice Series', *Environmental Management Plans* (IEMA 2008) provides comprehensive guidance on the purpose, benefits, design, implementation and review of EMPs in EIA practice in the UK. A model EMP structure is presented, which it proposes should comprise: a review of the project proponent's existing policies; details of the project team roles/responsibilities; summary of emergency procedures including in the event of a breach of EMP measures; EMP implementation cost estimates; record of consents and permissions; and a record of significant changes (including responsibilities for oversight). Since its initial publication, UK EIA practice has given increasing consideration to the use of draft EMPs in Environmental Statements to systematically set out the mitigation actions identified during the iterative EIA and design process. Recent IEMA guidance[2] on EIA, *Delivering Quality Development* (IEMA 2016), also makes extensive reference to the role of EMPs.

Table 20.3 Typical SIMP content (Vanclay et al. 2015)

Chapter title	Description of chapter contents
Cover	
Inside front cover	Publication statement indicating the authors (i.e. names of the individuals responsible for doing the work and writing the report), publisher, date of publication and other information to establish the nature and purpose of the document.
Executive summary	A short statement of key issues and findings.
Expert review statement	A letter/report from any expert or peer reviewer (or perhaps a joint statement if there were several reviewers) to indicate how the review was conducted, what constraints applied to the reviewers, and any comments, concerns and recommendations of the reviewers. A response from the authors to the review might also be appropriate.
Introduction	A general introduction to the report making the purpose of the report clear, perhaps including a short general statement about how the document connects to SIA literature/philosophy.
Project summary	A good description of the project and all ancillary activities so that readers can get a sense of the project. Where project alternatives or options exist, they could be explained here.
Methodology	A statement about the overall design of the SIA, what methods were used, what community engagement processes were used, and how ethical issues were considered and addressed. Perhaps definitions and/or a discussion of key concepts, and some link to the SIA and social research literature would be expected here. A discussion of the governance arrangements for the conduct of the SIA should be provided. Importantly, the limitations of the applied methodology would also be included, including decisions to narrow or expand the scope over the course of the SIA.
Applicable legal framework and standards	A discussion of the legal framework(s) and applicable legislation, regulations and guidelines that apply to the particular case. This would include not only local legislation/regulation, relevant institutions and their responsibility towards the project, but also mention of international standards, such as the IFC Performance Standards, guidance from international industry organisations, and reference to the IAIA guidance for SIA.
Community profile and social baseline	If an extended community profile and social baseline are to be included as appendices, then at least include a summary of key characteristics and key stakeholder groups here; alternatively include the community profile and baseline data here. Key historical issues should also be discussed. Key aspects of the physical environment that may be relevant to understanding the context should be included too.

Table 20.3 continued

Chapter title	Description of chapter contents
Scoping report	A statement of all potential social impacts considered in the assessment phase. The disposition of each impact considered should be made clear. Where this is presented as a separate report, a summary should be provided. Alternatively, this can be an appendix.
Prioritised listing of key social impacts	This is a listing of the residual impacts with a discussion of how different stakeholders are affected. There should be a particular focus on Indigenous peoples, women and vulnerable groups.
Resettlement (summary)	If resettlement is required, or if physical or economic displacement will occur, a short description of how the resettlement process will be undertaken, what compensation will be provided and how it will be determined, and what measures will be taken to restore and enhance livelihoods. A fully developed Resettlement Action Plan will be required as a separate document.
Summary of mitigation and management measures	A list of mitigation and other management measures to address social issues should be provided. There should be a costing and timeframe for implementation for proposed mitigation measures.
Monitoring plan and contingency plan (adaptive management)	A plan for how monitoring will be undertaken – what will be monitored, how monitored, how often, and who is responsible, as well as how the company will respond should an allowance threshold be exceeded – needs to be provided.
Benefit statement	This is a statement of the likely project benefits to the local communities, including of all proposed social investment actions, and local content and local procurement strategies.
Ongoing community engagement strategy and grievance mechanisms	A description of the intended ongoing community engagement processes. Also a description of what grievance mechanisms will be provided and what processes will be used for managing grievances.
Governance arrangements	A discussion of the governance arrangements that will apply to the ongoing community engagement processes, the grievance mechanisms, the monitoring process, and to ensure the ongoing acceptability of the social investment programme.
References	A list of all references used in the report, and any key references that informed the design of the SIA research.
Appendices	The appendices that are to be included will vary from project to project and will be affected by what is included in the body of the report but may include: questionnaires, interview schedules, consent form templates, an extended community profile, baseline data, and a scoping report (i.e. a listing of all issues considered as possible social impacts).

20.4 Developing an ESMP

As noted in §20.2, an ESMP is a document that sets out the actions needed to manage the environmental and social effects associated with the construction and operation of a development, including details on when each action should occur and who is responsible for its delivery. Due to the evolutionary path it has followed (§20.1) practice is diverse, terminology can vary, and there are a range of 'management plans' which might be used differently in different contexts or geographical settings.

The following section sets out a generic outline methodology for developing an ESMP and provides guidance on the likely structure and content. It builds on the work of the World Bank (1999), Lochner (2005), IEMA (2011 and 2012) and Vanclay et al. (2015).

20.4.1 Outline methodology and ESMP structure

Engaging the principal contractor (or their representatives) during the ESIA process and when drafting an ESMP is the most crucial action that needs to be undertaken if an ESMP is to be successful (IEMA 2012). The effectiveness of an ESMP in managing impacts is greatly enhanced when the actions proposed are built into contractual agreements and there is a clear indication of who will have overall responsibility for implementation. An effective *change management process* is also required to ensure that actions set out in the ESMP can be modified in order to remain relevant to any changes that take place post-consent, i.e. it needs to adopt a flexible and adaptive management approach to mitigation and enhancement (IEMA 2011).

Other key principles for successful ESMPs include (Lochner 2005, IEMA 2008, 2016):

- The cost of the mitigation actions and implementation of the ESMP (including estimates of recurring expenses, e.g. for training and environmental awareness raising; monitoring and auditing and corrective actions) should be identified during the pre-consent phase and built into the budget for the development.
- Competent environmental professionals should be involved in the design and specification of mitigation actions, and also in conducting audits during the implementation phase.
- There should be proactive collaboration with stakeholders, both internally within the project team (developer/designer/contractor/construction delivery teams) and externally (consenting authority and key stakeholders).
- Relevant stakeholders should be engaged during the ESMP development, implementation and revision stages.
- Roles and responsibilities need to be clearly defined, e.g. for the implementation of management actions, monitoring implementation, setting criteria/timeframes.

- There should be a document handling and control system, and all relevant legislation, standards, guidelines, permits, licences should be identified.
- Reporting procedures and practices should be identified, with management reviews scheduled for key stages in the life cycle of the project.
- The approach to non-compliance should be described in the EMP and also specified in relevant contractual agreements.

Table 20.4 provides a recommended structure for an ESMP. This structure is generic and should be adapted as necessary, as an ESMP should always be bespoke to the scale and location of the project.

20.4.2 Challenges

There are a number of challenges in using ESMPs. A key challenge is maintaining the adaptive element, particularly in relation to large and complex projects. Reflecting on the role of private sector investment in large-scale infrastructure projects, Faith-Ell and Arts (2009) observed that,

> as early market involvement will usually lead to longer-term contracts, these contracts will have to deal [with] changing contexts – new spatial and environmental developments, new techniques, new regulations and policies etc. As a consequence, there is need for some form of environmental management responsiveness.

They suggest adopting processes which are in essence those advocated as key by IEMA, i.e. the establishment of a system to manage environmental impacts, with careful monitoring of contract requirements and periodic evaluation of environmental performance measures, e.g. by the regulator/decision-making body or, in the case of projects funded by International Financial Institutions, by the funder.

Aspects of the knowledge transfer of essential information can also be problematic if the bodies responsible for the operational stage of the development are different to those involved at the construction stage. In the UK context, arguably there has been a tendency to concentrate the use of ESMP at the construction stage of the development. This was apparent in the early stages of development of the concept when, due to the nature of the projects being delivered by the Environment Agency, the EAPs concluded at the point when the development became operational. What is possibly still an issue, and which needs further consideration, are those developments where the ESMP needs to link into the long-term operation and ultimate decommissioning stage of a development. The World Bank (1999) guidance states that the ESMP should continue through to the operational stage of developments and be regularly reviewed. Sánchez (2012) highlights the importance of collective learning, stating that 'the best EMPs prepared today reflect, in first

Table 20.4 Recommended structure for an ESMP (based on IEMA 2008, Lochner 2005, and Vanclay et al. 2015)

Cover	To record issue number and date, revisions and reasons for changes, issuing authority/contact details.
Contents	
Executive summary	Key issues and findings.
Introduction	Including a summary of the project, the host environment (biophysical, economic and social components) and aims of the ESMP.
Project summary	This should cover project location, layout plans, project phases (e.g. design, construction, commissioning, operations and decommissioning), construction activities, operational processes and activities, employment and labour, directly associated infrastructure, and project schedule. A brief description of the affected social and environmental aspects should also be provided, particularly those elements of the environment that may be impacted upon by the project and which should be included in the monitoring programme. The environment in this context includes the biophysical, economic and social components. Key information about the environmental management context for the project (e.g. if the project environmental management is located within an overarching EMS, ESMS or SEMP) and the relevant legal and planning context should be provided. The information should be drawn from the ESIA Report/Environmental Statement and/or Scoping Report.
Summary of effects	A summary should be provided of the predicted positive and negative effects associated with the proposed project that require management actions (i.e. mitigation of negative effects or enhancement of positive effects). The necessary information should be drawn from the ESIA process.
Methodology for ESMP including limitations	
Project proponent's existing policies	
Project team roles and responsibilities	This will be particularly important where multiple organisations are involved in a project, e.g. several subcontractors. This section should also detail where queries should be directed within the team (including contact details), with procedures to escalate up to technical specialists as required.
Applicable legal framework and standards, consents and permissions	This should provide a record of the consents relevant to the project.

Table 20.4 continued

Cost estimates for ESMP implementation	It is the responsibility of the proponent/developer to fund the cost of ESMP implementation, however, the proponent/developer and organisation undertaking the ESIA should liaise closely.
Environmental and socially significant changes register	This should detail procedures to be followed if any significant changes are encountered once a project commences on the ground and which would result in a change to the ESMP, e.g. the use of alternative construction methods or design. A tabular format document to record changes to construction methods, design and mitigation and the implications of these changes and authorising personnel. This should also detail who has responsibility for overseeing changes and ensuring these do not conflict with any consenting or planning conditions.
Register of site-specific environmental and social actions including resettlement (if required)	This information, forming the core of the document, should be detailed for each action; a tabular format is often used to provide clarity and ease of reference. Also to include: • training of personnel • programme timeline – a visual indication of when measures should be implemented • monitoring – to detail monitoring equipment/methods and schedules.
Liaison, grievance mechanism and consultation requirements	Including any requirements for prior authorisation or provision of monitoring data, also key contacts (details for internal and external contacts).
Non-compliance and management reviews	The approach to non-compliance (whether based on penalties or incentives) must be described in the ESMP and specified, e.g. in tender documents and contracts. The actions specified by the ESMP must be enforced through the legal standing of the ESMP. Management reviews should be scheduled at key stages in the life cycle of the project and the ESMP revised accordingly.
References	
Appendices	To provide further detail on measures, e.g. monitoring methodologies to be followed, maps delineating boundaries/areas applicable to measures etc.

place, our collective learning about issues such as effectiveness of mitigation measures and how to design effective mitigation'. The UK DTHA (2011) has introduced the use of Handover Environmental Management Plans (see §20.3.3) to ensure that essential information and learning is transferred between the construction and operation phase of a development.

EIA/ESIA reports are required to describe the measures proposed to mitigate any 'significant' adverse effects of the development. These mitigation measures often include commitments that are influential in the decision to approve the project. Ensuring that sufficient measures are implemented is particularly important for the integrity of the ESIA process, particularly for those affected by a development project. However, research has shown the difficulties in linking the environmental and social mitigation within ESIA to development-consent obligations:

- ESIA reports are often variable in the degree to which the environmental and social mitigation measures they propose can be translated into consent obligations that are enforceable and precise (IEMA 2008, 2011; King 2016).
- Environmental and social mitigation measures proposed in EIA reports are often not required through planning consent obligations, and therefore often not implemented (Tinker et al. 2005, Mordue 2008). However, additional mitigation measures can be introduced through planning conditions and obligations that were not proposed in EIA reports, which points to influences other than the EIA findings (Tinker et al. 2005).

20.5 Conclusions

ESMPs are increasingly included in ES/ESIA Reports to summarise the mitigation, compensation, monitoring, and enhancement measures required to effectively manage the predicted environmental and social effects of a development if it gains development consent. Delivering successful ESMPs in practice can be the key to ensuring that developments cause no erosion to natural and social capital. The benefits of preparing an ESMP can be summarised as follows (based on IEMA 2008, 2016):

- To create a framework for ensuring and demonstrating conformance with:
  - legislative requirements
  - development-consent conditions, and
  - monitoring of mitigation measures set out in the ESIA report;
- To provide a link or 'bridge' between the design phase of a project, the construction and possibly operational phase;
- To ensure that effective communication and feedback systems are in place between the different actors involved in the project (such as operators on-site, the contractor, the environmental manager and/or consultant, stakeholders, regulators);
- To demonstrate a developer's commitment to protecting the environment through mitigation;
- To drive cost savings through improved environmental and social risk management.

ESMPs are a means of 'bridging the gap' between ESIA (typically at desk-based design stage) and 'real world' implementation of a project. Like any sound environmental management tool, to be effective an ESMP must be understandable (e.g. through the use of the vernacular of the language it is written in), practical, contain or require specific actions, include indicators/measurements of success, have responsibilities allocated, and be clear with respect to locations and timescales (where and when) for mitigation and monitoring.

The key to advancement of ESMP practice is to enhance learning from experience at all levels of application and practice. This chapter is only a starting point for this learning process and it is hoped that it will help to inform and improve ESMP practice.

Notes

1   UK practitioner vernacular is to refer to EIA/EMP rather than ESIA/ESMP.
2   The authors of this chapter led the production of early drafts.

References

ADB (Asian Development Bank) 2003. *Environmental assessment guidelines.* www.adb.org/documents/adb-environmental-assessment-guidelines

Anglo American 2012. *SEAT toolbox: socio economic assessment toolbox. Version 3.* www.angloamerican.com/~/media/Files/A/Anglo-American-Plc/docs/seat-toolbox-v3.pdf.

Arts J 1998. *EIA follow-up – the role of ex-post evaluation in Environmental Impact Assessment.* Groningen: Geo-press.

Arts J, P Caldwell and A Morrison-Saunders 2001. Environmental impact assessment follow-up: good practice and future directions – findings from a workshop at the IAIA 2000 conference. *Impact Assessment and Project Appraisal* 19(3), 175–185.

Bailey J 1997. Environmental impact assessment and management: an underexplored relationship. *Environmental Management* 21(3), 317–327.

Barnes J and D Lemon 1999. Life-of-project environmental management strategy: case study of the Confederation Bridge project, Canada. *Journal of Environmental Assessment Policy and Management* 1(4), 429–439.

Bennett S, S Kemp and MD Hudson 2016. Stakeholder perception of Environmental Management Plans as an environmental protection tool for major developments in the UK. *Environmental Impact Assessment Review* 56, 60–71

Broderick M 2012. Why EIA needs EMS. *The Environmentalist.* www.environmentalistonline.com/article/why-eia-needs-ems.

Broderick M and B Durning 2006. Environmental impact assessment and environmental management plans – an example of an integrated process from the UK. In *Geo-Environment and Landscape Evolution II*, JF Martin-Duque, CA Brebbia, D Emmanouloudis and U Mander (eds), *WIT Transactions Ecology and the Environment* Volume 89. Southampton: WIT Press.

Carroll B and T Turpin 2009. *Environmental impact assessment handbook: a practical guide for planners, developers and community*, 2nd Edn. London: Thomas Telford

DTHA (Department for Transport Highways Agency) 2011. *Design manual for roads and bridges Vol. 11*. London: DTHA.

DTHA 2014. *Interim Advice Note 183/14 Environmental Management Plans*. www.standardsforhighways.co.uk/ians/pdfs/ian183.pdf.

Durning B 2012. Environmental Management Plans – origin, usages and development. In *Furthering EIA – towards a seamless connection between EIA and EMS*, A Perdicoulis, B Durning and L Palframan (eds), 55–70. Cheltenham: Edward Elgar.

EBRD (European Bank for Reconstruction and Development) undated. *Performance requirements and guidance*. London: EBRD. www.ebrd.com/who-we-are/our-values/environmental-and-social-policy/performance-requirements.html%20.

EBRD 2014. *Environmental and social policy*. London: EBRD. www.ebrd.com/cs/Satellite?c=Content&cid=1395238867768&pagename=EBRD%2FContent%2FDownloadDocument.

EC (European Commission) 2009. Report from the Commission to the Council, the European Parliament, the European Economic and Social Committee and the Committee of the Regions on the application and effectiveness of the EIA Directive (Directive 85/337/EEC, as amended by Directives 97/11/EC and 2003/35/EC). http://eur-lex.europa.eu/legal-content/EN/TXT/?uri=CELEX:52009DC0378.

EC 2014. Directive 2014/52/EU of the European Parliament and of the Council of 16 April 2014 amending Directive 2011/92/EU on the assessment of the effects of certain public and private projects on the environment. Brussels: *Official Journal of the European Union*, L 124, 25 April 2014.

Eccleston C 1998. A strategy for integrating NEPA with EMS and ISO14000. *Environmental Quality Management* 7(3), 9–17.

Eco Power Holdings Ltd 2013. *RWIMI small hydro power project, Kassese, Uganda: Environmental & Social Management Plan (ESMP)*. www-wds.worldbank.org/external/default/WDSContentServer/WDSP/IB/2014/03/14/000442464_20140314104922/Rendered/PDF/E44440V50AFR0E080Box382173B00PUBLIC.pdf.

EMAS 2015. Regulation No 1221/2009 of the European Parliament and of the Council of 25 November 2009 on the voluntary participation by organisations in a Community eco-management and audit scheme (EMAS III). http://ec.europa.eu/environment/emas/documents/guidance_en.htm

EPD (Hong Kong Environmental Protection Department) 2002. The Operation of the EIA Ordinance. Hong Kong: EPD.

Equator Principles 2013. *The Equator Principles III*. www.equator-principles.com/resources/equator_principles_III.pdf.

Faith-Ell C and J Arts 2009. Public Private Partnerships and EIA: Why PPP are relevant to practice of impact assessment for infrastructure. Paper presented at the 'Impact Assessment and Human Well-Being' 29th Annual Conference of the International Association for Impact Assessment, 16 EPD (Hong Kong Environmental Protection Department) 2002. The Operation of the EIA Ordinance. Hong Kong: EPD. 22 May, Accra, Ghana.

Franks D and F Vanclay 2013. Social Impact Management Plans: innovation in corporate public policy. *Environmental Impact Assessment and Review* 43, 40–48.

Fuller K, C Vetori, B Munro and K House 2012. EIA-EMS link from the flood risk management sector. In *Furthering EIA – towards a seamless connection between EIA and EMS*, A Perdicoulis, B Durning and L Palframan (eds), 157–174. Cheltenham: Edward Elgar

Gibson RB 2012. In full retreat: the Canadian government's new environmental assessment law undoes decades of progress. *Impact Assessment and Project Appraisal* **30**(3), 179–188.

Glasson J, R Therivel and A Chadwick 2012. *Introduction to environmental impact assessment*, 4th Edn. Abingdon: Routledge

Hickie D and M Wade 1997. The development of environmental action plans: Turning statements into actions. *Journal of Environmental Planning and Management* **40**(6), 789–801.

Holling CS 1978. *Adaptive environmental assessment and management.* Chichester: John Wiley & Sons.

ICMM (International Council on Mining and Metals) 2010. *Good practice guidance for mining and biodiversity.* London: ICMM.

IEMA (Institute of Environmental Management and Assessment) 2008. *Practitioner best practice series vol 12: Environmental Management Plans.* Lincoln, UK: IEMA.

IEMA 2011. *Special report – The state of environmental impact assessment practice in the UK.* Lincoln, UK: IEMA. http://oldsite.iema.net/iema-special-reports.

IEMA 2012. IEMA workshop: delivering EIA's promises post-consent. Birmingham, 17 October.

IEMA 2016. *IEMA environmental impact assessment guide to delivering quality development.* Lincoln, UK: IEMA.

IFC (International Finance Corporation) 2012. *Performance standards on environmental and social sustainability.* Washington DC: IFC. www.ifc.org/wps/wcm/connect/ 115482804a0255db96fbffd1a5d13d27/PS_English_2012_Full-Document.pdf? MOD=AJPERES.

IFC 2013. *Environmental and social review procedures manual, version 7.* Washington DC: IFC. www.ifc.org/wps/wcm/connect/190d25804886582fb47ef66a6515bb18/ESRP%2B Manual.pdf?MOD=AJPERES.

IFC 2016. *Environmental, health and safety guidelines.* Washington DC: IFC. www.ifc.org/wps/wcm/connect/topics_ext_content/ifc_external_corporate_site/ifc+sus tainability/our+approach/risk+management/ehsguidelines.

ISO (International Organization for Standardization) 2015. *ISO14001 Environmental management systems – requirements with guidance for use,* 3rd Edn. Geneva: ISO.

King P 2016. Compliance and enforcement of EIA and EMMPs in Asia. Resilience and Sustainability: 36th Annual Conference of the International Association for Impact Assessment, Aichi-Nagoya, Japan.

Lochner P 2005. *Guideline for Environmental Management Plans.* CSIR Report No ENV-S-C2005-053 H. Cape Town, Republic of South Africa: Provincial Government of the Western Cape, Department of Environmental Affairs & Development Planning.

Marshall R 2004. Can industry benefit from participation in EIA follow-up? The Scottish Power experience. In *Assessing Impact: Handbook of EIA and SEA Follow-up,* A Morrison-Saunders and J Arts (eds), 118–153. London: Earthscan.

Mordue M 2008. The monitoring of mitigation measures in environmental impact assessment. MSc Dissertation in Environmental Assessment and Management, Oxford Brookes University.

Morrison-Saunders A and J Arts 2004. Introduction to EIA follow-up. In *Assessing Impact: Handbook of EIA and SEA Follow-up,* A Morrison-Saunders and J Arts (eds), 1–21. London: Earthscan.

Noble B 2010. *Introduction to environmental impact assessment. A guide to principles and practice,* 3rd Edn. Toronto: Oxford University Press.

Perdicoulis A, B Durning and L Palframan 2012. *Furthering EIA – towards a seamless connection between EIA and EMS*. Cheltenham: Edward Elgar.

Rankin M, B Griffin and M Broderick 2008. IFC raises the bar on environmental health and safety. *International Mining*, April.

Runhaar H 2011. Twenty five years of environmental impact assessment in the Netherlands, effectively governing environmental protection? In *Proceedings of Health in Environmental Assessment: Developing an International Perspective*. University of Liverpool, 28 June.

Sadler B 1996. *Environmental assessment in a changing world: evaluating practice to improve performance*. Final report of the International Study of the Effectiveness of Environmental Assessment. Canadian Environmental Assessment Agency and IAIA. www.ceaa.gc.ca/Content/2/B/7/2B7834CA-7D9A-410B-A4ED-FF78AB625BDB/iaia8_e.pdf.

Sánchez LE 2012. Information and knowledge management. In *Furthering EIA – towards a seamless connection between EIA and EMS*, A Perdicoulis, B Durning and L Palframan (eds), 19–38. Cheltenham: Edward Elgar.

Sanvicens G and P Baldwin 1996. Environmental monitoring and audit in Hong Kong. *Journal of Environmental Planning & Management* 39(3), 429–440.

Sheate WR 1999. Editorial. *Journal of Environmental Assessment Policy and Management* 1(4), iii–v.

Tinker L, D Cobb, A Bond and M Cashmore 2005. Impact mitigation in environmental impact assessment: paper promises or the basis of consent conditions? *Impact Assessment and Project Appraisal* 23(4), 265–280.

Vanclay F, AM Esteves, I Aucamp and DM Franks 2015. *Social impact assessment: guidance for assessing and managing the social impacts of projects*. International Association for Impact Assessment. www.iaia.org/uploads/pdf/SIA_Guidance_ Document_IAIA.pdf.

Wood C 2003. *Environmental impact assessment: a comparative review*, 2nd Edn. Harlow: Pearson Education.

World Bank 1993. *The World Bank and environmental assessment: an overview*. http://siteresources.worldbank.org/INTSAFEPOL/1142947-1116495579739/ 20507 374/Update1TheWorldBankAndEAApril1993.pdf.

World Bank 1999. *Environmental Management Plans*. http://siteresources.worldbank.org/ INTSAFEPOL/1142947-1116495579739/20507392/Update25Environmental ManagementPlansJanuary1999.pdf

World Bank 2013. *Operating manual: operating policy 4.01 – environmental assessment*. https://policies.worldbank.org/sites/ppf3/PPFDocuments/090224b0822f7384.pdf.

Glossary

· ·

The terms defined here are highlighted in **bold italics** at least the first time they appear in a chapter. Terms similarly highlighted within these definitions are defined elsewhere in the glossary.

abundance see **species abundance**

acid deposition Dry deposition (gravitational settling, impact with vegetation) and wet deposition (in precipitation) of acidic substances such as sulphates and nitrates. It is often called acid precipitation or 'acid rain', but these terms strictly refer to wet deposition only.

adaptive (monitoring and) management An approach to managing a project's impacts in the face of uncertainty, which involves temporarily managing the impacts while at the same time monitoring those impacts. Over time, the monitoring improves the knowledge base and reduces uncertainty, allowing the project's management to be fine-tuned or revised.

algae Mainly aquatic, unicelled or multicelled plants that lack true stems, roots or leaves. They include *phytoplankton*, filamentous 'pond scum' species and seaweeds.

algal blooms Rapid growth of *algae* in water bodies, facilitated by high nutrient levels and/or other physical and chemical conditions. They increase water *turbidity* (which inhibits light penetration and hence photosynthesis) and reduce dissolved oxygen levels at night and when the algae decay. Blooms of some algae and cyanobacteria also produce toxins that can affect fish and other wildlife, and present a hazard to human health.

alternatives (also options) Options considered in a project proposal, including: location/siting; alignment of linear projects; design (scales, layouts etc.); processes; procedures employed during the construction, operational and decommissioning phases; and the 'no action' option of the project not going ahead. Assessment may result in the selection of preferred options.

anthropogenic Generated and maintained by human activities.

appropriate assessment An assessment that must be carried out under the European Habitats Directive by a *consenting authority* when a project is considered likely (alone or in combination with other projects or plans) to have a significant effect on a Special Area of Conservation, Special Protection Area or Ramsar site.

aquifer A stratum of porous or fractured rock that contains groundwater and allows this to flow through.

audit trail A record of all analyses, decisions, etc. during a process such as ESIA, to assist in (a) explaining how alternatives were considered, why decisions were made, how the views of stakeholders were taken into account etc. and (b) reviewing the study, e.g. if conditions change.

authenticity (of cultural heritage) The truthful and credible expression of the cultural values of cultural heritage.

bathymetry Underwater depth of lake or ocean floors, the underwater equivalent to topography.

benthic (zone) The lowest ecological level of a water body, including the sediment surface and some sub-surface layers. Benthic organisms (the benthos, 'bottom dwellers') generally live in close relationship with the substrate, and may tolerate low oxygen levels.

bioaccumulation The process by which some pollutants accumulate in the tissues of living organisms.

biochemical oxygen demand (BOD) The quantity of dissolved oxygen in water (mg/l) consumed (under test conditions) by microbial degradation of organic matter during a given period (five days). It is one of the standard tests used to characterise effluent quality and to measure organic pollution in surface waters.

biodiversity The variety of life, globally or within any area – defined in the UN Convention on Biological Diversity (1992) as 'The variability among living organisms from all sources, including terrestrial, marine and other aquatic ecosystems and the ecological complexes of which they are part; this includes diversity within species, between species and of ecosystems'.

biomass The amount of organic matter in a community's living organisms at a given time, usually measured as dry weight per unit area (e.g. g/m^2) or (in aquatic systems) volume (e.g. g/m^3).

biomes The major climatic climax communities on a given continent, characterised largely by the vegetation and the governing climate. Similar biomes on different continents belong to global biome types, e.g. tropical rainforest, tundra.

biotope Usually defined as an area of uniform environmental conditions providing living place for a distinctive assemblage of species, i.e. a physical habitat with its biological community. Thus 'biotope' is almost synonymous with 'habitat' as used in habitat classifications but with emphasis on the whole biota (not just vegetation).

buffer zones/strips Vegetated strips of land designed to manage various environmental concerns, e.g. (a) to intercept waterborne pollutants and hence protect groundwaters and surface waters; (b) to slow *runoff* and enhance infiltration (within the buffer), so stabilising streamflows; (c) to reduce soil and streambank erosion; (d) to provide visual/noise/odour screens and landscape features; (e) to protect wildlife habitats/sites from pollution and/or disturbance, e.g. as 'recreational buffer zones'; and (f) to provide *wildlife corridors* and habitats/refuges. Buffer types include: wellhead protection zones, buffers along rivers or streams, grassed waterways, shelterbelts/windbreaks/snowbreaks, contour strips, roadside verges and field borders.

carrying capacity Can have various meanings, e.g. the population size of a species (including man) which a given environment can support; the ability of a habitat to support one or more given species; or the capacity of an ecosystem to tolerate a given stress such as a pollution level.

catchment A drainage area/river basin within which precipitation drains into a river system (and associated lakes and wetlands) and eventually to the sea. Catchment boundaries are generally formed by ridges, on different sides of which rainfall drains into different catchments. In the UK, these are usually called watersheds; but in the US, the term watershed is used in place of catchment.

climate change adaptation Ensuring that developments will be resilient to unavoidable climate change.

climate change mitigation Reduction of the causes of climate change (e.g. lowering greenhouse gas emissions).

competent authority See *consenting authority*.

connectivity The degree to which habitat patches in an urban or agricultural matrix are interconnected by *linear habitats* and/or *stepping stone habitats* between the main patches.

consenting authority (also competent authority) The organisation that determines whether a project should be permitted or not, often the land-use planning department in a regional/local government.

critical load/level A quantitative estimate of an exposure to one or more pollutants below which significant harmful effects on specified sensitive elements of the environment do not occur according to present knowledge. Critical load strictly refers to deposited pollutants (e.g. in soils or waters), while critical level refers to atmospheric concentrations. Exceedance of critical loads/levels may affect organisms directly, or indirectly, e.g. through increased dissolved aluminium concentrations associated with acidity.

cumulative impact Combined environmental or social impact caused by (a) a range of human activities, (b) activities of a given type such as land development, or (c) a new development in conjunction with other (past, present and future) developments and/or other activities.

diffuse (non-point source) pollution Pollution that cannot be attributed to discharges at specific locations. Typical causes are *runoff* to surface waters, or percolation to groundwater from farmland, roads, urban and industrial areas, or many minor point sources (e.g. land drains, leakages from sewers etc.). It is generally more difficult to control than *point source pollution*.

ecosystem A dynamic complex of plant, animal and micro-organism communities and their non-living environment interacting as a functional unit.

ecosystem services The direct and indirect contributions made by ecosystems to human well-being and project performance. They are typically classified into provisioning, regulating, cultural and supporting services.

environmental or social components Aspects of the natural or man-made environment (e.g. population, landscape, heritage, air, soils, water, ecosystems) that are individually assessed in an ESIA because they may be significantly affected by a proposed project, i.e. they are *receptors*.

environmental impact statement (EIS) The document that presents the findings of an EIA. See ESIA report.

ESIA report The document that presents the findings of an ESIA, including proposed mitigation measures, and is submitted (with the planning application) to the *competent authority* responsible for deciding if the proposal may proceed.

eutrophication The process or trend of soil or water enrichment by plant nutrients – especially nitrogen and phosphorus. It can occur naturally, but usually refers to *anthropogenic* enrichment (sometimes called enhanced eutrophication) which can lead to excessive nutrient loading and consequent ecosystem degradation.

evaporites Water-soluble mineral sediments, such as salt and carbonates, that are deposited as a result of the evaporation of surface water. They are considered sedimentary rocks.

evapotranspiration Total evaporative loss from a land area, including evaporation from soils and surface waters, and *transpiration* (which is the major component in well-vegetated terrestrial ecosystems).

field capacity (of a soil) The moisture content of a soil when water percolating downwards under gravity has drained out; usually expressed as cm^3 water per cm^3 soil (see also *saturation capacity*).

greywater Used water from sinks, showers etc. but not toilets.

hachure (plan) A way of showing steepness on a map, with closer, thicker lines indicating steeper slopes, and thinner lines that are further apart indicating gentler slopes.

heath/heathland A habitat/vegetation type usually dominated by dwarf shrubs. European heathland is an anthropogenic community that was created by forest clearance (often in the bronze age) and maintained by grazing, fire, and the use of materials for fuel etc. If unmanaged, it may quickly revert to woodland.

heavy metals Metals with atomic weight > 63.5 and specific gravity > 4.0. Some (e.g. cobalt, copper, iron, manganese, molybdenum, zinc) are essential nutrients, although more than trace amounts of most are toxic, especially to some taxa. Others, which have an atomic weight > 100 (e.g. silver, cadmium, mercury, lead) are highly toxic, and the term heavy metals is often restricted to these.

hydraulic conductivity The permeability of soil or rock, and hence the ease, and potential rate, of water flow through it, usually expressed at cm/hr.

hydraulic head A measure of the combined effects of elevation and water pressure at a point in an aquifer, representing the total energy of the water.

hydraulics Processes and regimes of water flow (velocities, volumes, duration, frequency etc.) in hydrological systems such as surface waters and groundwaters.

indicator species Species that can be used as biological indicators: (a) to define and identify community or habitat types, e.g. ancient woodland indicator species; (b) to assess the conservation value of habitats, e.g. protected species; or (c) to assess environmental quality and monitor change in this, e.g. lichens as atmospheric pollution (especially acid deposition) indicators.

indirect impact Impact that is not a direct result of the project, but that emerges through a complex impact pathway, possibly some distance or time from the project's direct impacts.

integrity (of cultural heritage) A measure of the wholeness and intactness of the cultural heritage and its attributes.

isoline map/plot A contour map where lines link places that share a common value, e.g. a topographic map where lines show areas at the same elevation, a noise map where lines show areas affected by the same noise level, or an air-quality map showing areas affected by the same air pollution levels.

keystone species A species having an important or vital influence on the structure and functioning of a community/ecosystem (e.g. with a key role in a food web), and/or that can be used (a) to identify genetic issues or (b) as an *indicator species* of habitat health/quality (with fluctuations in abundance indicating habitat change).

landscape effects Changes in the landscape, its character and quality (as compared to *visual effects*).

leachates Solutes, including pollutants, in water (or a non-aqueous liquid) that has leached from a 'solid' matrix (see *leaching*).

leaching The removal of soluble nutrients and other chemicals from a 'solid' matrix – such as a soil horizon, whole soil or landfill – by water percolating through it.

leakages (economic) The flows of money out of a national, regional or local economy, following from an initial injection of money into that economy. The most significant leakages are for taxation (direct and indirect), savings, and improved goods and services.

LiDAR ('Light Detection and Ranging') A type of remote sensing that measures distance using a laser.

life cycle analysis (or life cycle assessment) Assessment of the impacts associated with all stages of a project's or product's life, from raw material extraction to closure/disposal/recycling. Most ESIAs only deal with the impacts of construction and operation.

linear habitats Linear (much longer than wide) features that support biological communities. They can be valuable habitats in their own right and may also act as *buffer zones* and *wildlife corridors*. Examples are hedgerows, field margins/linear set-aside, road/railway verges, habitat edges, woodland rides/fire breaks, transmission line routes, urban green belts, avenues of trees, ditches, streams, river corridors, and lake/coastal shorelines.

macro-invertebrates Invertebrate animals that are large enough to be seen by eye or can be captured using a sieve of mesh 0.5–1.0mm.

macronutrients Nutrient elements needed by organisms in relatively large amounts. They are: carbon, hydrogen, and oxygen – which plants obtain by photosynthesis; and calcium, iron, magnesium, nitrogen, phosphorus, potassium and sulphur – which plants absorb in solution from soil or a water body (although some have root nodules in which nitrogen-fixing bacteria assimilate gaseous nitrogen) (see *micronutrients*).

macrophytes Plants large enough to be seen by eye. The term is most commonly applied to aquatic (freshwater and marine) species including *vascular plants* and seaweeds.

micronutrients (trace elements) Nutrient elements needed by organisms in small quantities, e.g. boron, chlorine, copper, manganese, molybdenum, and zinc. Some (e.g. copper) are toxic if present in more than small amounts (see *macronutrients*).

natural capital The stock of living and non-living environmental resources potentially available to generate *ecosystem services*.

negative feedbacks Homeostatic regulatory mechanisms that tend to maintain equilibrium in ecological (and other) systems by dampening the effects of perturbations. For example, factors such as predation and food supply normally prevent the growth of species populations beyond the **carrying capacity** of their habitats.

non-point source pollution See **diffuse pollution**.

notable (species/taxa and habitats) A general term, denoting some designation of high conservation value, that can include legally protected, and internationally, nationally, regionally, or locally threatened, rare or scarce. It can also be applied to **keystone species**.

options See **alternatives**.

pelagic Relating to or living in or on the open ocean. The pelagic zone commences at the low tide mark and includes the entire ocean water column.

pH Scale of 0–14 defining the acidity/alkalinity of solutions including those in soils and water bodies: 0 = extremely acid, 14 = extremely alkaline, and 7 = neutral (although soils and waters with pHs between c.6.5 and c.7.5 are often referred to as neutral).

phytoplankton The 'plant' component of **plankton**. They are the primary producers of open water communities (with water too deep for **macrophytes**).

plankton The small (often microscopic) freshwater or marine 'plants' (**phytoplankton**) and animals (**zooplankton**) that are suspended in, and drift with, a water body.

plastic limit (of a soil) The water content of a soil when the soil changes consistency from semi-solid to a more liquid, malleable 'plastic' consistency.

point source pollution Pollution from specific locations, e.g. into surface waters from sewage outfalls and industrial effluent discharge points; or into groundwaters from underground pipelines, wells or the bases of quarries and disposal sites. It is generally easier to control than **diffuse pollution**.

pollutant See **pollution**.

pollution Any increase of matter or energy to a level that is harmful to living organisms or their environment (when it becomes a pollutant). It thus includes **physical pollution** (e.g. thermal, noise and visual) and **biological pollution** (e.g. microbial or by non-native plants and animals), but most commonly refers to **chemical pollution**. Chemical pollutants can be: (a) man-made compounds such as pesticides; (b) toxic chemicals, such as **heavy metals**, harmful levels of which are not normally present in ecosystems; or (c) normally benign or even essential substances such as nutrients, either because these are **micronutrients** that are toxic in more than trace amounts, or because of excessive nutrient loading (**eutrophication**).

precautionary principle Taking avoiding action notwithstanding scientific uncertainty about the nature and extent of a risk, e.g. to respond to the possibility of a significant environmental impact without conclusive evidence that it will occur.

quadrat Strictly a four-sided, usually square, sampling plot; but can include shapes such as circles or rectangles, e.g. for sampling linear features. Quadrats can be any size, e.g. from portable frame quadrats (usually $\leq 1\text{m}^2$) to national grid squares.

receptor Any component of the natural or man-made environment that is potentially significantly affected by a development (sometimes referred to as *valued ecosystem component*).

remote sensing The science – and art – of taking measurements of the Earth without coming into direct contact with it. Typically involves the analysis of images collected via sensors mounted on drones, aeroplanes or satellites.

residual impact Impact remaining after mitigation.

resilience The ability of a system to absorb disturbance without crossing a threshold to a different (usually degraded) state, or to recover from disturbance.

resource efficiency Inputs needed to produce a required product or project (fewer inputs = more efficient).

return period/interval A period within which there is a given probability/risk of a design event occurring. For instance, a 1-in-100-year event is likely to occur once in any 100-year period. Return periods are based on long-term average time intervals between past (recorded) events, and it is statistically possible for a 1-in-100-year event to occur more than once within a year (or shorter period) or not for several hundred years – so they are often expressed as 'per cent chance', e.g. a 1-in-50-year event has a 2 per cent chance of occurring in any one year, a 45 per cent chance of occurring within any 30-year period, and a 76 per cent chance of occurring within any 70-year period.

runoff The part of precipitation that flows as surface water from a site, *catchment* or region and eventually reaches the sea. It is effectively the excess of precipitation over *evapotranspiration*, making allowance for storage in surface, soil and ground waters, and excluding groundwater seepage. Most runoff occurs in streams/rivers, and the term is often restricted to this.

saturation capacity (of a soil) The amount of water held by a soil when it is saturated, usually expressed as cm^3 water per cm^3 soil (see also *field capacity*).

scoping Decision on what to include and what to leave out of the ESIA, generally including geographical and temporal scale, possible methodologies to use, possible alternatives, and whom to consult.

screening Examination of a development proposal to determine whether it requires ESIA.

sediments Organic or inorganic material that has settled after deposition from suspension in water, ice, or air, usually as the water current or wind speed decreases. Aquatic sediments include those that accumulate on the floor of a water body (e.g. lake or ocean), water course or trap, or by deposition on a floodplain. Sediment commonly consists of silt, but can include coarser particulates and material such as calcium carbonate that has precipitated through chemical reaction. Suspended particulates that have not yet undergone *sedimentation* are usually called suspended solids or (incorrectly) suspended 'sediments'.

sedimentation The act or process of depositing *sediments*.

seed bank The accumulation of viable seeds in a soil (mainly the top 40 cm) which may germinate if conditions become suitable – often when the soil is disturbed.

semi-natural (ecological system) A habitat, ecosystem, community, vegetation type or landscape that has been modified by human activity – but largely consists of, or supports, native species (and/or has relatively undisturbed soils, waters and geomorphological features) – and appears to have a similar structure and function to a natural type. Very few completely natural systems now exist, so conservation is largely concerned with protecting semi-natural systems.

setting (of cultural heritage) The surroundings in which cultural heritage is experienced.

sewage treatment levels (in the UK) **Primary** – usually physical treatment to remove gross solids, and to reduce suspended solids by c.50 per cent, and *biochemical oxygen demand* (BOD) by c.20 per cent. **Secondary** – biological treatment to significantly reduce suspended solids, BOD and ammonia. **Tertiary** – additional treatment, e.g. nutrient removal/stripping or ultraviolet treatment to kill pathogenic bacteria.

species abundance The 'amount' of a species in an area or community, expressed by a quantitative measure such as number, density or cover.

species diversity A measure of both the number of species (*species richness*) and their relative abundance (proportion of the sum of the abundances of all the species) in a community. It is more informative than species richness because a community with a given number of species has higher diversity if the overall abundance is fairly evenly distributed between them rather than being concentrated in one or a few dominant species.

species richness The number of species in a biological community. Communities consisting of few or many species are often referred to as 'species-poor' and 'species-rich' respectively.

statutory consultee An organisation that must be consulted in ESIA, typically a government body with responsibility for environmental or social matters.

stepping stone habitats Small habitats that may be scattered and apparently isolated in a landscape, but which may assist in the dispersal of species by providing staging posts between larger habitats. Staging post habitats are also needed by long-distance migrants such as migratory birds, especially along their regular migration routes.

SuDS (sustainable drainage system) Drainage systems that aim to replicate natural systems by collecting, storing and naturally cleaning surface water *runoff* before slowly releasing it back into natural water bodies. SuDS can reduce surface water flooding, improve water quality, and enhance amenity and biodiversity. Examples include basins and ditches that are normally dry but fill up during rain events, wetland habitats, and gravel-filled drains.

sustainable development Defined in the 1987 Report of the World Commission on Environment and Development as 'Development that meets the needs of the present without compromising the ability of future generations to meet their own needs'.

synergistic impact Impact where the combined effect of two agencies, such as pollutants, is greater than the sum of their separate effects, i.e. the effect of one is enhanced by the other. Examples include: bioamplification (the increase in concentration of bioaccumulating pollutants along food chains, culminating in high concentrations in top carnivores), the ability of pests to develop resistance to biocides, the vulnerability of ecosystems to *eutrophication*, effects of species introductions or removals, and the rapidity of global warming and its consequences.

transpiration Evaporative loss of water from plants. When the plants are 'in leaf', it is normally the largest component of evapotranspiration from well vegetated terrestrial ecosystems, and can return > 50 per cent of precipitation water to the atmosphere.

turbidity The opacity of (and hence the degree of light attenuation in) water, due to the presence of suspended matter and *plankton*. High turbidities are harmful to aquatic life.

valued ecosystem component See *receptor*.

vascular plants 'Higher' plants which transport water and nutrients in a specialised structural system that is not present in simple (non-vascular) plants such as bryophytes (mosses and liverworts), *algae* and lichens. They include angiosperms (flowering plants), gymnosperms (mainly conifers) and pteridophytes (ferns, horsetails and clubmosses).

viewshed The area that can be seen from a specific viewpoint.

visual effect The appearance of landscape changes to people, and the resulting effect on visual amenity.

water balance The balance (or budget) between inputs and outputs. In many countries, for instance, there are seasonal patterns of the water balance: in wet seasons precipitation exceeds evapotranspiration, creating a water surplus.

weathering The physical and chemical breakdown of geological materials which contributes to erosion and soil formation.

wildlife corridors Linear habitats/landscape features, such as river corridors, hedgerows, field margins and roadside verges, that may increase connectivity by acting as routes between habitat patches, and hence: increasing the overall extent of habitat for animals with large range requirements; facilitating migration or dispersal of species between habitats; and facilitating access to, and colonisation of, new habitats. Together with *stepping stone habitats* they can be particularly important in areas in which there is severe habitat fragmentation, and may be the only remaining wildlife habitats in urban or intensively cultivated areas.

zone of theoretical visibility The area over which a development can theoretically be seen, also known as 'zone of visual intrusion'.

zooplankton The animal component of *plankton*, many of which graze on *phytoplankton* and are thus equivalent to the herbivores of terrestrial communities.

Index

••

critical habitat 179, 187, **198**, 202–3

cultural heritage 13, **432–75**, 485–6, 534, 588, 655; baseline 449–57; impact prediction 457–64; legislation/policy 176, 251, 444–9; mitigation 464–9; resettlement 553–4, 563–6; transport 380, 383, 386, 394

cumulative impacts (IA) **7–8**, 13–14, **649–77, 706**; acid rain 92; baseline 660–4; climate change 135, 140, 147; cultural 463–4; ecological 165, 200, 202, 207–9, 253, 265, 301, 311; landscape 421–2, 429; legislation/policy 549, 655–60; mitigation/management 130; see also bio-accumulation

dam see barrage

daylight/sunlight (impacts on) 412, 421–5

decommissioning/closure 8, 112, 143, 148, 273–4, 330, 486–7, 605, 619

deforestation 23–4, 31, 45, 64, 77, 88–90, 165, 272, 396–7; effect on ecosystem services 141, 298, 304–5, 314, 479, 488, 555, 568–9; see also forest

demography/human population 234, 325, 366, 590; changes over time 118, 155–6, 375, 396, 484; ecosystem services 307, 312, 325; in economic assessment 476, 478–9, 482–5, 492, 507; in social assessment 515–16, 521–2, 525–9, 532–6; see also vulnerable human groups

deprivation see distributional impacts, vulnerable human groups

design 2, 6, 8–9, 14, 681, 691, 694, 697–8; air/climate change 119, 128, 144, 151–2, 158–60; cultural/landscape 415, 424–5, 429, 467; ecology 202–4, 212–18, 220, 285; noise 355, 359, 361; resettlement 549, 557–8, 560, 562, 565–7; socio-economic/health 491, 524, 535, 584, 629–30; water/soil 32, 58–9, 78, 93; see also mitigation hierarchy, resource use

development/land use plan 74–5, 90, 346, 397, 445, 521, 557; cumulative impacts of 650, 665, 668; ecosystem services in 298–9; links to transport planning 368, 372, 375, 397

dewatering/drawdown 37, 45, 58, 454, 459, 619

diffuse/non-point source pollution 25, 39, 48, 51, 57, 268, **706**

digital terrain model 80–1, 357, 407–8

disaster see risk

disease/illness 517; air pollution/climate change 103, 105, 150, 393; ecosystem services protect from 301, 314; in health impact assessment 578–81, 587–90, 593–600; non-human species 199–200, 274; spread through transport 381, 384–5

dispersion: of air pollution 92, 119–24, 126–30, 382; of noise 337–9, 354–6; of sediment 51

distributional impacts/environmental justice/equality 10, 15, 483, 485, 487, 516, 519, 541, 553, 563, 596; see also Index of Multiple Deprivation

drainage 25; climate change implications for 152; impact of development on 43–7, 71, 91, 381, 387, 395; impact of 36, 38, 44, 208, 458–9; measuring 66–7, 73; river channel 33, 38; see also sustainable drainage system

dredging/dredgings 38, 43, 58, 250, 271–2, 281, 284, 645

drone 454, 557, 710

drought 23, 25; impact of climate change 137, 146, 151, 159–60; impact of development 43–4, 208

dust 102, 115, 129, 215, 380, 460, 465, 557, 597

earth mound see bund

ecology (IA) **164–233**; baseline 180–201; impact prediction 201–11; legislation/policy 15, 175–9; links to other environmental components 20–1, 29, 54–6, 80, 380, 383–4, 387, 394, 404, 427; mitigation 211–21; see also coastal ecology, ecosystem services

ecological succession 170, 172–3, 200, 220, 247

economic (IA) 21, **475–514**, 690, 608, 620, 624, 634; baseline 484–99; considerations in landscape/cultural heritage management 399, 439–40, 464; considerations in soil/land management 65, 75–6, 88; determinant of health 578–80, 589; ecosystem services 298–300, 303–4, 307, 310, 319; impact prediction 499–508; legislation/policy 477–84; mitigation 508–9; transport 365, 380, 382–3, 385–6, 389, 393; value of ecosystems/species 181, 192, 251–2,

change 131, 153; cultural heritage 453; ecology 183, 185, 201, 256, 260–2, 311–12, 316; flooding/water 29, 41–2, 50, 52–3, 250; landscape 403, 407–9, 412; noise 339, 349, 351; other impacts 453, 557–8, 565, 613, 629–30; soil 75, 80–4, 96; topographic 124, 255, 387, 452; see also hachure, isoline, GIS

marine farm see fish farm

marsh 173; impact on/of water regime on 23, 208; see also saltmarsh

mass/materials balance 605–7, 613

migration (of animals) 173, 191, 195, 203, 214–15, 259; climate change impact on 175, 259; dam impact on 38, 44; habitat removal impact on 203, 205–7; legislation 175; see also wildlife corridor

migration (of people) see in-migration

Millennium Ecosystem Assessment 298, 307, 312

mineral extraction/mining impact on: air/noise/water 102, 115, 120, 207, 331, 336, 350; cultural heritage 469–70; ecology 216, 218, 324; economy/social 496–9, 505, 508–9, 518, 535, 539–40, 560; health/risk 584, 592–3, 640–1, 645–6; soil/geology 45, 58, 64, 80, 84, 86–7, 90–1, 94

mitigation hierarchy 158, 212, 215, 322–3, 464, 466–7, 599, 619, 642, 672; in IFC PS1 and PS6 178, 202–3, 560; see also project design

modal shift (transport) 365, 369, 388, 390; see also public transport

monitor and manage see adaptive management

multiplier (economic/employment) 485, 503–6

natural capital 299–301, 305, 307, **708**; see also ecosystem services

negative feedback 170, **709**

net gain see enhancement

no net loss/impact neutrality 202–3, 219–20; see also compensation

noise (IA) 10, **330–64**, 459–61; baseline 347–52; cultural heritage impacts of 459–61, 465, 469; ecological impacts of 270, 273, 276, 284, 384; impact prediction 352–60; legislation/policy 339–47, 610; mitigation 11, 215, 359, 361–2, 426–7; transport impacts on 365, 380–1, 384–6, 392, 395

noise screening see noise mitigation

notable species 184–5, 191, 193–5, 197, 204, 209–10, 264, **709**

non-point source pollution see diffuse pollution

nuclear power station 114–15, 143–4, 486, 503, 507, 538

nutrient 807; impact on ecology 167–72, 248, 268–9, 272–3; in air 104, 107; in soil 66, 73–4, 90, 92–3, 284; in water 30, 37, 39, 41, 44–5, 49

odour 102–4, 110–16, 118, 120, 126–9

offset see compensation

oil see gas

option see alternative

perceived/perceptual impact 103; landscape 400, 403; noise 334–5, 348; risk 623, 632; social 484, 487, 493

piling 43, 215, 270, 273, 284, 341, 459–60, 467

pipeline/transmission line 44, 81, 146, 185, 204, 213–14, 274, 458

plankton 191, 248, 262–3, 268, **709**

point source pollutant 216, 268, **709**; air pollution 112–13, 119–23, 126–7; noise pollution 354–6; water pollution 31, 39, 51, 57

pollutant/pollution/contamination: chemical 41, 59, 636, **709**; cultural heritage impacts of 460, 465–6, 555; ecological impacts of 103–4, 199, 205, 207–8, 215–17, 251, 253, 268–9, 272, 274; groundwater 25–7, 87, 275; health impacts of 10, 13, 103, 127, 130, 555, 588, 596–7; interactions/linkages/transport of 8, 31, 74, 79, 83, 91–2, 94, 169, 384, 636; soil/land 64, 78–9, 83, 86–8, 91–3, 95; standards 8, 12, 34, 105–13, 688; thermal 31, 45, 57; transport related 39, 51, 122–3, 129, 380–2, 384–5, 393–5; water 30–1, 39, 41–3, 45–6, 48, 51–2, 56–7, 59, 387–8; see also air, bioaccumulation, critical dose, diffuse pollution, indicator species, light, noise, point source pollution, sustainable drainage system

polluter pays principle 220, 304

population (human) see demography

power station 7, 31, 39, 92, 115, 460, 486, 500, 502; see also nuclear power station, tidal power, windfarm

swamp *see* wetland
synergistic impact 8, **650–2**, 654, 670–2, **712**

threshold *see* standard
tidal power 270
tourism 269, 314, 380, 413, 446, 456–7, 464–5, 503; *see also* recreation
tranquility 346, 349, 403
transmission line *see* pipeline
transpiration/evaporation 22–4, 36, 44–5, 167, 285, **706**, **712**
transport (IA) **365–98**; air/climate change impact of 120, 135, 149, 380–2, 393; baseline 372–6; communities/resettlement impact of 380, 382–3, 385, 393–4, 535, 554, 570; cultural heritage impact of 380, 383, 394, 458; ecology impact of 181, 208, 213, 380, 383–4, 394; guidance 370–2; health impact of 380, 384–5, 394; impact prediction 376–8; landscape impact of 380, 385–6, 394, 406; mitigation 388–95; noise impact of 339, 381, 386, 395; resource use links to 604, 606, 611, 614, 619; soil/water impact of 386–8, 395
transport modes 366, 372; *see also* public transport, rail, road, ship
turbidity (water) 43, 58, 90, 248, 272, 284, **712**

uncertainty 6, 11–12, 126, 491; in archaeology 444, 468–9; climate change 142–3, 165; coastal change 279; cumulative impacts 665–6, 669; risk 623–4, 627, 631–2, 637–44, 646; use of multiple approaches/scenarios 52, 119, 145, 500; *see also* adaptive management, 'monitoring' under individual topic headings, precautionary principle, Rochdale envelope
urban area/development impacts on: air/climate change/land/water 89–90, 120, 149, 207, 234; cultural heritage 439, 457–8; ecology 174, 182, 204, 217, 234, 266, 268, 298; landscape 385–6, 399, 407, 423–4; transport/economy 368–71, 375, 377–8, 570; water 24, 28, 36, 39, 45, 47–8, 58–9, 250

valued ecosystem component (VEC) 181–2, 655, 659–64, **710**

vegetation 167–8, 170–1, 206–8; air/climate change impacts 104, 106–7, 110, 118, 154–5, 160, 393; coastal 243, 259, 261–2, 269, 285; cultural heritage links to 439, 458–9; in landscape 402, 408; mitigation of damage to 216, 218–9; soil/water effects on 37, 43, 58, 67, 90–3, 387; survey 183, 185, 189, 191, 198
vibration 330–1, 339–42, 362, 381, 386, 395, 460, 465, 467
visual *see* landscape and visual IA
visual amenity 399, 403–4, 411–12, 418, 426
vulnerable area/asset 26, 51, 92–3, 234, 405, 457–8; *see also* sensitivity of receptor
vulnerable human group 10, 519, 523–4; air quality impact on 105–6, 127, 381, 385; climate change impact on 137–9, 145–6, 150, 153–4; ecosystem service changes to 302, 309, 316–7; health impact on 590, 599; in resettlement 553, 561, 563–4, 568–71; *see also* Aboriginal people, Indigenous people, sensitivity of receptor
vulnerable ecosystems/species 174, 207, 276, 281–2

waste/disposal/landfill 30, 36, 45, 612; air pollution by 115; contaminated land as waste 94–5; health impact of 588, 596, 598; hierarchy of management 607–8; legislation 112, 216, 250–1, 609–10; odour from 103, 111; resource efficiency links to 603–7, 612–19; water impacts of 87, 207, 274; *see also* contaminated land, leaching
wastewater *see* effluent
Water Framework Directive 32, 34, 176, 249–50
water (IA) **20–62**; availability to communities 517–18, 553–5, 565, 568; baseline 40–2; climate change impacts on 134, 137, 149–52, 159–60; cultural links to 436–7, 439, 458, 466; ecology links to 167–8, 173, 199–200, 206–9, 215; ecosystem services of 300–1, 314, 316; geology/soil links to 64, 66, 68, 85, 90–2, 95; health impacts of 580–1, 588–9; impact prediction 35–40, 42–56; in the landscape 401–2; legislation/policy 34–5; mitigation 56–9; resource use 606–7, 612–13,

618–19; transport impacts on 381, 384, 387–8, 395; *see also* coastal ecology, flooding, groundwater, water quality

water balance 22–3, 44, 200, 607, 613, **713**

water quality 20, 29–31; data 41, 47–9; legislation 34–5; impact prediction 39, 51–2, 54–6, 270, 272, 274; indicators of 197; mitigation 57; scoping 35–6; *see also* ecosystem services, water

waterlogging 35, 43–4, 66, 72–3, 167, 459; *see also* flooding

wellbeing 578–9, 584–5; cultural/social/resettlement impacts on 445, 465, 516–17, 523, 546, 553–5, 566; ecosystem services benefits for 300, 315, 317, 320–3, 326; *see also* health

wetland/reedbed/swamp 173, 198, 234; as mitigation 59, 213, 216–17, 395; benefits of 307, 314, 316, 323; Ramsar 175–6, 251; relocation of 219; restoration of 218–19, 285, 323; water level/pollution impacts on 23–6, 43–4, 58–9, 207–8

Wild Birds Directive 176, 249–51, 654–6, 659

wildlife corridor/linear habitat 181–2, 204, 214, 218, 705, 708, **713**; hedge/hedgerow 213–14, 219–20

wildlife disturbance/mortality 206–7; from barriers to movement 204; from climate change 165; from mining 87; from new roads 284; from noise 330–1; from riverbank clearance 43; from windfarms 273

windfarm 491; climate change analysis of 140, 147–8, 154; compensation for 537; cumulative impacts of 658–9; ecological impacts of 273; geomorphological impacts of 271, 273–4; landscape/visual impacts of 406–7, 411–2

woodland *see* forest

World Bank 1, 13; cultural heritage guidance 436, 445, 457, 468–9; ecology guidance 205, 210; environmental management plan/monitoring guidance 59, 223, 680–3, 688, 695; resettlement guidance 547–8; social impacts guidance 479, 487–8; transport guidance 366, 370; water guidance 35

World Health Organisation (WHO) 105–6, 114, 330, 339–40, 345

zone of influence *see* scoping

zone of theoretical visibility/visual influence 407, **713**